ESO ASTROPHYSICS SYMPOSIA
European Southern Observatory

Series Editor: Philippe Crane

Springer-Verlag Berlin Heidelberg GmbH

Jacqueline Bergeron (Ed.)

The Early Universe with the VLT

Proceedings of the ESO Workshop
Held at Garching, Germany,
1–4 April 1996

 Springer

Volume Editor

Jacqueline Bergeron
European Southern Observatory
Science Division
Karl-Schwarzschild-Strasse 2
D-85748 Garching, Germany

Series Editor

Philippe Crane
European Southern Observatory
Karl-Schwarzschild-Strasse 2
D-85748 Garching, Germany

Cataloging-in-Publication data applied for

Die Deutsche Bibliothek - CIP-Einheitsaufnahme

The **early universe with the VLT** : proceedings of the ESO
workshop, held at Garching, Germany, 1 - 4 April 1996 /
Jacqueline Bergeron (ed.). - Berlin ; Heidelberg ; New York ;
Barcelona ; Budapest ; Hong Kong ; London ; Milan ; Paris ;
Santa Clara ; Singapore ; Tokyo : Springer, 1997
(ESO astrophysics symposia)

NE: Bergeron, Jacqueline [Hrsg.]; European Southern Observatory

ISBN 978-3-662-22488-5   ISBN 978-3-540-49709-7 (eBook)
DOI 10.1007/978-3-540-49709-7

Typesetting: Camera ready by editor/authors
Cover design: *design & production* GmbH, Heidelberg
SPIN: 10517740   55/3144-543210 - Printed on acid-free paper

Preface

This is the second ESO workshop in a series dedicated to science opportunities with the VLT. At the first workshop all areas of astronomical research were discussed. This second workshop is dedicated to research projects on the early Universe and has provided a forum for discussing strategies for studying faint distant objects in the optical and infrared spectral regions.

This field is evolving very rapidly. There are several new surveys of galaxies and clusters of galaxies at intermediate redshift and quasars at very high redshift. Major advances in the morphological studies of distant galaxies, surveys of galaxies at high redshift and searches for primeval galaxies have been rendered possible by the new facilities provided by the Hubble Space Telescope and the Keck Telescope.

Observational constraints on the evolution and formation of galaxies and large-scale structures as well as the cosmic chemical evolution were critically discussed with regard to theory and numerical simulations.

In this context, the VLT first generation instrument capabilities were presented comprehensively and their use as cosmological tools discussed. The concluding remarks of the workshop focussed on the analysis of various possibilities for the VLT second generation instrumentation.

Many of these topics were covered by invited reviews and talks, as well as some contributed talks. They are included in this volume together with the poster papers.

It is with great pleasure that I express my deep thanks to all participants for having contributed greatly to the success of this meeting especially through their numerous and lively discussions both during the formal sessions and at the various breaks. The very efficient organisational assistance of Christina Stoffer and the extensive contribution of Pamela Bristow to the editing process are also gratefully acknowledged.

Garching, 24 September 1996   J. Bergeron

Contents

4. Search for Highest Redshift Galaxies

6. First Active Objects

7. Concluding Remarks

8. Poster Papers

List of Participants

Name	Institution
ABEL, Tom	MPI für Astrophysik, Garching mak@mpa-garching.mpg.de
ADAM, Gilles	Observatoire de Lyon adam@obs.univ-lyon1.fr
ADORF, Hans-Martin	ST-ECF, Garching hmadorf@eso.org
AMICO, Paola	ESO, Garching pamico@eso.org
APPENZELLER, Immo	Landessternwarte Heidelberg iappenze@hp2.lsw.uni-heidelberg.de
BECKWITH, Steven	MPI für Astronomie, Heidelberg svwb@mpia-hd.mpg.de
BELLONI, Paola	Institut für Astronomie & Astrophysik, Universität München belloni@hal6.usm.uni-muenchen.de
BENDER, Ralf	Universitäts-Sternwarte München bender@usm.uni-muenchen.de
BERGERON, Jacqueline	ESO, Garching jbergero@eso.org
BJÖRNSSON, Claes-Ingvar	Stockholm Observatory bjornsson@astro.su.se
BLANCHARD, Alain	Observatoire de Strasbourg blanchard@astro.u-strasbg.fr
BÖHRINGER, Hans	MPI für extraterrestrische Physik, Garching hxb@mpe-garching.mpg.de
BÖRNER, Gerhard	MPI für Astrophysik, Garching grb@mpa-garching.mpg.de
BRAUN, Robert	NFRA, Dwingeloo rbraun@nfra.nl
CAPPI, Alberto	Osservatorio Astronomico di Bologna cappi@astbo3.bo.astro.it
CARLBERG, Ray	University of Toronto carlberg@astro.utoronto.ca
CAVALIERE, Alfonso	II Università di Roma, Dip. di Fisica cavaliere@roma2.infn.it
CHINCARINI, Guido	Univ. di Milano & Osservatorio Astronomico di Brera chincarini@astmim.mi.astro.it

CIMATTI, Andrea — Osservatorio Astrofisico di Arcetri
cimatti@arcetri.astro.it

CLEMENTS, David — ESO, Garching
dclement@eso.org

COLLESS, Matthew — Australian National University,
Mount Stromlo & Siding Spring Obs.
colless@mso.anu.edu.au

COMERON, Fernando — ESO, Garching
fcomeron@eso.org

COMTE, Georges — Observatoire de Marseille
comte@obmara.cnrs-mrs.fr

CÔTÉ, Stéphanie — ESO, Garching
scote@eso.org

COUCH, Warrick — University of New South Wales,
Dept. of Astrophysics
wjc@newt.phys.unsw.edu.au

CRANE, Philppe — ESO, Garching
crane@eso.org

CRISTIANI, Stefano — Università di Padova,
Dip. di Astronomia
cristiani@astrpd.pd.astro.it

D'ODORICO, Sandro — ESO, Garching
sdodoric@eso.org

DA COSTA, Luiz — ESO, Garching
ldacosta@eso.org

DE LAPPARENT, Valérie — Institut d'Astrophysique, Paris
lapparen@iap.fr

DE MARCHI, Guido — ESO, Garching
demarchi@eso.org

DELTORN, Jean-Marc — Observatoire de Paris-Meudon, DAEC
deltorn@gin.obspm.fr

DENNEFELD, Michel — Institut d'Astrophysique, Paris
dennefel@iap.fr

DI SEREGO ALIGHIERI, Sperello — Osservatorio Astrofisico di Arcetri
sperello@arcetri.astro.it

DICKINSON, Mark — Space Telescope Science Institute,
Baltimore
med@stsci.edu

DIETRICH, Matthias — Landessternwarte Heidelberg
M.Dietrich@lsw.uni-heidelberg.de

DOUBLIER, Vanessa — ESO, Garching
vdoublie@eso.org

DURRET, Florence — Institut d'Astrophysique, Paris
durret@iap.fr

EGAMI, Eiichi — MPI für extraterrestrische Physik, Garching
egami@mpe-garching.mpg.de

ELLIS, Richard — University of Cambridge, Inst. of Astronomy
rse@ast.cam.ac.uk

ELST, Eric — Royal Observatory of Belgium, Uccle
elst@oma.be

FALL, Michael — Space Telescope Science Institute, Baltimore
fall@stsci.edu

FELENBOK, Paul — Observatoire de Paris-Meudon, DAEC
felenbok@obspm.fr

FISHER, David — Kapteyn Astronomical Institute, Univ. of Groningen
fish@astro.rug.nl

FLORES, Hector — Observatoire de Paris-Meudon, DAEC
Hector.Flores@daec.obspm.fr

FOCKENBROCK, Ralf — MPI für Astronomie, Heidelberg
fock@mpia-hd.mpg.de

FONTANA, Adriano — Osservatorio Astronomico di Roma, Monteporzio
fontana@coma.mporzio.astro.it

FORT, Bernard — Observatoire de Paris, DEMIRM
fort@mesiom.obspm.fr

FOSBURY, Bob — ST-ECF, Garching
rfosbury@eso.org

FRANCESCHINI, Alberto — Università di Padova, Dip. di Astronomia
franceschini@astrpd.pd.astro.it

FRANCIS, Paul — University of Melbourne, School of Physics
pjf@physics.unimelb.edu.au

FRANSSON, Claes — Stockholm Observatory
claes@astro.su.se

FRANX, Marijn — Kapteyn Astronomical Institute, Univ. of Groningen
franx@astro.rug.nl

FRICKE, Klaus — Universitäts-Sternwarte, Göttingen
kfricke@uni-sw.gwdg.de

FRITZE - V. ALVENSLEBEN, Uta — Universitäts-Sternwarte, Göttingen
ufritze@uni-sw.gwdg.de

GALAZ, Gaspar — Institut d'Astrophysique, Paris
galaz@iap.fr

GEIGER, Bernhard — MPI für Astrophysik, Garching
bernhard@mpa-garching.mpg.de

GIACCONI, Riccardo — ESO, Garching
dg@eso.org

GIALLONGO, Emanuele — Osservatorio Astronomico di Roma, Monteporzio
giallo@coma.mporzio.astro.it

GIAVALISCO, Mauro — OCIW, Pasadena
mauro@ociw.edu

GILMOZZI, Roberto — ESO, Garching
rgilmozz@eso.org

GREGGIO, Laura — Universitäts-Sternwarte München
greggio@usm.uni-muenchen.de

GROSBØL, Preben — ESO, Garching
pgrosbol@eso.org

HAEHNELT, Martin — MPI für Astrophysik, Garching
haehnelt@mpa-garching.mpg.de

HAMMER, François — Observatoire de Paris-Meudon, DAEC
hammer@gin.obspm.fr

HIPPELEIN, Hans — MPI für Astronomie, Heidelberg
hippelei@mpia-hd.mpg.de

HOPP, Ulrich — Universitäts-Sternwarte München
hopp@usm.uni-muenchen.de

JÄGER, Klaus — Universitäts-Sternwarte, Göttingen
jaeger@uni-sw.gwdg.de

JØRGENSEN, Henning — Copenhagen University, Astronomical Observatory
henning@astro.ku.dk

KASAI, Masumi — MPI für Astrophysik, Garching
mak@mpa-garching.mpg.de

KAUFFMANN, Guinevere — MPI für Astrophysik, Garching
gamk@mpa-garching.mpg.de

KÄUFL, Hans Ulrich — ESO, Garching
hukaufl@eso.org

KOEKEMOER, Anton — Institut d'Astrophysique, Paris
koekemoer@iap.fr

KOLLATSCHNY, Wolfram — Universitäts-Sternwarte, Göttingen
wkollat@gwdg.de

KOO, David — Lick Observatory
koo@helios.ucsc.edu

KÜMMEL, Martin — Landessternwarte Heidelberg
mkuemmel@mail.lsw.uni-heidelberg.de

LA FRANCA, Fabio — Terza Università di Roma, Dip. di Fisica
lafranca@astrom.rm.astro.it

LAZZATI, Davide — Osservatorio Astronomico di Brera
lazzati@merate.mi.astro.it

LE FÈVRE, Olivier — Observatoire de Paris-Meudon, DAEC
lefevre@daec.obspm.fr

LEIBUNDGUT, Bruno — ESO, Garching
bleibund@eso.org

LENZEN, Rainer — MPI für Astronomie, Heidelberg
lenzen@mpia-hd.mpg.de

LEVEQUE, Samuel — ESO, Garching
sleveque@eso.org

LINDNER, Ulrich — Universitäts-Sternwarte, Göttingen
ulindner@uni-sw.gwdg.de

LOBO, Catarina — Institut d'Astrophysique, Paris
lobo@iap.fr

LUCCHIN, Francesco — Università di Padova,
Dip. di Astronomia
lucchin@astrpd.pd.astro.it

MACCAGNI, Dario — Istituto di Fisica Cosmica, Milano
dario@ifctr.mi.cnr.it

MACCHETTO, Duccio — Space Telescope Science Institute,
Baltimore
macchetto@stsci.edu

MANNUCCI, Filippo — CAISMI - CNR, Florence
filippo@arcetri.astro.it

MARANO, Bruno — Osservatorio Astronomico di Bologna
marano@astbo3.bo.astro.it

MATHEZ, Guy — Observatoire Midi-Pyrenées, Toulouse
mathez@obs-mip.fr

MCMAHON, Richard — University of Cambridge,
Inst. of Astronomy
rgm@ast.cam.ac.uk

MEISENHEIMER, Klaus — MPI für Astronomie, Heidelberg
meise@mpia-hd.mpg.de

MELLIER, Yannick — Institut d'Astrophysique, Paris
mellier@iap.fr

MENDEZ, Rene — ESO, Garching
rmendez@eso.org

MEYLAN, Georges — ESO, Garching
gmeylan@eso.org

MONNET, Guy — ESO, Garching
gmonnet@eso.org

MOORWOOD, Alan — ESO, Garching
amoor@eso.org

MOSCARDINI, Lauro — Università di Padova, Dip. di Astronomia — moscardini@astrpd.pd.astro.it

MÜLLER, Karen — ESO, Garching — kmueller@eso.org

NEUMANN, Doris — MPI für extraterrestrische Physik, Garching — don@mpe-garching.mpg.de

OMONT, Alain — Institut d'Astrophysique, Paris — omont@iap.fr

PAIN, Reynald — University of Paris VI & VII, LPNHE — rpain@lpnax1.in2p3.fr

PAPADEROS, Polichronis — Universitäts-Sternwarte, Göttingen — papade@usw008.dnet.gwdg.de

PETITJEAN, Patrick — Institut d'Astrophysique, Paris — petitjean@iap.fr

PICAT, Jean Pierre — Observatoire Midi-Pyrenées, Toulouse — picat@obs-mip.fr

PIERRE, Marguerite — C.E. Saclay – Service d'Astrophysique — mpierre@ariane.saclay.cea.fr

PRANDONI, Isabella — Istituto di Radioastronomia, Bologna — prandoni@astbo1.bo.cnr.it

PUGET, Jean-Loup — Institut d'Astrophysique Spatiale Université de Paris XI — puget@ias.fr

QUINN, Peter — ESO, Garching — pjq@eso.org

REIMERS, Dieter — Hamburger Sternwarte — dreimers@hs.uni-hamburg.de

RENZINI, Alvio — ESO, Garching — arenzini@eso.org

RICH, Michael — Columbia University, Dept. of Astronomy — rmr@carmen.phys.columbia.edu

RIGOPOULOU, Dimitra — MPI für extraterrestrische Physik, Garching — dar@mpe.mpe-garching.mpg.de

ROCCA-VOLMERANGE, Brigitte — Institut d'Astrophysique, Paris — rocca@iap.fr

RODRIGUEZ-ESPINOSA, José — Instituto de Astrofisica de Canarias — jre@iac.es

RÖNNBACK, Jari — ESO, Garching — jroennba@eso.org

ROSATI, Piero — Johns Hopkins University,
Dept. of Physics and Astronomy
rosati@stsci.edu

RÖTTGERING, Huub — Sterrewacht Leiden
rottgeri@reusel.strw.LeidenUniv.nl

ROWAN-ROBINSON, Michael — Imperial College, London
m.rrobinson@ic.ac.uk

RUPPRECHT, Gero — ESO, Garching
grupprec@eso.org

SAGLIA, Roberto — Institut für Astronomie & Astrophysik,
Universität München
saglia@hal1.usm.uni-muenchen.de

SAVAGLIO, Sandra — ESO, Garching
ssavagli@eso.org

SCARAMELLA, Roberto — Osservatorio Astronomico di Roma,
Monteporzio
kosmobob@coma.mporzio.astro.it

SCHINDLER, Sabine — MPI für extraterrestrische Physik,
Garching
sas@mpa-garching.mpg.de

SCHNEIDER, Peter — MPI für Astrophysik, Garching
peter@mpa-garching.mpg.de

SEITZ, Stella — MPI für Astrophysik, Garching
stella@mpa-garching.mpg.de

SEITZ, Carolin — MPI für Astrophysik, Garching
caro@mpa-garching.mpg.de

SHAVER, Peter — ESO, Garching
pshaver@eso.org

STEINMETZ, Matthias — Univ. of California, Berkeley
mhs@astro.berkeley.edu

SWINGS, Jean-Pierre — Institut d'Astrophysique, Liège
jpswings@vm1.ulg.ac.be

TARENGHI, Massimo — ESO, Garching
mtarengh@eso.org

TAYLOR, Keith — Anglo Australian Observatory
kt@aaoepp.aao.gov.au

THOMMES, Eduard — MPI für Astronomie, Heidelberg
thommes@mpia-hd.mpg.de

THOMPSON, David — MPI für Astronomie, Heidelberg
djt@mpia-hd.mpg.de

TYTLER, David — University of California, San Diego
dtytler@ucsd.edu

ULRICH-DEMOULIN, Marie-Helene — ESO, Garching
mhulrich@eso.org

VAN DOKKUM, Pieter — Kapteyn Astronomical Institute, Univ. of Groningen
dokkum@astro.rug.nl

VAN LOON, Jacco — ESO, Garching
jvloon@eso.org

VETTOLANI, Giampaolo — Istituto di Radioastronomia, Bologna
vettolani@astbo1.bo.cnr.it

VLADILO, Giovanni — Osservatorio Astronomico di Trieste
vladilo@oat.ts.astro.it

WAGNER, Stefan — Landessternwarte Heidelberg
swagner@mail.lsw.uni-heidelberg.de

WALSH, Jeremy — ST-ECF, Garching
jwalsh@eso.org

WEBSTER, Rachel — University of Melbourne, School of Physics
webster@physics.unimelb.edu.au

WHITE, Simon — MPI für Astrophysik, Garching
swhite@mpa-garching.mpg.de

WIEDEMANN, Günter — ESO, Garching
gwiedema@eso.org

WILLIGER, Gerard — MPI für Astronomie, Heidelberg
williger@mpia-hd.mpg.de

YAN, Lin — ESO, Garching
lyan@eso.org

ZAMORANI, Giovanni — Osservatorio Astronomico di Bologna
zamorani@astbo3.bo.astro.it

Part 1

Introduction

Cosmological Questions for the European Southern Observatory Very Large Telescope

David Tytler

University of California San Diego, 0111, La Jolla, CA 92093-0111, USA.

Abstract. The next decade promises an observational revolution which will change cosmology for ever. The precise measurement of the angular anisotropy of the cosmic microwave background should specify to a few percent all of the parameters of the cosmological model which effect astrophysics. The growth of structure will then be determined (but not yet observed) until gravitational collapse becomes highly non-linear and stars, galaxies and active galactic nuclei (AGN) form. Not all of the gas enters these collapsed objects, in part because of feed back: stars eject gas, and both stars and AGN emit ionizing radiation which heats the gas. Instead, the universe enters a prolonged out-of-equilibrium state, which we are in today, when its appearance is determined by the balance between three things: the parameters of the cosmological model, the efficiency with which matter enters various collapsed objects, the matter and energy released by those objects. The competition between these collapse and feedback processes determines the appearance of the universe today, and their study will replace the specification of the model as the long term focus of research in cosmology.

These processes are hard to model with basic physics because they are are complex and allow a rich variety of expression. Instead observations will determine when the first stars and quasars formed, and how and when galaxies assembled. These processes will be parameterized, and used to improve computer models, which will be tested in our own and other near-by galaxies. If we succeed, and reconcile the numerous contradictions which characterize the subject today, cosmology will be a mature subject, founded on the agreement between detailed, inclusive and realistic models, which make precise predictions, and the wealth of new data which will come from a wide variety of observations, at all wavelengths. This is an ambitious schedule, but nothing less is worthy of the outstanding capabilities of the 8 – 10 m telescopes, the next generation space telescope, the opportunities at millimeter to sub-millimeter wavelengths and advanced computer modeling. The ESO Very Large Telescope (VLT) should play a major role in this revolution.

1 What is Cosmology?

Cosmology, the study of the universe on the largest scales of space and time, began with ancient creation myths which described the beginning of the universe and the Earth, the differences between heaven and Earth. In this century General Relativity provided a dynamical framework for a variety of cosmological models, which allowed mathematical discussions of the growth of perturbations. Observations with the Mount Wilson reflector showed that galaxies are the building blocks, and that the universe is expanding. Experimental verification of the Big Bang came in the 1960's with the detection of the evolution of radio galaxies and QSOs, the discovery of the cosmic microwave background (CMB), and the prediction of the abundances of the light elements from big bang nucleosynthesis (BBNS). There followed much interest in the physics of the early universe, especially the growth of perturbations to form galaxies, and the prediction of how different types of perturbations could be distinguished in the anisotropy of the CMB and the motion and distribution of galaxies today. In the 1980s the suggestion of inflation at the GUTs energy scale emphasized that the model was incomplete, and stimulated intense interest in phenomena at the highest energies. The subject is now extremely active, on many frontiers, and changing rapidly.

From an observational perspective, the major questions in the subject can be grouped under a few headings:

Parameters of the Model. The standard big bang model is now specified in terms physical parameters, which can be measured. We believe that these parameters have an origin in the physics of extremely high energies, beyond those which are well understood, but perhaps they will be predictable with new physical theories.

Contents of the Universe. Which particles are important in the Universe? Where are the baryons, and how many of them are missing in dark matter? What are the main forms of non-baryonic matter? What media fill the universe (intergalactic medium, background radiation field, gravitational waves)? What are the main objects (large scale structures, clusters of galaxies, galaxies, active galactic nuclei, stars)?

Origins. How, why and when did the particles, media and objects form?

Evolution. How have they changed on cosmological time scales?

Processes. What are the main physical processes which affect origins and evolution?

Ecology. How do objects interact, merge and feed back on the media from which they form?

2   Much Cosmology Will Be Done Locally

The frontiers of astronomy, including cosmology, have moved out in distance (solar system, stars, our Galaxy, other galaxies, quasars, the cosmic microwave background) and back in time (solar system formation, oldest stars and QSOs, CMB, primordial nuclear synthesis, baryosynthesis, inflation, . . .) but in addition to direct observations of the most distant objects, much of the critical cosmological information will continue to come from the detailed study of local objects, and there could be laboratory detections of non-baryonic dark matter.

Oldest Stars. The ages, orbits and chemical compositions of the oldest halo stars are the best clues to the formation of our Galaxy, and are critical to our understanding of primordial element abundances, and chemical evolution.

Structure of Galaxies. Our Galaxy and other near by galaxies are the best places to understand the distribution of dark matter, the relative and absolute ages of stellar populations, and galactic chemical evolution (how much gas went into stars? what types of stars? what types of remnants did they leave?).

Local Objects and Structures. Detailed understanding of the different types of galaxies, of active galactic nuclei (AGN), clusters of galaxies, and of large scale structure, streaming motions and biases in the distribution of galaxy light relative to mass will continue to come from studies of the nearest examples.

Understanding Key Cosmological Processes. Again many, but not all of the key processes can be observed locally: star formation, galaxy mergers, supernovae.

3   Determination of the Cosmological Model

The apparent detection of the first doppler peak in the CMB (White, 1996) suggests that this radiation was last scattered at the recombination epoch, and still contains information from that time. In less than 10 years ground based interferometers, balloon born detectors, and especially the COBRAS/SAMBA and MAP satellites should together measure the angular power spectrum of the CMB on scales $l < 1000$ (1000 independent pieces of cosmological information) to within the limits of cosmic variance and the (benign) Galactic and extra-galactic sources. This data should specify the values of all the main cosmological parameters to high precision: COBRAS/SAMBA (1996) now promises H_o, Ω_o, and Λ to 1% and Ω_b, Q_{rms}, and n_s to few percent (Jungman et al. 1996; Hu, Bunn & Sugiyama 1995).

COBRAS/SAMBA will also measure the spectrum distortion parameter y in $> 10^4$ clusters of galaxies, giving estimated of H_o from X-ray data, the cosmological evolution of clusters, cluster bulk velocities to 50 km s^{-1} out to $z = 1$, large catalogues of IR galaxies, radio galaxies, AGN and counts of normal galaxies.

3.1   The Microwave Revelation

Precise measurement of the cosmological parameters will revolutionize all of cosmology. Models will make precise predictions, and observations will directly yield absolute physical quantities (redshift will yield distance, and look back time; apparent magnitude will yield luminosity; angles and redshifts will give linear separations). The classical cosmological tests from the 1960's, designed to measure the parameters of the model, will be more relevant as consistency checks and especially measures of evolution (Gunn 1977). Indeed, we predict that the model will be so well determined that evolution, especially of complex structures like galaxies, AGN and stellar populations, will replace the determination of the model as the main goal of cosmology. And we should expect surprises, perhaps coming from todays list of problems: the age problem, Λ, the types of dark matter, $\Omega \neq 1$, non-gaussian fields (strings), strong gravity waves, early reionization of the intergalactic medium (IGM), isocurvature perturbations, ...

4   Checking the Model

Before the main CMB data arrive (2002), and after, if they are complex to interpret, we will check the parameters of the model.

Λ can be determined from the number and distribution in z of gravitationally lensed QSOs (Kochanek 1996). The Sloan digital sky survey should provide the 100 - 1000 lensed QSOs.

Ω can be determined from supernovae out to $z \simeq 1$ (Goobar & Perlmutter 1995; Liebundgut 1996), and from lenses.

H_0 can be determined from a few simple lens systems and supernovae.

4.1   Time Scales

The CMB should give the age of the universe t_0 and $t(z)$: our first well defined cosmological time scale. It will then become critical that we know the ages of stars as precisely as possible, because they date the key events in galaxy formation. Globular cluster ages, now known to 10 percent, should be known to a few percent. One percent seems too hard at this time, because each of the following errors gives a 1 percent age error on its own: δ(distance modulus) = 0.01 mag, δ(Helium abundance) = 0.01, δ([Fe/H]) = 0.03 dex, δ([alpha elements/Fe])= 0.03 dex, δ(E(B-V)) = 0.003 mag. The first two may be reachable, but the last three items are beyond hope (Renzini, private communication, and Renzini et al. 1996).

Very complete luminosity functions, from the turnoff to lower giants, and improved data are needed. We can also use the VLTI to give 10 μarcsec parallax measurements for globular clusters, provided we find or obtain reference stars of known distance or position within the isoplanatic angle. This gives a 1 percent distances error at 1 kpc. Rough checks can be made with eclipsing binaries at or below the turnoff.

Improved ages will also be obtained from radioactive chronometers, especially Thorium in halo stars. Here we need to find tens of stars which have low metal abundances (to avoid blending with metal lines) and enhanced r-process elements. Only one is known today. It is not know if the VLT will uncover new cosmological clocks.

4.2   Thorough Tests of Big Bang Nucleosynthesis

The accurate measurement of the abundance of the light elements is an ideal project. The ratios of the abundances of different nuclei test the predictions of BBNS, while the variations of these abundances at later times are excellent tests of the predictive ability of our understanding of stellar interiors (especially mixing), and stellar and Galactic chemical evolution. These observations are well suited to the VLT, and there are many outstanding problems, which will not be solved for many years.

There are two observational approaches. First, and ideally we would measure abundances in primordial gas. QSO absorption systems often have low abundances [C/H] < −2, which are low enough, but there are no cases with [C/H] << −3, because such low abundances do not occur where total column densities are high enough to allow such low limits. Second, we can seek to understand how abundances change over time, using the ratios of various nuclei measured in different places.

Deuterium: This nucleus is made in BBNS (only) and destroyed in stars. The 1215Å line, 82 km s^{-1} to the blue of the H Lyman-α line has been seen in at least two QSO absorption systems which have low [C/H] (Tytler, Fan & Burles 1996; Burles & Tytler 1996). If the other possible detections (Songalia et al. 1995; but see Tytler & Burles 1996) are also real, then D/H may vary spatially. We believe that it can be seen in about 3 percent of QSOs at $z_{em} = 3$, so a search of 3000 QSOs would give 90 D/H measurements. If primordial D/H is high, there is a lot of destruction, which is not understood. We would see how D/H correlates with metal abundance, and measure the dispersion at a given abundance. The change in D/H as a function of metal abundance depends on the masses of the stars: high mass stars make more metals for each H atom returned to the interstellar medium (ISM). However if the first 10 high quality measurements show that primordial D/H is low, and homogeneous, then we would stop the project at that time.

Helium-3: This nucleus is made in BBNS and low mass stars, and destroyed in high mass stars. But this traditional description does not account for the basic features of the data, and must be wrong. The abundance in H II regions varies from 1 to 8×10^{-5}. Low mass stars are expected to make and release a lot of ^{3}He, but abundances in H II regions are an order of magnitude less than expected (Galli et al. 1995). It has been suggested that H II regions are not representative of the ISM because they contain winds from high mass stars which destroy ^{3}He (Olive et al. 1995).

Helium-4: Recent measurements of the abundance of this nucleus (Thuan, Izotov & Lipovetsky 1996) do not agree with either the low or high values for deuterium seen in QSOs. Something is wrong, perhaps large systematic errors which should be corrected.

Lithium-6: This nucleus is created by cosmic ray spallation on oxygen (and C, N) in the interstellar medium, and perhaps also in stellar flares. It is extremely hard to measure, since its lines are only 7 km s^{-1} to the red of those of the

6708Å doublet of ^{7}Li, and signal to noise of hundreds is required, but this is an excellent project for the VLT. It has been seen in only two stars. Measurement in others with different metallicities and mass would tell us more about mixing since this isotope is more readily depleted than ^{7}Li. It has been suggested (Copi, Schramm & Turner 1996) that ^{6}Li is the "strongest" argument against depletion of ^{7}Li, since ^{6}Li is more easily destroyed than ^{7}Li inside stars, but this argument is flawed. There may be other ways of making ^{6}Li in stars, and there may be ways of depleting ^{7}Li without removing all of ^{6}Li, such as stellar winds (Vauclair & Charbonnel 1995), or gas with no Li which is mixed to the surface.

Lithium-7: Multiple creation sites are considered: BBNS, cosmic ray spallation, AGB stars, supernovae, novae. Population I stars have more than we expect from BBNS, and ten times more than low metallicity population II stars. It is destroyed inside stars, for example in low temperature population II stars where convection extends deeper and brings depleted gas to the surface. Warm halo stars with [Fe/H] < -1 show approximately constant ^{7}Li, the Spite plateau, which some cosmologists would like to treat as the primordial abundance. But there are variations on the plateau, similar stars have different Li/H, and there may be a correlation of Li/H with T_e and possible [Fe/H] (Ryan et al. 1996, Molaro et al 1995). We must expect some depletion of Li because it either gravitationally sinks into a star if the atmosphere is non-turbulent, or it is convected down and destroyed inside (Vauclair & Charbonnel 1995). The question of how much depletion, and hence the determination of the BBNS abundance, will be answered with more data.

5   Astrophysical Input to the Model

Many important parameters, especially those determined by complex astrophysical processes, will not come from the CMB measurements, and must be determined by other observations.

First, we need to determine the distribution of baryonic dark matter, and the nature of the non-baryonic dark matter. From astrophysics alone we get phenomological descriptions of the various ways in which the dark matter acts in cosmological situations. If different particles behave in the same way in all astrophysical situations, then we will not distinguish between them, but we can pass these phenomological descriptions on to particle physics.

Second, we need an observational description of the various important nonlinear processes: large scale structure formation, galaxy formation, star formation and AGN formation. We also need to map out non-linear evolution and the relevant feed backs.

Third, we need to measure the radiation and other backgrounds, and determine the sources of radiation, and the extinction in dust (intergalactic and inside galaxies and QSOs).

VLT observations can be used to test and guide computer simulations, for example, by parameterizing star formation (initial mass function, efficiency, burst size and distribution) in terms of variables (ρ, T, background ionizing spectrum I_μ, abundances, magnetic fields) which should be followed in simulations.

6 Dark Matter

First we must determine how many types of dark matter are dynamically important on various scales. Candidates include hot and cold particles, hot diffuse baryons (hard to detect because the H is ionized and the density is too low for substantial X-ray emission), and baryons in collapsed objects (stellar remnants, MACHOs, black holes). We should attempt to measure the ratios of the densities of these dark components, together with luminous matter, and we must work out how the ratios vary with environment.

Radial velocities, proper motions and distances for selected stars, together with gravitational microlensing will trace dark matter in our Galaxy and a few near by galaxies, including local dwarfs and M31, where pixel lensing should be decisive. Weak lensing is powerful for clusters of galaxies, while proper motions of galaxies out to Virgo (VLT Interferometer; Tytler 1997) and improved galaxy distances will explore larger scales. The lensing of QSOs is one way to find massive dark galaxies.

7 Formation of Dense Objects

Understanding the formation of dense objects (galaxies, AGN, stars, Pop III objects) is central to cosmology after the cosmological model is determined. These objects dominate evolution of the observable universe. They release energy which heats the ISM and intergalactic medium (IGM), limiting their formation. They are the source of metals, which effect the cooling of gas, and they release the ionizing radiation, which controls the opacity of the ISM and IGM. These complex feed back processes produce highly inhomogeneous media, with orders of magnitude variations in density and temperature.

At early times, galaxies, clusters, QSOs all look different. We will need to find ways to place objects at different z on their correct evolutionary tracks.

7.1 Formation of Galaxies

Galaxy formation extends over a large range of redshift, and is not complete for at least 2 Gyr because of large dynamical time scales in the outer halos (> 100 kpc). The epoch of formation depends on the spectrum of primordial perturbations. For CDM models most mass in galaxies was accreted at low z, and virialized systems at $z = 4$ were of sub-galactic mass. This will be tested, for example with the splitting of QSO lenses (large samples are needed to get many lenses at high z), which measure σ_v, and with the velocity extent of Quasar absorption lines (Wolfe 1994).

Peebles (1988) has noted that different aspects of galaxy formation may happen at different times:

z_b: the assembly of 0.5 of the mass in bulges/halos today,

z_d: the assembly of 0.5 of the mass in disks today,

z_s: the time of formation of 0.5 of the mass in long lived stars seen today. The last can be measured with the images and spectra of high z galaxies. How do these epochs depend on galaxy morphology, mass, and environment (voids, clusters)? How much mass do galaxies eject in hot winds? What is the metal contents of those winds? Were the first pre-galactic objects Pop III stars?

7.2   First Stars

When do first stars form, especially when in relation to the formation of QSOs and galaxies? The recent discovery that the spectra of common galaxies at $z = 3$ can be obtained with the Keck telescope prompts us to prepare for major new investigations of the properties of young galaxies, perhaps reaching back to the earliest stars. Steidel and colleagues (Macchetto 1996; Giallongo 1996) find that galaxies with a surface density of 0.4 per arcmin^{-2}, and $R < 25$ have half the density of galaxies with $L > L^*$ today. Star formation rates are about $4 - 25\ h_{50}^{-2} M_0 yr^{-1}$ ($q_0 = 0.5$), and the majority have 0.7 arcsec ($8.5 h_{50}^{-1}$kpc) cores and half light radii of $0.2 - 0.3$ arcsec, similar to galactic bulges. A minority have exponential light distributions. There are immediate opportunities to make larger samples, explore higher and lower (HST for UV images) redshifts, and lower luminosities, to derive the luminosity function, masses, morphology, dust contents, clustering and large scale structure at these times. These techniques might find "proto-clusters".

A related question is the origin of the carbon seen in the high column density Ly-α forest. Was this carbon made by the first stars? Did those stars have a wider distribution sufficient to contaminate much of the volume of the universe, or did the carbon come from from stars which formed in the structures which we see in absorption? Alternatively, the carbon could be ejected from pre-galactic units which later merged to form galaxies.

Were the first stars all of high mass, so that none exist today, and are there other types of objects at high z, objects not seen today?

7.3   First Active Galactic Nuclei

Some QSOs at $z > 4$ are of such high luminosities that they have $R = 19$. Objects of this luminosity could be seen at much higher z if they exist and there is no obscuration.

Are QSOs the first collapsed objects? When do they form in relation to the stars in galaxies? If they form in the rarest of high density peaks, then we expect to see galaxies clustered near by.

What determines the rate of increase of the central mass (Haehnelt & Rees 1993; Umemura, Loeb, & Turner, 1993)? How efficiently does gas loose angular momentum and sink to the center of a galaxy? What determines luminosity, and are the first QSOs like those seen at $z < 4$?

How and why do QSOs evolve? How long are QSOs luminous? What determines if there is radio emission? How common are low luminosity AGN at high z? What fraction of galaxies contain massive black holes as a function of epoch? About 46% of all nearby galaxies having $B_T < 12.5$ mag have AGNs, if one includes LINERs in the AGN (Ho et al. (1995), Ho et al. (1995)). How does black hole mass relate to galaxy morphology and mass? What determines which galaxies are QSOs today – the fueling of the black holes by galaxy interactions?

QSOs at $z < 7$ may be found with wide field IR surveys (McMahon 1996). We need a wide field telescope with IR detectors across the focal plane, and 4-m telescopes to identify those objects with the colors of QSO candidates.

Why do QSOs cluster around local galaxies? Lensing apparently does not explain this.

8   Galaxy Evolution

I shall not discuss galaxy evolution, which is covered by several contributions to this volume, except for the following.

Population synthesis from integrated spectra can be used to determine the distribution of relative ages and abundances in high redshift galaxies. In local galaxies we can use HR diagrams (Ortolani et al. 1995).

Clusters of galaxies are laboratories for galaxy and star formation. Like other types of clusters, they are special targets, because many galaxies are observed at once, and all are at similar distances, although probably not of the same age. We need to find samples of clusters, and forming clusters at $z > 1$ to determine the star formation rate, galaxy morphology, cluster mass, and mass to light ratios all as function of z and environment. Samples at different z may be matched by comparing comoving densities.

Data now hint that the regions with the most active star formation may change in time: $z \simeq 3$ halos/bulges, $z \simeq 2$?, $z \simeq 1$ irregular blue galaxies, $z \simeq 0$ spiral disks. VLT images and spectra should correct this speculation, and show if there is a dependence on mass and environment.

9   Universal Chemical Evolution

Quasar absorption lines should provide the cosmological averages densities of various common elements: $\Omega_{element}(z)$. We can also measure element abundance ratios (e.g. $\Omega_O(z)/\Omega_C(z)$) which indicate the origins of the elements (Pettini, Lipman & Hunstead 1995; Lu, Sargent & Barlow 1996). Dispersion in ratios at a given time relates to the amount of mixing.

Most common elements have been seen in Quasar absorption lines: H, D, He, C, N, O, Na, Mg, Al, Si, S, Ca, Ti, Cr, Mn, Fe, Zn and Ni. In the most favorable cases, we may detect some of the following (Verner, Barthel & Tytler 1994): Li, Be, B, Cl, Co, Cu, Sc, V, Ga, Ge, Cd, Sn, Ba and Pb. We need damped Lyα systems with $\log N(H\ I) = 21.5$, 0.1 solar abundances, and signal to noise 1000 spectra. At lower abundances, which are more interesting, the lines are too weak to see.

Much more detailed studies of element ratios, and their origins can be made on halo stars. The distribution of halo stars as a function of metallicity tells us about the first stars, while their element abundances tell about the first supernovae (McWilliam et al. 1995). Perhaps the elements in some halo stars were made by individual supernovae, before the gas from many stars had mixed to yield average abundances.

10   Intergalactic Medium

Here, as in other areas, we have yet to determine the basic parameters. The mean density of the IGM could comprise a few percent of all baryons, or most baryons, with high values preferred by the ionization of the Lyα forest (Petitjean et al (1993), Bi & Davidsen 1996, Rauch et al. in preparation). The density determination is difficult because the IGM is highly ionized, and we see the small amount of gas which is neutral. When was the IGM re-ionized? We expect different epochs for H and He, depending on the spectrum of the ionizing energy. The reionization, and galaxy formation will lead to variations in density and temperature. We can also hope to determine the origin – galaxies, QSOs – of the background ionizing radiation using the proximity effect on QSO absorption lines. We may be able to measure how the intensity and spectrum of this radiation varies spatially and in time.

We expect galaxies and pre-galactic clumps to eject some, or even a lot of metal enriched gas. The gas in clusters of galaxies is very metal rich, perhaps in part because galaxies everywhere eject gas, not just those in clusters, where the gas is hot enough to see. How do abundances vary in the IGM, with epoch and especially with respect to the masses and types of the nearest galaxies, and past environment (what types of galaxies in this neighborhood in the past?).

One outstanding question is the determination of the amount of dust opacity on cosmological distances (Fall 1996). Are there dust obscured QSOs (Webster et al. 1995)? There are two ways to approach this question. First we can estimate the dust contents of $z = 3$ star forming galaxies (Steidel's galaxies are biased against dust). Second, we can use element ratios to measure dust in damped Lyα systems.

11   Large Scale Structure

We can distinguish three epochs by the mode of observation: $z \simeq 1000$ can be observed in the CMB, $z < 1$ at $r \simeq 21$ can be observed with existing instruments like the Two Degree Field of the Anglo-Australian Telescope and the Sloan Digital Sky Survey. The VLT should then plan to work on the intermediate redshifts, at $r >> 21$. We can also hope to use QSO absorption, QSOs and clusters of galaxies to supplement galaxies in the determination of structure at high z.

12   The Full Cast of Players

Thirteen mirrors with apertures > 6.5 m, at good sites, and with new instruments will be operating in a few years. Some will be specialized.

The Sloan Digital Sky Survey will cover 10,000 sq deg. with a 2.5-m telescope, giving the positions and images of 10^8 objects to $R = 23$ (5σ), the spectra of 10^6 galaxies to $B = 19$, and $< 10^5$ QSOs to $B = 20$.

The HST may continue to operate in no-repair mode past 2005 with much improved instruments, including NICMOS and STIS (1997) and the Advanced Camera for Surveys (1999).

The Next Generation Space Telescope, an 8-m optimized for 1-5 μ with a 2007 launch, should have a dramatic effect on astronomy. It should excel at its goals: the study of the first starlight, assembling pieces of galaxies, changes in morphology, interactions, ... indeed, most of the central cosmological projects for the ground based telescopes. It will have vastly superior angular resolution, and sensitivity in the IR, and for many projects it will easily out perform all ground based optical/IR telescopes. But in the decade before it is launched, we note that;   *If it can be done from the ground, it will be done from the ground before a space mission is launched,* (Beichman 1995).

New millimeter and sub-millimeter arrays (Shaver 1996) should also have a dramatic effect on cosmology since the peak of dust emission redshifts closer to those wavelengths, which allows the detection of thermal dust emission at $z < 20$ for $L > 10^{11} L_o$, far beyond what we expect to reach in the optical and infrared. The Millimeter Array could detect CO emission from ordinary galaxies to $z = 1$, and from ultraluminous galaxies to $z = 3$.

Computer simulations will also improve enormously once we know the cosmological model, and because of continued improvements in computing power, perhaps by a million fold by 2002 (Physics World (1996)). They will give much more precise predictions of what to expect.

13   Strengths of ESO and its VLT

The following strengths should be used to target particular projects in cosmology.

• The VLT is the largest telescope.

• It has the most varied complement of focal instruments.

• The Southern site is best for the Galactic center, the bulge, the Magellanic Clouds, and for gravitational microlensing.

• The community of astronomers is large. If there is interest in coordination and trying out new ways to conduct research, this could be turned into a major advantage, as data rates rise, and the pace of research in astronomy as a whole increases, as it must with all these new facilities.

• The ESO community has lead the way with adaptive optics and has the most experience in this area.

• The VLT Interferometer, with laser guide stars and AO on all four unit telescopes, and ≥ 3 outrigger telescopes would be the best IR/near-IR interferometer.

14   The Urgent Need to Find Targets

Keck is running out of ideal targets for some projects! It is absolutely essential to get new large samples, including:

• High z clusters of galaxies.

• 10^4 QSOs to find damped Lyα absorption, and measure the rise of metal abundance with epoch.

• 3000 QSOs at $z \simeq 3$ to make 90 high quality measures of D/H.

• A sample of 10^5 QSOs would yield about 1500 lens candidates and 400 lenses.

• A sample of 50 metal poor stars with [Fe/H] < -4 should come from spectra of 5000 stars with [Fe/H] < -2 (Beers et al. 1992).

• A sample of 100 – 1000 QSOs at $z > 5$ might require spectra of 10^5 candidates (McMahon, private communication, and this volume).

A variety of telescopes could be fully occupied finding targets.

† 2-m telescopes: z for 10^4 - 10^6 QSOs, and rough abundances for 5000 halo stars.

† 4-m telescopes:

    200 nights for 3000 QSO spectra to find candidates for D/H measurement.

    200 nights for 10^4 spectra to find damped Lyα systems.

    3000 nights to check 1500 lens candidates.

    1000 nights for 10^5 $z > 5$ QSO candidates.

    Light curves of SN Ia at $z = 1$ with $I = 25$ at max.

† Wide field telescopes: find QSOs, find lensed QSOs, find low abundance halo stars, and find and follow up microlensing events.

We note that it would be natural for ESO to lead in microlensing because Paranal is the best site for bulge observations and the VLTI will determine the distances and masses of individual lens objects (Tytler 1997, Miralda-Escude 1996).

15   Deep Wide Field Imaging Surveys

We should begin new deep wide field surveys which will open up new areas for research. Examples from the past include the Shane-Wirtanen Lick galaxy counts, which lead to an understanding of the distribution of galaxies (Peebles 1990), the Abell cluster catalogue, and the 3C and related radio catalogues which showed that the universe is evolving. These new surveys could include deep images, like the HST Deep Field, and spectra to produce deeper versions of the Sloan Survey. Others should be specialized, and cover large areas to find high z clusters, QSOs, lensed QSOs, and supernovae.

It might be most effective to devote one 8-m to such surveys. Many of the best projects are very large, and will require team work. We should practice all stages of these large projects: the planing, allocation of resources, distribution of work, coordination of analysis and the sharing of rewards. There will remain numerous opportunity for small projects because both the objects and the processes in cosmology are complex.

16   Cosmology Will Become Richer

In past centuries and decades the frontiers of cosmology have moved out in distance, and back in time. We predict that this trend will end, once we have detected the earliest objects, perhaps in the next decade. Instead the focus will move to the detailed understanding of know objects and their evolution. The subject will become richer, more demanding, and more mature.

COMPLEX OBJECTS. Like geology and biology, and unlike particle physics, the objects in cosmology are complex. Although we speculate that most types of objects have been discovered, there is great variety within each type of (neutron star, white dwarf, brown dwarf, giants, supernovae, ...) and this variety can help us understand.

COMPLEX PROCESSES. The processes in astrophysics are themselves complex, whether in dynamics, nuclear physics or gas dynamics. Consider the variety of effects which occur in dynamical systems: tides, tidal locking, orbit evolution, precession, stability, relaxation, mass segregation, evaporation, binary formation, mergers, accretion, fragmentation, ...

COMPLEX SYSTEMS. The processes which connect different types of objects make complex systems. The ISM is a rich example, with mass sources:– in fall, stellar winds, stellar ejecta; mass loss:– galactic fountains & winds, mass stripping, star formation; energy sources:– cloud-cloud collisions, supernovae shocks, star light, cosmic rays; variations in phase:– 3 stable phases of temperature and density. processes:– ionization, molecular formation, dust formation, radiative transfer, opacity. In contrast, our understanding of the IGM is primitive.

EVOLUTION. All objects had an origin, but there can be multiple origins for objects which are later identical: e.g. star formation. Objects and processes can be far from equilibrium, leading to increased variety. Most objects die, and there are multiple possible deaths: e.g. planets merge, fragment, fall into stars, and are ejected from systems. Many processes are irreversible and destroy information about the past. But even when the large scale picture does not reveal origins, the details, such as element and isotope ratios, orbits and spin axes may. For these reasons of complexity, cosmology has a long and rewarding future.

I should like to thank Chas Beichman, Arlin Crotts, Alex Filippenko, Chris Kochanek, Richard McMahon, Alvio Renzini, Michael Rich, Peter Schneider, Richard Simon, Ed Turner, Art Wolfe and especially Jacqueline Bergeron for the invitation to consider this topic, and for hosting this forward looking meeting.

References

Beers, T.C., Preston, G.W. & Shectman, S.A. (1992): AJ **103**, 1987-2034

Beichman, C. in *A Road Map for the Exploration of Neighboring Planetary Systems* http://techinfo.jpl.nasa.gov/www/exnps/roadmap.html

Bi, H.G & Davidsen, A.F. (1996): submitted to ApJ

Burles, S. & Tytler, D. (1996) submitted to Science (astro-ph 9603069)

COBRAS/SAMBA (1996): http:// astro.estec.esa.nl/ SA-general/ Projects/ Cobras/ cobras.html

Copi, C.J., Schramm, D.N. & Turner, M.S. (1995): Science **267**, 192

Faber, S.M. (1993): in *Cosmic Velocity Fields* (Editions Frontieres; Paris), 485–496

Fall, M.S. this volume

Filippenko, A. V. (1996) in *The Physics of LINERs*, ed. M. Eracleous, et al. (San Francisco: ASP), in press

Galli, D., Palla, F., Ferrini, F. & Penco, U, (1995): AA **443**, 536

Goobar, A. & Perlmutter, S. (1995): ApJ **450**, 14-18

Giallongo, E. this volume.

Gunn, J.E. (1997): in *Observational Cosmology* eds. Gunn, J.E., M. Longair, & Rees, M.J. 8th Advanced Course, Saas Fee (Sauverny: Observatoire de Geneve)

Haehnelt, M.G.; Rees, M.J. (1993): MNRAS **263**, 168

Ho, L.C., Filippenko, A.V., Sargent, W.L.W. (1996): ApJ submitted

Hu, W., Bunn, E.F. & Sugiyama, N. (1995): ApJ **447**, L59–L63

Jungman, G., Kamionkowski, M., Kosowsky, A. & Spergel, D.N. (1996): submitted to Phys. Rev. D.

Kochanek, C.S. ApJ in press (1996): (astro-ph 9510077; "Is there a Cosmological Constant")

Liebundgut, B. this volume

Lu, L., Sargent, W.L.W. & Batlow, T. (1996): submitted to ApJ

Miralda-Escude, J. (1996): ApJL in press

Olive, K.A., Rood, R.T., Schramm, D.N. & Truran, J., & Vangioni-Flam (1995): ApJ **444**, 680

Pettini, M., Lipman, K., Hunstead, R.W. (1995): ApJ **451**, 100-110

Petitjean, P., Webb, J.K., Rauch, M., Carswell, R.F., and others. (1993): MNRAS **262**, 499-505

Physics World (1996): **9**, 29-48

Macchetto, D. this volume.

McMahon, R. this volume.

McWilliam, A., Preston, G.W., Sneden, C. & Searle, L. (1995): AJ **109**, 2757–99

Molaro, P.; Primas, F.; Bonifacio, P. AA 295, L47-50

Ortolani, S. et al. (1995): Nature **377**, 701-4

Peebles, P.J.E. (1989): in *The Epoch of Galaxy Formation* (Kluwer Academic Dordrecht), 1–14

Peebles, P.J.E. (1990): *The Large-scale Structure of the Universe* (Princeton, N.J. : Princeton University Press)

Renzini, A. et al. (1996): ApJ **465**, L23

Ryan, S.G., Beers, T.C., Deliyannis, C.P. & Thorburn, J.A. (1996): ApJ **458**, 543-560

Shaver, P. ed. (1996): *Science with Large Millimeter Arrays* (Springer)

Thuan, T.X., Izotov, Y.I. & Lipovetsky, V.A. (1996): in "Interplay between Massive Star Formation, the ISM and Galaxy Evolution" eds. D. Kunth et al. (Gif-sur-Yvette: Editions Frontieres)

Tytler, D, Fan, F-X. & Burles, S. (1996), Nature, **381**, 207 (astro-ph 9603070)

Tytler, D, & Burles, S. (1996), in *Origin of Matter and Evolution of Galaxies* eds. T. Kajino, Y. Yoshii & S. Kubono (World Scientific Publ. Co.) (astro-ph 9606110)

Songali, A., Cowie, L.L., Hogan, C.J. & Rugers, M. (1994): Nature **368**, 599

Tytler, D. in *Science with the VLTI* eds. F. Paresce, this series.

Umemura, M., Loeb, A., Turner, E.L. (1993): ApJ **419**, 459

Verner, D.A., Barthel, P. D. & Tytler, D. (1994): AAS **108**, 287-340

Webster, R. L., Francis, P.J., Peterson, B.A., Drinkwater, M.J. & Masci, F.J. (1995): Nature **375**, 469–471; and this volume

White, S.M. this volume

Wolfe, A.M. in *QSO Absorption Lines*, Proc. ESO Workshop, eds. J. Bergeron, G. Meylan & J. Wampler (Springer: Heidelberg).

Vauclair, S., & Charbonnel, C. (1995): AA **295**, 715-24

Part 2

The VLT First Generation
Instrument Capabilities
as Cosmological Tools

ISAAC – IR Spectrometer and Array Camera

Alan F.M. Moorwood and Jason Spyromilio

ESO, Karl - Schwarzschild - Str. 2, D-85748 Garching, Germany

Abstract. ISAAC covers the 1-5μm wavelength range and will provide for both imaging over a field of up to 2.5x2.5' and long slit spectroscopy at resolving powers in the range 300-5000. It is now in the integration phase at ESO in Garching and is scheduled to be installed at one of the Nasmyth foci of the first VLT unit telescope in mid-1998. Our main purpose here is to report the latest performance estimates made with a recently developed ISAAC software simulator and to discuss them in the context of studies of high redshift galaxies.

1 Instrument Concept and Status

ISAAC comprises two cameras optimized for the 1-2.5μm and 2-5μm ranges which can be used directly for imaging or to re-image the intermediate spectrum formed by a grating spectrometer. Each camera contains an objective wheel allowing scale changes, two filter wheels providing for 26 positions in each arm and a 2D array detector of up to 1024x1024 pixels. The spectrometer is based on the use of a compact 3 mirror collimator/camera and two back to back mounted gratings used in Littrow mode. All functions are integrated into a casted optical support structure which can be pre-cooled with a continuous flow of liquid nitrogen and maintained at cryogenic temperature by means of two closed cycle coolers. The entire instrument is housed in a $\simeq$ 1.5m diameter vacuum tank attached to the Nasmyth adapter/rotator. Further technical details, layout drawings and photographs can be found in Moorwood (1995) and the ESO WWW pages.

Present status is that the vacuum vessel and most of the instrument functions have been manufactured, integrated and tested (cryogenic operation, flexure etc.) and system integration to the extent necessary for first tests is expected around September 1996. Final objectives and detectors will be delivered somewhat later. Detectors baselined at present are the Rockwell 1024x1024 Hg:Cd:Te array for the short wavelength arm and the SBRC 256x256 InSb array for the long wavelength arm.ESO is, however, planning a foundry run of SBRC 1024x1024 InSb arrays with the intention of using one of these to upgrade the LW arm. Control software and electronics for both the detectors and instrument is nearing completion and a software simulator of the instrument has been developed and will be described in more detail below.

2 Observational Capabilities

ISAAC will offer the following basic modes:

- ## Channels
  - SW 1-2.5μm: Scales (0.07), 0.15, 0.27"/pixel
  - LW 2-5μm: Scales 0.07, 0.15, 0.5"/pixel

- ## Imaging
  - 2.5x2.5' Max. field
  - Seeing improvement by M2 tip-tilt
  - Broad, intermediate and narrow band filters
  - Polarizing analyzers (or Wollaston prisms)

- ## Spectroscopy
  - R $\simeq$ 500 and 3000 (two pixels, 1" slit)
  - Slit widths 1, 0.6, 0.3"
  - Slit length 2' max.

3   The Simulator

Each VLT instrument will have an exposure time estimator for use in the preparatory phases of observations. Traditionally, such aids were simple spreadsheet style calculators. For ISAAC we have developed a simulator with functionalities between that of a full ray-tracing model of the instrument and a simple calculator. The emphasis has been on the physical processes that affect the quality of the data rather than blind modeling of every parameter of the system.

The sky brightness in the near IR may be more than 8 magnitudes brighter than the source observed with an 8-m telescope. Therefore, when modeling the performance of an IR instrument more attention needs to be expended on following variations in the background. There is little or no point in looking at 1 or 2 per cent effects that affect the source counts unless you can remove the background to two or three orders of magnitude higher accuracy.

The ISAAC simulator is a throughput measuring machine which convolves the source spectrum with the transmission/reflection characteristics of the optical elements between the source and the detector and then distributes the flux on the detector according to the seeing and instrument mode. The atmosphere is treated as an optical element with a transmission and emission spectrum. The telescope emissivity is also taken into account. Input parameters for the celestial source may be a FITS image, point or extended sources to which an observed or calculated spectrum (e.g black - body of given temperature) can be assigned. The atmosphere is also considered as an extended source of emission with a time variable spectrum. All observations result in a final image of the source as observed on the detector. Signal to noise estimates are made by extracting the spectrum/image from the final image rather than calculating the signal to noise and then making the appropriate image. This approach allows fully 2 dimensional computations of instrumental effects including scattered light and dark current (e.g amplifier glow).

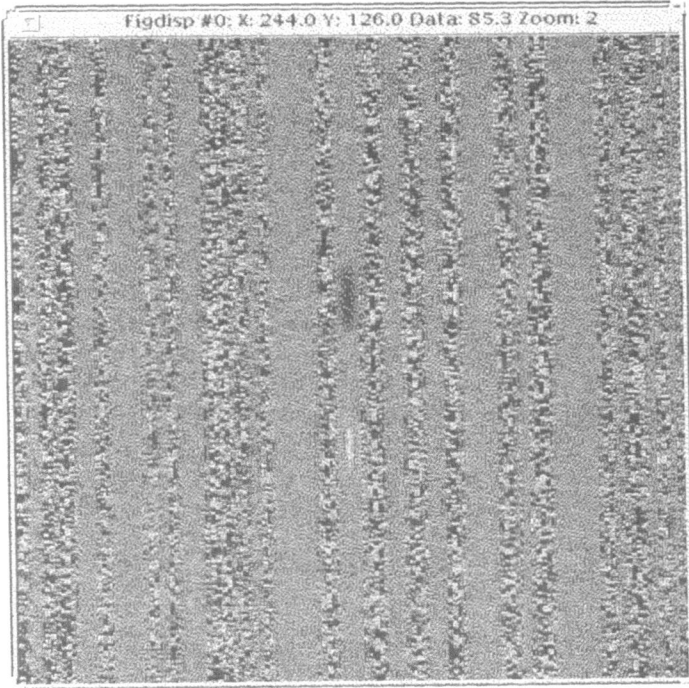

Fig. 1. Simulated H band spectrum of a spatially extended emission line (see text).

The simulator is generic such that optical elements can be replaced or changed with no impact on the code. Generic functions to calculate the distribution of light on the detector as well as the throughput have been coded such that the same code could easily be used for optical instrumentation. Such a port has already been made for the EMMI instrument currently operating on the NTT. Multiple, short exposures (as is applicable to high background situations) may be combined to create the final image. Broad and narrow band imaging as well as imaging polarimetry (using a Wollaston prism) are supported. In addition, spectroscopic observations at a variety of resolving powers are also available. Slit losses as well as the effect of the slit width on the resolution and throughput are fully taken into account. Chopping, nodding, jitter and mosaic observations are all supported for the modes where they are applicable.

3.1 Uses of the Simulator

Although the large number of instrumental parameters necessary to run the simulator are known or are in the process of being determined the critical parameters needed to determine the success or otherwise of a planned near-IR observation will not and cannot be known with sufficient accuracy until the moment of execution. What the simulator allows the observer/applicant to do is a *what if?* scenario and therefore to place the constraints on the conditions under which the observation may be executed.

Moreover, the simulator can be used as a research tool for instrumentation specialists to diagnose the instrument by playing various scenarios to tune parameters of the instrument before and during construction as well as when operating on the mountain. Here the simulator is used to explore the parameter space relevant to the observation of high redshift galaxies.

As an example, Fig. 1 shows a simulated H band spectrum at R $\simeq$ 2500 of a spatially extended galaxy emitting a 1.7×10^{-17} erg s^{-1}cm^{-2}arsec^{-2} emission line. Frames corresponding to total integration times of 1 hr with the object at different slit positions have been subtracted yielding positive and negative images of the line plus vertical stripes due to the shot noise remaining on the cancelled OH lines. Because the astronomical line falls between the OH sky lines the s/n achieved is much higher than in simulated narrow band (1.5% filter) images of the same source where this line is close to the detction limit.

3.2 Performance Summary

The following limits have been obtained with the simulator and correspond to 3σ detections in $\simeq$ 1hr but assuming no sky variations.

- ## Photometry
  - J$\simeq$ 24.5, H$\simeq$ 23.5, K$\simeq$23.5 mag. arcsec^{-2}

- ## Narrow band (1.5%) Line Imaging
  - $\simeq 2 \times 10^{-17}$ erg s^{-1}cm^{-2}arcsec^{-2}

- ## Spectroscopy
  - R$\simeq$ 500: J$\simeq$21.5, H$\simeq$20.5, K$\simeq$20.5 mag arcsec^{-2}; lines $\simeq 2 \times 10^{-17}$ erg s^{-1}cm^{-2}arcsec^{-2}
  - R$\simeq$3000:J$\simeq$19.5, H$\simeq$18.5, K$\simeq$19 mag arcsec^{-2}; lines $\simeq 5 \times 10^{-18}$ erg s^{-1}cm^{-2}arcsec^{-2} (between OH).

4   Studies of the Early Universe

ISAAC offers a powerful capability for both the discovery and study of high redshift galaxies through its combination of large field (for an 8m telescope) broad and narrow - band imaging and possibility for follow - up spectroscopy.

For broad - band observations the natural backgrounds are such that ISAAC and FORS will have comparable sensitivity to galaxies with V-K $\simeq$ 5. In the case of ellipticals this means that ISAAC should become more sensitive at z $\simeq$ 1. At z=2, V-K$\simeq$ 8 and the predicted K band limit is $\simeq$ 3 mag fainter than an L* galaxy at this redshift. In the range z $\simeq$ 2-5 the 4000Å(H and K) break falls in the near infrared and provides a means for crudely estimating the redshift. The Keck telescope has already pushed infrared galaxy counts to K $\simeq$ 24 mag but only has a small field of 38x38 arcsec. ISAAC will provide larger area covereage to similar limits leading to improved statistics for cosmological tests and a larger sample of objects for detailed study. A specific application to fundamental plane studies of cluster ellipticals is being investigated by Renzini et al., within the framework of the VLT Reference Proposals.

For bluer disc galaxies it will be possible to measure the spectral energy distributions from the UV to 2.5μm using FORS and ISAAC and to estimate redshifts from the Lyman and H+K breaks and ages from comparison with models. Medium resolution spectra over the same range will be possible for galaxies which are bright enough. As Hα moves into the infrared at z = 0.5 ISAAC will be *the* instrument for 'visible' spectroscopy of high redshift galaxies. For individual galaxies found by any technique it should be a major tool for determining redshifts; velocity dispersions; star formation rates and for detecting AGN activity. Starburst, radio, ultraluminous infrared and active galaxies will be obvious targets both in their own right as individual objects and as samples to probe the evolution of activity to earlier epochs. Radio galaxies in particular feature strongly in the on-going elaboration of VLT Reference Proposals (Fosbury et al.). Around the peak of AGN activity at z=2-2.6 it should be noted that Hα falls in the infrared K band while Lyα can be observed in the UV-visible allowing estimates of dust extinction and a test of the relative merits of these lines for detecting starburst activity.

The perspectives for high redshift studies with the VLT have improved dramatically during just the last year following the successes already achieved in detecting galaxies at z $\geq$ 2 in Lyα and Hα emission line surveys and by colour selection of objects showing the Lyman break in deep images obtained both on the ground and with the Hubble Space Telescope (for a review of recent results see Moorwood, 1996). Most of the detected galaxies still require extensive spectroscopic and other follow-up studies with large groundbased telescopes. These results show starburst activity underway at z=3.5 but, provisionally, little evolution between z=1 and 3.5. At higher redshifts, most objects out to z=5 are radio galaxies and quasars and beyond that nothing is yet known. Deeper surveys over larger areas are therefore still required to measure the evolution of starburst activity and to continue the search for primeval galaxies undergoing their first bursts of star formation. This will require a multi-wavelength approach.

One contribution expected from ISAAC is a continuation of recent narrow band imaging searches for redshifted Lyα, [OII]3727, [OIII]5007, Hα (z=0.5-2.6) and other lines as described at this Workshop (Meisenheimer; Mannucci, Thompson and Beckwith; van der Werf et al.). The largest area surveyed to data with adequate sensitivity to detect Hα at z $\simeq$2.3 is 42 arcmin^2 using IRAC2 at the ESO/MPI 2.2m telescope on La Silla. Three candidate Hα emitting galaxies in addition to three known radio galaxies have been detected and the results already tend to exclude models of a strongly evolving starburst population (van der Werf et al., 1996). ISAAC with the same instantaneous field (2.5x2.5') but on an 8m telescope will clearly push these limits and, hopefully, establish the volume density of starburst galaxies. Recent theoretical predictions (Mannucci and Beckwith, 1995) indicate that at least one primeval (spheroidal) galaxy/field should be detected at the narrow band imaging limits estimated here. Also, as shown by the simulations, follow-up medium resolution spectroscopy at higher s/n should be possible of any line detected by narrow - band imaging.

References

Mannucci, F., Beckwith, S. (1995), ApJ., **442**, 569.

Moorwood, A.F.M. (1996): Starburst Galaxies, *Space Science Reviews*, (in press)

Moorwood, A.F.M. (1995): *Infrared Detectors and Instrumentation for Astronomy* (ed. A. Fowler), SPIE Vol. **2475**, 262–268.

van der Werf, P.P, Bremer, M.N., Moorwood, A.F.M., Röttgering, H.J.A., Miley, G.K.(1996): in preparation

Observing the Early Universe with CONICA

Rainer Lenzen

Max-Planck-Institut für Astronomie, Königstuhl 17, 69117 Heidelberg

Abstract. The high resolution NIR camera CONICA being attached to one of the Nasmyth foci of the first VLT unit is designed to offer diffraction limited imaging and spectroscopic observational capabilities in combination with the Nasmyth Adaptive Optics System. Several modes are available to overcome the seeing limited resolution. In combination with these high resolution capabilities the variety of observing modes available with CONICA is described, emphasizing those applications which are of special interest for extragalactic observations.

Several fields of cosmological studies and high redshift research programs are addressed for which the observational capabilities of CONICA will contribute important and hopefully astounding new data. Especially the high angular resolution direct imaging mode and the opportunity of combining these capabilities with slit spectroscopy will be a powerful tool for studying early type high redshift galaxies, quasars or in general faint distant objects at near infrared wavelengths.

1 Introduction

In addition to the multi-mode infrared-instrument ISAAC, the first VLT unit will be equipped with a high resolution NIR camera that will provide some complimentary observational capabilities: For the same wavelength region from 1 to 5 μm CONICA is specialized in providing high spatial resolution for direct imaging, spectroscopic imaging (using a cryogenic Fabry Perot étalon), long-slit spectroscopy and polarimetry down to the diffraction limit of an 8m-telescope.

CONICA will be attached to the second Nasmyth focus of the first VLT unit right opposite or face to face to ISAAC. There is the possibility to mount it directly to the Nasmyth flange, but normally it will be used in combination with the Adaptive Optics System NAOS.

In general, no guiding star near enough to the studied object and bright enough to make use of the full NAOS correction mode will be available, thus, CONICA is not specialized in doing in general extragalactic research. For several applications however the diverse observational modes and capabilities offered by CONICA will provide a powerful tool for studying far distant high redshift objects, especially emphasizing the diffraction limited spatial resolution and those modes that are unique to this instrument.

2 Observational Capabilities of CONICA

2.1 General Overview

CONICA will be equipped with a 1024^2-InSb detector with a pixel size of 27 μm. A variable optics is offering different image scales between 110 and 13 mas/pixel, the highest resolution of which is provided for the short wavelength region only. Beside a number of broad band and interference filters a cryogenic Fabry Perot etalon is implemented offering a spectral resolution around 1800 within the K-band. There are several grisms to use CONICA in a low spectral resolution mode. This mode is especially interesting in combination with the high spatial resolution provided by the Adaptive Optics System.

In summary, the following different observational capabilities are available:

- Diffraction limited direct imaging (NAOS, Speckle, TADC,...)
- Spectroscopic imaging using cryogenic Fabry Perot
- Low resolution long slit spectroscopy with grisms
- Polarimetry with Wollaston prisms and Wiregrid analyzer
- Fokal plane (slits, coronogr.) and pupil (apodization) masking

2.2   Diffraction Limited Resolution

In 1998 the Hubble Space Telescope is planned to be equipped with a near infrared detector array. Nevertheless, CONICA in combination with NAOS can compete in the domain of high resolution infrared astronomy because the fundamental limitation of spatial resolution by diffraction strongly favours the use of large telescopes: The main mirror diameter of the HST of 2.4m is limiting the spatial resolution to 190 mas at K (2.2μm) while the 8m-class telescope offers a resolution of 57mas at the same wavelength.

Thus, ground based high spatial resolution infrared astronomy will be of high astrophysical interest even at the end of this century.

To overcome the disadvantage of ground based astronomy as given by atmospheric turbulences CONICA will provide several modes:

- The main high spatial resolution mode is the long exposure mode in combination with the adaptive optics system NAOS. All instrumental configurations of CONICA can be combined with this mode, especially long slit spectroscopy, chronographic masking or Wollaston-polarimetry with high spatial resolution is possible.
- CONICA is equipped with a fast cryogenic shutter that may be used for best seeing selection without reading out the detector.
- Speckle modes are offered for those cases where no guiding star is available or as an additional mode of improving the NAOS image for the short wavelength region.
- Deconvolution algorithms like LUCY and CLEAN or a Fourier image processing package are available to improve the image resolution even further by off-line data reduction.
- To correct for atmospheric dispersion effects a special TADC can be inserted and tuned depending on the given zenith distance and wavelength.

This TADC can be switched into the observing beam if required, that is for large zenith distances and/or for small wavelengths. Beside the entrance window (see Fig. 1) it is the only optical component that remains at room temperature. The

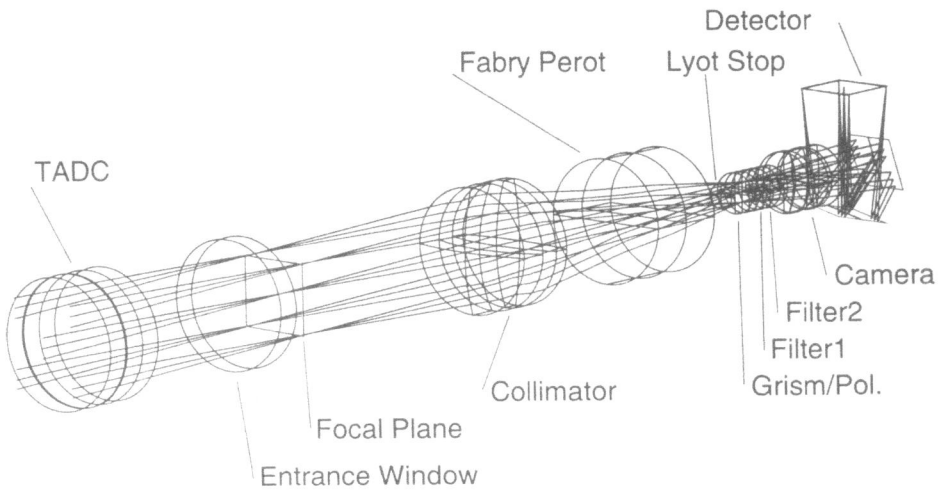

Fig. 1. schematic drawing of the imaging optical ray path

collimator is an achromatic doublet that forms an image at the Lyot stop plane. Between the Lyot stop wheel and the camera wheel, that is within the collimated beam there are two filter wheels and one grism/polarizer wheel.

In Tab. 1 an overview over the camera date is given: The pixel scale can

Camera	f-ratio	magn.	Pixels/AD 1.0 - 1.9 μm	Pixels/AD 2.0 - 5.0 μm	Scale [mas/pxl$_{27}$]	FOV$_{256}$ [arcsec]	FOV$_{1024}$ [arcsec]
C50	51.0	3.4	1.89-3.60	——	13.6	3.5	14.0
C25	25.5	1.7	0.94-1.79	1.89-4.72	27.3	7.0	28.0
C12	12.75	0.85	0.47-0.90	0.94-2.36	54.6	14.0	56.0
C06	6.375	0.425	0.24-0.45	0.47-1.18	109.2	28.0	Ø75.6

Table 1. Camera data for CONICA

be modified between 13.6 and 109 mas/pixel corresponding to f-ratios between 6.4 and 51. Using a 1024^2-detector the resulting field of view for the highest resolution is 14x14 arcsec. The FOV for the fastest camera is limited by the entrance aperture and does not illuminate completely the detector.

The importance of high angular resolution for the study of the early universe that is for long distance objects is evident: At 2.2μm the angular resolution is improved by a factor of about 10 from seeing to diffraction limited resolution. Less evident, however, is the improvement of sensitivity for point-like objects: For faint objects NIR-broad band imaging in general is background limited. Decreasing the pixel scale by a factor of 10 from typical seeing limited application of

1/3 arcsec to diffraction limited application (25 mas) will reduce the background radiation per pixel by a factor of 100, the signal-to-noise ratio is improved by a factor of 10.

Thus, especially for long distance objects which remain unresolved at seeing limited application, there is a certain advantage of sensitivity for shorter wavelengths. In Fig. 2 the limiting magnitudes for one hour total integration time (object and sky) for different modes is shown. The most sensitive mode is the

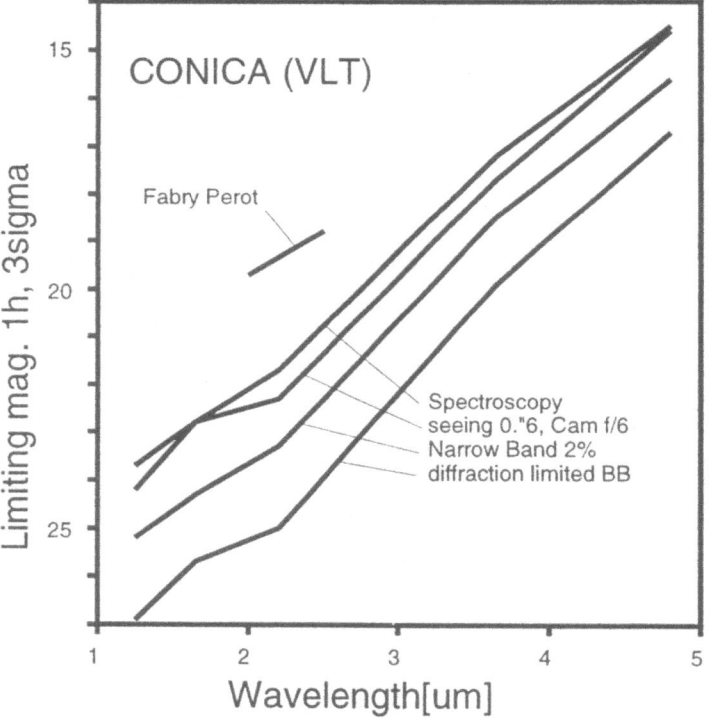

Fig. 2. 3 σ-limiting magnitudes of CONICA for different observing modes, 1 hour total integration time

broad band direct imaging mode for diffraction limited resolution down to K (A Strehl ratio of 70% is assumed). For higher spectral resolution using narrow band filters or grism spectroscopy the limiting magnitude is reduced by 1.5 and 3.0 magnitudes, respectively. In addition, using the fastest camera available for CONICA the limiting magnitude for seeing limited resolution is shown. Even the spectroscopic imaging mode using the Fabry Perot can be used down to about 19 magnitudes if full diffraction limited resolution is achieved.

2.3   The Cosmic Distance Scale

One interesting application of the high resolution direct imaging mode with cosmological implication is the calibration of cosmological distance indicators in the infrared: Anticipating the performance of high resolution (≤ 0.1 arcsec) and high sensitivity (K $\geq$ 25mag) the reach of several primary distance indicators could be extended considerably: Especially for the RR Lyrae and Cepeids the long debated influence of extinction could be almost avoided in the K-band. In Tab. 2 the main distance indicators and the expected maximum distance module are summarized. Such studies have been carried out at optical wavelengths

Distance Indicator	M_K	Max. dist. mod.	D[Mpc]	Comments
RR Lyrae	0.3	24.7	0.9	
δ Cepeids	-6.0	31.0	15.8	Virgo cluster
LF peak of GC	-8.7	33.7	55.0	Pegasus, Pisces cluster
Brightest GC	$\geq$ -14	39	631	3C 273
SNIa	-19.5	44.5	8000	

Table 2. Reach of cosmological distance indicators ($H_0 = 50$)

using the HST (see Saha et al. 1994 and Saha et al. 1995)with distance modules of (m-M)=28.36 (IC4182) and 28.08 (NGC5253). Long time monitoring of distant galaxies to calibrate SNIa will be an interesting application of CONICA to improve the calibration of these distance standard candles.

2.4   Filters and Polarization Analyzers

CONICA is equipped with two filter wheels offering 40 positions in total (see Tab. 3). One wheel is completely equipped with blocking filters for the Fabry Perot, covering the K-window with 3% interference filters. These filters can be used within the K-window for any line observation of any redshift, as well. In addition, there are the standard broad band filters including the somewhat shorter and narrower K' and a list of narrow band filters, which are optimized for special emission lines from galactic sources. There are two modes of measuring linear polarization:

Broad Band Filters:	J, H, K, K', L, M, continuum1.040
Narrow Band Filters:	Ice2.90, Ice3.07, Ice3.30, Ice3.60, HeI1.083, P_γ1.094, OII1.237, FeII1.257, P_β1.282, FeII1.644, H2(1-0)S(7)1.748, Pf$_\gamma$3.74, Br$_\alpha$4.051, Pf$_\beta$4.653
Blocking Filters:	NB1.97, NB2.00, NB2.03, NB2.06, NB2.09, NB2.12, NB2.15, NB2.18, NB2.21, NB2.24, NB2.27, NB2.30, NB2.33, NB2.36, NB2.39, NB2.42, NB2.45, NB2.48, NB2.51
Wollaston Prism:	1–5 μm, 3arcsec beam separation, 0 deg. and 45 deg.
Wire Grids:	1–5 μm, 0 deg., 45 deg., 90 deg., 135 deg.

Table 3. List of available filters and polarizer

- A set of four individual exposures through wire-grids at several position angles is taken to analyze large extended objects.
- For very compact polarized sources like BL-Lacs e.g. a second mode is recommended: Using Wollaston prisms within the collimated beam double images can be taken observing simultaneously the ordinary and extraordinary beam. This technique may be especially adequate for long distance extragalactic objects. The study of synchrotron radiation originated within jets like that in 3C273 or M87 e.g. will be an interesting application of this observational capability.

The beam separation is slightly wavelength dependent being well represented by the formula:
$\delta[arcsec] = a + b(\lambda[\mu m] - 3) + c(\lambda[\mu m] - 3)^2 + d(\lambda[\mu m] - 3)^3$
with a=3.161, b=-0.20, c=-0.035 and d=-0.0041

2.5 Long Slit Spectroscopy with Grisms

The combination of the high angular resolution mode with long slit spectroscopy is a very attractive tool for extragalactic research application: For a redshift of z = 0.2 a spatial resolution of 0.1arcsec will allow to measure the IR-profile of host galaxies of quasars as close as 300pc to the center of activity.

CONICA will be equipped with four NIR-grisms that can be used at different orders (see Fig. 3):

For doing long slit spectroscopy there is a mask wheel within the focal plane that offers three different slit widths of 25, 50, and 100 μm. In addition, several focal masks are provided, field limiting masks as well as special coronographic mask that allow to increase the sensitivity near luminous point sources.

2.6 Imaging Spectroscopy Using the Cryogenic Fabry Perot

A second spectroscopic mode is available in terms of slitless imaging spectroscopy: CONICA is equipped with a cryogenic Fabry Perot which offers the

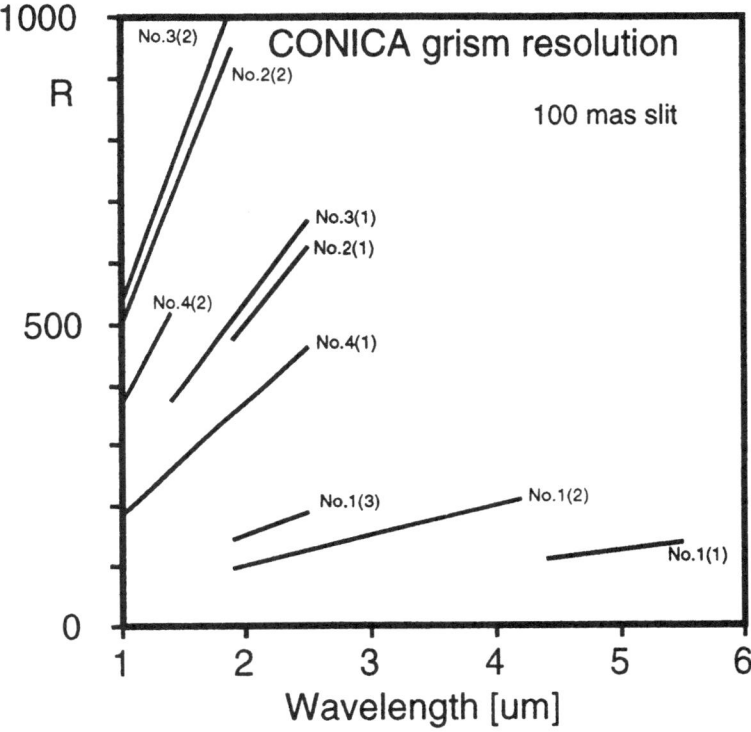

Fig. 3. Graphical presentation of the long slit spectroscopic mode for different grisms and orders assuming a slit width of 100 mas

resolution power of 1800 within the K-band (a set of narrow band blocking filters is provided). This unit has been built by Queensgate Instruments and is currently being tested at our institute. For galactic application it has been combined with MAGIC, the near infrared camera of the Max-Planck-Institut für Astronomie at the 3.5m-telescope of Calar Alto.

This device may be used to study galaxies at high redshift by observing lines like Hα, β, γ or the oxygen lines which coincides with the K-window. The search for primeval galaxies at least in the redshift range between 4 and 5 can be effectively done as well as re-confirmation studies of candidates found by other survey programs.

3 Examples of Scientific Programs Using CONICA

I would like to close this presentation with a small list of examples of some extra-galactic applications for CONICA which is by no means complete but should give a rough impression of what can be done using the infrared instrument CONICA for high redshift objects:

- **Structure of High-Redshift galaxies:** Study of stellar population of high-z galaxies and the galaxy evolution.
- **Quasar Fuzz and Environment of BL Lacs:** Study of properties of Quasars and BL Lac host galaxies. Possible role of gravitational lensing in BL Lac objects. Satellite galaxies and clustering environment.
- **Narrow-band Imaging of Fields around Quasars:** search for emission line galaxies at high redshift (Fabry Perot).
- **Search for Black holes in centers of galaxies** Direct images and spectra (R=1000) at 0.1 arcsec resolution to study the light distribution close to the center and the velocity dispersion as a function of distance from the nucleus
- **Are there obscured quasars?** The hypothesis of powerful radio galaxies containing a quasar which is hidden behind several ten magnitudes of visual extinction is easily tested by highly resolved infrared images (L-band) of the galactic cores.
- **Cosmic distance scale:** Using combination of CONICA with NAOS the reach of several primary distance indicators could be extended considerably. The long debated influence of extinction could be almost avoided in the K-band.
- **Search for Primeval Galaxies:** Search for primeval Galaxies beyond z=5. Candidates found by survey programs can be studied by imaging and slit spectroscopy using CONICA.
- **CONICA Deep Field:** Selection of any appropriate high latitude guide star to take very deep IR JHK images.

References

R. Lenzen: SPIE Vol 1946, Infrared Detectors and Instrumentation, Orlando, 14-16 April 1993, p635

R. Lenzen, R. Hofmann: SPIE Vol 2475, Infrared Detectors and Instrumentation, Orlando, 18-21 April 1995, p268

Saha, A., Labhardt, L., Schwengler, H., Macchetto, F.D. Panagia, N., Sandage A. Tammann, G.A., 1994 ApJ 425, p14

Saha, A., Sandage A., Labhardt, L., Schwengler, H., Tammann, G.A., Panagia, N., Macchetto, F.D., 1995 ApJ 438, p8

G.A.Tammann, Proceedings of ESO Conference 24-27 March 1981 "Scientific importance of High Angular Resolution at Infrared and Optical Wavelengths"

Spectroscopy of Faint Distant Objects with FORS

I. Appenzeller[1], O. Stahl[1], S. Kiesewetter-K.[2], R.-P. Kudritzki[2], H. Nicklas[3], and G. Rupprecht[4]

[1] Landessternwarte, Königstuhl, D-69117 Heidelberg, Germany
[2] Universitäts-Sternwarte, Scheinerstraße 1, D-81679 München, Germany
[3] Universitäts-Sternwarte, Geismarlandstraße 11, D-37083 Göttingen, Germany
[4] ESO, Karl-Schwarzschild-Str. 2, D-85748 Garching, Germany

Abstract. The two FORS instruments will allow imaging, photometry, low-dispersion spectroscopy, polarimetry, and spectropolarimetry of faint astronomical objects at the ESO VLT. The instruments have been designed to match the field available at the Cassegrain foci of the VLT unit telescopes. They are optimized to reach the faintest objects accessible with these telescopes. In the field of early-universe studies the FORS instruments will have the potential to contribute new information to questions of the cosmic distance scale, the mass in the universe, and of the formation, chemical evolution and kinematics of distant galaxies and active nuclei.

1 Basic Objectives of FORS

The **FO**cal **R**educer **S**pectrographs FORS I and II have been designed to carry out imaging, photometry, and low-dispersion spectroscopy at the Cassegrain foci of the ESO VLT Unit Telescopes 1 and 3. FORS I has additionally polarimetric and spectropolarimetric capabilities, while FORS II will contain additional grisms for echelle-format spectroscopy.

In all observing modes a wavelength coverage corresponding to the spectral region accessible to groundbased CCD observations had been specified. Therefore, the FORS optics was designed to be usable in the range 330 - 1500 nm, although with present detectors the instruments will not be operated at wavelengths longward of 1100 nm.

Mounting the FORS instruments at the Cassegrain foci has the advantage of having minimal light losses and low instrumental polarization. On the other hand, since the VLT optical system is designed primarily for the F/15 Nasmyth foci, and since the same secondaries are used for the F/13.6 Cassegrain foci (cf., e.g., Enard 1988), aberrations and vignetting limit the FOV at the Cassegrain foci to about 12 arcminutes. Moreover, this area has to be shared with the Cassegrain guide probe. These constraints resulted in a field size of FORS (in the standard operating mode) of about 7×7 arcminutes (corresponding to a 10 arcminutes diagonal).

The main design goal of the FORS project was to reach very faint objects near the feasibility limit of groundbased 8-meter class telescopes. At the same time the FORS instruments had to be sufficiently versatile to allow accurate quantitative photometric and spectroscopic observations of very different types of faint galactic and extragalactic objects.

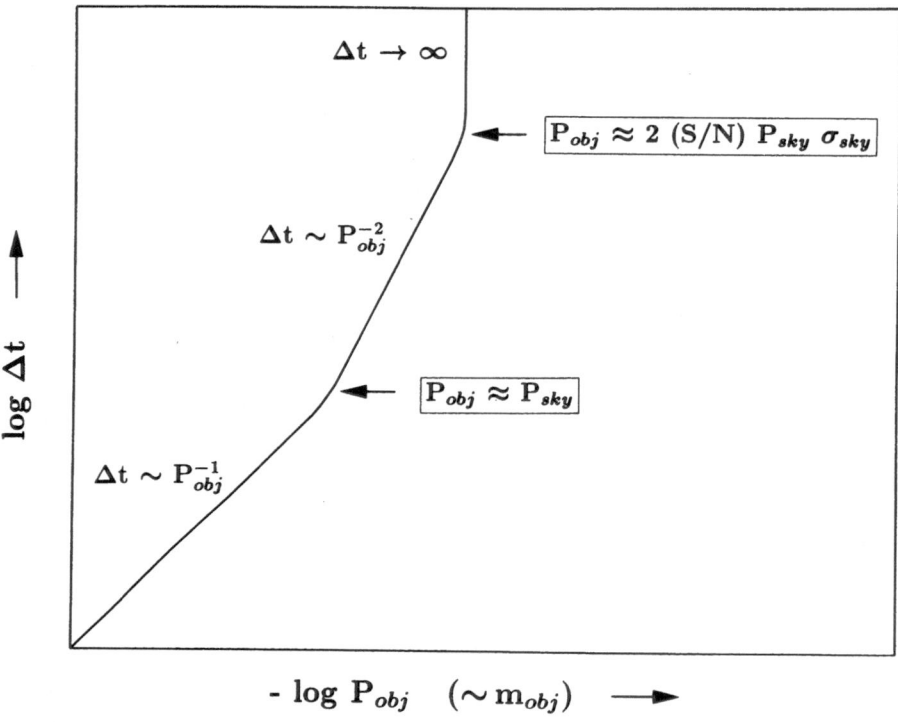

Fig. 1. Schematic relation between the observing time Δt required to achieve a fixed S/N value and the logarithm of the object photon rate P_{obj} (or the apparent magnitude m) for groundbased astronomical observations

For groundbased CCD observations the signal-to noise ratio S/N can be approximated by the relation

$$\frac{S}{N} = \frac{\Delta t P_{Obj}}{\sqrt{\Delta t P_{Obj} + \Delta t P_{Sky} + \Delta t I_{Dark} + RON^2 + \Delta t P_{Sky}\sigma_{Sky}}} \tag{1}$$

where Δt is the integration time, P_{Obj} and P_{Sky} are the photoelectron rates (photoelectrons recorded per second) from the object and the sky, respectively, I_{Dark} is the dark current (in electrons per second), RON is the readout noise, and σ_{Sky} is the relative systematic accuracy of the sky background determination. If the instrumental efficiency is known the photon rate of a source with a visual magnitude m_V can be estimated from the telescope aperture and the photon flux in the V-band outside the atmosphere which is approximately

$$P(m_V) \approx 10^{3-0.4m_V} \text{photons s}^{-1}\text{cm}^{-2}\text{\AA}^{-1} \tag{2}$$

Hence one VLT unit telescope collects in the V-band at most about 5 photons $s^{-1}cm^{-2}\overset{\circ}{A}^{-1}$ at $m_V = 20$ and about 2 photons $hour^{-1}cm^{-2}\overset{\circ}{A}^{-1}$ at $m_V = 30$.

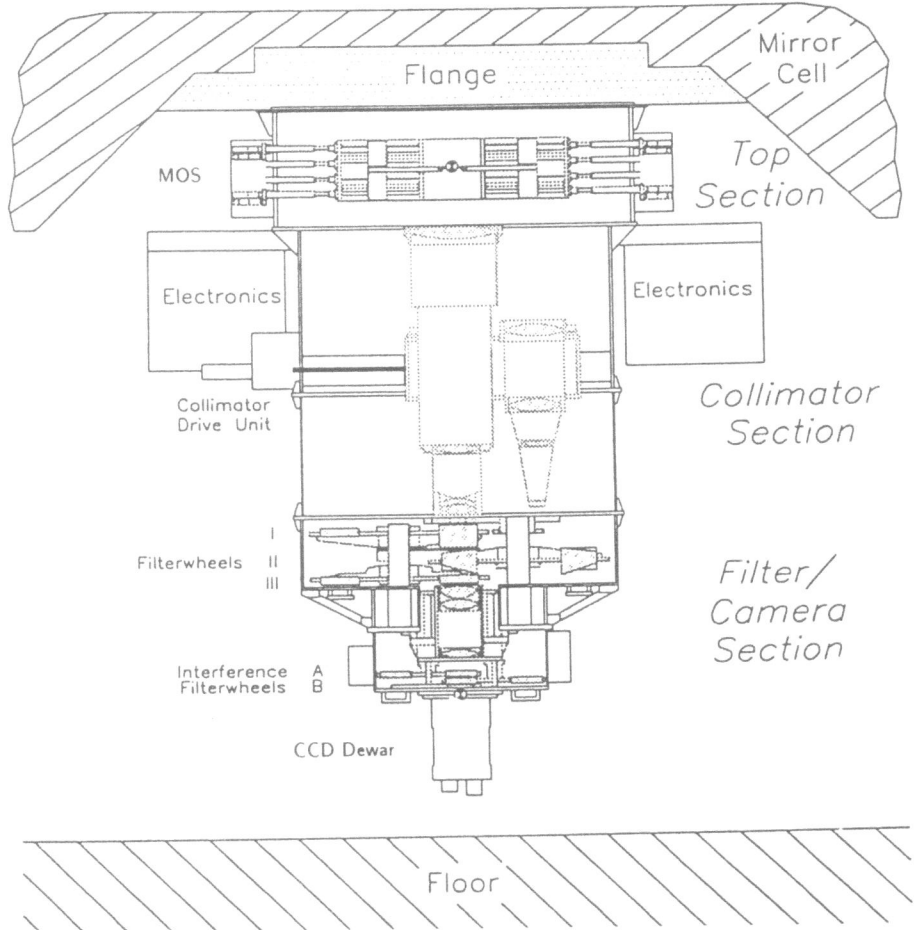

Fig. 2. Schematic cross section of the FORS instruments

As illustrated schematically by figure 1, if a fixed value S/N has to be reached and if the dark current and the readout noise are negligible compared to the object and sky photon rates (which can be safely assumed for practically all FORS applications) equation (1) results in three different domains of the relation between the required integration time and the object photon rate (or the object apparent magnitude). For sufficiently bright objects where P_{Obj} is the dominant term in the denominator of equation (1) Δt increases proportinal to P_{Obj}^{-1}. When the sky background flux dominates, the slope becomes steeper and we have

$\Delta t \sim P_{Obj}^{-2}$. Finally, if

$$P_{Obj} < 2(S/N)P_{Sky}\sigma_{Sky} \qquad (3)$$

the slope approaches infinity. Hence relation (3) characterizes the limiting flux level or (limiting magnitude) of groundbased instruments. Note that this limiting flux level is independent of the spectral resolution. For a fixed S/N (usually determined by the astrophysical problem) the magnitude limit of an instrument depends only on the sky contribution and on the relative error of the sky background subtraction. Hence, the main design considerations of FORS was to keep these two quantities as low as possible.

2   Design Details and Performance

Technical details of FORS has been described in various earlier papers (Appenzeller and Rupprecht 1992, Appenzeller et al. 1992, Seifert et al. 1994, Böhnhardt et al. 1995, Möhler et al. 1995, Seifert et al. 1995). The basic design is illustrated by the instrument cross section reproduced in figure 2. The main components are a multiple slit unit (MOS) in the telescope focal plane, a set of two interchangeable collimators, a parallel beam section with "filter wheels" for filters, grisms, and for the polarization optics, followed by a camera with a CCD detector.

The standard resolution (SR) collimator gives a pixel scale of 0.2 arcseconds and a FOV of 6.8×6.8 arcminutes. The "high resolution" (HR) collimator, designed for use under exceptionally good seeing conditions, provides a 0.1 arcsecond/pixel scale for a correspondingly smaller FOV.

In the case of FORS I the spectral resolution R $= \lambda/\Delta\lambda$ will be ≤ 1800 for a 1 arcsecond slit width (or about ≤ 4500 for the 2-pixel sampling limit, corresponding to 0.4 arcseconds). In the case of FORS II we expect R ≈ 3000 for a 1 arcsecond slit (R ≈ 7500 for the 2-pixel limit). At present we are investigating the feasibility of grisms of high refractive index optical materials (such as ZnS or KRS6). If such grisms can be realized, the spectral resolution of FORS II may possibly be extended to about 5400 for the 1" slit, or about 13 000 in the 2-pixel limit.

In the case of direct imaging the sky contribution and the accuracy of the sky background subtraction depend on the image quality, the flatfielding accuracy of the detector, the uniformity and reproducibility of the background, and on the brightness and behaviour of ghost images. The FORS optics was, therefore, designed to produce an intrinsic point spread function which even under the best expected seeing conditions is much narrower than the seeing disk. Hence, we expect that the actual PSF will be determined fully by seeing effects without a significant instrumental contribution. In both imaging modes 80 % of the light of a point source was to be concentrated within one pixel (of 0.2" (SR) or 0.1" (HR), respectively). Laboratory tests of the the completed imaging optics of both FORS instruments show that this goal has been reached. The level of ghost images and other background effects also appear to be very low, although a reliable measurement of these quantities will be possible only at the telescope with the original FORS CCD system. On the basis of the test carried out so far, we expect that with carefull flatfielding FORS should be able to reach $\sigma_{Sky} \approx 10^{-3}$, corresponding to a limiting B-magnitude of about 30.

In the case of spectroscopy the sky contribution and σ_{Sky} depend mainly on the size and quality of the slits used. With present technologies, it is difficult to produce slit masks giving sufficiently low values of σ_{Sky} for the spectroscopy of the faintest objects accessible to 8-m class telescopes. Moreover, because of operational considerations ESO's "Call for Proposals for the VLT Cassegrain Low-dispersion Spectrograph" (on which the FORS concept was based) excluded explicitly the use of slit masks produced outside the instrument. Therefore, we use in FORS a set of 19 pairs of conventional slit jaws which can be moved individually with very high precision. The individual slitlets are of sufficient quality to reach nearly the same sky subtraction accuracy (and hence nearly the same limiting magnitudes) as for direct imaging at the same S/N. The more typical application of FORS will probably be spectroscopy of less faint objects at a higher S/N. However, according to relation (3) this will require equally low values of σ and hence an equally high slit quality.

An important advantage of individually movable slit jaws is the possibility of adjusting the slit widths on a short time scale to time variations of the seeing and the sky background conditions. In this way the individual slit widths can always be optimized and adjusted for each object to reach the best S/N achievable at the conditions prevailing during a given exposure.

As an alternative to the individual slit jaws the FORS (MOS) slit unit allows to introduce masks into the telescope focal plane. At present masks with a set (lower quality but fully reproduceable) long slits and a strip mask for surface polarimetry can be introduced into the MOS unit by means of slide mechanisms.

In principle these mask drives could also be used to introduce externally produced slit masks into the focal plane. Such masks could be of advantage for spectroscopic survey programs where the simultaneous observation of a larger number of objects is more important than reaching a high S/N or a very faint limiting magnitude. The use of such externally produced masks in FORS would require that the operational constraints described above are waived by ESO and that mechanical components for loading and storing the masks were added to FORS. Since the MOS unit itself is inside the mirror cell (and therefore not accessible) when FORS is mounted the loading mechanism (which had been considered at an earlier stage of the project) would require modifications of the lower instrument housing sections.

Predicted exposure times of FORS have been listed, e.g., by Möhler et al. (1995). Since FORS will normally be used to observe objects below the sky brightness, there is a strong dependence of the exposure times on the seeing when faint unresolved or small objects (such as very distant galaxies) are observed and very good seeing conditions will be a prerequisite for observing close to the limiting magnitude. According to relation (1) for a seeing of 0.5 arcsec (FWHM) we can expect, e.g., for B-Band photometry with $S/N = 5$ minimum exposure times of 6 seconds for $m_B = 25$, 30 minutes for $m_B = 28$, and 20 hours for $m_B = 30$. To just detect a $m_B = 30$ object ($S/N = 3$) we will need about 7 hours,

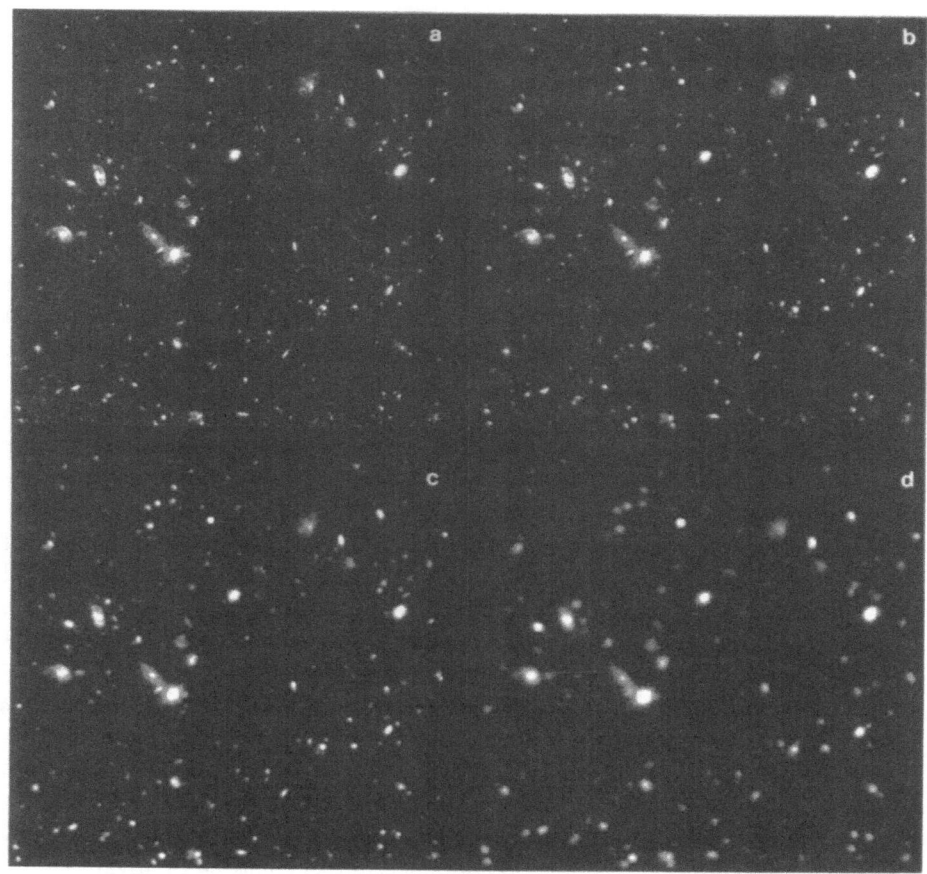

Fig. 3. Section of the Hubble Deep Field observed in the blue spectral range (f450 filter). (a) is the original HST image, (b), (c), and (d) are simulated groundbased FORS images assuming, respectively, 0.3, 0.5, and 1.0 arcseconds seeing

For a 0.3 arcsec seeing (according to the ESO VLT Report No. 62 the best seeing occuring occasionally on Paranal) we get, respectively, 2 seconds ($m_B = 25$), 10 minutes ($m_B = 28$), and 8 hours ($m_B = 30$). A $m_B = 30$ detection will require about 2.5 hours under these conditions.

Again assuming 0.5 arcsec seeing spectroscopy with a spectral resolution R $= 300$ and a $S/N = 10$ will require about 2.4 hours for $m_B = 25$, and 15 hours for $m_B = 26$. At 0.3 arcsec seeing the exposure time for $m_B = 26$, is reduced to about 6 hours. With the same exposure time one may $m_B = 27$ if the S/N and the resolution are reduced by a factor 0.5.

The predictions from relation (1) are confirmed by model experiments which we carried using HST "Deep Field" images. For direct imaging this is illustrated by figure 3 where we present, as an example, in addition to an original blue (f450)

HST image three simulated FORS images of the same region (which, being in the Northern sky, will actually never be observable with FORS). The simulated images were obtained assuming for the FORS frames an exposure time of two hours (which due to the larger aperture of the VLT results in about the same number of photons per object as the much longer ST exposure). Furthermore, we added the noise resulting from the higher sky background of groundbased observations (using ESO sky brightness data for a moonless night) and we convolved the image with various seeing disks.

The simulated images demonstrate very clearly the strong dependence of the information content of FORS frames on the seeing. At 0.3 arcsecond seeing most of the information of the HST image is also present in the FORS image. (The fact that the superior resolution of HST does not result in a larger difference may be due to predominance of extended objects, i.e. galaxies, in the image). With increasing seeing disk fainter objects disappear progressively in the background noise (as predicted from relation (1)) and the morphology of the objects becomes undetectable.

From the above considerations and experiments it is obvious that flexible seeing dependent scheduling will be extremely important for faint object studies with FORS.

3   Project Status

The two FORS instruments are being developed by a consortium consisting of the Landessterwarte Heidelberg and the University Observatories of Göttingen and München, in close cooperation with ESO. FORS I is scheduled to be installed at the VLT Unit Telscope 1 late in 1998. FORS II is expected to follow at the Unit Telescope 3 in about 2000.

At present both instruments are in the manufacturing phase. In the case of FORS I almost all components and subsystems have by now been completed and the subsystem tests have been started. Assembly of FORS I will begin in September 1996, followed by a series of system tests on a Telescope and Star Simulator during the rest of 1996 and the first half of 1997. After FORS I has been delivered to Paranal FORS II will be assembled and tested at the same test facility.

4   Opportunities for Early-Universe Studies

From the properties described above it is clear that the FORS instruments will be particularly useful for observing very faint and distant objects. Under excellent seeing conditions and if the same amount of observing time is invested the direct imaging mode of FORS may (at least in the blue and visual range) reach similar limiting magnitudes as the HST. Since (in spite of the limitations of the VLT Cassegrain focus) FORS has a larger FOV, for many applications FORS imaging may be an economical alternative to the HST. For spectroscopy of moderately faint objects the large aperture of the VLT unit telescopes gives FORS a clear advantage relative to space instruments.

Among the extragalactic problems where FORS has significant potential for important new results are the spectroscopy of individual objects in extragalactic systems, which can be used as distance indicators and for mapping the velocity field and thus the gravitational potential of galaxies. Examples are variable and supergiant stars, planetary nebulae, and globular clusters. For more distant galaxies FORS may help to derive velocity fields and population information from integrated spectra. The high quality MOS unit of FORS will be of advantage for all quantitative investigations of the chemical content and the physical conditions in distant galaxies and AGN. The polarimetric capability of FORS will allow making use of the light collecting power of the VLT to extend polarization measurements to much fainter and highly interesting objects, including distant dusty galaxies and nonthermal optical emitters. Finally we note that the expected high optical quality should make FORS particularly valuable for gravitational lensing studies for the derivation of the mass and mass distribution of galaxy clusters and the derivation of cosmic parameters.

Acknowledgements: This work has been supported by the German Ministery of Education and Science (BMBW grants nos. 052HD50A, 052GO20A and 052MU104).

References

Appenzeller, I., Rupprecht, G. (1992): *The Messenger* 67, 18

Appenzeller, I. et al. (1992): in "Progress in Telescope and Instrumentation Technologies", ed. M.-H. Ulrich, ESO Conference Proceedings No 42, 577

Böhnhardt, H., Möhler, S., Hess, H., J., Kiesewetter, S., Nicklas, H. (1995): in "Scientific and Engineering Frontiers for 8 -10 m Telescopes", ed. M. Iye, T. Nishimura, Universal Academy Press, Tokyo, p. 199

Enard, D. (1988): in "Very Large Telescopes and their Instrumentation", ESO Conference and Workshop Proceedings No. 30, ed. M.-H. Ulrich, p. 17.

Möhler, S., Seifert, W., Appenzeller, I. , Muschielok, B., (1995): in "Calibrating and Understanding VLT and ESO Instruments", ed. P. Benvenuti, ESO Conference Proceedings No 53, 149

Seifert, W., Mitsch, W., Nicklas, H., Rupprecht, G. (1994): in "Instrumentation in Astronomy VIII", SPIE Proc. Vol. 2198, 213

Seifert, W., Fürtig, W., Böhnhardt, H., Nicklas, H. (1992): in "Tridimensional Optical Spectroscopy Methods in Astrophysics", eds. G. Comte, M. Marcelin, Proc. IAU Colloquium No 149, 18

FUEGOS: Multi-Objects and 2-D Spectrograph for the VLT

Paul Felenbok

Observatoire de Paris-Meudon, 92195 Meudon Principal Cedex, France

Abstract. FUEGOS, a multi-object spectrograph is built by a consortium of the following institutes: Observatoire de Paris-Meudon, Observatoire de Toulouse, Observatoire de Geneve, Osservatorio Astronomico di Bologna. Its goal is to provide the European community with a first class equipment allowing high and low resolution spectroscopy, in a large field, for stellar and extragalactic observations. This large field of view, 26 arcmin, provided by the VLT, is quite unique on 8m class telescopes and it is fully exploited by the use of optical fibres. The multi-object spectrograph, fibre fed, will be implemented on the Nasmyth platform of one 8m telescope. FUEGOS is made of two parts: a fibre positioner attached to the Nasmyth rotator and a spectrograph sitting on the Nasmyth platform. The two components are linked together with optical fibres. For the MEDUSA mode, the positioner is of the robot type, made of two heads travelling simultaneously, on a X-Y carriage. It locks magnetically 86 independent optical fibre bundles at object coordinates, on a metallic focal surface. 2-D or integral field spectroscopy is provided through a 600 fibres anamorphoser, called the ARGUS mode. An atmospheric dispersion corrector is supplied for this mode. The spectrograph is of classical type, based on a catadioptric design, with spherical optical correctors. It uses a premonochromator as orders selector. Spectral resolutions are ranging from 2 500 to 45000 and the detector is equivalent to a 4k × 4K CCD. The spectral range extends from 370 nm to 900 nm. Examples of applications of this instrument to Early Universe investigations as well in stellar as in extragalactic research are presented. The first light is expected at the dawn of the third millennium.

1 Introduction

When ESO launched a call for proposals to build a fibre spectrograph for the VLT, our first approach was to propose an instrument that would fulfil the scientific needs of all the astronomical communities, as well stellar as extragalactic. The idea to build a fibre spectrograph for a Nasmyth focus of the VLT was an excellent one, due to the fact that there the field of view is 30 arcmin., very rare on a 8m class telescope and that only fibres could sweep such a surface. This led us to design an instrument fed with 400 individual fibres, with high and low spectral resolution. As a consequence of this choice, we had to build two high resolution spectrographs, each one with a large CCD. The other point was the positioning of 400 fibres on a huge spherical focal plate. A single and even a double positioner was insufficient to lock all the fibres in the focal plane in a reasonable time and exchanging focal plates, as it is done for the 2dF, not possible for such a large equipment. The complexity of the 400 fibres positioner and the cost of two large spectrographs led to a budget in excess of what was foreseeable and this approach had to be revised. The alternative was either to build a 200 fibres instrument with only low spectral resolution or a 80-100 fibres instrument with high and low resolution. The 200 fibres version, in the landscape of the international competition, was felt as weak compared to 2dF and especially to Hectospec on the 6.5m refurbished MMT. Also, for deep redshift surveys, it was thought that a mask instrument was more adequate. So the final choice was to build a 86 fibres instrument, with high and low spectral resolution, capable to use quite all the Nasmyth field, namely 26 arc min. In addition, an integral field unit, called the Argus mode, was implemented in the centre of the field and uses the same spectrograph as the Medusa mode. The selected scientific programmes need a large field and exhibit a low object density. In addition, access to as high spectral resolution as 45 000, makes FUEGOS unique in this niche. Stars and galaxies clusters and QSO's are best candidates for the Medusa mode and galaxies core kinematics, galaxies rotation curves and gravitational lenses are first rank targets for the Argus mode.

2   The Overall Design

FUEGOS is made of two parts, a fibre positioner and a spectrograph, which are linked together with optical fibres. Its general presentation is found in Felenbok et al 1994. The positioner is bolted to the Nasmyth adapter rotator (NAR) and the spectrograph is fixed on the Nasmyth platform. Both elements are linked together with optical fibres. The general layout is displayed on **Figure 1**. The optical fibres allow the relative motion between the mobile focal surface and the fixed Nasmyth platform. The optical fibres are taking care of this relative motion. The field of view (FOV) is 30 arc min. but due to a strong vignetting and a high image quality degradation at the field edge, we limited the FUEGOS FOV to 26 arc min. which is, at F/15, a physical surface with a diameter close to one meter.

3   The Positioner

The positioner function is to lock, on astrometric coordinates, 86 fibres connected to the spectrograph slit. The positioner is fixed on the NAR, and made of a X-Y carriage, on which two robot heads are travelling. **Figure 2** shows the positioner design. This double head design is made to gain a factor of two on the configuration time, that we expect to held between 5 and 10 minutes. Saving on this time is direct benefit on observing time. We choose an X-Y carriage because it is the simpler and the most tested system. It is affected by some defects, as the sag of the rails and backlash in the motion, but all that can be corrected to a satisfactory level. The global positioning precision is foreseen to be $0''15$ with a goal of $0''1$. The focal surface is spherical and not telecentric. To overcome this situation, the focal surface is approximated by four circular steps perpendicular to the incoming beam and crossing at best the focal surface. This leads to a surface lost below 8%.

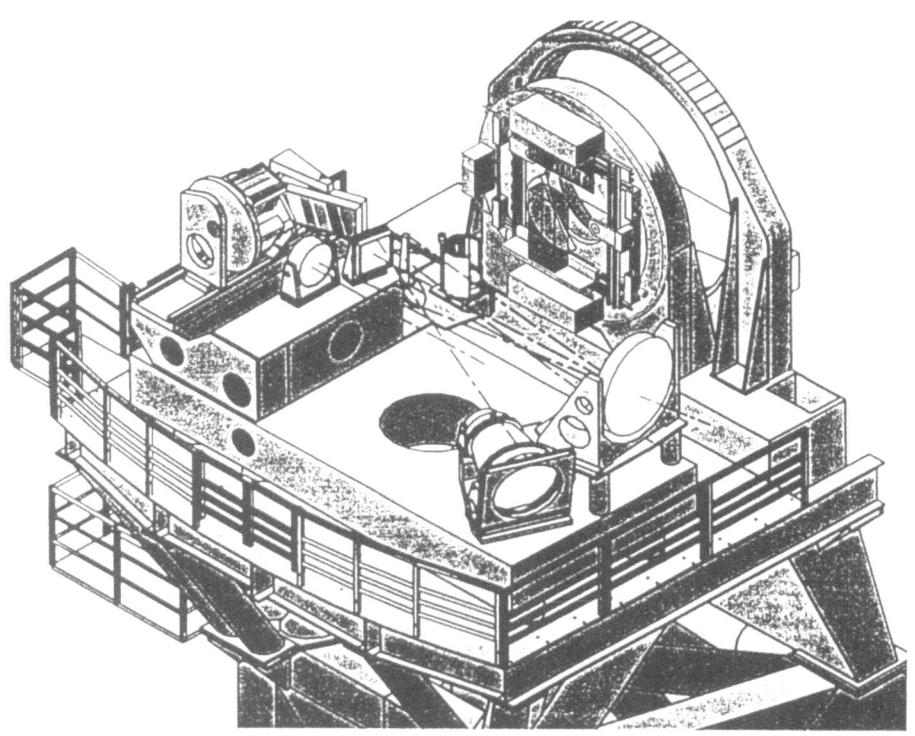

Fig. 1. The overall display of FUEGOS on the Nasmyth platform

The robot heads are picking up the fibres in a circular parking position and dropping them at the dedicated location. To do this, the robot head have to position itself perpendicular to the focal annulus and use a z motion to go in contact to the surface, before opening the gripper jaws. This is a three motions mechanism to add to the gripper pickup action. The robot heads are fitted with a gripper vision system that is able to visualise on a CCD the astronomical target and the fibre output. The match between the object and the fibre is checked before the lunching and can be corrected when the fibre button is still held by the gripper. This is a few seconds exercise, which guarantees the best position for a long integration time. The lunching procedure is over precise and if the gripper position is correct, the locked one will be to.

The assignation of the right fibre to every object is done through a software which is able to handle a full observation programme, divided in single exposures, with a large number of selection and priority criterion. The closest approach of two fibre buttons is 15″.

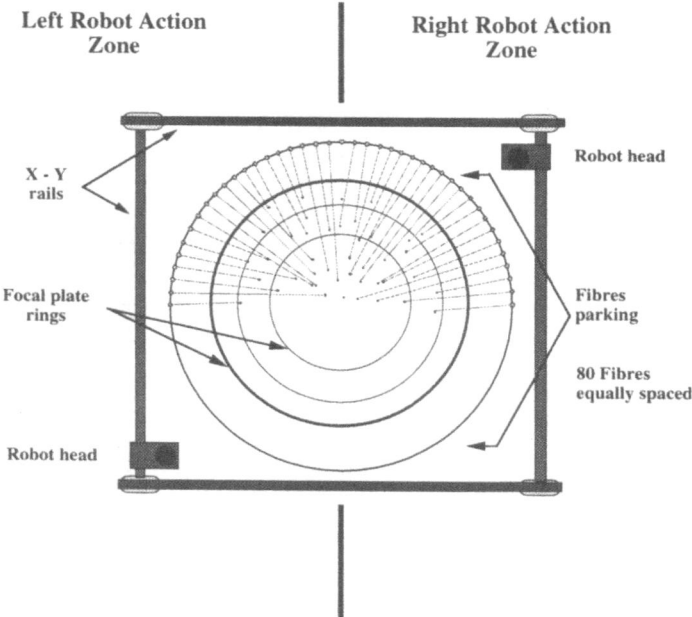

Fig. 2. The double gripper positioner

4   The Medusa Fibre Link and the Integral Field Unit

Each Medusa fibre link is in fact made of 7 fibres to make an image slicer that gives access to high resolution spectroscopy with out the use of a huge grating. A precise description of the design is described in Baudrand et al 1994. The fibre link does not use bare fibres but microlenses at the input and at the output, coupled to the fibres at both ends. At the input, microlenses are needed for several reasons. At first, the VLT focal ratio at Nasmyth is F/15. This is a very poor configuration to feed fibres, because it is subject to focal ratio degradation (FRD), which increases artificially the image size on the sky. To minimise the FRD, we have to try to use as an open input beam as possible, compatible with the maximum admittance angle of the fibre. We choose a design of seven hexagonal microlenses that are seen on **Figure 3**. The microlenses are also artificially baffling the direct telescope light and are illuminating the fibres in a pupil mode. This pupil mode is increasing the spectral stability, because any guiding defect is generating a total light modulation in the fibre without changing its centre of gravity. At the output, where the seven fibres are aligned to form the spectrograph slit, again microlenses are used to exchange the open beam coming out from the fibre to a F/10 beam, well suited for spectrograph collimator design. The input lenses are 390 μ in diameter, the output lenses 460 μ and the silica-silica fibre core, 100 μ.

In alternation with the Medusa mode, an Integral Field Unit (IFU) can be inserted in the focal plane. This unit is designed for 2-D spectroscopy and uses the same spectrograph, with the same spectral resolutions as the Medusa mode. The input and output microlenses and fibre design are the same as for the Medusa mode excepted that the input is made of 600 microlenses feeding the same number of fibres. At the output, the 600 fibres are divided in groups of seven to mimic the Medusa mode on the spectrograph slit. In front of the IFU are implemented three focal enlargers that supply three different samplings.

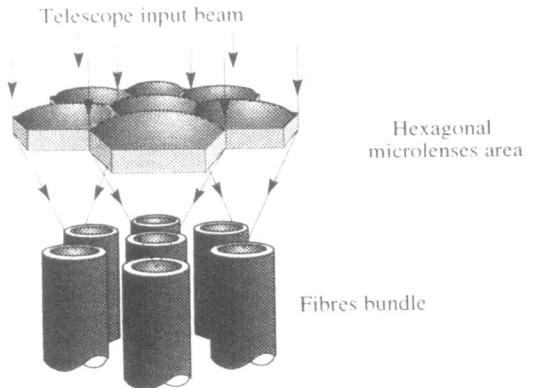

Fig. 3. The microlenses design

5   The Spectrograph

The main constraints on the spectrograph design are coming from the need to achieve high resolution spectroscopy. Two designs were in competition, one of the "white pupil" type and one of the classical type. The white pupil design was able to accommodate more fibres on the slit height but had a lower transparency and could only be associated with a filters orders sorting system. On the opposite, the classical design is only able to handle 86 fibre bundles but with a higher throughput and a monochromator for orders selection. Finally, the classical design was selected and the premonochromator gives a continues wavelength coverage. The spectrograph design is shown on **Figure 4**. Four spectral resolutions are supplied, 45 000, 11 000, 5 000 and 2 500. The lowest resolution is 2 500, because it covers the full spectral domain usable in a single grating order. Lowering the resolution at this level would lead to an uncompleted detector coverage. If lower resolution is needed, it is achieved by binning. Special care is taken to improve spectral stability and our goal is to be able to work on a 50 m/s radial velocity level on exposure of the order of 30 minutes. This performance could be even better, if simultaneous wavelength calibration is done during all the exposure, which is possible by leaving some Medusa fibres in the parking position, facing the calibration beam.

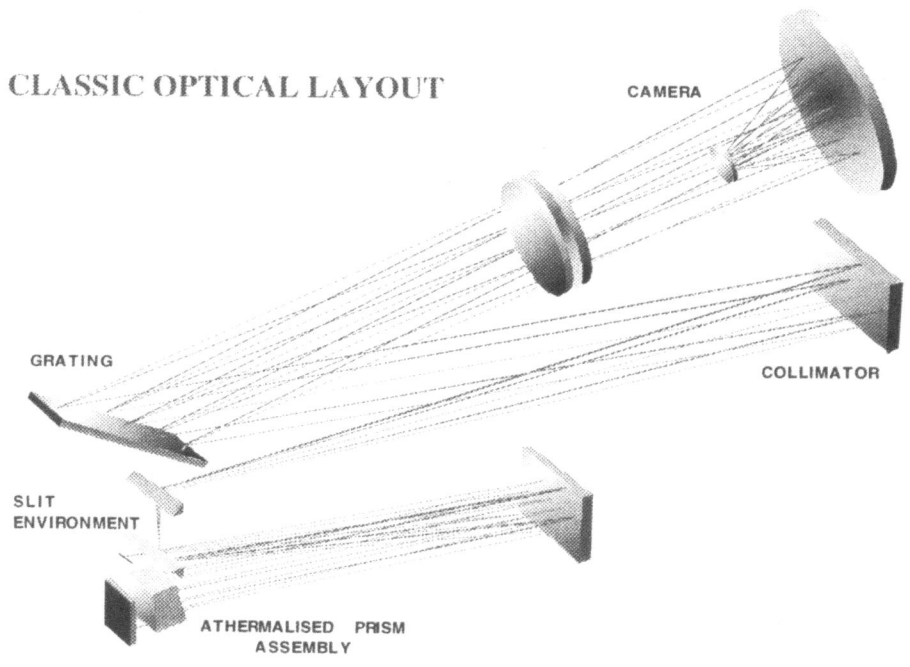

Fig. 4. The spectrograph design

6   Fibre Calibration and Sky Subtraction

This is a major problem with fibre spectroscopy in the Medusa mode. The fibre throughput is affected by fibre motion. When a fibre is changing its curvature, some beams could leave the core path and the transmission is changed. Standard calibration techniques, summarised in Cuby and Mignoli 1994, are leading to a sky subtraction level of 1% and redshift galaxies measurements on 4m class telescopes to a limit magnitude of B = 22 in less than two hours. This is done by fibres calibration on the sky or by beam switching. When a dome screen is available, flat field could be done without moving the fibres from the observing position.

In the FUEGOS case, no dome flat field is supplied and the limiting magnitude that we expect to reach is B = 23. This needs a sky subtraction of few ‰, and a fibre calibration with a high S/N. We are investigating the possibility to use a flat field lamp under permanent flux control that we will move in front of each object and sky fibre, kept in its observing position. The flat field calibration unit is travelling at a speed of 20 cm/s and the exposure in front of each fibre is of the order of one or two seconds. This gives a total calibration time of three to five minutes.

In the IFU mode, fibre calibration of one fibre with respect to the others is quite easy, the fibre bundle being compacted with no relative motion possible between individual fibres. Remains the sky subtraction problem, and two procedures are selected to solve it. If the sky recording could be done close to the object, fixed sky fibres inside of a one arc minute field are used. If the extended object is covering all the one minute field, six Medusa fibre bundles, that is to say 42 individual fibres of the IFU type, are able to record the sky in the 26 arc min. Medusa field, the one arc minute field around the IFU being excluded.

7   Scientific Programmes in Relation with the Early Universe

Collecting scientific information on the physical status of the early universe can be done in different manners, each one with its specific contribution. Stellar spectroscopy of external galaxies as well as the study of the bulge of our own galaxy are supplying high quality information. Of cause, quasars, gravitational lenses and arcs spectroscopy is of prime importance. I will recall some programmes previously presented in different conferences, papers and quote some extracts, that will make benefit from the FUEGOS features.

7.1   The Medusa Mode

At first, we have to recall the density of objects of different type in a FUEGOS field. This can be clearly seen in Figure 5, extracted from a report made by R. Ellis to ESO, March 1992, ref. VLT-TRE-DUR-13300-0237, p. 15. We notice that 20 to 30 QSOs are present in that field, depending the magnitude limit fixed.

Detailed Analysis of Stars in the Galactic Bulge with the VLT
R. Cayrel, P.E. Nissen, 1994, Science with the VLT, ESO Astrophysics Symposia, eds J.R. Walsh and I.J. Danziger, p. 171

In this paper it is shown that the VLT, in combination with the planned Multifibre Area Spectrograph, FUEGOS, can bring a bulk of new basic information on the Galactic Bulge, unobtainable in a reasonable time on 4m class telescopes. Such information would be of great interest for extragalactic astrophysics too, because of the deep connection between bulges, starbursts and QSO phenomenon.

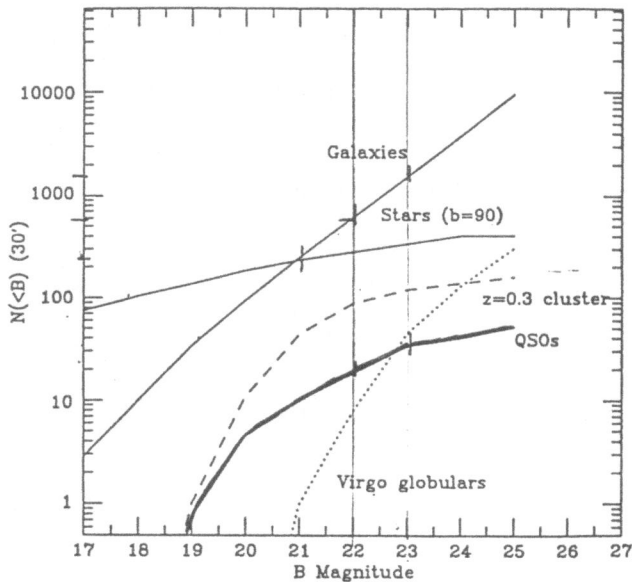

Fig. 5. Density of different objects in a FUEGOS field

Lithium Abundance of Solar Type Stars in Open Clusters
D.R. Soderblom, 1995, IAU Highlights of Astronomy, ed. I. Appenzeller, p. 447

Li is not easily studied because only one spectroscopic features available and that feature is often weak. Thus high spectroscopic resolution and good signal to noise are needed. But Lie is still the most easily studied of the light elements pentateuch (D, He, Li, Be, B). Much recent work on Li in stars has been for Population II objects because of the importance of Li for cosmology... Third, to apply that understanding of stellar physics to Pop II stars so we can correctly infer the Big Bang Li abundance.

FUEGOS and the VLT: A tool to Study the Depletion of Lithium in the Old Globular Cluster Stars
P. François, M. Spite, F. Spite, 1994, Science with the VLT, Poster papers Supplement, ESO Astrophysics Symposia, eds J.R. Walsh and I.J. Danziger, p.17

Lithium is one of the rare elements created during the Big Bang. The determination of the primordial Lithium abundance has important consequences for cosmology. It is possible to check the theories, and thus finally to determine the primordial value of the Lithium by studying the variation of the Lithium abundance in globular cluster stars. With FUEGOS (the VLT multi-fibre spectrograph) it will be possible to make an almost complete survey of a cluster in a few exposures. We propose to determine the variation of the Lithium abundance in Pop II clusters with different metallicities: NGC 6397, 47 Tuc, NGC 6752, M 22 and ω Cen.

Using Binaries to Split the Age Dichotomy of the Universe
M. Mario, Lin Yan, NOAO Newsletter No 45, March 1996, p. 6

An outstanding problem in modern cosmology is the persistent discrepancy between absolute ages of globular clusters and the age of the Universe. Recent ground and space-based measurements of the Hubble constant imply an expansion age of about 10 Gyr for the Universe, while the ages of globular clusters are typically estimated to be 12–16 Gyr. The globular cluster distance scale is crucial in properly defining the luminosities of turnoff stars in globulars and is therefore crucial in defining the age scale. Virtually every well-studied globular cluster contains eclipsing systems. The positive experience with M 71 suggests that it should be possible to do effective follow-up spectroscopy of many of these systems. The ultimate goal is to test critically the reliability of the absolute ages implied by modern evolutionary models and, at the same time, provide a new distance scale for globulars based on the binaries themselves.

Spectroscopy Survey of Faint Quasars
G. Zamorani, FUEGOS Scientific Report, 1993, Ref. VLT-PLA-FUE-13300-0012, p. 20

The faintest complete samples of spectroscopic Active Galactic Nuclei (quasars and Seyfert galaxies) which exist today reach B $\sim$ 22.0. These samples have been obtained at 4m class telescopes either with multifibre systems (AAT $B_{lim} \sim$ 21.8) or with slit spectroscopy (Kitt Peak, ESO). For two different reasons it is almost impossible to improve these results with 4m class telescopes. First, the intrinsic limitations in S/N induced by the number of photons collected by a 4m class telescope makes it very hard to obtain complete quasars samples at B $\geq$ 22. Secondly, the surface density of spectroscopilly confirmed AGNs at B $\sim$ 22.0 ($\sim$ 115 per sq.deg.) is still to low to use in an efficient way the multislit capabilities on small fields of view. For this reasons, a program for a spectroscopic survey of complete quasars samples would be perfectly suited for a high efficient multifiber system with a relatively large field of view and a large number of fibres on a 8m telescope (e.g. FUEGOS).

QSO Absorbers
P. Boissé, P. Petitjean, FUEGOS Preliminary Scientific Report of the phase A, 1993, Ref. VLT-PLA-FUE-13300-0003, p. 70

A large number of absorption lines are detected in QSO spectra blueward the Ly α emission line from the QSO. Most of this lines are identified as redshifted Ly α absorption from neutral hydrogen. The study of their distribution in space is of major importance in the way of understanding the distribution of matter in the Universe. By now, generally, only QSOs with magnitudes smaller than 19 have been extensively observed at the minimum resolution suitable for this studies, of R $\sim$ 5000. Decisive clues for this issue will be derived when QSOs with magnitudes larger than 21 will be routinely observed at the same time in the same field.

7.2   The Integral Field Spectroscopy Mode

The following examples are taken from the CFHT which is the only 4m class telescope used routinely with an integral field spectroscopy capacity. The examples are underlining the limit attained on this size of telescopes.

Gravitational Lensing on QSOs and Related Objects
M.C. Angonin, C. Vanderriest, J. Surdej, FUEGOS Scientific Report, 1993, Ref. VLT-PLA-FUE-13300-0012, p. 30

To summarise, we will limit ourselves to the gravitational mirages on quasars or on others point like sources. The part that we can see of such objects through different optical paths allows several applications:
– by measuring the time delay between the images, it is possible to measure directly the distance to the source
– the light beam cross eventual absorbing clouds between us and the source at slightly different positions; this allows to estimate the size of such clouds
– microlensing can also affect the (macro-)images; as the resulting amplification depends on the size of the source, it is possible to explore the structure of the quasar with a resolution of the micro arc sec. This needs a bidimensional spectroscopy facility on a large telescope, the limit on a 4m class one is presently reached.

Argus Images of 3C254
C.S. Crawford, C. Vanderriest, 1996, CFHT Bulletin No 34, p. 3

Two dimensional spectroscopy is an efficient use of telescope time, allowing simultaneous observation of the ionisation state and kinematic behaviour of the whole EELR. Interpretation of the physical properties of the immediate environment of a quasar is more straightforward than for a radio galaxy, as a better understanding of the ionisation source enables photo-ionisation modelling of the emitting gas. It is clear that both the radio source and the immediate environment of 3C254 play important roles in defining the observable properties of the quasar. We have further ARGUS data for a sample of radio loud quasars $(0.2 < z < 1)$, and intend to investigate the evolution and importance of these influences for a range of orientations, environments and epochs.

Discovery of a QSO companion at $z = 4.7$ with TIGER
P. Petitjean, E. Pecontal, D. Valls-Gabaud, 1996, CFHT Bulletin No 34, p. 2

Quasars at high redshift show absorption lines produced by gas concentrations on the line of sight. Damped absorption systems (for which the absorption is complete), have for origin clouds with high column densities that could be protogalactic disks. The search of this kind of disks is a very active field of observational cosmology, because one can observe this type of absorption at redshifts as high as 4 or 5, which corresponds to the most distant galaxies ever seen. TIGER advantage is to allow in a single shot to observe $3''5$ radius around the quasar in a spectrographic bidimensional mode.

8   Conclusion

As it is clearly seen from the above programmes, FUEGOS is a first order tool for the study of the Early Universe. With its two modes, the Medusa and the Integral Field, it offers a large variety of observational tools. Its high spectral resolution in a large field makes of it a unique instrument for a precise study of distant objects, especially when they are of low density, as QSOs.

Table 1. Table of performances

Number of fibres in Medusa mode	82	
Minimum proximity of two fibres	15 arc sec.	
Number of fibres in the IF mode	574, sky included	
Sampling and field in the IF mode	Sampling	Field
	0.7″	21.6″ × 12.6″
	0.45″	13.9″ × 8.1″
	0.2″	6.2″ × 3.6″
Spectral domain	370 nm	900 nm
Spectral resolution and wavelength range	45 000	17.5 nm 42.5 nm
	11 000	52 nm 104 nm
	5 000	92 nm 185 nm
	2 500	324 nm
Expected sensitivity. S/N=10, exp.=1h	45 000	mV = 17 – 18.5
	11 000	mV = 18.5 – 20
	5 000	mV = 19.5 – 21
	2 500	mV = 21 – 22

References

Felenbok, P., Cuby, J.G., Lemonnier, J.P., Baudrand, J., Casse, M., Andre, M., Czarny, J., Daban, J.B., Marteaud, M., Vola, P., 1994, Instrumentation in Astronomy, SPIE **2198**, 115

Cuby, J.G., Mignoli, M., 1994, Instrumentation in Astronomy, SPIE **2198**, 98

Baudrand, J., Casse, M., Jocou, L., Lemonnier, J.P., 1994, Instrumentation in Astronomy, SPIE **2198**, 1071

The ESO UV-Visual Echelle Spectrograph (UVES)

S. D'Odorico

European Southern Observatory, K. Schwarzschild Strasse 2, D-85748 Garching, Germany

Abstract. UVES is the cross-dispersed, echelle spectrograph built at ESO for installation at the VLT in 3Q 1999. It has two separate blue ($\lambda\lambda$ 300-500nm) and red ($\lambda\lambda$ 420-11000nm) optical paths and two CCD detectors to maximize efficiency. The slit-resolving power product is 40000, with the possibility to reach resolving powers of 80000 (blue) and 110000 (red) with a 2 pixel sampling of the instrumental profile with narrow slits or image slicers. At the typical resolution, the limiting magnitude is $\sim$ 19.5 ($S/N = 10$ in 2hrs for the spectral range 400-650nm). The instrument features an image derotator for observations of extended objects, an atmospheric dispersion corrector and exposimeters to optimize the exposure time to the desired S/N ratio. The expected advantages of UVES in the study of the absorption lines in the spectra of faint QSOs are presented.

1 The VLT Instrumentation Plan and UVES

The VLT Instrumentation Plan includes a first generation of 7 instruments which are now under construction : an overview of their characteristics is given by D'Odorico (1995). Details can be found in the VLT page of the ESO www site. One of the observing modes in which the increase of collecting area provides the largest gain is high resolution spectroscopy. In this mode the final cross-dispersed spectrum is distributed over many pixels and the detector read-out noise is comparable to the photon noise for observations of faint objects. In this regime the gain to be expected in the S/N ratio of the extracted spectrum can go as fast the square of the ratio of the two apertures leading to a very significant improvement with respect to the capabilities of a 4m telescope. This well-known prediction has been demonstrated by the rich harvest of scientific results obtained with the HIRES spectrograph (Vogt,1994) at the Keck telescope (see e.g. the contribution of Tytler in this volume). UVES is the VLT instrument dedicated to high dispersion spectroscopy. Like HIRES it is a cross-dispersed echelle spectrograph and it will be installed at a Nasmyth focus of the second 8m unit telescope. It is one of the two first generation instrument built under direct responsibility of ESO, the other being the infrared imager spectrometer ISAAC (see the contribution by A. Moorwood in this volume). Hans Dekker of ESO is the UVES project manager and optical engineer, the author is the instrument scientist. Table 1 summarizes the past and future milestone of UVES. The progress of the project is being regularly updated in the UVES site (http://www.eso.org/vlt/instruments/uves). Figure 1 shows a 3D view of the instrument.

Table 1. UVES Milestones

History:

- Project kick-off: 5/92.
- Freezing of the layout configuration 5/94
- Preliminary Design Review: 10/94
- Delivery of the two echelle gratings 9.96
- All contracts for main optical components placed 10.96
- Final Design Review 6.96

Planned:

- Start instrument integration: 7/97
- Testing, optimization and calibration of the complete instrument in the Garching Lab (12.97-12.98)
- UT2 available for instrumentation ,UVES installation 8.99
- UVES available for scientific operation 1Q.2000

2   UVES Concept and Modeled Performance

The instrument consists of two main parts: a group of a few functions is attached directly to the Nasmyth rotator and the main body is mounted on a steel table fixed to the floor of the Nasmyth platform and it is covered by a light-tight enclosure which also provides thermal insulation and protection from dust. The light beam from the telescope is eventually focussed on the blue arm slit aperture or is directed to the red arm slit by an high efficiency mirror. Two dichroics are also available for parallel work in the two arms. The blue arm ($\lambda\lambda$ 300-500 nm) and the red arm ($\lambda\lambda$ 420-1100 nm) have an identical layout. They are folded and cross each other to minimize the size of the table on the platform.

The two-arm solution gives the highest efficiency because it permits to optimize coatings, gratings and detectors. The design of both arms is of the white-pupil type (Baranne, 1988). With a beam of 200mm, the off-axis parabolic collimators illuminate the echelle gratings of $214 \times 840 \times 125$ mm with a large blaze angle ($76°$) for enhanced resolution. The echelle gratings are mosaics of two replicas on a single block of ZERODUR, the largest ever made of this type. They are operated in quasi-Littrow mode, that is with the angle of incidence and diffraction equal but in a different plane, to maximize efficiency.

Gratings have been chosen as cross-dispersers to obtain an order separation larger than 12 arcsec over the full spectral range. Advantages of this relatively large separation are the possibility to perform spatially-resolved slit spectroscopy of extended objects or to use image slicers and to properly sample the sky emission background and the scattered light in the spectrum. Each grating unit

Table 2. UVES Predicted Capabilities

	Blue	Red
Wavelength range	300 - 500 nm	420 - 1100 nm
Resolution-slit product, nm / pixel	40 000 , 19 x 10^{-4} at 450nm	40 000 , 25 x 10^{-4} at 600 nm
Max. resolution (2 pix)	80 000	110 000
Pixel scale ($\perp$ to disp)	0.215"	0.155"
Throughput (telescope + UVES,no slit, order aver.)	12% at 400 nm	13% at 600 nm
Limiting magnitude (2 h. exp. time, S/N =10)	19 (R = 50 000) at 400nm	19.5 (R = 60 000) at 600nm
Camera	dioptric F/1.8 field 43.5 mm diam.	dioptric F/2.5 field 87 mm diam.
Baseline CCD 15μm pix (optional format)	2048 x 2048, (2800 x 2048)	4096 x 2048, 4096 x 4096(mosaic of 2)
Echelle (blaze angle ≈ 75^0.8)	41.59 g/mm	31.6 g/mm
Crossdispersers	CD1: 1000 g/mm, λ_b 330 nm CD2: 660 g/mm, λ_b 420 nm	CD3: 600 g/mm, λ_b 510 nm CD4: 316 g/mm, λ_b 800 nm
λλ/frame (typ.)	80 nm at 350nm	100 nm at 600nm
Order separation	≥ 12"	≥ 15 "

mounts two cross-dispersers back-to-back again to maximize efficiency and order separation at all wavelengths.

The cameras are dioptric to provide an external focal plane for easy detector interfacing and upgrading during the lifetime of the instrument. The baseline CCD detectors have a format of 2048×2048 and 4096×2048, 15 μm pixels in the blue and red arm, respectively. The camera field size permits an upgrade to a 2800×2048 format in the blue and a 4096×4096 format in the red. Fig. 2 shows the distribution of the main functions on the table. A short description of the UVES subsystems can be found in the draft version of the User Manual (D'Odorico and Kaper,1996). Table 2 gives a self-explanatory overview of the UVES capabilities and expected performance. These have been computed on the basis of the physical model of the instrument, of the specified performance of the components under procurement and in the case of the echelle of the measured parameters. Table 3 gives more detailed information on the CCDs. At present, we foresee to mount on the blue arm and on the red arm CCDs of the specified properties under development for ESO by EEV and Lincoln Lab. The values in parenthesis in the table indicate best effort, not guaranteed values.

Table 3. Properties of UVES scientific CCDs

	Blue	Red
Format (15μm square pixels)	2800 x 2048 as windowed 4096x2048	4096 x 2048 (4096x4096, mosaic of 2)
Quantum Efficiency	>22 (50) % at 320nm >50 (65) % at 350nm >72 (78) % at 400nm >85 % at 500nm	>60 % at 450 nm >85 % at 600nm >70 % at 800nm >40 % at 900nm
Read-out-noise : at 50 kpix/port/sec at 500 kpix/port/sec	<2 e⁻ rms <6 e⁻ rms	<2 e⁻ rms <6 e⁻ rms
Read-out ports	2	2 (4 on mosaic)
Dark current (at -110°C)	<2 (0.2) e⁻/pix/h	<2 (0.2) e⁻/pix/h
Full Well	>10^5 e⁻	>10^5 e⁻
CTE	>0.99995	>0.99999
Flatness (peak to peak)	< 20 μm	<20 μm

3 Design Highlights

UVES will be the second high-dispersion spectrograph coming into operation on a 8-10m class telescope after HIRES at Keck. We enphasize below those design choices which lead to enhanced capabilities and operational advantages. To be fair, one has to stress that in case of UVES the capabilities are based on model calculations and on the specified but in most cases not yet delivered components, while in the case of HIRES they represent the actual performance obtained at the telescope. The layout based on two **separate blue and red arms** permits to optimize the coatings, the choice of echelle and cross-disperser gratings and CCD detectors in the two spectral ranges and thus maximize efficiency. This is illustrated in Figure 3. The use of **image slicers** combined with the high pixel sampling of the entrance apertures on the detectors gives resolving powers up to 1.1×10^5 in the red and 0.8×10^5 in the blue arm without significant slit losses while in HIRES a slit 0.6 arcsec wide has to be used to reach a resolution of 6×10^4. An **image derotator** can be used for the observation of extended objects and for close pairs up to an angular scale of 15 arcsec in the blue-visual region and up to 20 arcsec beyond 600nm. The **atmospheric dispersion corrector** permits observations over a wide spectral range (e.g. with the dichroics) without additional slit losses even at large zenith angles. The esposimeters on line on both arms will be used to optimize exposure times and to recover the time-dependance of the detected photons even in the presence of variation of the atmospheric transmission.

4   The Study of QSO Absorption Lines with UVES

One of the main science drivers of UVES is the determination of the properties of the universe at high redshift through the study of the intergalactic absorption lines of QSOs. One "plus point" of UVES for work in this field is the high efficiency down to the atmospheric transmission limit which will permit the study of the Ly α forest in QSOs at redshifts as low as $z = 2$ down to $m_B = 19$. A second plus is the high resolving power with good sampling of the instrumental profile which will give optimal results in the study of crowded Ly α forests and of the intervening metal systems . Finally, UVES has an operational advantage being initially the only instrument at the second unit telescope. This will facilitate programs requiring a statistically significant and homogeneous sample of observations.

We have been considering the feasibility of "reference programs" as a way to tune the final parameters and the operational mode of the instrument. The study of the relative abundance of SiIV and CIV in intervening metal systems in the redshift range $1.4 < z < 4.5$ is one of such cases. In the hypothesis of photoionization, the abundance ratios of the two ions as derived from their resonance absorption lines can be used to constrain the spectrum of the UV background light and hence to discriminate between QSOs and galaxies as light sources in the early universe (Miralda-Escudé and Ostriker, 1990). Recent observations do suggest a possible dependance on redshift, with a marked transition at $z \sim 3$ (Songaila and Cowie, 1996; Savaglio et al. 1996) but a large sample of high quality measurements are needed to reach a robust conclusion. The advantage provided by UVES in this type of observations is illustrated in Figure 4. At a resolving power of 60000, the thermal widths of most metal lines and the velocity structures are well resolved and the high S/N ratio in the spectra allows an accurate model fitting of the different components.

5   Acknowledgements

I am grateful to Hans Dekker, the project manager, for a large fraction of the information quoted in this paper, to Guy Hess, the mechanical designer, for the CAD view of the instrument and to Sandra Savaglio for computing the C IV absorption line simulation.

Fig. 1. A CAD view of UVES mounted on the Nasmyth platform of one of the 8m telescope. The cover of the instrument enclosure is partially lifted to give a view of the table where the main functions of the instrument are mounted. The approximate dimensions of the enclosure are 380x380x400 cm

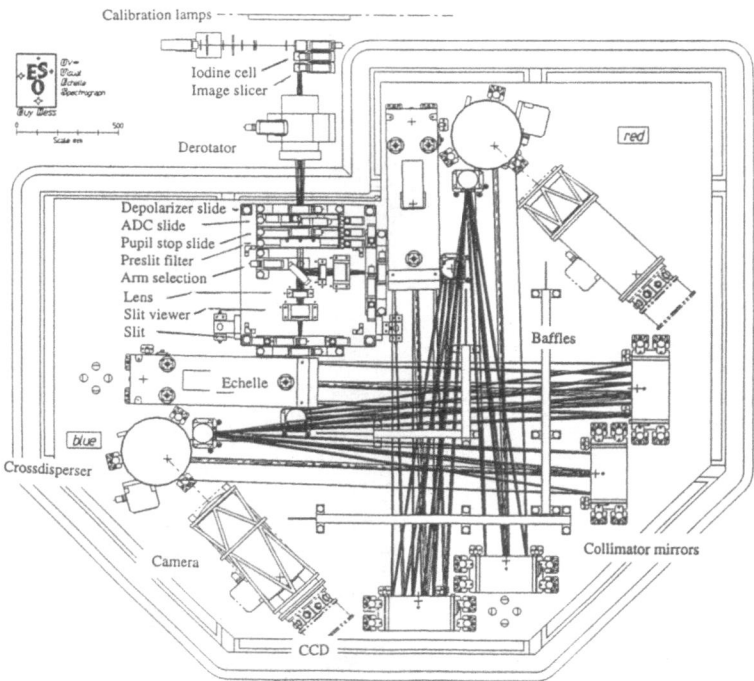

Fig. 2. The layout of UVES with the main functions and the outline of the optical beams in the blue and red arms.

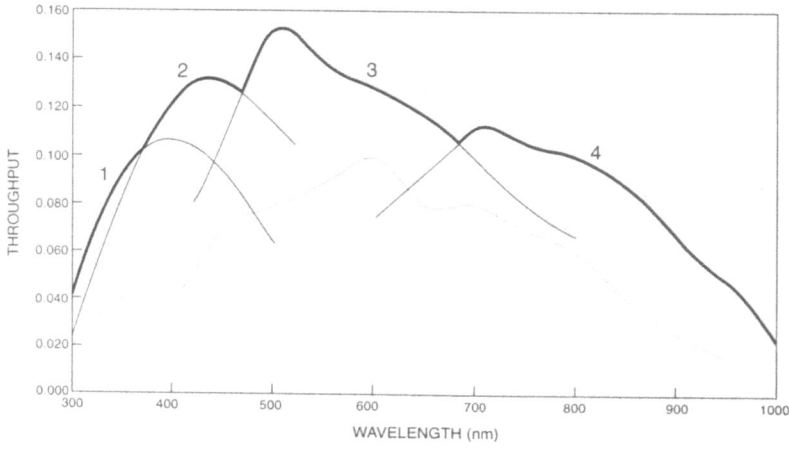

Fig. 3. The predicted efficiency of the telescope (three aluminium reflections) + UVES is shown as heavy line with the numbers identifying the four different cross-dispersers. No slit losses and 0.75 of the peak efficiency of the echelle orders are assumed. The broken lines gives the HIRES efficiency as quoted in the instrument User Manual, upgraded to the use of a CCD with the same QE of those foreseen for UVES.

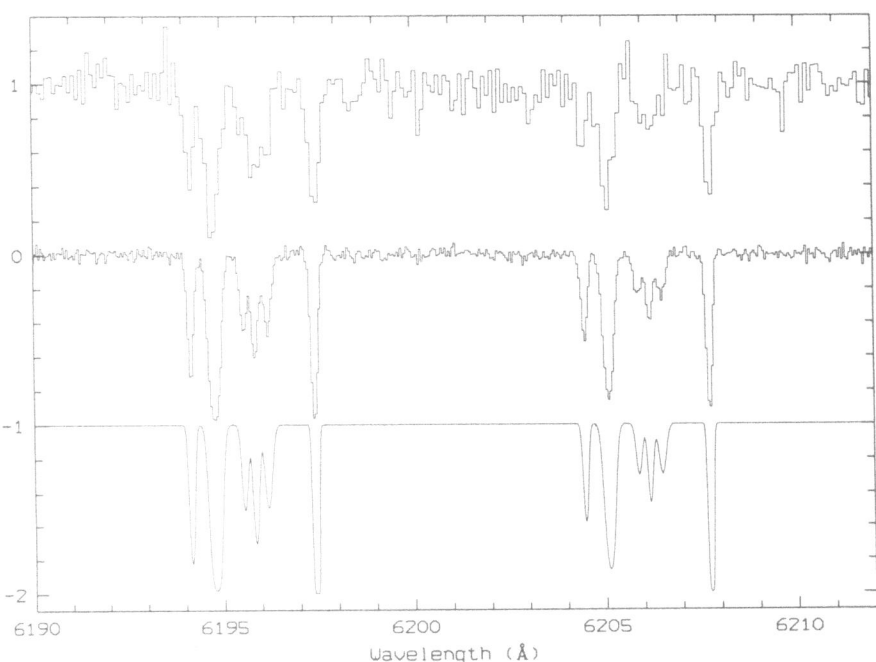

Fig. 4. A comparison between the simulated performance of an echelle spectrograph at a 3.5m telescope (EMMI at the NTT) and that of UVES for the observations of a CIV doublet absorption line system in the spectrum of a $m = 18$ QSO. The input normalized spectrum (lower plot) is that produced by six clouds at $z = 3$ spanning over 150 km/sec in the rest frame, with $logN(HI)$ between 12.9 and 13.9 and values of the b parameter between 3 and 7 km/sec. Upper plot: spectrum from a 4 hours EMMI integration at $R = 25000$. The S/N ratio of 5 is representative of the limit of todays observations at a 4m class telescope for objects of that magnitude. Middle plot: simulated spectrum with UVES, 3 hrs integration, R=60000, predicted $S/N = 40$.

References

Baranne, A. (1988), Proceedings of the ESO Conference on Very Large Telescope and their instrumentation, M.-H. Ulrich editor, pg 1195

D'Odorico, S. (1995) in "Scientific and Engineering Frontiers for 8-10m Telescopes" M.Iye and T. Nishimura edts., Universal Academy Press, Inc. Tokio, pg.51

D'Odorico, S., Kaper, L. (1996) UVES User Manual, Draft Version 1/11/96

Miralda-Escudé, J., Ostriker, J.P.,(1990) ApJ. 350,1

Savaglio, S., Cristiani, S., D'Odorico, S., Fontana, A., Giallongo, E., Molaro. P. (1996) A&A in press

Songaila, A., Cowie, L.L.,(1996) AJ. in press

Vogt, S. (1994)Proceedings of SPIE , D.L.Crawford and E.R.Craine edts. Vol 2198,pg.362

Part 3

Galaxy Evolution at $z \lesssim 1$

Evolution of Normal Galaxies:
HST Morphologies and Deep Spectroscopy

Richard Ellis

Institute of Astronomy, Madingley Road, Cambridge, UK

Abstract. I review progress in understanding the evolution of normal field and cluster galaxies through the combination of HST imaging and ground-based spectroscopy. These data suggest that the bulk of the star formation producing the present-day galaxy population occurred at accessible redshifts, $z < 2$. Furthermore, a surprising amount of the detailed processing that shaped the Hubble sequence and morphology-density relation occurred surprisingly recently. The stage is thus set for a concerted attack on these questions with the present generation of 8-10 metre large telescopes. An important step forward will be the development of efficient survey techniques for the systematic exploration of the $z > 1$ Universe. Some possible approaches are briefly discussed.

1 Introduction and the Role of Instrumentation

These are exciting times in the fields of observational cosmology, galaxy formation and evolution. The rate of publications on bulletin boards like SISSA is one (albeit possibly unreliable) indicator of interest. At the time of writing almost a third of all 1996 preprints thus far are in some way connected with these topics. One is also struck by the optimism in many minds that we are close to resolving one of the outstanding questions of modern cosmology, namely 'When did galaxies form?'. Lest we be over-optimistic about the progress we have apparently made, some of which I review below, it is salutory to recall Zel'dovich's enthusiastic remarks when he summarised the 1977 IAU Symposium on 'Large Scale Structure' in Tallinn. He claimed '..extrapolating to the next symposium in the early eighties, one can be pretty sure that the question of the formation of galaxies and clusters will be solved'!

Although it pays to be cautious in reviews, with results in abundance from the refurbished HST (not least from the remarkable Hubble Deep Field), and the first deep spectroscopic studies emerging from the Keck telescope, it is a timely moment to look ahead and discuss, in the context of the VLT, the respective roles of HST and large ground-based telescopes.

In planning programmes with the VLT, one must surely now take into account the quite remarkable capabilities of HST. HST already offers $\simeq 1$ kpc image resolution at all redshifts in most conventional cosmologies and to this will soon be added long-slit spectroscopy and infrared imaging provided STIS and NICMOS are delivered according to schedule.

HST is *not*, as one often sees quoted, a 'small aperture telescope'. At wavelengths where the ground-based sky is dominated by OH emission, HST offers a significant gain. For background-limited work, HST has an aperture equivalent to a 6.5m ground-based instrument. Furthermore, the wavelength range where this advantage accrues ($0.8 < \lambda < 1.6$ μm) is particularly crucial in studies of the $z > 1$ Universe. Likewise, although HST does suffer from having a small field of view, it is not yet clear whether large aperture telescopes using adaptive optics will recover resolution approaching that of HST over a big enough field to make an appreciable gain. And in the longer term, HST's Advanced Camera will offer an important step forward.

Therefore, I would contend that the traditional 'complementarity' between HST and ground-based telescopes has been overstated and that ground-based telescopes should, at least in this scientific area, respond with imaginative instrumentation. Their big advantage over HST is the ability, at least in principle, to respond rapidly to both technological progress and scientific developments. Rapid interplay is important in both directions. One example of science driving instrumentation might be the discovery that the remarkably small angular sizes of the majority of star forming sources in the Hubble Deep Field has significant implications for ground-based instruments designed to study them. An example of technology driving scientific capability is the recent availability of large format HgCdTe arrays which opens up the prospect of survey spectroscopy at high redshifts. Such interaction should ensure ground-based telescopes maintain a cutting edge in years to come.

Yet, with a few exceptions, much of the instrumentation being developed for the 8 metre telescopes coming online is of the 'monumental' variety. By this I mean large $3-5M general purpose instruments which have taken many years to develop and, because of the cost and associated wide community involvement, often become the brainchilds of committees preoccupied primarily with technical reliability and financial management. Although important considerations, I'm confident many would agree that we should ensure at least some funds are set aside for instrumentation that can be developed quickly to exploit what is, after all, a rapidly moving subject.

This article, therefore, addresses the progress made in understanding the evolution of normal galaxies from both HST and ground-based telescopes. §2 discusses the progress made in the study of cluster spheroidal galaxies together with the important implications these results may have on the visibility of distant primordial sources. I also make some remarks on the environmental effects occurring in clusters. §3 discusses results on the evolution of field galaxies, particularly in the context of the faint blue galaxy problem and deep imaging from HST. Finally, in §4 I speculate briefly on ways in which we might systematically explore the $z > 1$ Universe.

2   The Star Formation History of Cluster Galaxies

Ellipticals were traditionally imagined to be simple stellar systems whose stars were formed in a single burst of star formation at high redshift (Tinsley & Gunn 1976, Sandage & Visvanathan 1978, Bower et al 1992). The popularity of this hypothesis is easy to understand. Such systems are convenient for theorists to model and, as their ancestors should be luminous primaeval galaxies, the hypothesis produce exciting opportunities for observers too!

However, in the past decade, the simple picture has been under concerted attack. Many local ellipticals show evidence for intermediate-age stellar populations (O'Connell 1980) and dynamical peculiarities seen in many (dust-lanes, shells etc) can be readily explained if they formed more recently via the merger of gas-rich systems (Toomre 1977, Quinn 1984). Such arguments suggest a continued formation of ellipticals to quite low redshifts.

The conflict might be resolved if some ellipticals were old single burst systems, whereas the remainder formed via merging of gas-rich disk galaxies. In this case one might expect an environmental and/or mass dependence in the rate of occurrence of intermediate age populations. Reasonably good evidence is emerging that recent star formation is more prevalent in low density environments than in clusters. Rose et al (1994) find the mean stellar dwarf/giant ratio is higher in environments with low virial temperatures. This would be consistent with other environmental trends which indicate accelerated star formation histories in clusters (Oemler 1991). Kauffmann et al (1996) have suggested deep field redshift surveys indicate a paucity of high z red spheroidals, although the reliability with which such systems can be identified using ground-based colours needs to be verified with HST data.

The sensitivity of the U-band light to small numbers of hot, young stars enabled Bower et al (1992) to conclude that no more than 10% of the current stellar population in present-day E/S0s could have been formed in any subsequent activity in the past 5 Gyr as might be the case if merging of gas-rich systems had been involved. This result presents an important challenge for hierarchical theories of structure based on dark matter halos since these predict relatively recent formation eras for massive galaxies. Kauffman (1996) and Baugh et al (1996) have addressed the question quantitatively using a simple prescription for merger-induced star formation. They find that the homogeneity of Bower et al's colour-magnitude (c-m) data can be satisfied if the merging of disk galaxies that produce spheroidals was largely complete by a redshift $z \simeq 0.5$.

Although good progress has been made in tracking the UV-optical c-m relation to higher redshift (Ellis et al 1985, Aragón-Salamanca et al 1991, 1993), without morphological information a major uncertainty remains. The scatter of the photometric c-m relation may be underestimated if some spheroidal galaxies lie blueward of the c-m sequence. This could well be the case if the timescale for dynamical evolution is shorter than that for main sequence evolution as indicated in numerical simulations (Mihos 1995, Barger et al 1996a).

The MORPHS team (Dressler et al 1996) have recently extended the analysis of Bower et al to a sample of three $z \simeq 0.54$ clusters, taking advantage of HST to morphologically classify a sample of 177 faint spheroidal galaxies (Ellis et al 1996a). The clusters cover a range of optical richnesses and X-ray luminosities within a narrow redshift interval specifically chosen so that observed colours are close to rest-frame $U - V$. Overall, the morphological selection of Es appears reliable to $I=23$. However, the distinction between E and S0 galaxies becomes somewhat uncertain fainter than $I=21$-22.

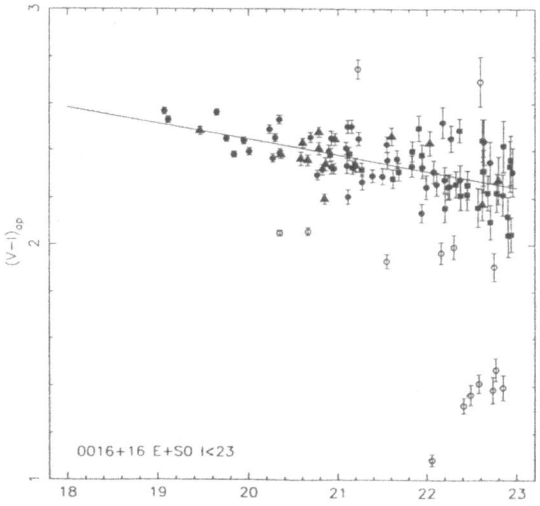

Fig. 1. Colour-magnitude diagram for morphologically-classified galaxies in 0016+16 ($z=0.54$) from the 'MORPHS' project (Ellis et al 1996a). Es are indicated by filled circles, S0's by triangles and E/S0s by squares. Those spheroidals and compact objects known to be field galaxies or discounted from the analysis are indicated by open circles. The small scatter in the rest-frame $U - V$ colours of the spheroidal population in this and other clusters argues that the bulk of the stars formed before a redshift 3.

The rest-frame $U - V$ colour-magnitude relations (Figure 1) for the morphologically-confirmed spheroidals in these clusters show remarkably small scatter (<0.1 mag rms) and there is no evidence the scatter for S0s is any larger than that for Es. After accounting for photometric errors, the intrinsic scatter is about 0.07 mag uniformly to $I=23$ ($M_V = -17.8 + 5 \log h$). Moreover, the combined sample shows little evidence of cross-cluster differences at a level greater than the internal scatter. The most straightforward interpretation is that the bulk of the star formation in cluster spheroidals occurred at least 5 Gyr before a redshift of $z=0.54$, i.e. $z > 3$ unless H_o is low or $\Lambda \neq 0$.

Although this result is consistent with analyses of larger samples of distant clusters (Aragón-Salamanca et al 1993), it does not necessarily apply generally to *all* elliptical galaxies, even those in clusters. Franx & van Dokkum (1996) warn of selection effects that might operate if ellipticals are identified morphologically in ways that guarantee they are least 2-3 Gyr old at any redshift. The most robust statement that can be made is that the stars that form the dominant proportion of red light in 3 $z=0.54$ clusters most likely formed before $z \simeq 3$. Thus there is every incentive to search for the star-forming ancestors of these galaxies.

Some of the above caveats might be minimised by examining the evolution of spheroidal galaxies in terms of mass/light ratios rather than broad-band luminosities. Impressive progress has been made of late in measuring stellar velocity dispersions (Franx & van Dokkum 1996, Bender et al 1996) and HST scale sizes (Pahre et al 1996, Barger et al 1996b) for high redshift galaxies. Preliminary results indicate only modest evolutionary changes consistent with passive evolution from a burst of star formation at high z. However, the selection biases above will only ultimately be overcome with a comprehensive sample of field spheroidals studied in a variety of ways. That Kauffmann et al (1996) and Lilly et al (1995) should come to rather different conclusions from analyses of the same CFRS dataset on the rate of evolution of field ellipticals is an indication of the degree of uncertainty inherent in the presently-available small samples.

How does the above help us to understand the physical origin of the morphology-density relation (Dressler 1980) which, according to observational evidence, was produced at quite low redshifts by environmental effects (Butcher & Oemler 1978, Allington-Smith et al 1993)? Morphological surveys of distant clusters such as those discussed above (Couch et al 1994, Dressler et al 1994) delineate a clear change in the morphological mixture in the sense that the proportion of disk galaxies was much greater in the past, apparently at the expense of a declining S0 population.

One traditional explanation for this evolution, viz. the transformation of spirals to S0s, goes a long way towards explaining the HST results (Dressler et al 1996). The cluster ellipticals provide a backbone of stability over a large range in redshift. By contrast, gas-rich spirals enter the cluster potential and are stripped to produce the abundant S0s we see in present day clusters. The evidence of radial gradients in diagnostic spectral features is particularly convincing support of this picture (Abraham et al 1996a). On the other hand, the small scatter seen in the S0 population at all epochs thus far studied is puzzling and, at least in the core regions, there appear to be some genuine ellipticals which have surprisingly strong $H\delta$ absorption lines indicating recent star formation (Barger et al 1996a).

A worry with all these studies thus far is the absence of a clear understanding of how the clusters were selected. Kauffmann (1996) argues that, by selecting the richest clusters at a given redshift, we are unlikely to be studying the precursors of present-day clusters. An X-ray flux-limited sample may not be much better given our limited physical understanding of the evolution of the X-ray luminosity (Castander et al 1995). Ultimately, one might contemplate undertaking a comprehensive survey using gravitational lensing to locate mass in a well-defined manner. At that stage of complexity, it is probably simpler to undertake very large field surveys if the primary goal is to understand the galaxy population.

3   Evolution of Field Galaxies

The surveys of Bergeron & Boissé (1991) and Steidel et al (1994) based on the identification of the galaxies responsible for Mg II absorption in QSO spectra indicate little change in the overall luminosity function (LF) of regular field galaxies to $z \simeq 0.7$. However, several details remain unclear with the interpretation of such samples in the context of galaxy evolution. These include the weak correlation between impact parameter and Mg II equivalent width (Churchill 1996) and the apparent absence of prominent absorption from gas-rich dwarf galaxies. Although viewing galaxies via their absorbing effects provides a valuable complement to the more traditional redshift surveys, it may be that complex selection biases operate in such samples.

On the other hand, it certainly is reassuring that the LFs of the absorbers can be reconciled with the results emerging from the deep redshift surveys. Impressive progress has been made in the past 2 years from the comprehensive surveys of the CFRS group (Lilly et al 1995), the LDSS/Autofib team (Colless et al 1990, Glazebrook et al 1995a, Ellis et al 1996b) and at the Keck (Cowie et al 1996). Collectively, the number of faint (>20 mag) redshifts is now over 1000 (Table 1) and each provides a complementary insight into the distant population.

The CFRS survey is I-selected and well-suited for sampling the evolving population of massive galaxies to $z \simeq 1$. In contrast, the LDSS/Autofib survey is b_J-selected and particularly tuned to address the nature and distribution of the faint blue population which lies around or fainter than L^*. The wide apparent magnitude range of this survey makes it ideally suited for exploring changes in the *shape* of the luminosity function with redshift. The unusually good spectral resolution also makes it appropriate to examine evolutionary trends as a function of spectral class (Heyl et al 1996). The Keck survey by Cowie et al is K-selected and thus at high redshift is least affected by uncertainties in k-corrections. Moreover, by using multicolour data, Cowie et al have extended their survey in order to construct B and I surveys to slightly deeper limits than has been possible on 4-m telescopes.

Table 1. Deep Redshift Surveys

Reference	N_{gal}	Selection
CFRS Lilly et al (1995)	591	$I < 22$
LDSS Ellis et al (1996b)	1726	$17 < b_J < 24$
Keck Cowie et al (1996)	346	$K < 20$
	203	$B < 24.5$
	130	$I < 22.5$

The empirical trends found by all 3 survey teams agree remarkably well in the sense that the evolutionary changes seen are strongest in the star-forming population which progressively occupy the fainter part of the LF at lower redshift. Given the different perspectives and survey strategies of the teams this is encouraging! Lilly et al (1996) characterise the global evolution in terms of the mean rest-frame luminosity density at various wavelengths and claim this corresponds to an order of magnitude decrease in the volume-averaged star formation rate since a redshift $z \simeq 1$.

The difficulty lies in the physical interpretation of the declining star formation rate in terms of the various populations and, in particular, the question of whether number evolution is required. The traditional 'faint blue galaxy' problem has been sold as requiring an 'excess population' which fades or merges by $z \simeq 0$ (e.g. Ellis 1996). Is it possible to directly identify such an excess population from the redshift surveys?

The CFRS team claim that the LF evolves such that the most rapid change occurs for those galaxies with rest-frame colours bluer than a typical Sbc. They discuss various galaxy populations whose characteristic evolutionary timescale differ. Number density evolution is not invoked. Although a very deep and well-controlled survey (the median redshift is $z \simeq 0.6$), the *time* baseline is fairly modest since there are few galaxies below a redshift of 0.3.

The LDSS team place greater emphasis on the changing *shape* of the LF in the sense that the faint end slope steepens with increasing redshift. Such a behaviour is not unexpected, at least qualitatively, in hierarchical merging. By subdividing their large sample according to [O II] strength and spectral class (Heyl et al 1996), they conclude that the bulk of the evolution can be characterised by a strong luminosity and/or number density evolution of the late type population. Crucial to the need for number evolution is the assumed form for the local LF. A flat (Schechter $\alpha=-1$) local LF would imply fairly dramatic changes have taken place between $z \simeq 0.5$ and today, and thus it is reasonable to question the reliability of the local LF before accepting this conclusion (McGaugh 1994).

Glazebrook et al (1995a), Ellis et al (1996b) and Cowie et al (1996) have each placed constraints on the faint end slope of the local LF from the absence of low z galaxies in their deep B-selected surveys and reject the hypothesis (Gronwall & Koo 1995) that the counts can be understood solely in terms of a local LF with a steep faint end slope. On the other hand, it is clearly worrying that the same authors admit a LF normalisation (ϕ^*) markedly higher than traditional estimates (e.g. Loveday et al 1992). Part of the difficulty may be the degree to which brighter photographic photometry can be effectively tied to that of the faint surveys (Bertin & Dennefeld 1996) and this is closely tied to the important role that surface brightness plays in isophotal surveys (Ferguson & McGaugh 1994).

What evidence is there for short-term star formation activity such as might be expected if merging is an ingredient driving the evolution? Broadhurst et al (1988) first suggested that the faint blue star-forming sources had spectral characteristics indicating short-term bursts rather than a gradual decline in the star formation rate. More recently, Heyl et al (1996) have found a similar effect in the more extensive LDSS and Autofib survey (Figure 2). In the blue-selected samples, a high proportion of the spectra are unlike those of local spirals. For this class of object there is also a marked increase in the median [O II] 3727 Å equivalent width with redshift (Figure 3).

Assuming galaxy morphology is a marker with some degree of permanence,

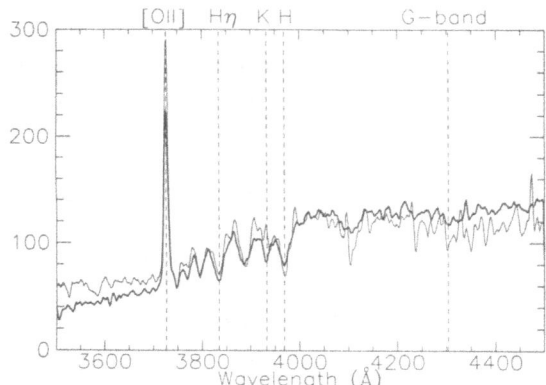

Fig. 2. Coadded spectra for late-type spiral galaxies in the LDSS/Autofib survey (Heyl et al 1996). The bold curve is the coaddition of those with z <0.2 while the light curve is for those with 0.2< z <0.5. The higher redshift sample shows absorption lines whose increased strength is indicative of recent ($\simeq$1 Gyr) star formation.

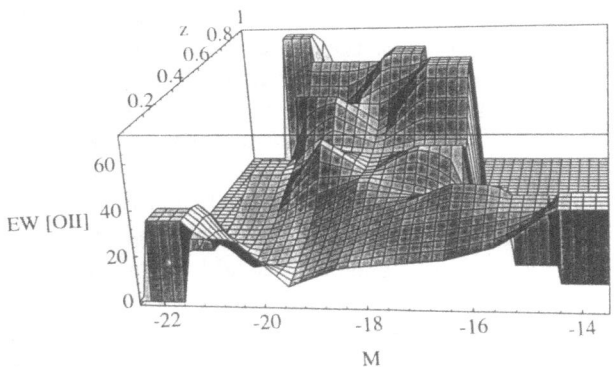

Fig. 3. Evolution of the median [O II] equivalent width for late-type spirals in the LDSS/Autofib survey (Heyl et al 1996).

HST galaxy counts can provide valuable insight into the galaxies that are responsible for the LF changes discussed above. Glazebrook et al (1995b) and Abraham et al (1996b) provide type-dependent counts to I=25 from the Medium Deep Survey and the Hubble Deep Field and claim a remarkable overabundance of irregular galaxies compared to local samples. Many are certainly suggestive of merging, although firm quantitative proof is difficult to establish (Neuschaefer et al 1995). Although these counts probe much deeper than the current redshift survey limits, the basic conclusion is that the number of regular spheroidal and disk galaxies to $I \simeq$22-23 is fairly close to no evolution expectations, whereas the bulk of the excess population seems confined to the irregular sources.

But how reliable are the morphological assignments in these faint samples? Abraham et al (1996c) have addressed this question via the development of more quantitative classification criteria based on assymmetry and light concentration and via simulations that take account, on a pixel-by-pixel basis, of differential k-corrections and surface brightness dimming. For z<1, where the bulk of the I <22 galaxies lie, the biases are quite small and amount, for example, to confusion only between Sdm, Irregular or merging systems rather than gross misclassification such as movement from Sbc to Irr. As local irregulars and late type spirals should not be difficult to see if they are still actively producing stars, their abundance in the HST counts compared to a paucity in local data is an important result.

The HST and ground-based data thus both assign a high proportion of the evolution to a single class, namely late-type spirals and irregulars. The mean luminosity and perhaps number density of this class of sources is rapidly decreasing with time. This is not to say there is not room for some evolution in the intermediate spirals or ellipticals. The CFRS and LDSS teams have joined forces and will soon have over 300 galaxies for which HST images and spectra will be available to I=22 and b_J=24. Schade et al (1996) have already claimed quite significant evolution in the surface brightness of disk galaxies with redshift and this work is being extended to the larger CFRS+LDSS database. Additionally, the luminosity and redshift distribution of morphologically-distinct samples is being analysed in the context of the question of whether number evolution is required.

The most significant conclusion thus far from the redshift surveys is the global evolution of the population (Lilly et al 1996) rather than results based upon dissection into individual types whose physical significance remain unclear. Nonetheless, for a detailed understanding of the origin of disk galaxies and a resolution of the 'faint blue galaxy problem' it is clear this is the way to go although possibly very large joint HST and ground-based spectroscopic samples will need to be gathered.

4   Exploring $z > 1$

The Hubble Deep Field has provided an exciting first glimpse into the distant Universe. By extending the galaxy counts to a regime affected by ground-based confusion, it is now clear that the surface density of very faint sources exceeds that predicted on the basis of most reasonable local luminosity functions, suggesting either: (i) galaxies are not conserved, viz. in the HDF we are seeing sub-units destined to become larger systems, or (ii) we live in a world model where $\Lambda \neq 0$.

Although Λ-dominated models remain popular in theoretical circles, the observational constraints are getting tighter. If the excess counts were produced primarily by huge volumes, one would see similar effects in the K-band (Djorgovski et al 1995). Spatially-flat models with $\Lambda \neq 0$ would also produce accelerating universe which seem in conflict with the constraints emerging from the distant supernovae programmes (Perlmutter et al 1996, Leibundgut, this volume).

There is growing evidence that the bulk of the star formation that made the present day population occurred between 1< z <3. Firstly, the steep rise in star formation rate to redshift $z \simeq 1$ is a major pointer to a low redshift of mean star formation (Fall et al 1996). Secondly, a key result, suggested initially by Guhathakurta et al (1990), is the rarity of $R < 25$ star-forming sources whose U band flux indicates a strong Lyman limit consistent with $z > 3$. The same result has been derived by Steidel and collaborators with the important advance that candidate Lyman limit sources beyond $z \simeq 2.3$-3 have now been spectroscopically verified using the Keck telescope both in the HDF (Giavalisco et al 1996) and elsewhere (Steidel et al 1996).

If the bulk of the star formation occurred at low redshift, could the high z star-forming galaxies recently identified be the ancestors of the present day spheroidal galaxies? HST images of those HDF galaxies satisfying Abraham et al's (1996c) Lyman-limit criteria are shown in Figure 4 and are highly suggestive of hierarchical merging of sub-units in the manner predicted by Kauffmann (1996) and Baugh et al (1996). However, a crucial pointer in this regard would be an estimate of the mass of such distant systems, either from resolved spectroscopy or absorption line widths (c.f. Giavalisco, this volume).

It is important to recognise that the Lyman-limit method, although remarkably effective, provides only a limited view of the high redshift Universe, namely star-forming sources within a narrow redshift range. Our inability to immediately 'connect' this interesting population with low z counterparts exemplifies the need to provide complete redshift coverage so the evolution of the various samples can be directly tracked.

How are we going to *systematically* explore the galaxy population beyond $z \simeq 1$ in a manner similar to that which has been so successful with the 4-m telescopes for $z < 1$? If our hypothesis concerning the star formation history is correct, the redshift interval $1 < z < 2.5$ is particularly important yet, paradoxically, this is a range which will be the hardest to systematically explore with the first tranche of 8-m instruments. The basic difficulty is the absence, at optical wavelengths, of any of the familiar diagnostic features. Although one is encouraged by the detection of absorption lines in the Keck spectra to faint limits, I am confident those same sources would be far easier to study at infrared wavelengths where their emission line spectra would be quite prominent.

At this conference we have witnessed three promising techniques. Firstly, the weak lensing signals seen in a variety of clusters provide valuable information on the mean statistical distance to an objective sample of very faint images viewed through the cluster lens (Fort, this volume). In certain clusters, the mass models are sufficiently tightly constrained (Kneib et al 1996) that the modelling can be used to estimate distances to very faint sources viewed through the lens. The process can be iteratively improved via arclet redshift measurements to make quite precise statements about extremely faint galaxies. A dramatic verification of this 'inversion' technique is the $z=2.515$ arc in Abell 2218 (Ebbels et al 1996, Figure 5) originally predicted to be a $R_{true}=24.1$ galaxy at $z \simeq 2.8 \pm 0.3$. This is just the beginning of several approaches which, through calibration spectroscopy and HST imagery, will provide mean distances to a variety of galaxy populations at very faint limits. As these methods are geometric, they avoid many of the

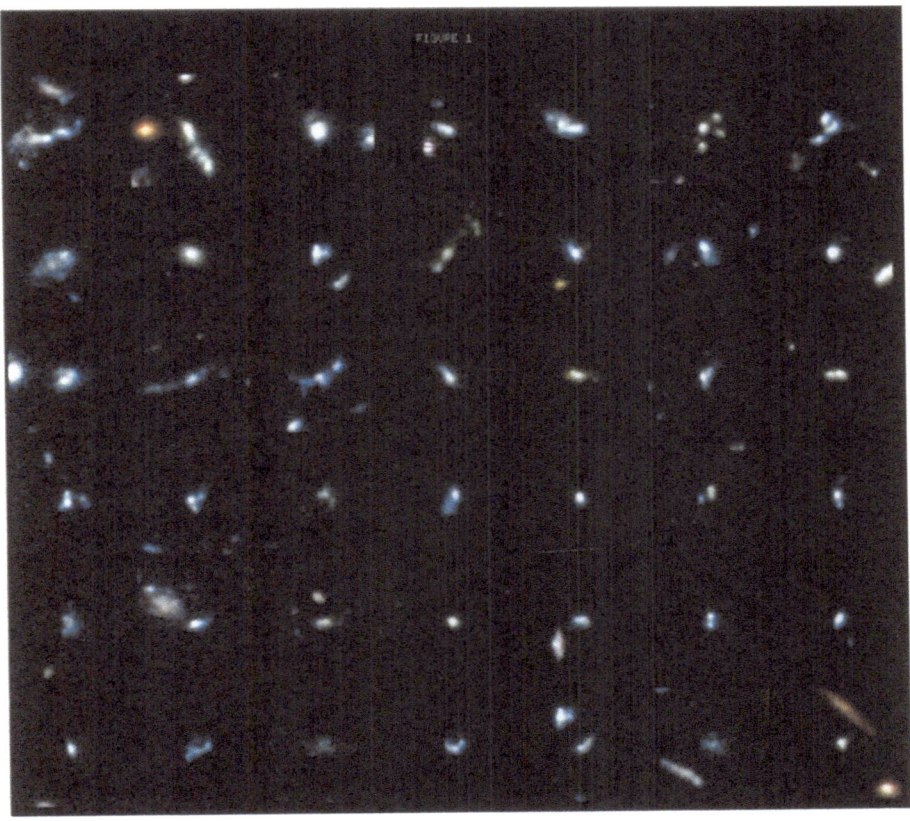

Fig. 4. Hubble Deep Field images of those galaxies selected by Abraham et al (1996b) to have Lyman limit drop outs suggesting they lie at redshifts $z > 2.3$. A significant fraction have since been spectroscopically confirmed (Giavalisco et al 1996).

biases inherent in traditional surveys.

Secondly, a high priority must be the effective use of the panoramic infrared spectrographs on 8-10m telescopes to sample the wavelength range where redshifted [O II] and $H\alpha$ lie. The troublesome OH background necessitates high dispersion which is costly in detector pixels. The VIRMOS project (LeFevre, this volume) is an imaginative and ambitious solution to this problem. The third technique is the highly complementary CADIS narrow band imaging programme (Meisenheimer, this volume). On the relative merits of these two techniques, I

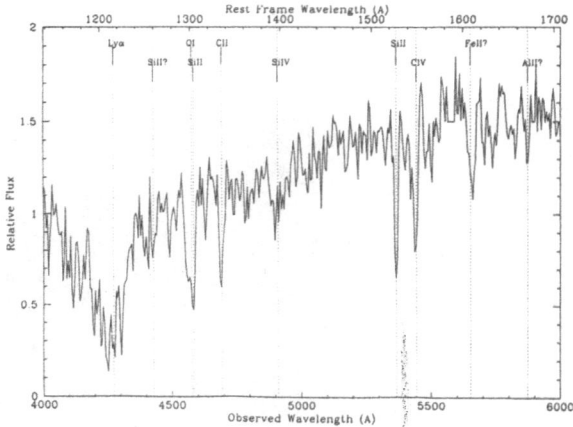

Fig. 5. Spectrum of a faint arc (#384) in the rich cluster Abell 2218 obtained with LDSS-2 on the WHT (Ebbels et al 1996). The spectroscopic redshift of 2.515 agrees closely with that inferred from the lensing inversion method developed by Kneib et al (1996) and illustrates the potential of determining the mean redshift of a very faint population of galaxies viewed through a lensing cluster.

believe it will be important to execute some scouting missions to determine the likely distribution of emission line strengths before finalising the design parameters of a major commitment like VIRMOS. One would hardly contemplate embarking on 2dF or the Sloan Digital Sky Survey without having gathered a representative set of optical galaxy spectra.

5   Acknowledgements

I thank Jacqueline Bergeron and Christina Stoffer for their generous financial assistance which made it possible for me to attend this enjoyable meeting. I acknowledge useful conversations with Carlos Frenk, Simon Lilly, Guinevere Kauffmann and Alvio Renzini. I thank all my collaborators on the lensing and cluster and field galaxy programmes for allowing me to discuss our work in the context of this review.

References

Abraham, R.G., Smecker-Hane, T.A., Hutchings, J.B., Carlsberg, R., Yee, H.C., Ellingson, E., Morris, S., Oke, J.B. & Rigler, M. 1996a, Ap J, in press (astro-ph/9605144)
Abraham, R.G., Tanvir, N.R., Santiago, B.X., Ellis, R.S., Glazebrook, K. & van den Bergh, S., 1996b, MNRAS 279, L47.
Abraham, R.G., van den Bergh, S., Glazebrook, K., Ellis, R.S., Santiago, B.X., Surma, P. & Griffiths, R.E. 1996c, Ap J, in press (November).
Allington-Smith, J.R., Ellis, R.S., Zirbel, E. & Oemler, A. 1993, Ap J, 404, 521.

Aragón-Salamanca, A., Ellis, R.S. & Sharples, R.M. 1991, MNRAS, 248, 128.

Aragón-Salamanca, A., Ellis, R.S., Couch, W.J. & Carter, D. 1993, MNRAS, 262, 764.

Barger, A., Aragón-Salamanca, A., Ellis, R.S., Couch, W.J., Smail, I. & Sharples, R.M. 1996a, MNRAS, 279, 1.

Barger, A., Aragón-Salamanca, A., Butcher, H., Couch, W.J., Dressler, A., Ellis, R.S., Oemler, A., Sharples, R.M. & Smail, I. 1996b, in preparation.

Baugh, C., Cole, S. & Frenk, C.S. 1996, MNRAS, submitted (astro-ph/9607056)

Bender, R., Ziegler, B. & Bruzual, G. 1996, Ap J, 463, L1.

Bergeron, J. & Boissé, P. 1991, A&A, 243, 344.

Bertin, E. & Dennefeld, M. 1996, A&A, in press (astro-ph/9602110)

Bower, R.G., Lucey, J.R. & Ellis, R.S. 1992, MNRAS 254, 601.

Broadhurst, T.J., Ellis, R.S. & Shanks, T. 1988, MNRAS 235, 827.

Butcher, H. & Oemler, A. 1978, Ap J, 219, 18.

Castander, F.J., Bower, R.G., Ellis, R.S., Aragón-Salamanca, A., Mason, K.O., Hasinger, G., McMahon, R.G., Carrera, F.J., Mittaz, J., Pérez-Fournon, I. & Lehto, H.J. 1995, Nature, 377, 39.

Churchill, C. W. 1996, in *Young Stars & QSO Absorbers*, eds. Viegas S.M. et al, ASP in press (astro-ph/9604127)

Colless, M., Ellis, R.S., Taylor, K. & Hook, R. 1990, MNRAS 244, 408.

Couch, W.J., Ellis, R.S., Sharples, R.M. & Smail, I. 1994, Ap J, 430, 121.

Cowie, L., Hu, E. & Songaila, A. 1996, AJ, in press (astro-ph/9606079)

Dressler, A. 1980, Ap J, 236, 251.

Dressler, A., Oemler, A., Butcher, H. & Gunn, J.E. 1994, Ap J, 430, 107.

Dressler, A., Couch, W.J., Oemler, A., Ellis, R.S., Smail, I., Butcher, H. & Sharples, R.M., in preparation.

Djorgovski, S., Soifer, B.T., Pahre, M.A., Larkin, J.E., Smith, J.D., Neugebauer, G., Smail, I., Matthews, K., Hogg, D.W., Blandford, R.D., Cohen, J., Harrison, W. & Nelson, J. 1995, Ap J, 438, L13.

Ebbels, T.M.D., LeBorgne, J-F., Pellò, R., Ellis, R.S., Kneib, J-P., Smail, I. & Sanahuja, B. 1996, MNRAS, in press (astro-ph/9606015)

Ellis, R.S. in *Unsolved Problems in Astrophysics*, eds. Bahcall, J.N. & Ostriker, J.P., Princeton University, in press (astro-ph/9508044)

Ellis, R.S., Couch, W.J., MacLaren, I. & Koo, D.C. 1985, MNRAS, 217, 239.

Ellis, R.S., Smail, I., Dressler, A., Couch, W.J., Oemler, A., Butcher, H., Sharples, R.M. 1996, Ap J, submitted (astro-ph/9607154)

Ellis, R.S., Colless, M., Broadhurst, T.J., Heyl, J. & Glazebrook, K. 1996, MNRAS 280, 235.

Fall, S.M., Charlot, S. & Pei, Y.C., 1996, Ap J, 464, L43.

Ferguson, H. & McGaugh, S. 1994, Ap J, 440, 470.

Franx, M. & van Dokkum, P.G. 1996, *New Light on Galaxy Evolution*, eds. Bender, R. & Davies, R.L., in press (astro-ph/9603029)

Glazebrook, K., Ellis, R.S., Colless, M., Tanvir, N. & Allington-Smith, J. R. 1995a, MNRAS, 273, 157.

Glazebrook, K., Ellis, R.S., Santiago, B.X. & Griffiths, R.E., 1995b, MNRAS, 275, L19.

Giavalisco, M., Steidel, C.C. & Macchetto, F.D., 1996, Ap J Lett, in press (astro-ph/9603062)

Gronwall, C. & Koo, D. 1995, Ap J, 440, L1.

Guhathakurta, R., Majewski, S. & Tyson, A.J. 1990, Ap J, 357, L9.

Heyl, J.S., Colless, M., Ellis, R.S. & Broadhurst, T.J., MNRAS, submitted.

Kauffmann, G. 1996, preprint (astro-ph/9502096)

Kauffmann, G., Charlot, S. & White, S.D.M., 1996, MNRAS, in press (astro-ph/9605136)

Kneib, J-P., Ellis, R.S., Smail, I., Couch, W.J. & Sharples, R. 1996, Ap J, in press (astro-ph/9511015)

Lilly, S.J., LeFévre, O., Crampton, D., Hammer, F. & Tresse, L. 1995, Ap J, 455, 50

Lilly, S.J., LeFévre, O., Hammer, F. & Crampton, D. 1996, Ap J L, in press (astro-ph/9601050)

Loveday, J., Peterson, B.A., Efstathiou, G. & Maddox, S.J. 1992, Ap J, 390, 338.

McGaugh, S. 1993, Nature, 367, 538.

Neuschaefer, L., Ratnatunga, K.U., Griffiths, R.E., Casertano, S. & Im, M. 1995, Ap J, 435, 559.

O'Connell, R.W. 1980, Ap J, 236, 340.

Oemler, A. 1991, in *Clusters & Superclusters of Galaxies*, ed. Fabian, A.C., Kluwer, p29.

Pahre, M, Djorgovski, S., de Carvalho, R. 1996, Ap J, 456, L79.

Perlmutter, S. et al, in *Thermonuclear Supernovae*, eds. Canal, R et al, in press (astro-ph/9602122)

Quinn, P. 1984, Ap J, 279, 596.

Rose, J.A., Bower, R.G., Caldwell, N., Ellis, R.S., Sharples, R.M. & Teague, P., 1994, A J, 108, 2054.

Sandage, A. & Visvanathan, N. 1978, Ap J, 228, 81.

Schade, D.J., Lilly, S.J., Crampton, D., Hammer, F., LeFevre, O. & Tresse, L. 1996, Ap J, 451, L1.

Steidel, C., Dickinson, M. & Persson, E. 1994, Ap J, 437, L75.

Steidel, C.C., Giavalisco, M., Dickinson, M. & Adelberger, K.L., 1996b, Ap J. Lett, in press (astro-ph/9604140).

Tinsley, B.M. & Gunn, J.E. 1976, Ap J, 302, 52.

Toomre, A. 1977, in *Evolution of Galaxies & Stellar Populations*, eds. Tinsley, B.M. & Larson, R., Yale Univ. Press., p401.

The Canada France Redshift Survey

F. Hammer[1], S. Lilly[2], D. Crampton[3], O. Le Fèvre[1], H. Flores[1] and D. Schade[2]

[1] DAEC, Observatoire de Paris-Meudon, F-92195 Meudon, France
[2] Department of Astronomy, University of Toronto, Canada,
[3] DAO, University of Victoria, Canada

Abstract. The properties of field galaxies are investigated based on a sample of $\sim$ 600 galaxies from z=0.05 to z=1.3. The strongest evolutionary features are found in the luminosity function of the blue galaxies, in the emission line properties and in the disk surface brightness. It is argued that the star formation rate was higher in the past than today but other mechanisms (such as evolution of the opacity) are also likely contributing to the observed evolution.

1 Introduction

During the last decade, considerable effort has gone into obtaining spectra of faint field galaxies. In their pioneering work, Broadhurst et al. (1988) and Colless et al. (1990) gathered several hundreds of galaxy spectra selected in B. They found a surprisingly low average redshift, and their data were apparently consistent with a uniform increase in the apparent number density of galaxies. A basic problem in working with B-band selected galaxies is that, at high-z, it corresponds to selection in the ultraviolet. Such samples are likely dominated by the most active objects, and their understanding requires precise knowledge of the local galaxy properties in the UV. Selecting objects in the I band substantially reduces the problems of working at high redshift, including minimizing K corrections up to z = 0.9, where the I light correspond to the B band at rest.

The Canada-France Redshift Survey (CFRS) has produced a unique sample of spectra of 591 field galaxies, 6 QSOs and 200 stars with $17.5 < I_{AB} < 22.5$ (Lilly et al. 1995a, hereafter CFRS I; Le Fèvre et al. 1995, hereafter CFRS II; Lilly et al. 1995b, hereafter CFRS III; Hammer et al. 1995a, hereafter CFRS IV). The completeness achieved is high (85%, Crampton et al. 1995, hereafter CFRS V), and deep imaging in I was carried out for all the sample in order to avoid selection biases against low surface brightness galaxies (CFRS I). Deep B, V and K photometry is also available for most of the galaxies.

2 Redshift Distribution and Luminosity Function

Redshift distribution of the 591 CFRS galaxies is between z=0.05 to z=1.3, with a median at z=0.56. The sample is enough large to split the galaxy population in various redshift, luminosity and color bins. For example it allows to investigate the properties of the B band at rest up to z=1, and to compare them to the current surveys of local galaxies. To study the luminosity function and its evolution, it is useful to account for several redshift bins and to compare the galaxy population properties within the CFRS sample. This avoid any undesirable effects related to comparisons with external references, such as different selection criteria or different observational technics. We have also chosen to split the sample in two color bins, one containing galaxies redder than Sbc at rest, the other the bluer objects.

We find (Lilly et al, 1995c, hereafter CFRS VI) that the luminosity function of the population of red galaxies shows virtually no change up to z=1. This suggests that the population of massive, reasonably quiescent galaxies, was to a very large degree "in place" at $z \sim 1$. Interestingly, we see no evidence for significant luminosity evolution (of order 0.8 magnitudes) that might be expected from the passive evolution of the stellar population. The luminosity function of bluer galaxies shows a substantial enhancement at $z > 0.5$ in the numbers of galaxies with roughly present-day L* ($M_{AB}(B) \sim -21.0$). This evolving blue population, now clearly revealed in the luminosity function is responsible for steepening the galaxy number counts down to $B \sim 24$. The observed changes in the blue luminosity function between $z \sim 0.3$ and $z \sim 0.7$ could be due to either (a) a uniform brightening of all blue galaxies by about 1.5 magnitudes, (b) an increase in the number density of galaxies of constant luminosity by a factor of three, (c) a combination of both effects, or (d) an even more complicated scenario involving galaxies crossing the red-blue divide. Further information on the environments, morphologies, kinematics, and stellar populations of the galaxies (see below) should allow us to discriminate between these scenarios and will lead to an understanding of the physical evolution of individual galaxies as well as the statistical evolution of the population as a whole.

3   Spectral Properties

3.1   Emission Line Galaxies

We have measured emission line intensities and equivalent widths as well as color indices, based on 200Å wide spectral regions which were calibrated by photometric measures (Hammer et al, 1996, hereafter CFRS XIV). The fraction of galaxies with significant emission lines ($W_0(OII) > 15$Å) increases from $\sim$ 13% locally to over 60% at z> 0.5 (Figure 1).

At all redshifts we find that a significant fraction of emission lines galaxies (40%) have large Balmer break indices, which would correspond to $W_0(H\delta) >$ 5Å (Figure 2). These galaxies harbour a significant population of A stars, and their large number suggests that most of the emission lines galaxies in the CFRS are experiencing relatively long duration bursts (at least 0.6 Gyr).

At redshifts larger than 0.7, most of the galaxies have rest-frame colors close to the ones of irregulars. We also identify a population of high redshift galaxies with very small D(4000) indices while having large Balmer break (Figure 2), which cannot be fitted by any kind of Bruzual and Charlot (1993,1995) models,

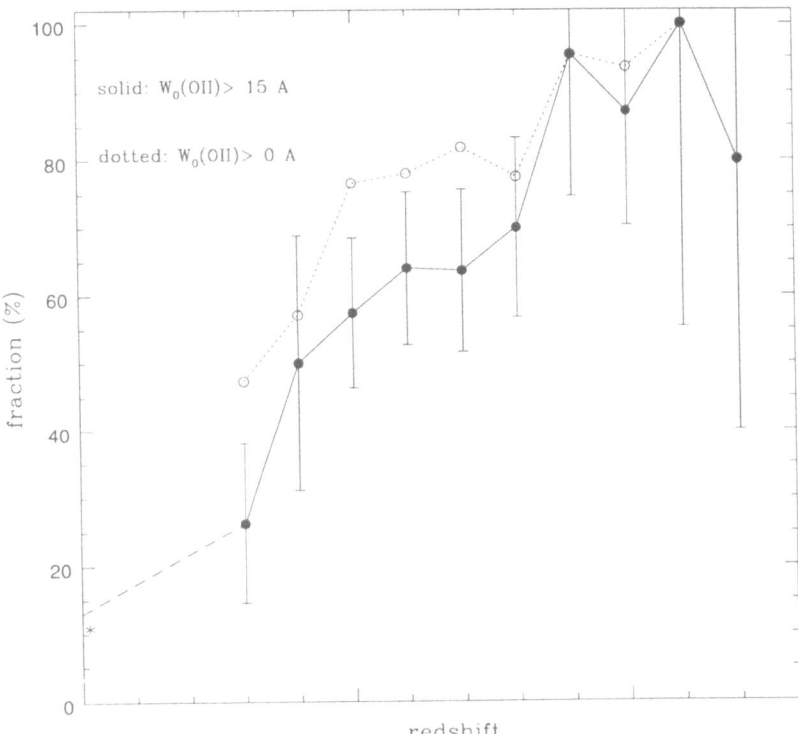

Based on 272 galaxies with $M_B < -20$

Fig. 1. fraction of objects with $W_0(OII) > 15$Å (solid line) and with $W_0(OII) > 0$Å; they increase by large factors especially when compared to similar galaxies at low redshifts (Vettolani et al, these proceedings).

and this even with various IMFs and burst scenarios. These galaxies, referred as "D(4000)-deficient objects", have continuum properties very similar to the ones of the Magellanic young star clusters (Bica et al, 1994), and their continuum can be fitted by population synthesis models with low metallicities (Bruzual and Charlot, 1996). There is a considerable evidence for the emergence at $z > 0.7$, of a population of metal-deficient objects ($Z/Z_\odot < 0.2$) representing to 1/3 of the emission line galaxies in that redshift range. Support for the emergence of a low metallicity population at high redshift is provided by the fact that among the four spectra of galaxies from the Cowie et al (1995) sample with $1 < z < 1.2$, three have similar properties than the "D(4000)-deficient objects". The low metallicities in the latter imply that the star formation took place several 10^8 years ago in an almost primordial medium.

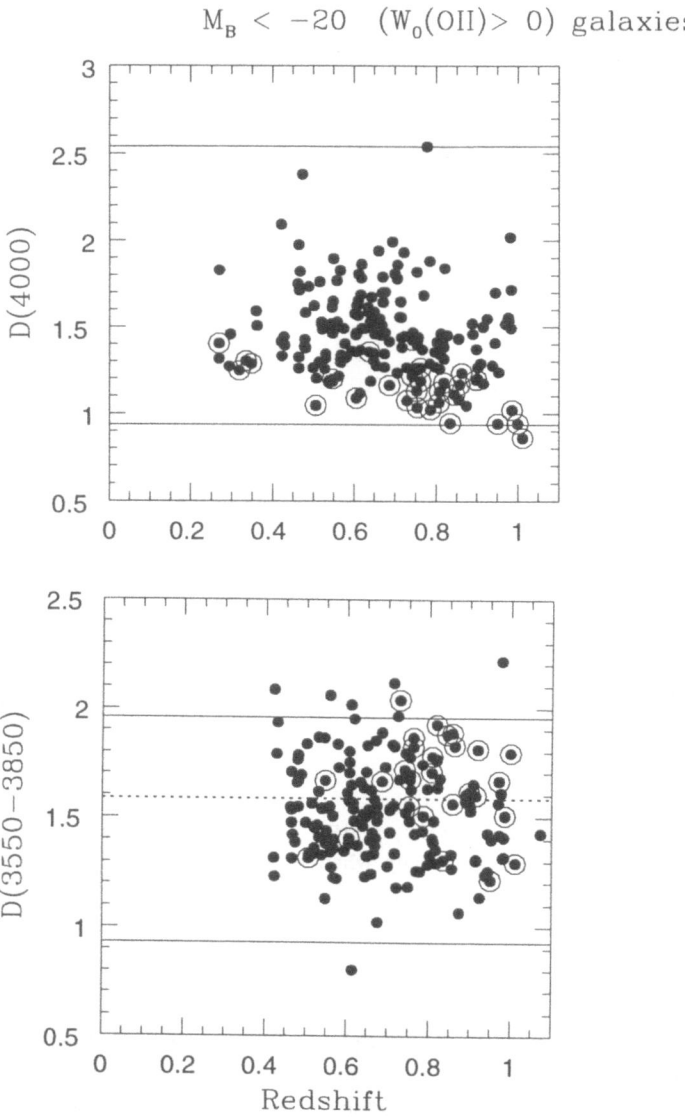

Fig. 2. D(4000) and Balmer (D(3550-3850)) indices versus redshift for the $M_B <$ -20 emission lines galaxies. Full lines shows the limits from the Bruzual and Charlot (1995) code (bottom lines=young population). In the bottom panel, the dotted line represents a Balmer index which corresponds to $W_0(H\delta)$=5Å. Circled points are the D(4000)-deficient objects (see text).

3.2 Quiescent Galaxies

Beyond z=0.5, the fraction of quiescent galaxies (i.e. having no [OII] emission line) become ~ 25% (Figure 1), and similar to the one of elliptical/lenticular galaxies found in our HST subsample (Schade et al, 1995, hereafter CFRS IX). We believe that the further increase of the rate of emission line objects beyond z=0.8 is mostly due to selection effects against red and faint objects. A large fraction have D(4000) indices and Mg2 indices smaller than expectations for a large redshift of formation. This supports the fact that a majority of early type galaxies have experienced relatively recent bursts prior to z=2 or even to z=1.

4 Other Major Results

4.1 HST Observations of 32 CFRS $z > 0.5$ Galaxies

To date, we have analysed images in F450W and F814W of four WFPC2 fields from the CFRS and a resulting sample of 32 galaxies at $z > 0.5$ (CFRS IX). The 32 galaxies at $z > 0.5$ show the full range of Hubble types from ellipticals through spirals to irregulars. Spiral arms and bars are clearly visible in many galaxies and 4 of the 32 clearly look as if they are merging/interacting galaxies. As expected, the galaxies are considerably less regular on the shorter wavelength images.

Our analysis has centered on two-dimensional fitting of the galaxies using bulge+disk models and examination of the subsequent residual images (CFRS IX). About 70% of the galaxies form a regular Hubble sequence of red bulges and blue disks, most (70 %) of which are blue disk-dominated galaxies with B/T ratios less than 0.5. In the 50% of galaxies that are both well-represented by bulge-plus-disk models and have $B/T < 0.5$, the disk surface brightnesses are significantly brighter (by about one magnitude) than seen in local samples, indicating a higher star-formation rate per unit area.

The remainder of the sample are what we have termed "blue nucleated galaxies" - galaxies with concentrated blue components that nearly always have large, highly asymmetric, residuals from the fitting process. The fit parameters indicate that these are "bulge-dominated" galaxies, but given the asymmetrical nature of the galaxies, we have prefered to interpret these as either irregular galaxies with off-centered star-formation or as systems that are the result of mergers or interactions.

Thus our present interpretation of the imaging data is that at least two processes are occuring to produce the changes in the blue galaxy luminosity function, including the disk brightening in normal regular galaxies and the emergence of a population of relatively bright "irregular" galaxies. As our sample of galaxies with both HST imaging and spectroscopically determined redshifts grows (we already have 250 galaxies since the end of HST Cycle 5) we should be able to study statistical distribution functions such as the disk size function, the B/T distribution, the bulge luminosity function, etc.. This will enable us to constrain much better the range of allowable scenarios.

4.2   Optical Counterparts of μ-Jy Radiosources

During the preliminary deep imaging phase of our large spectroscopic survey of faint field galaxies (CFRS), one of our fields (10 arcmin × 10 arcmin) was chosen to coincide with the Fomalont et al (1991) radiosource field, including 36 S~ 16μJy radiosources of their complete sample. All sources but two have been identified to V < 25 and/or I_{AB} $\leq$24, and/or K_{AB} $\leq$21.

The microJy population is mainly constituted of three distinct populations of galaxies with different redshift regimes: early-type galaxies at z > 0.75 with inverted radio spectra (possibly related to a low powered AGN in their cores), post-starburst galaxies at intermediate redshifts (z = 0.375 to z = 0.8 or slightly > 1), and emission-line galaxies at z < 0.45 containing AGNs. The fraction of μJy sources with z > 1 could be as high as 30%. Most of the μJy radiosources (> 50%) are likely associated to AGNs (including the early-type galaxies), conversely to what is found at mJy levels (mostly starburst galaxies, Benn et al, 1993). Only one galaxy in our sample has a classical starburst spectrum.

The strong decrease of the radio spectral index from sub-mJy to μJy counts appears to be due to a combination of three factors: (1) the emergence of an elliptical population at high redshifts with moderate radio emission (2) an increasing fraction of narrow emission-line AGNs (Seyfert 2 and LINER); (3) a higher contribution of the thermal radiation to the radio emission from spirals, and the almost complete disappearance of starburst galaxies. Details of the results summarized here can be found in Hammer et al (1995, hereafter, CFRS VII).

HST images of 19 μ-Jy radio sources shows that more than 50% of them are related to bulge dominated galaxies. Latter have often a small companions withing 25 kpc, while the spiral and post-starburst galaxies are apparently experiencing a major interaction with a neighbour of almost the same brightness (Hammer et al, 1996).

4.3   The Clustering Correlation Function

The evolving amplitude of the correlation function $\xi(r, z)$ reflects both the growth of large scale structure and the nature of the faint galaxy population, since the population at high redshift may be differently clustered (intrinsically) to that at low redshift. The strength of galaxy clustering at high redshift has been studied with the CFRS and we find a net decrease with the redshift of the physical correlation length $r_0(z)$ (Le Fèvre et al, 1996, hereafter CFRS IX). Latter reaches value of $1.15h^{-1}$ Mpc at z=0.85, which could be compared to the local CfA sample, with $r_0 \sim 5.5h^{-1}$ Mpc (Davis and Peebles, 1983), or the 60-μm-selected IRAS sample of star-forming galaxies, with $r_0 \sim 3.7h^{-1}$ Mpc (Fisher et al, 1994). For $q_0 \sim 0.5$ a scenario involving no growth of structure would produce a local population that would be much more weakly clustered than *any* local population. For $q_0 \sim 0$, then less evolution is required and evolution to an IRAS-like sample is consistent with the "stable clustering" growth case.

4.4 Population of Low-Luminosity Galaxies at $z < 0.3$

We have analysed in detail the emission line properties of the $z < 0.3$ galaxies, in order to include $H\alpha$ in our spectral range. Essentially all of the galaxies in the bluest half of the sample are emission line objects. Emission line ratios have been plotted in diagnostic diagrams, in order to identify the nature of the ionization source. We find that up to 33% of the faint blue galaxies (or 8 to 17% of our whole sample with $z < 0.3$) have AGN-like spectra rather than being associated with starburst galaxies (Tresse et al, 1996, hereafter CFRS XII). These objects would have spectral properties intermediate between Seyfert2 and LINERs and can contribute to the diffuse X-ray emission.

5 Conclusion

CFRS has provided us the largest complete sample of galaxy spectra up to z = 1 which is suitable for several analyses for a better understanding of the galaxy evolution. It is likely that galaxy evolution is a complex process involving stellar evolution, evolution of physical properties such as extinction or metallicity, and mechanisms operating within the galaxies and between galaxies and their environments.

There are considerable evidences of an increase of the star formation in most of the late type galaxies from z=0 to z=0.7. The fraction of galaxies with significant $W_0(OII)$ increases from 13% at z=0 to more than 60% at z=0.5-0.7. Increase of star formation is followed by a significant increase of the disk surface brightness and by the emergence of the blue nucleated galaxies (CFRS IX). These properties appears consistent with the fact that luminosity evolution is the main contributor to the changes with the epoch of the luminosity function. In this paper, we have not calculated star formation rates (SFR) at various epochs. Such calculations are currently based on a calibration of the SFR to the OII luminosity density. Such a calibration has been made by Kennicutt (1992) for a sample of local galaxies spanning the whole range of morphological type. Gallagher et al (1989) have found much lower values of SFR for a given OII luminosity, on the basis of a sample of local irregulars. The discrepancy between the two studies is mainly due to the difference in average opacity in the two samples (A_V=1.26 for Kennicutt and A_V=0.57 for Gallagher et al) and at a lower level to the different IMF used. Beyond z=0.7, we find that most of the emission line galaxies (i.e. > 75% of the whole population) have colors close to the ones of local irregular galaxies. Added to the fact of the emergence of a metallic deficient population, this led us to consider seriously that the opacity was much smaller at higher redshift than locally. This would considerably lower the apparent increase of the SFR, if based on the [OII] or UV luminosity density, as well as to contribute to the luminosity function evolution and to the reported increase of the disk surface brightness.

Further works would include analysis of the larger sample of galaxies with HST deep imagery, future planned observations at C.F.H.T. and ESO (kinematics, emission lines, strucutres etc...) and soonly expected ISO observations from 5 to 90 microns.

References

Benn, C,R., Rowan-Robinson, M., McMahon, R.G., Broadhurst, T.J., Lawrence, A., 1993, MNRAS, 263, 98.

Broadhurst, T.J., Ellis, R.S., Shanks, T., 1988, MNRAS, 235, 827

Bruzual, G. and Charlot, S. 1993, ApJ, 405, 538.(BC93)

Bruzual, G. and Charlot, S. 1995, preprint.

Bruzual, G. and Charlot, S. 1996, in preparation.

Colless, M.M., Ellis, R., Taylor, K., Hook, R.N., 1990, MNRAS, 244, 408

Cowie, Hu and Songalai, 1995, Nature, 377, 603.

Crampton, D., Le Fèvre, O., Lilly, S. 1995, ApJ, 455, 96 (CFRS V)

Davis, M., Peebles, P.J.E., 1983, ApJ, 267, 465

Fisher, B.F., Davis, M., Strauss, M., Yahil., A., Huchra, J., 1994, MNRAS, 266, 50.

Fomalont, E.B., Windhorst, R., Kristian, J.A., Kellerman, K., 1991, AJ, 102, 1258

Gallagher, Bushouse, Hunter, 1989, AJ, 97, 700

Hammer, F., Crampton, D., Le Fèvre, O., Lilly, S., 1995a, ApJ, 455, 88 (CFRS IV)

Hammer, F., Crampton, D., Lilly, S., Le Fèvre, O., Kenett, T., 1995b, MNRAS, 276, 1085 (CFRS VII)

Hammer, F., Flores, H., Lilly, S., Crampton, D., Le Fèvre, O., Rola, C., Mallen, G., Schade, D., Tresse, L., 1996, ApJ, submitted (CFRS XIV)

Kennicutt, 1992, ApJ, 388, 310

Le Fèvre, O., Crampton, D., Lilly, S., Hammer, F., Tresse, L., 1995, ApJ, 455, 60 (CFRS II)

Lilly, S., Le Fèvre, O., Crampton, D., Hammer, F., Tresse, L., 1995a, ApJ, 455, 50 (CFRS I)

Lilly, S., Hammer, F., Le Fèvre, O., Crampton, D., 1995b, ApJ, 455, 75 (CFRS III)

Lilly, S., Tresse, L., Hammer, F., Crampton, D., 1995c, ApJ (CFRS VI)

Lilly, S., Le Fèvre, O., Hammer, F., Crampton, D., 1996, ApJ, 460, L1

Schade, D., Lilly, S., Crampton, D., Hammer, F., Le Fèvre, O., Tresse, L., 1995, ApJL, 451, L1 (CFRS IX)

Tresse, L., Rola, C., Hammer, F., Stasinska, G., Le Fèvre, O., Lilly, S., Crampton, D., 1995, MNRAS, in press (CFRS XII).

Galaxy Evolution: Luminosities and Linewidths

Matthew Colless

Mount Stromlo and Siding Spring Observatories, Australian National University, Weston Creek, ACT 2611, Australia

Abstract. The results of recent redshift surveys which yield the evolution of the galaxy luminosity function out to z~0.75 are presented, along with a preliminary study of the luminosity–linewidth relation for galaxies at z~0.3.

1 Introduction

The underlying goal of the work described here is an understanding of galaxy evolution—both a descriptive, statistical understanding of the observed galaxy population and a physical understanding of the mechanisms through which galaxies evolve. It is only in the last decade that such goals have begun to be observationally feasible, largely through the advent of better instrumentation: multi-object spectrographs for deep redshift surveys and high-resolution imaging from the ground and from space. Future advances will depend on yet more powerful telescope/instrument combinations, in particular the step up to VLT-class telescopes and out into the near-infrared.

I cover two topics in this paper. The first is the evolution of the galaxy luminosity function (LF) as revealed by recent redshift surveys of faint galaxies. I briefly review current issues regarding the galaxy LF before describing the results from new redshift surveys. In particular I will look at the considerable uncertainties in the local LF, the evolution of the LF out to z~1, and the LF evolution of different spectral types. The second topic is some preliminary work investigating a new window on galaxy evolution beyond redshift surveys—the study of luminosity evolution as a function of galaxy mass by means of the linewidth-luminosity relation (the Tully-Fisher relation).

2 Current Issues Concerning the Galaxy LF

The luminosity function (LF) is the space density of galaxies as function of luminosity. Since a galaxy's luminosity is the most basic observational datum, the LF is a fundamental characterisation of the galaxy population. There are several current issues concerning the galaxy LF which are being hotly debated and intensively studied. These include the normalisation of the local (i.e. low-redshift) LF, the faint end slope of the LF and the number of dwarf and low surface brightness galaxies, the variation of the LF spectral and morphological type, and the existence and form of LF evolution with redshift.

Concern about the normalisation of the local LF stems largely from the unexpected steepness of the number-magnitude relation, $n(m)$, for bright galaxies.

Local determinations of the LF, when put into standard models for predicting the number counts of galaxies, fail to return the near-Euclidean slope of the $n(m)$ relation. Since this should in principle be simply a consistency check, there is clearly something wrong with either the existing number counts or the local LF determinations.

Detailed knowledge of the LF at low redshifts is based on bright (B<17) samples. There are a number of deficiencies in these samples. For faint dwarf galaxies the samples encompass only small volumes and are subject to significant errors due to clustering, while low surface brightness galaxies may be differentially excluded from different surveys by varying isophotal detection and photometry thresholds. Many of the bright surveys depend on photographic magnitudes, about which concerns are being raised regarding the photometric calibrations. Finally, almost all current samples are too small or have insufficient additional data to yield LFs as a function of morphological or spectral type.

The problems of LF evolution, as determined from faint redshift surveys, are (not surprisingly) even greater. Until recently the fundamental problems were the small size of deep redshift survey samples and the narrow magnitude ranges over which they were selected, which made them quite unsuitable for direct determination of the LF over a range of redshifts. Faint galaxy samples generally also have different selection criteria to bright samples, making intercomparisons difficult. Finally, there are the intrinsic difficulties of K-corrections and incompleteness in faint, high-redshift surveys.

3   Recent Redshift Surveys

In order to determine the LF over a broad range of redshifts it is essential to have a sample covering a very broad range of apparent magnitudes (so that a wide span of luminosities are sampled at all redshifts), and also as many galaxies as possible in several independent fields (in order to reduce both shot noise and clustering variations).

To meet these requirements our group combined a number of earlier redshift surveys we had carried out (DARS, BES, LDSS-1; see respectively Efstathiou et al. 1988, Broadhurst et al. 1988, Colless et al. 1990, 1993) with two more recent surveys: (i) the *Autofib redshift survey* of ~1000 galaxies with b_J=17–22 obtained with the 64-fibre Autofib spectrograph on the AAT (Ellis et al. 1996), and (ii) the *LDSS-2 redshift survey* of ~100 galaxies with b_J=22.5–24 obtained with the LDSS-2 multislit spectrograph on the WHT (Glazebrook et al. 1995).

Table 1 lists the main properties of these redshift surveys. The important features to note are the very broad magnitude range spanned by the combined surveys (b_J=11.5–24) and the large number of survey fields (67 in 9 widely-separated areas on the sky). The total sample includes 1700 galaxies with a redshift completeness among the fainter surveys of 70–85%, and allows us to investigate both the local LF and the evolution of the LF out to z~0.75.

Table 1. Redshift survey parameters.

Survey	b_J range	Area (sq.deg)	# fields	Comp -lete	# Gals
DARS	11.5–16.8	70.8	5	96%	326
BES	20.0–21.5	0.50	5	83%	188
LDSS-1	21.0–22.5	0.12	6	82%	100
AF bright	17.0–20.0	5.5	16	70%	480
AF faint	19.5–22.0	4.7	32	81%	546
LDSS-2	22.5–24.0	0.07	5	71%	60
					1700

4   The Local Luminosity Function

Using the combined redshift survey sample (excluding DARS; hereafter referred to as the Autofib survey) we have constructed the local LF from the 291 galaxies with $0.02<z<0.1$. The reason for excluding the DARS survey is readily apparent in Figure 1a, which shows the large difference in the normalisations of the DARS and combined LFs. Taking the well-known Stromlo-APM redshift survey (Loveday et al. 1992) as the canonical local LF determination, we see that the DARS LF lies below it (especially around the 'knee' at M*), while the Autofib LF approaches the Stromlo-APM LF at the bright end but is a factor ~2.5 higher at the faint end. The Autofib LF is in fact very well-fitted over the range $-20<M<-14$ by a Schechter function with parameters $M^*=-19.3+5\log h$, $\alpha=-1.1$ and $\phi^*=0.026\ h^3\ \mathrm{Mpc}^{-3}$.

Comparing this new result with other local LFs (see Marzke et al. 1994 and Lin et al. 1996), we find that brighter than M~-18 it agrees (within the errors) with all the other surveys except CfA. However at the faint end there is considerable disagreement, with our survey largely agreeing with the CfA and SSRS surveys in giving a density of faint galaxies 2–3 times higher than found in the Stromlo-APM, DARS or LCRS surveys. Our result is almost identical to the LF obtained from the ESO Slice Survey (see Vettolani, these proceedings).

Given that our goal to determine the evolution of the LF out to z~1, it is somewhat distressing that there are such significant differences in the LFs obtained at z~0. What causes these differences? Apart from particular deficiencies in a couple of the surveys that may be affecting their results (clustering in DARS and the use of Zwicky magnitudes in CfA), there are four main possibilities: (i) magnitude scale errors in the photographic magnitudes (a rather severe error of ~0.2mag/mag would be needed); (ii) different isophotal thresholds in selection/photometry of the samples (this would imply many low surface brightness galaxies, and it's not clear that the thresholds differ sufficiently for this to be plausible); (iii) large scale structure affecting the different surveys (this

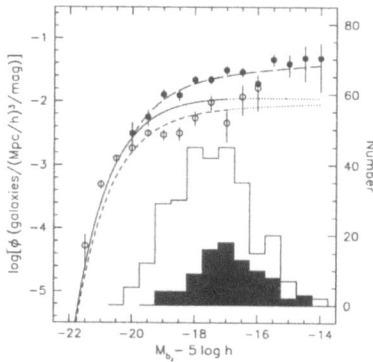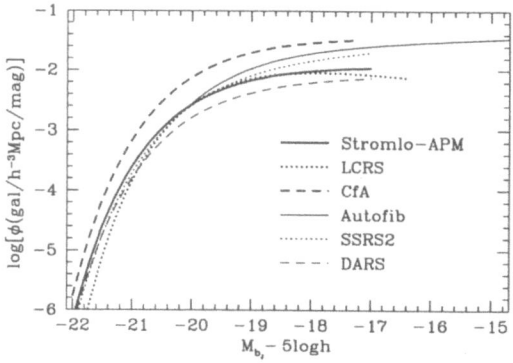

Fig. 1. The local luminosity function. (a) The local luminosity functions from the DARS (open circles) and combined (filled circles) redshift surveys. The curves are Schechter function fits to the Stromlo-APM (solid), DARS (short dash) and combined (long dash) LFs. (b) A comparison of the Schechter function fits to the local LFs from various surveys.

structure would need to have large amplitude over scales $\sim 150\,h^{-1}$ Mpc in order to explain why the CfA survey, with $\langle z \rangle \sim 0.02$, has a mean density 3 times that of the LCRS survey, with $\langle z \rangle \sim 0.1$); (iv) rapid galaxy evolution (hard to argue for, since one of the surveys with higher mean redshift, LCRS, gives a low normalisation while the most nearby survey, CfA, gives the highest normalisation; moreover the timescale for any evolution is implausibly short, less than a Gyr).

   None of the possible solutions to the puzzling variations in determinations of the local LF is entirely plausible, and further work clearly needs to be done to clarify the situation.

5   Evolution of the Overall Luminosity Function

Turning from the local LF to the question of *evolution* of the LF with redshift, we find that the broad magnitude range and faint limiting magnitude of the combined redshift surveys gives just sufficient luminosity coverage to allow us to obtain LFs in three redshift ranges from $z \sim 0$ to $z \sim 0.75$. These LFs are shown in Figure 2, along with the best-fitting Schechter functions.

   There is clear evidence for evolution—with redshift there is a gradual steepening of the faint-end slope of the LF. We have carried out a number of tests and simulations to ensure this result is not due to inadequate K-corrections or misclassifying spectral types. A χ^2 test rejects the possibility of a non-evolving LF with a probability formally less than 10^{-20}.

   Significant evolution around the knee of the LF at L^* appears to set in at $z > 0.3$. Only fainter than $M_B = -17$ (i.e. $L < 0.1L^*$) has there been evolution at

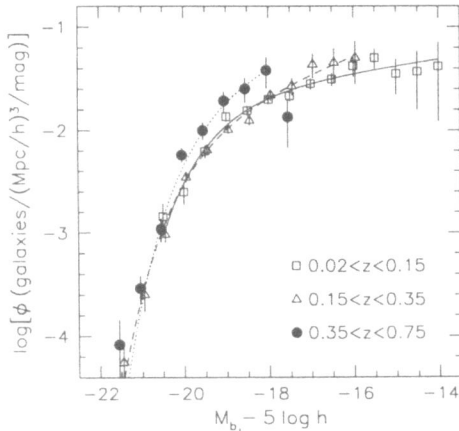

Fig. 2. The luminosity function in three redshift ranges: 0.02<z<0.15, 0.15<z<0.35 and 0.35<z<0.75. The solid, dashed and dotted curves are the respective best-fit Schechter functions.

lower redshifts. Claims for evolution at z<0.3 based on the number counts are tied to the problem of the numbers of faint galaxies in the local LF.

6   Evolution by Spectral Type

Given the observed changes in the overall LF with redshift, it is clearly of immediate interest to discover which components of the galaxy population are involved and whether we are witnessing luminosity evolution resulting from an increased rate of star-formation in the population as a whole or only in some particular morphological type.

The first step towards this goal can be made by using the equivalent width (EW) of the [OII] 3727Å emission line as an indicator of star-formation rate. We choose an EW in [OII] of 20Å as a somewhat arbitrary demarcation line between high-EW 'star-forming' galaxies and low-EW 'quiescent' galaxies. Because of the ease of obtaining redshifts for strong-lined galaxies, the high-EW galaxies form a complete subsample, simplifying the interpretation.

Comparing the LFs of the 'star-forming' and 'quiescent' samples, we find first that the star-forming objects have a steeper LF at all redshifts than the quiescent galaxies. Furthermore, the high-EW galaxies show more pronounced LF evolution than the low-EW objects—not only were there more star-forming sub-L^* galaxies at z~0.3 than there are now, but beyond z~0.5 the number of star-forming galaxies is ~5× the present number for all $L \leq L^*$.

This behaviour can also be seen in the distribution of [OII] EWs with redshift. The median [OII] EW of the galaxy population increases with redshift at fixed L, implying higher rates of star-formation in all galaxies. However bright galaxies show this increase at higher redshift than faint galaxies (this reflects the steepening of the LF with redshift).

Since we derive a spectral type for each galaxy, either directly from its spectrum or from its colour, we are able to recover the LF evolution of each spectral type separately. Dividing up the sample this way significantly reduces the precision with which the LFs are determined, but the overall trends are clear: (i) bright E/S0 galaxies show little evolution, while faint ones may actually decrease in numbers at higher redshifts; (ii) early-type spirals show a steepening of the faint-end slope of their LF, though $\phi(L^*)$ does not change by z~0.5; (iii) late-type spirals evolve in the same way as early-type spirals but more strongly, with even L^* galaxies increasing in number by z~0.5.

7   Linking Luminosity to Mass

A potentially very powerful approach to characterising the physical processes involved in galaxy evolution is to measure linewidth velocities for faint star-forming galaxies. Such measurements can in principle establish the mass scale for this evolving population and quantify the changes in galaxy luminosity as function of mass, thereby distinguishing rapid star-formation in dwarfs from mild evolution in brighter galaxies. Given that in any but the most radical merging scenarios the mass of an evolving galaxy changes far less than its luminosity, this approach may allow us to identify the low-redshift counterparts of high-redshift star-forming galaxies.

In a pilot study following this approach (Rix et al. 1995, 1996), spectra with σ=50 km/s resolution for 54 faint blue galaxies were obtained using the Autofib fibre spectrograph on the AAT. This sample of galaxies had B=21.25–22.0 and B−R<1.2, corresponding to the blue half of the population at these magnitudes. The wavelength range covered was $\lambda\lambda$=4300-5100Å, which for the [OII] 3727Å doublet which was our target corresponds to z=0.16–0.37 (close to the peak of the redshift distribution at B~22). At these redshifts the sample galaxies span the luminosity range $L_B^*>L>0.1L_B^*$, and have L([OII])=10^{40}–10^{41} erg s^{-1} cm^{-2}.

The [OII] doublet was detected in 24 cases, and a double Gaussian profile fitted to obtain redshift, flux, line ratio and line width. Excluding 5 galaxies with $R_{[OII]} \ll 1.3$ (implying gas densities larger than 100 cm^{-3}) which are possible AGN, the remaining 19 galaxies are all found to have $\sigma < 100$ km/s. Simulations show that the detection efficiency in these observations is approximately uniform for all $\sigma < 200$ km/s, so that the observed upper limit is real and not a detection bias. Both the high significance at which the [OII] lines were detected and the fact that the fraction detected is very close to that predicted for this redshift range argue that our sample is complete.

A maximum likelihood comparison of the σ–$\log L_{[OII]}$ relation with that found for giant HII regions and HII galaxies by Melnick et al. (1989) shows that the linewidths we obtain are ~ 2× too large to be due to turbulent motions within individual star-forming regions. We are thus justified in assuming that the linewidths reflect rotational motion, as in local spirals.

However relating linewidth σ to the circular rotation velocity v_c is complex. A wide variety of effects must be accounted for, including fibre size, seeing, rotation curve shape, spatial distribution, disk inclination and so on. Since not all of this information is available we can only recover v_c in a statistical sense. To do so, Fabry-Perot data for 3 local galaxies with matched properties was used to create simulated ensembles of galaxies at $z \sim 0.3$ from which linewidths were 'measured'. From these simulations the conditional probability $P(\sigma|v_c)$ of measuring a particular value of σ given v_c is derived. Not surprisingly it turns out to have a broad distribution, with $\langle\sigma/v_c\rangle=0.65$.

$P(\sigma|v_c)$ can then be used to obtain a maximum likelihood fit to the relation between linewidth, luminosity and colour (a 'Tully-Fisher' relation), which is assumed to have the simple form

$$\log[v_c(M_B, B-R)] = log[v_c(-19,1)] - \eta(M_B+19) + \zeta[(B-R)-1].$$

We find $v_c(-19,1)=66\pm8$ km/s, $\eta=0.07\pm0.08$ and $\zeta=0.28\pm0.25$. Thus the zero-point of the relation is well-determined, although the luminosity and colour dependences are quite uncertain (the luminosity dependence is consistent both with v_c independent of L and with the local value, $\eta=0.12$).

The zeropoint obtained in this sample at $z\sim0.3$ can be compared with that found in local B-band Tully-Fisher samples: Bothun et al. (1985) find a zeropoint $v_c(-19)=118$ km/s, while Fukugita et al. (1993) find $v_c(-19,1)=118$ km/s. Thus the Tully-Fisher relation zeropoint is 40% smaller at $z\sim0.3$ than at $z\sim0$ (at least 25% smaller at 95% confidence). The most straightforward interpretation is that faint blue galaxies were ~1.5 mag brighter at $z\sim0.3$ than they are today (>1 mag at 95% confidence).

Further investigations are underway. An obvious deficiency in the pilot study is the limited luminosity and and colour range spanned by the sample, which results in poorly determined dependences on these quantities and hampers the interpretation of the result (e.g. has the slope of the Tully-Fisher relation changed with redshift?). Another important check is extend the Fabry-Perot imaging to these redshifts so as to establish that the rotation curves and the spatial distribution of emission line flux are similar to those of low-redshift galaxies (Guhathakurta et al. 1995).

8   Acknowledgements

The redshift survey studies of the luminosity function described above were carried out in collaboration with Richard Ellis, Tom Broadhurst, Karl Glazebrook and Jeremy Heyl (Ellis et al. 1996; Heyl et al. 1996); the linewidth–luminosity work was performed in collaboration with Hans-Walter Rix, Raja Guhathakurta and Kristine Ing (Rix et al. 1995, 1996; Guhathakurta et al. 1995).

References

Bothun G.D., Aaronson M., Schommer B., Mould J., Huchra J. & Sullivan W.T., 1985, ApJS, 57, 423

Broadhurst T.J., Ellis R.S. & Shanks T., 1988, MNRAS, 235, 827

Colless M.M., Ellis R.S., Taylor K. & Hook R.N., 1990, MNRAS, 244, 408

Colless M.M., Ellis R.S., Broadhurst T.J., Taylor K. & Peterson B.A., 1993, MNRAS, 261, 19

Efstathiou G., Ellis R.S. & Peterson B.A., 1988, MNRAS, 232, 431

Ellis R.S., Colless M.M., Broadhurst T.J., Heyl J.S. & Glazebrook K., 1996, MNRAS, in press

Fukugita M., Okamura S. & Yasuda N., 1993, ApJ, 412, 13L

Glazebrook K., Ellis R.S., Colless M., Broadhurst T.J., Allington-Smith J.R. & Tanvir N.R., 1995, MNRAS, 273, 157

Guhathakurta P., Ing K., Rix H.-W., Colless M.M. & Williams T., 1995, "New Light on Galaxy Evolution", IAU Symposium 171, in press

Heyl J.S., Colless M.M, Ellis R.S. & Broadhurst T.J. 1995, in preparation (Paper II)

Lin H., Kirshner R.P., Shectman S.A., Landy S.D., Oemler A., Tucker D.L. & Schechter P.L., 1996, preprint

Loveday J., Peterson B.A., Efstathiou G. & Maddox S.J., 1992, ApJ, 390, 338

Marzke R., Huchra, J.P. & Geller, M., 1994, ApJ 428, 43

Melnick, J., Terlevich, R. & Moles, M., 1989, MNRAS, 235, 297

Rix H.-W., Colless M.M. & Guhathakurta P., 1995, "New Light on Galaxy Evolution", IAU Symposium 171, in press

Rix H.-W., Guhathakurta P., Colless M.M. & Ing K., 1996, in preparation

Distant Supernovae and Cosmic Deceleration

Bruno Leibundgut and Jason Spyromilio

European Southern Observatory
Karl-Schwarzschild-Strasse 2
D-85748 Garching
Germany

Abstract. Distant supernovae can now be detected routinely. To date 34 supernovae at $z > 0.1$ have been discovered. Among them are 12 Type Ia supernovae confirmed spectroscopically and suited to measure the cosmic deceleration when appropriately employed as standard candles. However, peak magnitudes have been determined for only two objects so far and a determination of q_0 is not yet possible.

We describe the current status of the searches and possible pitfalls of the method which rests on few basic assumptions. The importance of sufficient information on the distant events is stressed and the observations of SN 1995K are used as an example of the detailed procedures employed in the analysis. Only spectroscopic classification and light curves in at least two filter bands provide the basis to use correction schemes for the luminosity which have successfully been established in nearby samples.

Time dilation has been detected acting on the light curve of SN 1995K at a redshift of 0.478, providing clear evidence of universal expansion. The observations are fully consistent with local Type Ia supernovae in an expanding universe but incompatible with the expectations from a static universe.

The contributions of the new, large telescopes to this research area are described. The extension of the observations to even more distant objects will provide a better leverage to distinguish between the possible decelerations and the inclusion of Type II supernovae into the sample add an independent check on the cosmological distances.

1   Introduction

Observational cosmology has made use of supernovae for several decades now (Zwicky 1965, Kowal 1968). Their large luminosity makes them suitable for detection at the distances required for the measurement of cosmological parameters. The main drawback of supernovae is the brief time they shine at this luminosity and that they are intrinsically very rare events.

In principle, all supernova types can be used for cosmological purposes. Explosions of massive stars, which are normally recognized by strong hydrogen lines in their spectrum (Type II supernovae), can be physically understood when the radiation hydrodynamics are solved (Eastman et al. 1996) and yield distances through the expanding photosphere method (Schmidt et al. 1992, 1994). The explosive carbon incineration of white dwarfs produces very luminous events with no hydrogen in their spectra - Type Ia supernovae (SNe Ia). These objects display a fairly uniform behavior which led to their use as standard candles (Kowal 1968, Tammann & Leibundgut 1990, Branch & Miller 1993, Sandage et al. 1996). The simple adoption of a unique peak luminosity, however, has recently been challenged by SN Ia samples with accurate relative distances and a dispersion of 0.6 magnitudes in maximum luminosity (Phillips 1993, Hamuy et al. 1995). The uniformity of the class is further eroded by correlations found in the light curve shapes and spectral appearances (Phillips 1993, Hamuy et al. 1995, Nugent et al. 1995, Vacca & Leibundgut 1996). Local samples can be corrected by the proposed light curve shape fitting and only a small residual scatter (<0.2 mag.) remains (Hamuy et al. 1995, Riess et al. 1996).

Despite the variety of SNe Ia in the nearby universe we are still confident that the mechanisms are in place to use them at large distances for the determination of cosmological parameters, provided sufficient observations of each distant event are available.

The value of the Hubble constant has been the main focus of SN Ia applications (Sandage et al. 1996, Branch et al. 1996, Hamuy et al. 1995, Riess et al. 1996). The discrepancies in the reported results can be traced to the absolute nature of the measurement. The discussion of Riess et al. (1996) exemplifies the positions and assumptions adopted by several groups and resolves some of the conflicts rather convincingly.

Measuring the cosmic deceleration is fundamentally simpler, since it is done through relative measures. It is based on the comparison of the apparent brightness of objects with known relative luminosity at largely different redshifts. The ability to find the luminosity of a distant SN Ia relative to a nearby twin rests purely on observables (Schmidt et al. 1996, Perlmutter et al. 1996).

In this paper we present the basic ingredients necessary for successful detection and observation of distant supernovae and describe the current status of projects which use SNe Ia to measure q_0 (§2). In section 3 we develop the time dilation test for universal expansion as applied to SN 1995K. The prospects for and the contributions of the VLT to experiments like this one are presented together with the conclusions.

2 Observing Distant Supernovae

Finding distant ($z > 0.3$) supernovae is hampered by their variability and the rareness with which they occur. To overcome these obstacles a massive approach is needed. This has become possible with the development of sensitive, panoramic CCD detectors. Faint brightness levels have to be reached to increase the period a supernova is detectable. At the same time a large number of galaxies has to be surveyed to increase the chance of catching a supernova in the act.

There are a few pitfalls in the use of supernovae for accurate cosmological distance determinations. The first one - to find the objects - has been overcome by the concerted efforts at 4m telescopes to image many galaxies in a short time period. Typically prime focus cameras with their large field of view are employed. The search fields, some including known clusters at $z > 0.3$, are observed after new moon to be re-imaged at the beginning of the next dark period. This ensures that all events detected during the second run are still on their rising branch and their light curve peak can be observed (Perlmutter et al. 1996). The currently employed strategy detects several supernovae (5–10) in such a search period (Garnavich et al. 1996a,b, Perlmutter et al. 1995b, 1996). All candidate objects for which it was possible to obtain a spectrum have been confirmed as supernovae at large distances.

Spectroscopic classification is of paramount importance as only Type Ia supernovae should be used for this method. Furthermore, on has to discriminate against SNe Ia which show clear peculiarities (SN 1991bg: Filippenko et al. 1992a, Leibundgut et al. 1993, Turatto et al. 1996; SN 1992K: Hamuy et al. 1994; SN 1991T: Filippenko et al. 1992b, Phillips et al. 1992, Ruiz-Lapuente et al. 1991; SN 1992bc: Maza et al. 1994) to avoid contamination of the sample. Since a SN Ia reaches about R=22.5 at z=0.5, the required spectroscopy stretches the capabilities of the current 4m telescopes to the limit. The ability to schedule the spectroscopic follow-up observations ahead of time has proven to be essential for their success. Thanks to the combined efforts by several observatories has it been possible to obtain a spectrum for a fair fraction of the events (see below).

A further complication is the correction required to compare the observed flux evolution to the rest frame light curves of local supernovae. This K-correction, which depends on the phase of the supernova, can be overcome in two ways. One possibility is to use special filters corresponding to redshifted B and V passbands to minimize the corrections and also largely remove the dependence on phase. The other approach is to use standard broad-band filters which are closest to B and V in the rest frame of the supernova. Such K-corrections for regular SNe Ia have been calculated (Hamuy et al. 1993, Kim et al. 1996). In practice, this problem does not contribute significantly to the error budget.

Finally, photometry in two filters delivers the color of the supernova. The multi-color light curve shape method (Riess et al. 1996) makes use of this information to correct for possible absorption. Potentially absorption is one of the major contributors to the uncertainty in the derivation of q_0.

The first distant supernova was discovered in a dedicated search already in 1988. The object located in a galaxy cluster at z=0.31 (SN 1988U, AC 118) was discovered about two weeks after maximum light and the spectrum suggested a Type Ia classification (Nørgaard-Nielsen et al. 1989). The same search reported a second event at a somewhat smaller redshift (SN 1988T, z=0.28; Hansen et al. 1989). The classification of this object is uncertain.

In recent years two major efforts to find and observe distant supernovae with the goal to determine the geometry of the universe have emerged. SN 1992bi was reported at a redshift of 0.458 by the Berkeley Cosmology Project and proved that it is possible to obtain an accurate light curve (Perlmutter et al. 1995a). The High-Z Supernova Team (Schmidt et al. 1995, 1996) has reported its observations of SN 1995K which include light curves in two filters and a spectrum near maximum light. The event occurred in a galaxy at a redshift of 0.478 and appears superposed on a spiral arm. The galaxy spectrum displays Hα, Hβ, [O II] and [O III] lines in emission and is consistent with an Sbc classification.

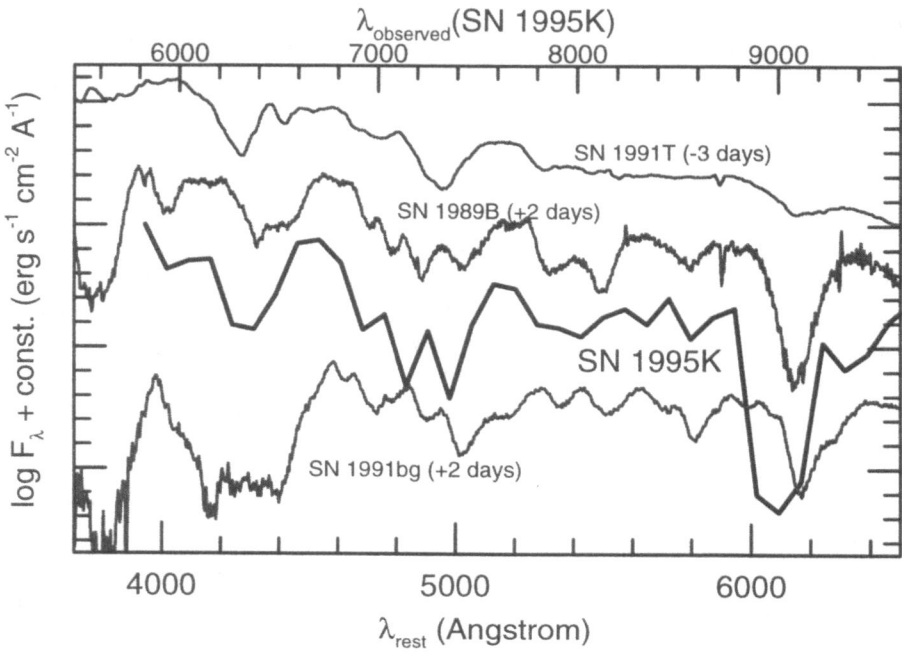

Fig. 1. Comparison of the spectrum of SN 1995K with local SNe Ia near maximum light. The spectra are plotted in rest frame wavelengths.

The supernova spectrum itself had to be corrected for galaxy contamination and heavily smoothed. The result is a spectrum which displays the usual absorption and emission features of a Type Ia supernova near maximum light (Figure 1). All lines are shifted to the same redshift as determined for the galaxy confirming the association. A comparison of the SN 1995K spectrum with nearby events clearly shows the close relation to regular SNe Ia. The closest match is the near-maximum spectrum of SN 1989B. The comparison with the overluminous SN 1991T exemplifies the importance of spectroscopic classification. All spectral features of SN 1995K are stronger than in SN 1991T. In particular, the deep absorption due to Si II is characteristic of 'normal' SNe Ia (Nugent et al. 1995) and distinguishes SN 1995K from SN 1991T. Similarly, the absorption ascribed to Ti II in low-temperature supernovae as observed in SN 1991bg near maximum is missing in SN 1995K. The phase of the spectrum is independently set by the light curve (see below) and is fully consistent with the spectral observation.

Light curves have been obtained in special B and V filters which correspond to a redshift of 0.45 and the Kron-Cousins R and I filters. At the redshift of SN 1995K (z=0.478) the transmission curve of the R filter is almost identical with the redshifted B passband and K-corrections reduce to a constant describing the zero-point offset between the B and R passbands (see also Kim et al. 1996). We hence combined the B45 and R data sets as well as the V45 and I filter observations, respectively. The photometry covers the maximum phase quite adequately

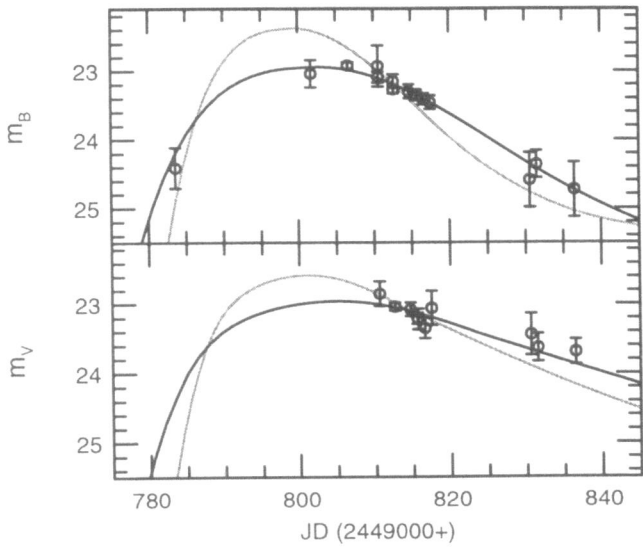

Fig. 2. B and V rest frame light curves of SN 1995K. The best χ^2 fits for a simultaneous fit to the dilated (dark) and non-dilated (grey) light curves of SN 1991T are shown.

(Figure 2). The light curves span from about 18 days before until 35 days after maximum in B and from about peak light to 30 days thereafter in V. In fact, the first two measurements are from pre-discovery search observations. The supernova was discovered close to maximum light. The detection was not triggered in the earlier observations due to the faintness of the object ($R_{max}\approx22.2$) and its projection onto the high surface-brightness disc of the galaxy (e.g. Leibundgut et al. 1995). The pre-maximum points are essential for the determination of accurate light curve fits (Leibundgut et al. 1996) and a reliable measurement of the peak brightness.

As of June 1996 28 supernovae at redshifts larger than 0.3 have been reported. Of those at least 9 have been confirmed spectroscopically as being of Type Ia (Nørgaard-Nielsen et al. 1989, Schmidt et al. 1995, Perlmutter et al. 1995b, Garnavich et al. 1996a, b). Several supernovae without spectra have light curves typical of SNe Ia. The peak is covered in most cases, but modeling is still required for an accurate determination of the maximum brightness (cf. Perlmutter et al. 1996). The histogram of all supernovae discovered in these campaigns shows a clear concentration to the range of $0.3 < z < 0.5$ with 70% of all events detected so far (Figure 3). Monte Carlo simulations indicate that with the accuracy achieved for SN 1995K it will take about 30 SNe Ia to decrease the uncertainty on q_0 to about 0.1 (1 σ, Schmidt et al. 1996). The remaining error is mostly due to the uncertainty in the normalization by the local supernovae. The application of corrections like the multi-color light curve shape method (Riess et al. 1996) require the availability of at least two filter light curves and suffi-

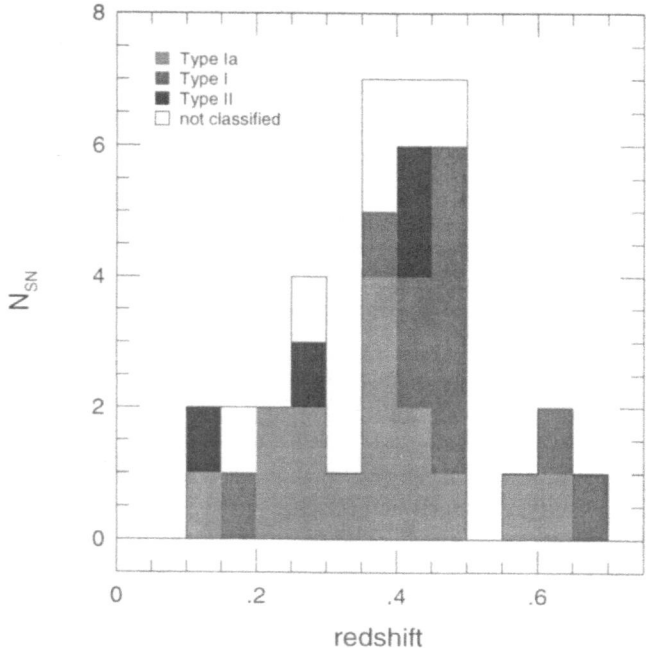

Fig. 3. Histogram of distant supernovae reported before June 1996.

cient photometry coverage. Of course, it also implies that distant supernovae are indeed relatives of the nearby events and the training set employed to find the 'corrections' according to light curve shape and color is applicable.

The current status of the determination of q_0 is summarized in Figure 4 (cf. also Perlmutter et al. 1996). There are three SNe plotted at redshifts larger than 0.3. Two have spectroscopic SN Ia classifications (SN 1988U and SN 1995K) and the third (SN 1992bi) displays a light curve typical of SNe Ia. SN 1995K and SN 1992bi have been observed through their maximum and have fairly secure peak magnitudes, while SN 1988U has been extrapolated to maximum. Clearly it is too early to estimate the size of any cosmic deceleration with these three objects. Nevertheless, it is encouraging that the two distant objects fall between the lines for $1/2 > q_0 > 0$. Note that we have used the normalization of the 'uncorrected' nearby sample of Hamuy et al. (1995). No correction for the light curve shape has been applied to the distant supernovae either. Light curve shape corrections change the zero-point of the normalization and have to be applied carefully. This further exemplifies the importance of a large knowledge base for the distant supernovae to confirm their conformity with the local samples.

The redshifts of additional supernovae already discovered in the two projects are indicated in Fig. 4. It appears that quite a few events have been recorded to date and that we should be able to find the value for the deceleration fairly soon. However, not all objects are indeed SNe Ia. Some have no spectroscopic

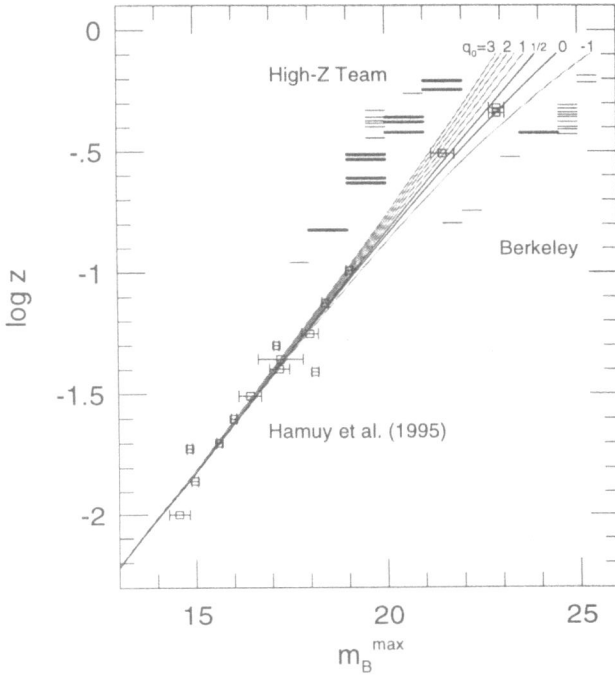

Fig. 4. Hubble diagram of supernovae with the lines of constant luminosity in different mass universes. The horizontal bars show the redshifts of known distant supernovae. Spectroscopically classified SNe Ia are indicated by thick bars, SNe II and unclassified objects by short, thin lines.

confirmation while others are known to be Type II. The importance of sufficient light curve coverage is another limiting factor. Figure 4 thus presents a very optimistic view and we will not be able to fill the diagram with a point for each line indicating a SN redshift. In the next year a similar amount of distant events will be discovered and soon we should be able to measure q_0.

3   Time Dilation

A significant result, which can be derived from the single event currently available to us, is the direct observation of time dilation due to universal expansion. The light curve of a distant supernova acts as a clock, the ticks of which can be directly compared to wavefront stretching as observed in the spectra. In an expanding universe the two have to go together while other theories of redshift invoke energy dissipation of photons or similar mechanisms (Arp 1987, Arp et al. 1990). Two distant SNe Ia have been observed at least 18 days before maximum (SN 1994am, Perlmutter et al. 1996, and SN 1995K). No local event has been observed at such an early phase. The earliest observations of nearby SNe Ia are reported in the range of 14 days before peak (e.g. SN 1990N; Leibundgut et al. 1991). Moreover, the brightness evolution of SN 1995K is the slowest ever observed for any SN Ia (Leibundgut et al. 1996). In Figure 2 we have overplotted the best fits as determined by minimization of χ^2 to the combined B and V rest frame dilated and non-dilated light curves. For the comparison we have chosen one of the slowest local supernovae known, SN 1991T. The non-dilated curves cannot fit the observations, while the photometry of SN 1995K is fully consistent with the dilated template. Analyses with comparison light curves of other local events confirm this result (Leibundgut et al. 1996). The combination of regular color, typical spectrum, a luminosity in the range of reasonable deceleration parameter values, and the conforming light curve of SN 1995K strongly favor the dilated case. In a non-expanding universe SN 1995K would represent a weird event displaying a regular spectrum but deviating strongly from the locally established luminosity-decline rate relation, and displaying the slowest decline of all known SNe Ia. Malmquist bias could not explain this supernova as it would be less luminous than local SNe Ia in a static universe. The same conclusions was found based on light curves in a single filter for a set of seven distant SNe ($0.3 < z < 0.5$) three of which with a spectroscopic Ia classification (Goldhaber et al. 1996).

4   Conclusions

Finding supernovae at redshifts above 0.3 has become a routine enterprise over the last two years. There are currently two groups vigorously following this route to find the value of the deceleration parameter (Perlmutter et al. 1995a, 1996, Schmidt et al. 1996, Leibundgut et al. 1996). Several supernovae are found each semester in the scheduled search runs. However, the currently available data set is not quite large enough to make a serious attempt to measure q_0. Only a dozen objects has the required spectral classification and sufficient light curve coverage to provide the basis for an accurate measurement of the peak brightness and the correction to the luminosity from the light curve shape. There also remains too much slack in the zero-point of the expansion lines in the Hubble diagram as defined by local supernovae. This uncertainty directly goes into the estimate of the cosmic deceleration.

Another error source which has to be carefully excluded are systematic differences of distant events from the nearby ones. Thus the sample has to be tightly checked against possible evolutionary effects. The theoretical understanding of SNe Ia is not sound enough to exclude evolutionary effects which could change the luminosities of the explosions (Canal et al. 1996). Not all models predict such changes but some do. By a careful analysis of the distant sample we should be able to distinguish among the various current candidate models of SN Ia explosions.

Sufficient statistics and, eventually, detailed spectroscopic follow-up observations should be able to detect such differences.

The large aperture telescopes will provide the necessary spectroscopy for the classification of the supernovae. Their rôle will be in the detection and follow up observations of more distant events. SNe Ia at maximum should be around I≈25 at $z=1$ and the difference between an empty and a critical universe is 0.54 magnitudes providing more leverage for the determination of q_0. The complication of a cosmological constant can also be explored with standard candles at large distances (Goobar & Perlmutter 1995). The infrared capabilities of the large telescopes are of paramount importance as the rest frame V light is shifted into the J band at $z=1$.

Important cosmological tests also remain for the VLT and its 8m-class partners. The time dilation as observed in the spectral evolution of a distant SN Ia would provide an even more stringent proof of cosmic expansion. The distances derived for Type II supernovae through the expanding photosphere method (Eastman et al. 1996) are based on the combination of a luminosity distance with an angular diameter distance. These two distances are related in all cosmological models (Carroll, Press, & Turner 1992). Since these explosions are typically less luminous than SNe Ia only the large telescopes will be able to deliver the signal needed to measure the expansion velocities accurately.

Acknowledgment: It is a pleasure to thank our collaborators in this experiment. Some of the data presented in this review has been kindly provided before publication.

References

Arp, H. C. 1987, Quasars, Redshifts and Controversies, Interstellar Media, Berkeley

Arp, H. C., Burbidge, G. R., Hoyle, F., Narlikar, J. V., & Wickramasinghe, N. C. 1990, Nature, 346, 807

Branch, D., & Miller, D. L. 1993, ApJ, 405, L5

Branch, D., Romanishin, W., & Baron, E. 1996, ApJ, 465, 73

Canal, R., Ruiz-Lapuente, P., & Burkert, A. 1996, ApJ, 456, L101

Carroll, S. M., Press, W. H., & Turner, E. L. 1992, ARA&A, 30, 499

Eastman, R. G., Schmidt, B. P., & Kirshner, R. P. 1996, ApJ, in press

Filippenko, A. V., et al. 1992a, AJ, 104, 1534

Filippenko, A. V., et al. 1992b, ApJ, 384, L15

Garnavich, P., et al. 1996a, IAU Circ. 6332

Garnavich, P., et al. 1996b, IAU Circ. 6358

Goldhaber, G., et al. 1996, Thermonuclear Supernovae, eds. R. Canal, P. Ruiz-Lapuente, & J. Isern, (Dordrecht: Kluwer), in press

Goobar, A & Perlmutter, S. 1995, ApJ, 450, 14

Hamuy, M., Phillips, M. M., Wells, L. A., & J. Maza 1993, PASP, 105, 787

Hamuy, M., Phillips, M. M., Maza, J., Suntzeff, N. B., Schommer, R. A., & Avilés, R. 1995, AJ, 109, 1

Hamuy, M., et al. 1994, AJ, 108, 2226

Hansen, L., Jørgensen, H. E., Nørgaard-Nielsen, H. U., Ellis, R. S., & Couch, W. J. 1989, A&A, 211, L9

Kowal, C. T. 1968, AJ, 73, 1021

Leibundgut, B., Kirshner, R. P., Filippenko, A. V., Shields, J. C., Foltz, C. B., Phillips, M. M., & Sonneborn, G. 1991, ApJ, 371, L23

Leibundgut, B., et al. 1993, AJ, 105, 301

Leibundgut, B., et al. 1995, The Messenger, 19

Leibundgut, B., et al. 1996, ApJ, 466, L21

Maza, J., Hamuy, M., Phillips, M. M., Suntzeff, N. B., Avilés, R. 1994, ApJ, 424, L107

Nørgaard-Nielsen, H. U., Hansen, L., Jørgensen, H. E., Salamanca, A. A., Ellis, R. S., & Couch, W. J. 1989, Nature, 339, 523

Nugent, P., Phillips, M. M., Baron, E., Branch, D., & Hauschildt, P. 1995, ApJ, 455, L147

Perlmutter, S., et al. 1995a, ApJ, 440, L41

Perlmutter, S., et al. 1995b, IAU Circ. 6270

Perlmutter, S., et al. 1996, Thermonuclear Supernovae, eds. R. Canal, P. Ruiz-Lapuente, & J. Isern, (Dordrecht: Kluwer), in press

Phillips, M. M. 1993, ApJ, 413, L105

Phillips, M. M., Wells, L. A., Suntzeff, N. B., Hamuy, M., Leibundgut, B., Kirshner, R. P., & Foltz, C. B. 1992, AJ, 103, 1632

Riess, A. G., Press, W. M., & Kirshner, R. P. 1996, ApJ, in press

Ruiz–Lapuente, P., Cappellaro, E., Turatto, M., Gouiffes, C., Danziger, I. J., Della Valle, M., & Lucy, L. B. 1992, ApJ, 387, L33

Sandage, A., Saha, A., Tammann, G. A., Labhardt, L., Panagia, N., & Macchetto, F. D. 1996, ApJ, 460, L15

Schmidt, B. P., Kirshner, R. P., & Eastman, R. G. 1992, ApJ, 395, 366

Schmidt, B. P., Kirshner, R. P., Eastman, R. G., Phillips, M. M., Suntzeff, N. B., Hamuy, M., Maza, J., & Avilés, R. 1994, ApJ, 432, 4

Schmidt, B. P., et al. 1995, IAU Circ. 6160

Schmidt, B. P., et al. 1996, in preparation

Tammann, G. A., & Leibundgut, B. 1990, A&A, 236, 9

Turatto, M., Benetti, S., Cappellaro, E., Danziger, I. J., Della Valle, M., Gouiffes, C., Mazzali, P. A., & Patat, F. 1996, MNRAS, in press

Vacca, W. D., & Leibundgut, B. 1996, Thermonuclear Supernovae, eds. R. Canal, P. Ruiz-Lapuente, J. Isern, (Dordrecht: Kluwer), in press

Zwicky, F. 1965, Stellar Structure, eds. L. H. Aller & D. B. McLaughlin, (Chicago: University of Chicago Press), 367

Evolution of Elliptical Galaxies up to $z \approx 1$

R. Bender, R.P. Saglia, B. Ziegler

Universitäts–Sternwarte, Scheinerstraße 1, D–81679 München, Germany

Abstract. We review the observational evidence showing that luminous cluster elliptical galaxies are old stellar systems, undergoing mostly passive stellar evolution up to redshift $z \approx 1$, with approximate coeval epoch of formation. This scenario is supported by observations of local early–types, collected by 2m – 4m class telescopes, and fits the recently gained high resolution imaging (given by the refurbished HST) and deep spectroscopic data coming from 4m telescopes of $z \approx 0.4$ cluster ellipticals. Up to $z \approx 1$, luminosity functions, colors and surface brightnesses provide further evidence for mild evolution of massive cluster ellipticals.

The new 8m class telescopes and in particular the VLT with its imaging and multiplex spectroscopic instruments will give further essential information on early–type galaxy evolution over the full galaxy mass–spectrum and as a function of environment, disk–to–bulge ratio and other parameters. The VLT will also allow to explore the possibility to use evolution–calibrated massive ellipticals as cosmological standard candles/rods.

1 Introduction

What was the major epoch of galaxy formation? How extended is this epoch? Are elliptical galaxies old? When did the last episodes of star formation happen in early–type galaxies? Can elliptical galaxies be used as cosmological standard candles? These are some of the most discussed questions in the last 20 years of studies of early–type galaxies. A coherent picture is now slowly emerging, which may be able to answer these questions and trace the history of star formation in these galaxies.

In Section 2 we review formation scenarios and age estimates for elliptical galaxies based on the observations of local field and cluster samples. In Section 3 we discuss the constraints on the evolution of the stellar populations of ellipticals up to redshifts $z \approx 1$ coming from the study of their colors, luminosity function, surface brightness, Fundamental Plane and $\mathrm{Mg} - \sigma$ relation and we illustrate how cluster ellipticals can possibly be calibrated and used as standard candles. In Section 4 we highlight a number of projects which involve the study of the properties of early–type galaxies at high redshift and which will be best performed with the VLT and its imaging and spectroscopic instruments. In Section 5 we summarize our conclusions.

2 2m Science: Constraints from Local E/S0

The large amount of observational material accumulated over the last decade with medium–class telescopes (2m – 4m telescopes) allows one to put some important constraints on the merging and star formation (SF) history of early–type galaxies in the "local" universe.

About 1/3 of luminous nearby ellipticals show kinematically decoupled cores or otherwise peculiar kinematics. Such configurations point to formation scenarios which involve major merging events between progenitor objects with low gas–to–star ratio (see, e.g., Bender 1996, Barnes 1996). In addition, the statistics of counter–rotating gaseous disks present in E and S0 galaxies indicate that each early–type galaxy experiences accretion of a low luminosity galaxy or of intergalactic gas at least once in its life time (Bertola et al. 1992). Finally, the presence of shells, tails, x–shaped structures etc., mostly in field galaxies, is further indication for at least minor merging or accretion events taking place today in low density environments (see review by Schweizer 1990).

On the other side, the small scatter observed in the colour–velocity dispersion (σ) and $Mg_b - \sigma$ relations indicates that the bulk of the stars of *luminous cluster* ellipticals must be rather old and must have an almost coeval epoch of formation, with $\Delta t/t \lesssim 0.15$ (Bower et al. 1992; Bender, Burstein, Faber 1993), in agreement with recent calculations in the context of CDM models (Kauffmann 1996). The high Mg_b absorption values and the large [Mg/Fe] ratios imply also large ages and presumably short star formation time scales for luminous Es (Matteucci 1994, Bender 1996, Greggio 1996). In addition, the small scatter observed in the mass–to–light ratio (M/L) derived from the "Fundamental Plane" analysis sets tight constraints on the relative variations in dynamics, IMF and ages of the stellar populations (Renzini & Ciotti 1993). Finally, the absorption–index diagrams between H_β, Fe and Mg_b confirm the old age for luminous Es, but imply smaller ages or extended SF histories for low luminosity early–type galaxies (Faber, Worthey, Gonzalez 1992, Gonzalez 1993). Since most low luminosity ellipticals are likely to contain disks contributing up to 30% to the total light (Bender et al. 1989, Rix and White 1990, Scorza and Bender 1995), there is still the possibility that younger *mean* ages may solely be associated with extended star formation in disks while bulges may still be old. Some low luminosity Es (especially compact Es) may also be "disk–less" bulges or bulges with stirred disks (see Bender et al. 1992). Independent from this uncertainty, ellipticals in low density environments show more peculiarities and may be genuinely younger than cluster ellipticals (Schweizer et al. 1990, Kauffmann 1996, Gonzalez 1993), though the evidence is conflicting (De Calvalho & Djorgovski 1992, Lucey & Guzmán 1993, Burstein 1989).

Summarizing, the above observational facts suggest the following conclusion. Luminous cluster ellipticals formed rapidly in early merging events. Luminous field ellipticals may be younger and could be late mergers. Faint ellipticals and S0 galaxies may have had extended SF histories.

3   4m Science + HST: Redshift Evolution of Elliptical Galaxies

Confirmation and even tighter constraints on stellar population ages in early–type galaxies come from the combined use of 4m (spectroscopic) telescopes and the imaging capabilities of the refurbished HST. In order of increasing accuracy, the following tests of the evolutionary history of early–type galaxies can be listed.

The median colours of cluster E galaxies evolve only slowly with redshift, consistent with mostly passive evolution up to $z \approx 1$ (Aragon–Salamanca et al. 1993, Stanford et al. 1995). The galaxy counts, divided as a function of morphological type, show that the number density of E/S0 galaxies does not evolve with redshift within current errors (Driver et al. 1996). The luminosity function of red galaxies and K–band selected galaxies changes only very weakly with redshift up to $z \approx 1$ (Glazebrook et al. 1995, Lilly et al. 1996, Ellis et al. 1996).

The surface brightnesses of E/S0s decrease with redshift following closely the Tolman relation ($I \approx (1+z)^{-4}$) and passive evolution models (Franx 1993, 1995; Dickinson 1995, Pahre et al. 1996), up to redshifts $z \approx 1$.

More recently, pushing to the limit the current generation of 4m telescopes and instrumentation, it has been possible to investigate problems that will be best tackled with the future 8m telescopes. The evolution of the Fundamental Plane (and the related M/L) has been followed up to $z \approx 0.4$. It is consistent with mostly passive evolution, but smaller SF events cannot be excluded (Franx 1993, 1995, see also this conference).

On the same line of research, the evolution of the $Mg - \sigma$ relation (Bender, Ziegler, Bruzual 1996) demonstrates that luminous cluster ellipticals are very old. Figure 1 shows the $Mg_b - \sigma$ relation for the clusters A370 and MS1512+36, at redshift $z \approx 0.375$, together with the local ellipticals of the Virgo and Coma cluster. An aperture correction has been applied to put the two samples on the same scale. Distant ellipticals also show a correlation between Mg_b and σ as local ellipticals do. However, there is clear evidence for evolution: at any given σ, the mean Mg_b of Es at $z = 0.375$ is lower than at $z = 0$. The evolution is very small for massive Es and likely stronger for faint Es, on average it is about 0.3 Å. Using Worthey's (1994) population synthesis models, the Mg_b weakening at a given σ can be translated into a relative age difference. This results in the conclusion that the bulk of the stars in the luminous cluster ellipticals must have formed at redshifts $z > 2$. Moreover, the same models can be used to translate the observed Mg_b weakening into a –0.4 mag change of the B–band luminosity, assuming a Salpeter IMF. The fact that the slope of the $Mg_b - \sigma$ relation at $z = 0.375$ appears to be slightly steeper than today indicates that less luminous ellipticals may be generally younger (consistent with Faber et al. 1992 local $H_\beta - Mg-Fe$ measurements, see previous section). Note that, in contrast to luminosity and surface brightness evolution tests, the $Mg_b - \sigma$ test is independent of the slope of the IMF in elliptical galaxies (see discussion in Bender et al. 1996).

We have imaged some of the ellipticals observed spectroscopically by Bender et al. (1996) with HST and determined the structural parameters R_e (the half–

luminosity radius) and $\langle SB_e \rangle$ (the effective average surface brightness) in the F675W filter, using the two–component fitting algorithm developed by Saglia et al. (1996), with HST psf convolution tables. The surface brightness term has been transformed to rest–frame B–band, corrected for cosmological dimming $((1+z)^4)$ and for passive evolution using the $Mg_b - \sigma$ relation described above. Figure 2 shows the Fundamental Plane of the clusters A370 and MS 1512+36 (filled/open squares before/after evolution correction) together with the data points of the Coma ellipticals (open circles), all having $\sigma > 150$ km/s. The distances are computed using $H_0 = 50$ km/s/Mpc and $q_0 = 0.25$. The figure shows that the passive evolution correction fully explains the observed luminosity evolution and allows for q_0 values in the range $0 - 0.5$.

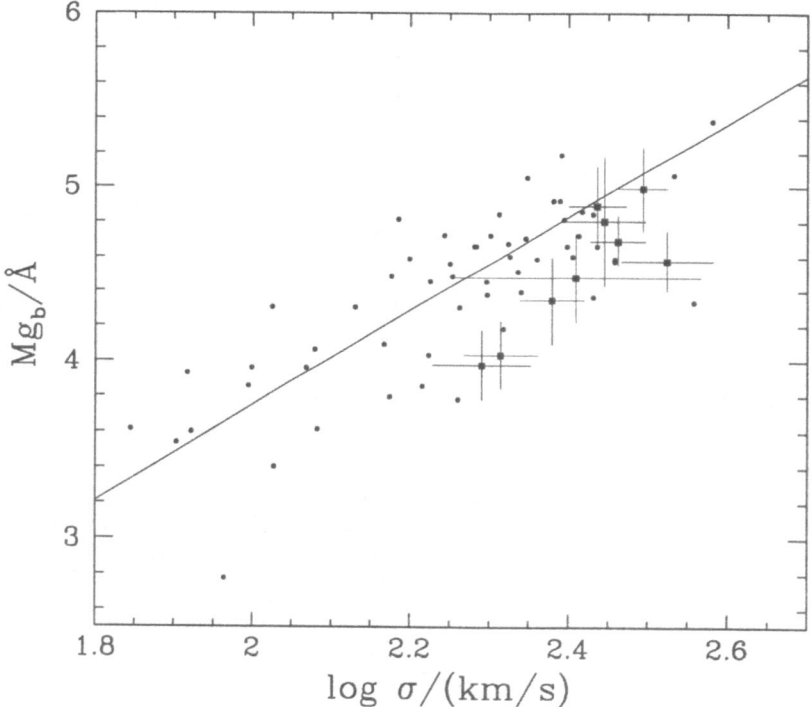

Fig. 1. The $Mg_b - \sigma$ relation for ellipticals in the clusters A370 and MS1512+36 at redshift $z = 0.375$ (filled squares), compared to the one of Coma Cluster ellipticals (small filled circles).

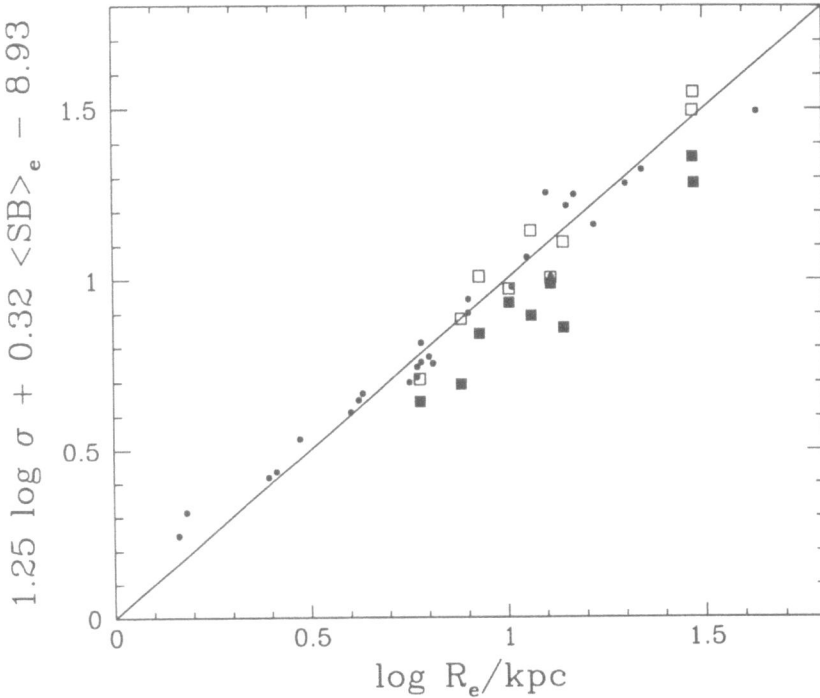

Fig. 2. The Fundamental Plane at redshift $z = 0.375$ for the same elliptical galaxies as in Figure 1 (filled/open squares: before/after correction for luminosity evolution). The filled small circles show the position of Coma Cluster ellipticals. We used $H_0 = 50$ km/s/Mpc and $q_0 = 0.25$.

In contrast to cluster ellipticals, blue members and E+A galaxies are known to show strong evolution (Butcher and Oemler 1978, Dressler and Gunn 1983). However, these objects are unlikely to end as ellipticals. Hubble Space Telescope imaging of blue and E+A galaxies indicates that presumably most of them are infalling spirals experiencing tidal shaking or 'harrassment' (Moore, Katz, Lake 1996) with only a small percentage undergoing merging events. Since the outer parts of the disks are stripped during these processes and/or star formation is likely to enhance the central stellar densities, large disk–to–bulge ratios are transformed into low disk–to–bulge ratios. The end products of this process are therefore likely to resemble S0 galaxies or disky ellipticals. In fact, S0s, and not ellipticals, are the dominant galaxy population in clusters at lower redshifts (e.g., Saglia et al. 1993, Jørgensen et al. 1994).

Summarizing, the conclusions based on the redshift evolution of elliptical galaxies are in full agreement with the analysis of the properties of local samples of ellipticals (see Section 2).

4   8m Science (+HST)

Following the discussion of the previous sections, it is natural to highlight a few tasks that the VLT (and other 8m telescopes) will be able to perform (the list is of course very incomplete).

The search for the oldest ellipticals at high redshifts is connected to the determination of the epoch of first star formation, a crucial test of cosmological models. Some of these objects may have been found already (see Giavalisco, this conference), but larger statistics and better information about these objects is still needed. As preparatory work to the spectroscopic observations with the VLT, a (photometric) survey of distant clusters ($z \gtrsim 0.6$) in the southern hemisphere of the kind described by Da Costa et al. (1996) is needed.

An extremely powerful test of cosmological models of structure formation is the determination of the evolution of the *potential function* (or *dispersion function*) of E, S0 and spiral galaxies via the measurement of their internal kinematics as a function of redshift. This will allow to follow the evolution of the potential depth of dark matter halos directly, bypassing the uncertain steps related to the ill–known processes of star formation needed to compute the luminosity function.

The VLT, in combination with HST photometry, can allow the accurate determination of the redshift evolution of the luminosities, stellar population parameters, internal kinematics and structural parameters of galaxies. This will be the essential information to test galaxy formation models in detail, i.e., beyond dark halo evolution. In addition, this information can possibly be used to calibrate elliptical galaxies as standard candles and cosmological tools, to be used to constrain q_0 via the Fundamental Plane relations and the Volume test.

Spectroscopic instruments with multiplex capabilities and low/medium resolution in the optical and near infrared spectral range such as FORS, ISAAC, NIRMOS/VIRMOS will play a decisive role in the above outlined projects. The imaging capabilities offered by FORS and especially CONICA plus Adaptive Optics in the NIR will complement the HST high resolution with the power of collected flux.

5   Conclusions

Observations of local massive ellipticals obtained with 2m – 4m class telescopes over the last decade demonstrated that these objects, though most likely being formed in mergers, exhibit only small scatter in color/line–strengths vs. velocity-dispersion diagrams and in mass–to–light ratios (from 'fundamental plane' analysis). The small amplitude of the scatter indicates *approximately coeval formation of massive ellipticals at high redshift.* Further support for this conclusion comes from H_β vs. metal–linestrength analysis and from the high overabundance of light elements relative to iron (indicating short star formation time scales). Age and formation constraints on lower luminosity early–type galaxies and also field ellipticals are less tight and lower mean ages or extended star formation histories are possible.

Observations of redshifted luminous cluster ellipticals with 4m–class telescopes and the Hubble Space Telescope up to $z \approx 1$ confirm this picture. Luminosities, surface brightnesses and mass–to–light ratios, as well as the fundamental plane relation evolve only slowly with redshift z. The most reliable measurements show a rest–frame B–band evolution following the simple relation $\Delta B(z) \approx -z$. Again, our information on low luminosity ellipticals and field ellipticals at higher redshift is much more uncertain.

These results put strong constraints on the formation of luminous cluster ellipticals and may further offer a route to calibrate these objects as cosmological standard candles or standard rods.

Acknowledgments: Financial support by the Deutsche Forschungsgemeinschaft under SFB 375 is gratefully acknowledged. We thank our collaborators Drs. Paola Belloni, Laura Greggio and Ulrich Hopp for discussions and comments. Part of the work presented here is based on observations made with the NASA/ESA Hubble Space Telescope, obtained at the Space Telescope Science Institute, which is operated by the Association of Universities for Research in Astronomy, Inc., under NASA contract NAS 5-26555.

References

Aragon–Salamanca, A., Ellis, R.S., Couch, W.J., Carter, D. (1993): MNRAS 262, 784

Barnes, J.E. (1996): IAU Symp. 171, p.191

Bender, R. (1996): IAU Symp. 171, p. 181

Bender, R., Surma, P., Döbereiner, S., Möllenhoff, C., Madejsky, R. (1989): A&A 217, 35

Bender, R., Burstein, D., Faber, S. (1992): ApJ 399, 462

Bender, R., Burstein, D., Faber, S. (1993): ApJ 411, 153

Bender, R., Ziegler, B., Bruzual, G. (1996): ApJ 463, L51

Bertola, F., Buson, L.M., Zeilinger W.W. (1992): ApJ 401, L79

Bower, R.G., Lucey, J.R., Ellis, R.S. (1992): MNRAS, 254, 601

Burstein, D. (1989): in The World of Galaxies, Eds. H.G. Corwin, L. Bottinelli, Springer Verlag, Berlin, p. 547

Butcher, H., Oemler, G. (1978): ApJ 219, 18

Da Costa, L., Baker, A., Beletic, J., Clements, D., Coté, S., Freudling, W., Huizinga, E., Méndez, R., Roennback, J. (1996): Proposal for the ESO imaging survey

De Carvalho, R.R., Djorgovski, S. (1992): IAU Sump. 194, p. 400

Dickinson, M. (1995): ASP Conference 86, p. 283

Dressler, A., Gunn, J. (1983): ApJ 270, 7

Driver, S.P., Windhorst, R.A., Griffiths, R.E. (1996): IAU Symp. 171, p. 221

Ellis, R.S., Colless, M., Broadhurst, T., Heyl, J., Glazebrook, K. (1996): preprint

Faber, S.M., Worthey, G., Gonzalez, J.J. (1992): IAU Symp. 149, p. 225

Franx, M. (1993): PASP 105, 1058

Franx, M. (1995): IAU Symp. 164, p. 269

Glazebrook, K., Peacock, J.A., Miller, L., Collins, C.A. (1995): MNRAS 275, 169

Gonzalez, J.J. (1993): PhD Thesis, University of Santa Cruz

Greggio, L. (1996): MNRAS, submitted

Kauffmann, G. (1996): MNRAS in press

Jørgensen, I., Franx, M., Kjærgaard P. (1994): ApJ 433, 533

Lilly, S., Le Fevre, O., Hammer, F., Crampton, D., Schade, D.J., Hudon, J.D., Tresse, L. (1996): IAU Symp. 171, p.209

Lucey, J., Guzmán, R. (1993): ESO/EIPC Workshop "Structure, Dynamics and Chemical Evolution of Elliptical Galaxies", Eds I.J. Danziger et al., ESO Garching, p. 171

Matteucci, F., (1994): A&A 288, 57

Moore, B., Katz, N., Lake, G. (1996): IAU Symp. 171, p. 203

Pahre, M.A., Djorgovski, S.G., de Calvalho, R.R. (1996): ApJ 456, L79

Renzini, A., Ciotti, (1993): ApJ 223, 707

Rix, H.-W., White, S.D.M. (1990): ApJ 362, 52

Saglia, R.P., Bender, R., Dressler, A. (1993): A&A 279, 75

Saglia, R.P., Bertschinger, E., Baggley, G., Burstein, D., Colless, M., Davies, R.L., McMahan, R.K Jr., Wegner, G. (1996): submitted to ApJS

Schweizer, F. (1990): in Dynamics and Interactions of Galaxies, Ed. R. Wielen, Springer Verlag Heidelberg, p. 232

Schweizer, F., Seitzer, P., Faber, S.M., Burstein, D., Dalle Ore, C.M., Gonzalez, J.J. (1990): ApJ 364, L33

Scorza, C., Bender, R. (1995): A&A 293, 20

Stanford, S.A., Eisenhardt, P.R.M., Dickinson, M. (1995): ApJ 450, 512

Worthey, G. (1994): ApJS 95, 107

Constraining the Evolution of Galaxy Masses and the M/L Ratios out to $z=1$

Marijn Franx and Pieter G. van Dokkum

Kapteyn Astronomical Institute, P.O. Box 800, 9700 AV Groningen
The Netherlands

Abstract. Galaxy evolution is probably a complex process. Mergers, infall, and starbursts may change galaxy properties systematically with time. As a result, information on the mass evolution of galaxies is needed. Such information can be retrieved from the evolution of the Tully-Fisher relation, Faber-Jackson relation, or the Fundamental Plane with redshift.

Observations of this kind have recently become possible. We present the Fundamental Plane relation measured for galaxies in the rich cluster CL 0024+16 at $z=0.391$. The galaxies satisfy a tight Fundamental Plane, with relatively low scatter (15 %). The M/L is 31 $\pm$ 12 % lower than the M/L measured in Coma, which is consistent with simple evolutionary models. Hence, galaxies with very similar dynamical properties existed at a $z=0.4$.

An important goal of new, large optical telescopes will be to extend this type of work to larger samples, fainter magnitudes, and higher redshifts.

1 Measuring the Evolution of Mass $F(M_\star, z)$, or $F(v_c, z)$

Galaxy evolution may be a complex process, with possibly a large role for mergers, interactions, infall, and starbursts triggered by these events. Such processes complicate the interpretation of observations of high redshift galaxies, as galaxies can change rapidly in luminosity (due to starbursts), and can change morphology due to mergers, and infall of gas. The progenitors of certain types of galaxies at some redshift may be of different type at some other redshift, and their luminosities may be quite different.

In order to quantify these effects, more information is needed than the evolution of luminosity and color of galaxies, such as measured by the evolution of the luminosity function. Detailed information on the morphological evolution, and the evolution of the mass function is essential. The evolution of the mass function is possibly the most important, as it gives direct insight into the mass evolution of individual galaxies, and can directly determine when typical galaxies were assembled.

Unfortunately, the total masses of galaxies are notoriously difficult to measure. However, there exist good relations between circular velocity, and velocity dispersion, and photometric parameters: the Tully-Fisher relation for spirals (Tully & Fisher 1977), the Faber-Jackson relation (Faber & Jackson 1976), and the Fundamental Plane for early-types (Djorgovski & Davis 1987, Dressler et al. 1987). These relations are very suitable for evolutionary studies, because their intrinsic scatter is low at $z=0$.

The general purpose of these studies is to obtain the evolution of the Tully-Fisher relation, Faber-Jackson relation, and Fundamental Plane with redshift. The combination of such observations with the evolution of the luminosity function of galaxies, can be used to constrain the evolution of the distribution of circular velocities $F(v_c, z)$, which is less sensitive to starbursts than the equivalent $F(L, z)$. Similarly, the stellar mass locked up in early type galaxies can be measured in a similar way. Such results will provide direct constraints on theories of galaxy formation and evolution.

2   An Example: Evolution of the Fundamental Plane

Here we present new results on a program to measure the evolution of the Fundamental Plane relation with redshift. Earlier results can be found in Franx (1993a,b, 1995). The Fundamental Plane is a relation between effective radius r_e, effective surface brightness I_e, and central velocity dispersion σ of the form $r_e \propto \sigma^{1.24} I_e^{-0.82}$ (e.g., Bender et al. 1992, Jørgensen et al 1996, JFK). Its scatter is low, at 17% in r_e (Lucey et al. 1991, JFK). The implication of the Fundamental Plane is that the M/L ratio of galaxies is well behaved (e.g., Faber et al. 1987). Under the assumption that galaxies are a homologous family, the implied M/L scaling is $M/L \propto r_e^{0.22} \sigma^{0.49} \propto M^{0.24}$. Such scaling is sufficient for the existence of the Fundamental Plane, and vice versa. The cause of the variation in M/L with mass is not well understood, but it is thought to be mainly due to variations in metallicity (see also, e.g., Renzini & Ciotti 1993).

Observations at higher redshifts will yield the evolution of the M/L ratio as a function of redshift. Below we explore the expected variation of M/L.

2.1   Models for the Evolution of the M/L Ratio

The luminosity of a co-eval stellar population is expected to evolve with time. Tinsley (1980) showed that the luminosity evolves like

$$L \propto 1/(t - t_{form})^\kappa$$

where $\kappa = 1.3 - 0.3x$, and x is the slope of the IMF. The Miller–Scalo IMF implies $x=0.25$, and $\kappa \approx 1.2$. Recent studies indicate that the value of κ depend on passband and metallicity (Buzzoni 1989, Worthey 1994). These authors find $0.6 < \kappa < 0.95$ for the V band.

To first order, this evolution implies that the M/L ratio evolves like

$$\ln M/L(z) = \ln M/L(0) - \kappa(1 + q_0 + 1/z_{form})\, z,$$

where z_{form} is the formation redshift (Franx 1995). Hence the logarithm of the M/L ratio is expected to decrease linearly with redshift, and the coefficient depends on κ(IMF), q_0, and z_{form}. This equation is valid for $q_0 \approx 0$, and high z_{form}. The rate at which the M/L ratio decreases is a function of several unknown variables, and a direct interpretation of the observed decrease of the M/L ratio may not be very straightforward.

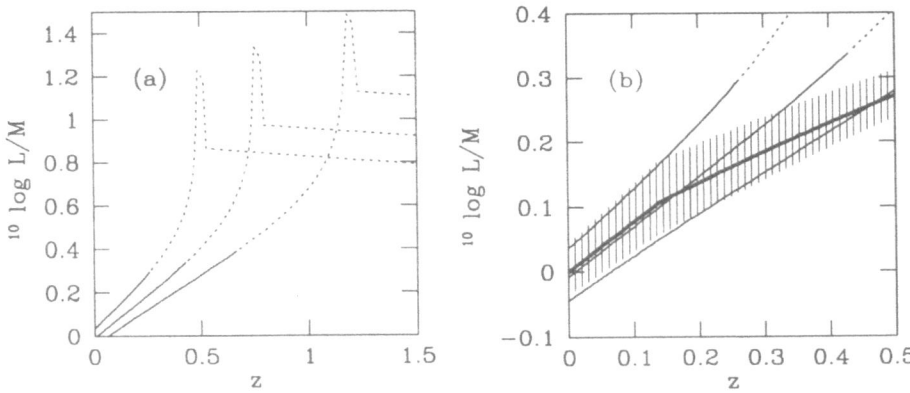

Fig. 1. The evolution of galaxies which undergo three distinct phases: I regular star formation in a disk, II starburst, III quiescent evolution. a) shows the luminosity evolution for three such galaxies. The galaxies are classified as regular early-types after 1.5 Gyr after the burst. This is indicated by the drawn curve. b) the evolution of the mean L/M ratio for a sample of early-types which formed in this complex way. The starburst is assumed to occur at a random time between $z=0.5$ and $z=2$. The line indicates the median L/M, the shaded area is bounded by the upper and lower quartile of the sample. The median L/M ratio bends at $z=0.2$, as more and more galaxies drop out from the sample. The sample becomes more and more biased to the oldest early-types at higher redshifts.

2.2 Complex Evolution

There is no good reason to assume that all early-type galaxies formed at the same redshift. Furthermore, a single galaxy may have had a complex formation history, with starformation extending over a long time. The evolution of the M/L ratio will be much more complex if such age differences are taken into account.

We have created models in which early type galaxies form by transformation of galaxies with continuous starformation. It is assumed that the progenitors form stars in a continuous way, until a burst of starformation occurs, and the galaxies are transformed into non-starforming galaxies. These will appear as post starburst galaxies for another 1.5 Gyr, and then appear to be early types.

This type of evolution implies that the morphologies of galaxies evolve with time, from spiral, to post star burst galaxy, to early-type. This has important consequences, since the set of early-types at higher redshifts will be a special subset of the set of early-types at $z=0$. If we select early-types at higher and higher redshift, we are selecting a subsample that is more and more biased towards the oldest early-types. In short, we may be selecting the oldest galaxies, and find that they are old.

The problem is illustrated in Fig. 1. The typical evolution of 3 galaxies is shown. The drawn curve is the phase in which they appear as early-types. Clearly, the oldest early-types appear as early-type for the longest time. Fig. 1b demonstrates the effect on the observed L/M ratios of a large sample. At low redshifts, all galaxies appear as early types, and the evolution of the median L/M ratio remains normal. The scatter around the mean increases rapidly with redshift. Around $z = 0.2$, some of the galaxies appear as post star burst galaxies, and they would be excluded. The median L/M ratio is biased towards low values. This effects increases at higher redshifts. The bias is as strong as 30 % at $z=0.5$. As galaxies disappear from the sample, the scatter in the L/M ratio may decrease at higher redshifts.

2.3   The Fundamental Plane in CL 0024+16 at $z=0.39$

CL 0024 is a rich cluster at $z=0.39$, and has been extensively observed (e.g., Dressler et al. 1985). We have obtained a deep, 19 hour integration at the MMT to measure the internal velocity dispersions of luminous galaxies in the cluster. HST images were used to measure the structural parameters of the galaxies. Full details of the observations and the analysis can be found in van Dokkum and Franx (1996).

Fig. 2a shows the resulting Fundamental Plane. There is a very clear relation, with relatively low scatter (15 %). The slope is very similar to that for nearby cluster galaxies (e.g., JFK). In short, *early-type galaxies exist at $z=0.4$ which are very similar to galaxies at $z=0$.*

Fig. 2b shows the observed M/L ratios for Coma and CL 0024 against the parameter $r_e^{0.22}\sigma^{0.49}$. The Fundamental Plane implies that galaxies lie along a line in the plot. We see a clear offset between the two data sets. The lines indicate fits to both data sets. The mean difference in the M/L ratio is 31 %. The error is dominated by systematic effects, and is estimated at 12 %. It is clear that the sample for CL0024 is biased towards the most massive galaxies, and this selection bias is partly the cause for the systematic uncertainty.

Fig. 2c shows the evolution of the mean M/L ratio with redshift. The current results are in good agreement with a large formation redshift, and small values of κ. The resulting constraint is $\kappa(1 + q_0 + 1/z_{form}) = 0.84 \pm 0.32$. Obviously, many combinations of q_0 and z_{form} are allowed, given the uncertainty in κ and the observed change in M/L ratio.

3   Discussion

We have shown that the equivalent of the Fundamental Plane exists at $z=0.39$, and we have derived the mean evolution of the M/L ratio. The evolution is low at 31%, but is in agreement with models. More data are needed to constrain the formation redshift better.

These observations demonstrate that information on galaxy masses can be obtained from deep, ground based spectroscopy. The next step is to extend this work to higher redshift, and to the field. Furthermore, similar studies have demonstrated that the Tully-Fisher relation can be used (Vogt et al 1996, Rix, Colless, & Guhathakurta 1996). The new generation optical telescopes will allow

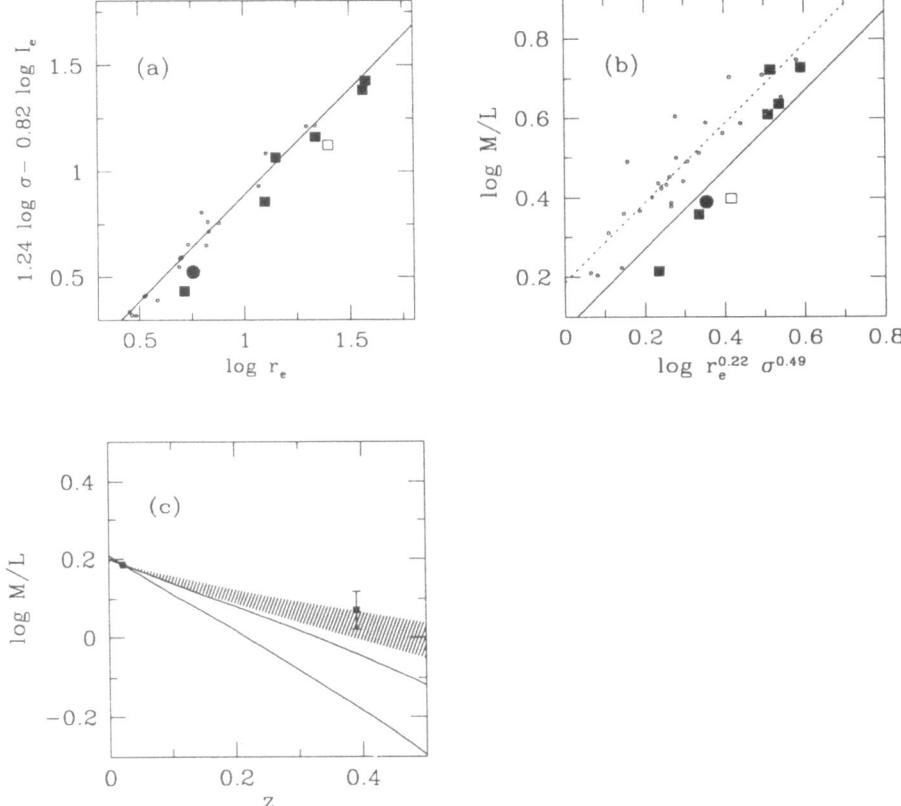

Fig. 2. a) The Fundamental Plane for galaxies in CL 0024+16 at z=0.391 in the red-shifted V band. The small symbols are galaxies in Coma. The Fundamental Plane in CL 0024 is very similar to that in Coma, with similar low scatter (15%). b) The M/L ratio against $r_e^{0.22} \sigma^{0.49} \propto M^{0.24}$, for CL 0024 and Coma. The lines are fits to the data points. The M/L ratio in CL 0024 is lower by 31± 12 %. c) The evolution of the M/L ratio against redshift. The two points are Coma and CL 0024. The hatched area is the area allowed by models with formation redshift $z_{form} = \infty$. The two lines delineate the area allowed by models with $z_{form} = 1$. Such models are marginally consistent with the data.

a rapid extension of this type of work to larger samples, higher redshifts, and lower galaxy masses.

Eventually, these observations can be used to determine the evolution of the distribution of circular velocities for galaxies $F(v_c, z)$, and the evolution of stellar masses locked up in early type galaxies $F(M_\star, z)$. This requires accurate observations of the evolution of the luminosity function and the "mass relations" in the field. The final outcome of such a program can be expected to provide strong constraints on the models of galaxy formation and evolution.

It is a pleasure to thank the organizers for a stimulating workshop.

References

Bender R., Burstein D., Faber S. M., 1992, ApJ, 399, 462

Buzzoni A., 1989, APJS, 71, 817

Djorgovski S., Davis M., 1987, ApJ, 313, 59

Dressler A., Gunn J. E., Schneider D. P., 1985, ApJ, 294, 70

Dressler A., Lynden-Bell D., Burstein D., Davies R. L., Faber S. M., Terlevich R. J., Wegner G., 1987, ApJ, 313, 42

Faber S. M., Dressler A., Davies R. L., Burstein D., Lynden-Bell D., Terlevich R., Wegner G., 1987, Faber S. M., ed., Nearly Normal Galaxies. Springer, New York, p. 175

Faber S. M., Jackson R. E., 1976, ApJ, 204, 668

Franx M., 1993a, ApJ, 407, L5

Franx M., 1993b, PASP, 105, 1058

Franx M., 1995, van der Kruit P. C., Gilmore G., eds, Proc. IAU Symp. 164, Stellar Populations. Kluwer, Dordrecht, p. 269

Jørgensen I., Franx M., Kjærgaard P., 1996, MNRAS, 280, 167 [JFK]

Lucey J. R., Guzmán R., Carter D., Terlevich R. J., 1991, MNRAS, 253, 584

Renzini A., Ciotti L., 1993, ApJ, 416, L49

Rix, H. W., Colless, M. M., Guhathakurta, P., 1996, Bender, R., Davies, R. R., eds, Proc. IAU Symp. 171, New Light on Galaxy Evolution. Kluwer, Dordrecht, p. 241

Tinsley B. M., 1980, Fundamentals of Cosmic Physics, 5, 287

Tully R. B., Fisher J. R., 1977, AA, 54, 661

van Dokkum P. G., Franx M. 1996, MNRAS, 281, 985

Vogt, N. P., et al, 1996, ApJ, 465, L15

Worthey G., 1994, APJS, 95, 107

2D Spectroscopy of Remote Radiogalaxies with TIGER

Gilles Adam[1]

C.R.A.L., Observatoire de Lyon, F-69230 Saint Genis Laval

Abstract. After a short reminder of the need for remote radiogalaxies studies, their promises and their problems, we emphasize the two-dimensional spectroscopy achievements. The TIGER concept is briefly described. As a first example, 3C435A (z=0.471) results are presented; a cocoon model seems adequate to fit the observed spectral characteristics, and old elliptical galaxy population are observed for two components. The latest TIGER observation of a radiogalaxy was made on 4C41.17 (z=3.8). A complete Lyα mapping was obtained, and the ideas which are suggested by the data are presented. Regarding the future prospects, a good indication is given by the first spectrum obtained, with TIGER, of the z=4.7 resolved companion to the quasar BRI1202-0725. To end this paper, the CFHT adaptive optics OASIS spectrograph is presented, and an extrapolation to 8-metre class telescopes is evaluated.

1 Baby Universe and Data Cubes

1.1 First Hopes ...

Nearby elliptical galaxies are characterised by an old stellar population and a low gas content; this suggests that the bulk of their stars were formed in an explosive burst at a remote epoch. As low redshift radiogalaxies are identified with "normal" ellipticals, studying the stellar evolution over cosmological times was first thought to be straightforward: identify high-z radiogalaxies counterparts, assume that they are normal ellipticals, and compare their SED's to local ones.

1.2 And Immediate Disapointment ...

A growing body of data soon showed that high-z radiogalaxies, that is high radio-luminosity ones, are not so "normal", and that nuclear activity is strongly influencing their optical properties (Dunlop and Peacock, 1993). The first consequence is that the observed spectrum is a mix of various effects: stellar continuum, strong emission lines, due either to stars ou to nuclear activity, nuclear non thermal continuum (see for instance Chambers et al., 1990), dust scattering and absorption (for an example, Dunlop et al., 1994). The second consequence is a complex morphology, with nuclear jets propagating their effects to the external regions. This raises well known questions about the alignment effect, and the effects of interactions between the jet and the IGM: star formation triggering (Rees, 1989, de Young, 1989, Bithell and Rees, 1990, Begelman and Cioffi, 1989), nebular emission (Eales et al., 1993), various dust or electrons scatterings mechanisms (see for instance Daly, 1992a or Daly, 1992b). So, the spectral properties are strongly non-uniform over the surface of the objects, each emitting region showing a specific Doppler shift. This is further strengthened by the possibility that some z>3 objects may still be in a coalescence phase, with numerous sub-galactic sized blobs accreting to eventually form a galaxy (see several contributions to this colloquium).

1.3   Data Cubes

To understand this 3D picture, we need to know the positions and kinematics of the various components; so, we need, for each picture element, a position and a spectrum. Finally, the only way to disentangle the superposition of spatial and spectral effects, is to get complete (α,β,λ) **data cubes**. This is the goal of the 2D spectrographs currently developped.

2   From Stars to Stripes: TIGER

Among the various possible designs, TIGER was a pioneer, proposed by G. Courtès and used at CFHT since 1987. The instrument is fully described elsewhere (Bacon et al., 1995). The main advantages over other designs are the filling factor of the lens array, essentially equal to 1, an overall transmission near to that of a classical spectrograph and lack of spectral cross-talk between spatially adjacent picture elements. The drawback is the limited wavelength range (at best 2500Å) on a single exposure. For the observer, important points are the number of picture elements (572), the spatial sampling (0.39″ to 0.61″), the field of view (8″ to 12″), the spectral domain (4000Å to 10000Å), and the spectral sampling (1.8Å to 8Å), with a maximum of 470 spectral elements. The data reduction process has some very specific aspects; for instance, night sky spectrum may be obtained by statistical reconstruction, even in the case of a fairly crowded field, cosmic rays removal is done by median comparison with neighbouring spectra. The many published papers show that TIGER is a very efficient concept, used in fields ranging from planets to remote extragalactic objects.

3   Observation of $z < 1$ Radiogalaxies

3.1   The Data

The radiogalaxy 3C435A (z=0.471) has been observed with TIGER at the CFHT, with 0.6″ and 8Å spatial and spectral samplings. In 2 hours (V) and 3.2 hours (R) exposure times, the S/N ratio attained is 20 on the [OII] line, 10 on the continuum, for the main component (V=21.5, R=20, summation over 3.6 square arcseconds).

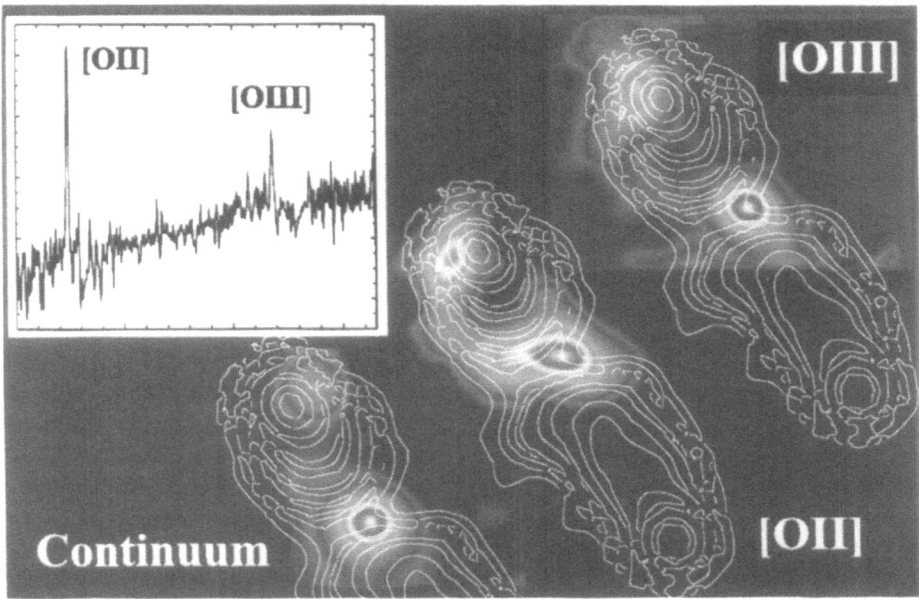

Fig. 1. Reconstructed spectral maps of 3C435A (grey scale) with 1.4GHz VLA map (McCarthy et al. 1989) superimposed.

3.2 The Results

Six distinct components have been identified, and SED have been detected and fitted with synthetic spectra (Rocca-Volmerange and Guiderdoni, 1988) for two of them; the fits indicate old elliptical populations (8.5 Gyr to 11 Gyr at $z = 0.471$). Two other components have marginally detected continua; one seems also to host an old stellar population, the last one is a young object lying outside the radio axis. The gas kinematics, from [OII]3727Å and [OIII]5007Å lines, as well as the [OII]/[OIII] ratio map, are consistent with the overpressured cocoon model (Begelman and Cioffi, 1989). The radio/optical alignment, in this object, is mainly due to nebular [OII] and [OIII] emission. See Rocca-Volmerange et al., 1994 for details.

4   High Redshift Radiogalaxies

4.1   The Data

The $z = 3.8$ radiogalaxy 4C41.17 has been observed at CFHT, for a total integration time of only 1.86h, with 0.6″ and 8Å samplings, over a 12″ field. Twin-peaked Lyα emission was observed, with a patchy structure over the whole field. The stronger peak (2.7×5.4 kpc^2) reaches $4.8\ 10^{43}$ erg s^{-1}. The two peaks are aligned on the radio axis defined by the two strong 15GHz components B2 and

B3 (Chambers et al., 1990), the flat-spectrum N component (Carilli et al., 1994), and the high resolution HST components H2 and H4 (Miley et al., 1992). A general curvature of the Lyα axis is noticeable, and matches the radio axis twist noted by Miley et al. Filaments extend out of the central region, especially towards South-East; the very inner parts of these extensions are conspicuous on HST images (HST archive, 1993).

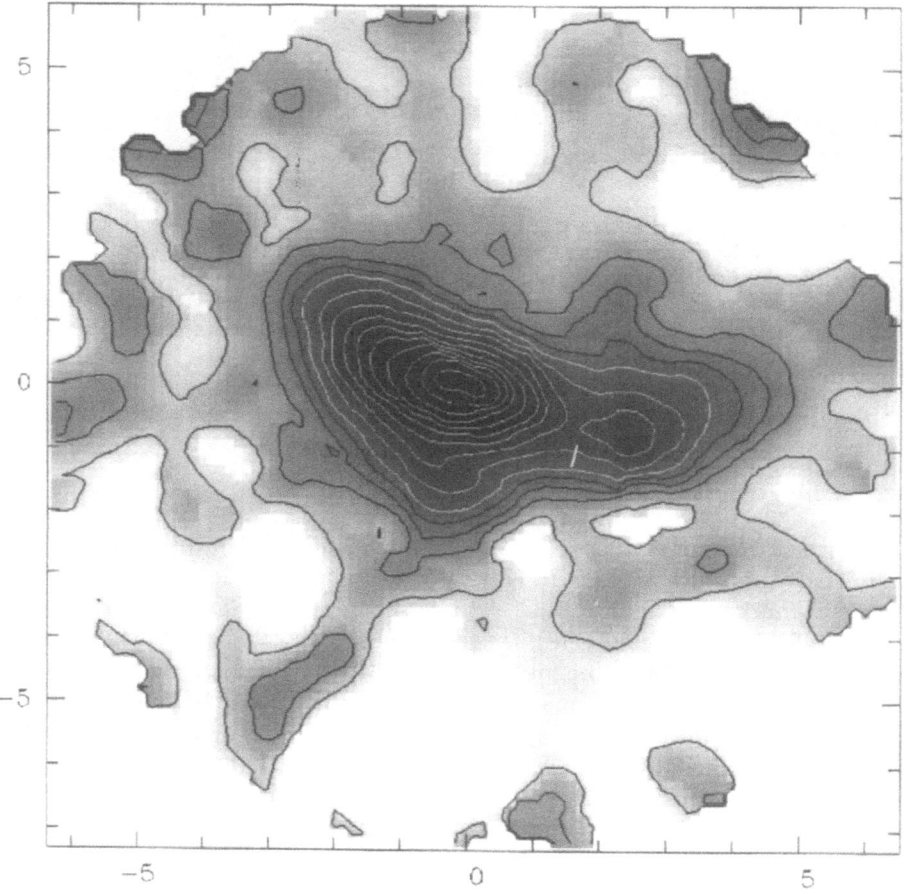

Fig. 2. Lyα image of 4C41.17, smoothed with a 0.8″ fwhm gaussian filter. Scale units are arc seconds, and the greyscale is logarithmic. The black contours run from 8% to 20%, with a 4% step, the white ones from 24% to 94% with a 8% step. The (0,0) point indicates the Lyα peak, where the brightness is 5.9 10^{-8} erg cm^{-2} arcsec^{-2}, that is 100%.

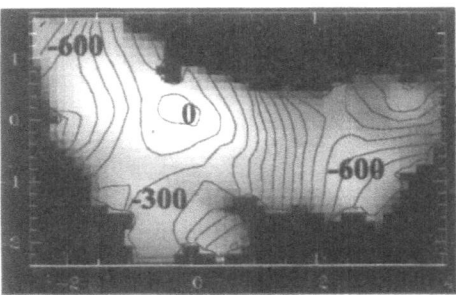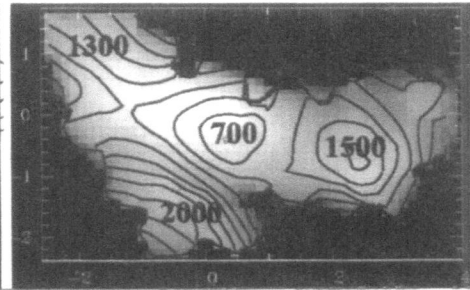

Fig. 3. The velocity (left) and velocity dispersion (right) fields, superimposed over the Lyα image. Isovelocities are plotted from -650 to 0 km s^{-1}, with a 40 km s^{-1} step. Isodispersions are plotted from 300 to 1700 km s^{-1} with a 100 km s^{-1} step.

4.2   The Results

A detailed analysis will be presented elsewhere (Adam et al., 1996, Rocca-Volmerange et al., 1996), but some strong points may already be emphasized: For H_0=75 km s^{-1} Mpc^{-1}, q_0=0.5, the dense central part has a size of 35×18 kpc^2: if a merging process in on the way, the bulk of the galaxy seems to be already formed at $z = 3.8$. The curved radio-jet follows the Lyα isophotes, over at least 18 kpc; this would favor a star formation process against scattering to explain the alignment effect (Miley et al., 1992). The high velocity dispersion values (up to $\approx$ 2000 km s^{-1}) could derive from the jet/IGM turbulent interaction at the cocoon frontier (Begelman and Cioffi, 1989), but no emission enhancement is detected where the frontal bow shocks are expected. The dip between the two Lyα peaks may be the trace of an intervening absorber (Hippelein and Meisenheimer, 1993), or of a large dust component (Dunlop et al., 1994), acting as an efficient Lyα killer (Chen and Neufeld, 1994). This point is presently studied (see Rocca-Volmerange et al., 1996), as the existence of large amounts of dust in a $z = 3.8$ object would be a significant constraint on evolutionary models. Last, the crowded Lyα neighbouring of the radiogalaxy, and the high velocities observed, may trace the last stages of a merging process.

5   Future Prospects

5.1   The Companion to BRI 1202-0725, at $z = 4.7$

This object have been observed at CFHT, during a 2.3h total integration time producing the first spectrum of this companion to a z=4.7 quasar. The luminosity is 3.10^{-1} erg cm^{-2} Å^{-1} in one spatial element (0.3 $''$ ^{2}). The Lyα shows exactly the same blue cut-off as the quasars's one, indicating that the same absorber is involved. The emission comes most certainly from gas ionised by the quasar UV, but an additional continuum source is needed to explain the measured broadband luminosity (see Petitjean et al., 1996).

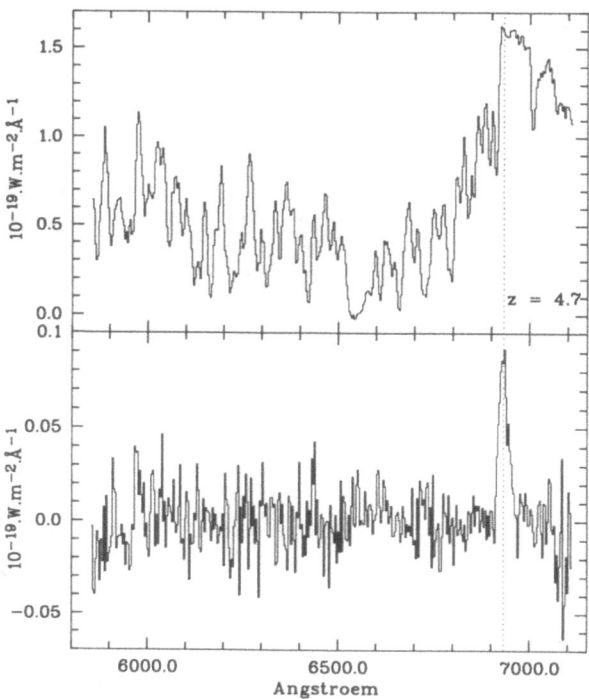

Fig. 4. TIGER observation of the companion to BRI 1202-0725, $z = 4.702$. Upper spectrum: quasar, lower spectrum: companion.

5.2   The Adaptive Optics OASIS Spectrograph

The nearby future of TIGER spells OASIS. This is the integral field spectrograph to be put into regular operation at the adaptive optics focus of the CFHT in 1997. It will expand TIGER capabilities, with spatial resolutions ranging from $0.52''$ to $0.04''$, over a $20'' \times 15''$ to $1.6'' \times 1.2''$ field, and spectroscopic resolutions ranging from 1600 to 3350.

5.3   Very High Resolution Studies of Distant Radiogalaxies with 8m-class Telescopes

Clearly, the present observations do not allow to answer the questions we asked about early stars and galaxies evolution, as well as about the interactions of jets and IGM. We need more detailed kinematical and spectral analysis, and/or deeper data. The new 8m-class telescopes will give this oportunity; they must be as time-efficient as possible, and two-dimensional spectroscopy definitely IS (the most immediate advantage beeing the absence of time lost to position an almost invisible object on a slit). In the TIGER design, the spatial sampling, done by the micro-lens array, cannot be impaired in any way by the spectrograph characteristics: there is no sensible cross-talk between spectra, and the photometric quality remains high, with or without adaptive optics. Both catadioptric or dioptric designs are feasible, with a large range of sampling. The limiting magnitude is nearly the same as that of a long-slit spectrograph, with much cleaner data, as present TIGER observations show. Magnitude limits have been extrapolated, from the OASIS documents, for a continuum source giving $s/n=5$ in one hour on an 8m telescope, in the V domain (5000-6000Å) or in the R domain (6000-9000Å), for both resolved and unresolved objects. As an example, $m_R=17.6$ per $''^2$ will be reached for an extended source sampled at 0.1″, with a spectral resolution of 3040, or $m_V=23.2$ per $''^2$ for an extended source sampled at 0.5″, with a spectral resolution of 400. So, a TIGER IFS will be well adapted to high spatial resolution / low-to-medium spectroscopic resolution observations of the distant faint and small objects of the early universe.

References

Adam, G., Rocca-Volmerange, B., Gérard, S., Ferruit, P. and Bacon, R. (1996): submitted to *Astron. Astrophys.*.

Bacon, R., Adam, G., Baranne, A., Courtès, G., Dubet, D., Dubois, J.P., Emsellem, E., Ferruit, P., Georgelin, Y., Monnet, G., Pécontal, E., Rousset, A.,Sayède, F. (1995): *Astron. Astrophys. Suppl. Ser.*, **113**, 347.

Begelman, M.C., Cioffi, D.F. (1989): *Astrophys. J.*, **345**, L21.

Bithell, M., Rees, M.J. (1990): *MNRAS*, **242**, 570.

Carilli, C. L., Owen, F.N., Harris, D.E. (1994): *Astron. J.*, **107**, 480.

Chambers, K.C., Miley, G.K., and van Breughel, W.J.M. (1990): *Astrophys. J.*, **363**, 21.

Chen, W.L. and Neufeld, D.A. (1994): *Astrophys. J.*, **432**, 567.

Daly, R.A. (1992a): *Astrophys. J.*, **386**, L9.

Daly, R.A. (1992b): *Astrophys. J.*, **399**, 426.

Dunlop, J.S., Peacock, J.A. (1993): *MNRAS*, **263**, 936.

Dunlop, J., Hughes, D., Rawlings, S., Eales, S., Ward, M. (1994): *Nature*, **370**, 347.

Eales , S., Rawlings, S, Puxley, P., Rocca-Volmerange, B., Kuntz, K. (1993): *Nature*, **33**, 14.

Hippelein, H., Meisenheimer, K.,(1993): *Nature*, **362**, 224.

HST archive, exposures from proposal 2438, Miley G. et al., (1993).

McCarthy, P.J., van Breughel, W., Spinrad, H., (1989): *Astron. J.*, **97**, 3.

Miley, G., Chambers, K., van Breugel, W., Macchetto, F., (1992): *Astrophys. J.*, **401**, L69.

Petitjean, P., Pécontal, E., Valls-Gabaud, D., Charlot, S. (1996): *Nature*, **380**, 411.

Rees, M.J. (1989): *MNRAS*, **239**, 1P.

Rocca-Volmerange et al., *in preparation*.

Rocca-Volmerange, B., Adam, G., Ferruit, P., and Bacon, R. (1994): *Astron. Astrophys.*, **292**, 20.

Rocca-Volmerange, B., Guiderdoni, B. (1988): *Astron. Astrophys. Suppl. Ser.*, **75**, 93.

de Young, D.S. (1989): *Astrophys. J.*, **342**, L59.

Spectro-Imaging of Starburst Galaxies: Present and Future

Georges Comte[1], Tigran Movsessyan[2], and Artashes Petrosian[2]

[1] Observatoire de Marseille and Institut Gassendi, 2, pl. Le Verrier,
   F-13248 Marseille Cedex 4
[2] Byurakan Astrophysical Observatory, Ashtarak Region, Armenia

Abstract. Recent advances in spectro-imagery enable to reconstruct emission line and continuum images of extended objects at both high spatial and spectral resolutions. The applications of such instruments to investigation of blue compact dwarf galaxies is illustrated with a few examples. The combination of such instruments with very large telescopes in the next future will allow a deep understanding of the starburst phenomenon supposed to occur at the onset of any galaxy formation.

1 Introduction

Dwarf starburst galaxies are intrinsically low luminosity objects that are seen in an active star formation episode, producing an increase in surface brightness due to the rich main-sequence population created during the burst, generally associated with a blue and ultraviolet continuum excess. The interstellar gas ionized by the newborne massive stars emits an HII region-like emission spectrum. The increase in surface brightness and blue color occurring in small physical sized systems leads to the well-known "blue compact dwarf" galaxies (hereafter BCDGs).

Previous studies of BCDGs have revealed a number of features that could lead to the idea that some of these objects may be considered as reasonable analogs at $z = 0$ of primeval galaxies: high to very high star formation rates with average low metallicities, and presence of relaxed dynamical stellar components. Therefore, their detailed study could open clues to the comprehension of the problem of galaxy formation and early evolution, especially as regards the triggering mechanisms of the starburst, that are still very largely unknown, and its early evolution in terms of stellar population, internal dynamics and chemical abundance distribution. The investigation of BCDGs by methods of 3D spectroscopy (also called spectro-imaging) allows to obtain at once spatial and spectral characteristics of extended objects with an excellent separation of line emission from continuum.

2 The Instruments

The table below summarizes the instrumental compromises that are available for spectro-imaging of active star-forming galaxies with large optical telescopes. Integral field spectrometers (IFS) using channeled spectra produced when crossing a scanning Fabry-Perot with a grating are still under development (Le Coarer

	Scanning FP	Grating IFS: TIGER (microlenses) or ARGUS (fibres)	Interferential IFS scanning FP crossed w. a Grating
Spectral range	Narrow typ. 1 nm	Broad typ. 100 nm	Broad typ. 100 nm
Spatial resolution	Moderate	High	High
Spectral resolution	High typ. 15000	Low to moderate typ. 1500	High typ. 15000

et al. 1996). Note that IFS are especially well suited for use at an adaptive optics fully corrugated focus on a large or very large telescope, because the separation of spatial and spectral information by means of spatial sampling at the entrance of the instrument preserves the image quality given by the telescope/atmosphere combination.

As an illustration of what is currently accessible to state-of-the-art auxiliary instrumentation on 4 to 6 m telescopes, in the following we present studies of BCDGs performed with scanning Fabry-Perot and with a TIGER-type IFS.

3   Interferometric Study of the Ionized Gas in IZw18

The extreme metal-poor BCDG IZw18 (Mk 116) has been observed at the Canada-France-Hawaii 3.6 m telescope with a Fabry-Perot spectrometer installed at the Cassegrain focus. The instrumental configuration gave a focal reduction of f/8 to f/2. The piezoelectrically scanned Queensgate Instruments Fabry-Perot etalon had a free spectral range of 375 km.s^{-1} at Hα with a velocity sampling of 18.8 km.s^{-1}. The 6.2 * 6.2 arc minutes field was sampled at 0.73".pixel^{-1} on a photon-counting detector. The Hα emission line map and a red continuum (660 nm) map were constructed from the data cube. The distribution of Hα emissivities, barycentric radial velocities and profiles were subsequently mapped across the object.

The main results that are discussed in detail in a forthcoming paper (Petrosian et al. 1996) are: i) discovery of a systematic offset of some 80-120 pc between the red continuum peaks and the Hα ones these latter being themselves offset from the HI densest clumps, ii) discovery, besides the two main compact components, of a population of small angular diameter individual HII regions whose properties are quite similar to those observed in dwarf Magellanic irregulars, iii) a velocity field with peculiar motions superimposed over a solid-body general rotational pattern, iv) asymetric profiles of the Hα line across most of the area of the galaxy, with blue and redshifted secondary components of various amplitudes, suggesting a complex pattern of accretion and / or expulsion of gas onto / from the central compact cores, v) some evidence of recent gravitational interaction with the extreme dwarf component called "Zwicky's flare".

4   Multi-Pupil Spectrograph Observations of BCDGs

4.1   The MPS Spectrograph of the 6m Telescope

The Multi-Pupil Spectrograph (MPS) of the Special Astrophysical Observatory of the Russian Academy of Science (Afanasiev et al. 1990), is operated at the prime focus of the Zelenchuk 6 m telescope. The MPS is an integral field spectrograph very similar to the TIGER spectrograph (Bacon et al. 1995). The field is sampled by means of a microlens array illuminated by an enlarger. The spectrograph itself, a classical slitless reflexion grating instrument, produces on the detector (photon-counting camera or CCD), the spectra of elementary images of the primary mirror of the telescope, (micro-pupils) each one containing the integrated light from the field spatial element subtended by the corrresponding microlens. Spectra are densely packed on the detector, in a way designed to avoid confusion between spectra arising from adjacent spatial field elements and to obtain the maximum number of spectra with the broadest possible spectral range within the finite size of the detector.

The reduction process, designed by V. Vlasiuk, basically similar to that designed for TIGER (Rousset et al. 1996) consists in building a data cube (x, y, λ) after careful extraction and calibration of the spectra. The following operations are usual manipulations of the data cube, as map building in the continuum, which traces the projected stellar density, and in emission lines whose local peaks of brightness are supposed to indicate the location of the most recently born stellar population. Profile analysis and velocity mapping are also performed.

4.2   Mapping the Starburst Galaxies

21 BCDGs from the First and Second Byurakan Surveys for ultraviolet excess galaxies (Markarian, 1967, Markarian and Stepanian, 1983), have been observed in several runs, in 1994-1995. A parallel imaging and photometric study is conducted (Doublier et al., 1996) on the same galaxies. Two detectors were used, a photon-counting camera with the 9 * 10 microlens array, and a CCD with a 8 * 11 microlens array, the spatial sampling being 1.25 arcsec.microlens^{-1}. The spectral ranges studied were: 460–530 nm (with Hβ and [OIII] lines and 630–680 nm (with [OI], Hα, [NII] and [SII] lines) with a spectral resolution of about 1500.

Three kinds of BCDGs could be distinguished:

- Class 1 objects have an apparently single starburst component, usually located at or very near the approximate geometrical center of the isophotes in broadband visible light. Some of them obey a $r^{1/4}$ law in their projected brightness distribution (Doublier et al., 1996). The emission lines brightness peaks generally coincide with the continuum peaks. This implies that the starburst observed at the present epoch is located at the point of maximum stellar density, therefore consistent with the hypothesis that the gaseous fuel of the stellar formation has accumulated at the bottom of the baryonic potential well of the object (typical example : Mk 1450). The radial behaviour of emission line ratios is consistent with the presence of a unique ionizing star cluster at the center, and a progressive dilution of the ionizing radiation field outwards.

- Class 2 objects have "double-nucleus" structure, that is, regular external isophotes and double bright core. Several objects of this class also obey a $r^{1/4}$ projected brightness distribution law (Doublier et al., 1996). They are highly suspected to be spheroidal systems in final phase of formation through a merging event involving two smaller bodies, one at least being gas-rich so as to provide the fuel necessary for the starburst ignition. Examples are Mk 324, Mk 900.

- Class 3 objects have multiple starburst components (appearing as blue knots with emission spectra). They look similar to dwarf Magellanic Irregulars, simply exhibiting an especially high star formation rate. Their emission line brightness peaks are almost systematically offset (on scales of 100–200 pc) from the continuum peaks. This can be easily explained by recurrence in the starburst phenomenon across timescales of a few million years, the "hot spot" of stellar formation migrating from a place where we observe to-day an evolved cluster of stars to a place where a new ionizing cluster is just born (this is also suggested in the case of IZw18, see above). Typical examples are SBS 1154+534 SBS 1331+493, Mk 1426, SBS 1707+565 etc... Some Class 2 objects also show this phenomenon (Mk 324 and Mk 900).

The velocity fields of these BCDGs show peculiar motions superimposed on a regular gradient, consistent, in most objects, with a quasi–solid–body rotation.

5 Perspectives for the Future

A big challenge for the future is the observation of galaxy formation itself, out of primeval protocloud material. Increasing evidence supports the idea that galaxy formation may have occurred across a large range of redshift, and could perhaps still take place, although being extremely rare at $z = 0$. The case of IZw18, with its exceptionally low metallicity for a star–forming system and its possibly quasi-primeval surrounding gas cloud (Lequeux et al. 1995) is very intriguing in this respect. On the other hand, galaxy population evolution is now claimed to be directly observed, especially through the occurrence of frequent merging within dense structures.

The combination of very large optical telescopes like the VLT, featuring adaptive optics providing regular sub-arcsecond image quality and visible and near-infrared spectroimagers will allow decisive progress in understanding the starburst phenomenon at low redshift, and will shed light on the similarities between high z early evolving galaxies still in their initial starbursting epoch and the present epoch active star-forming systems.

Acknowledgments: We are deeply indebted to SAO Telescope Allocation Comittee for generous granting of telescope time and to SAO staff for excellent welcome and friendly help with the instrument and data analysis software. CNRS gave an invaluable support to this research through PICS France-Armenie. INTAS provided the financial support for the installation of a modern computer network at Byurakan Observatory and for travel and stays at Zelenchuk.

References

Afanasiev, V.L., Vlasiuk, V.V., Dodonov, S.N., Sil'chenko, O.K. (1990): Preprint S.A.O. no 54

Bacon, R., Adam, G., Baranne, A. *et al.* (1995): A&AS **113**, 347

Doublier, V., Comte, G., Petrosian, A., Surace, C., Turatto, M. (1996): A&AS (submitted)

Le Coarer, E., Bensammar, S., Comte, G., Gach, J.-L., Georgelin, Y. (1995): A&AS **111**, 359

Lequeux, J., Kunth, D., Mass-Hesse, J.M., Sargent, W.L.W. (1995): A&A **301**, 18

Markaryan, B.E. (1967): Afz, **3**, 55

Markaryan, B.E., Stepanyan, D.A. (1983): Afz, **19**, 631

Petrosian, A.R., Boulesteix, J., Comte, G., Kunth, D., Le Coarer, E. (1996): A&A (submitted)

Rousset, A., *et al.* (1996): A&AS (in press)

Part 4

Search for Highest
Redshift Galaxies

Keck Spectroscopy of Redshift $z \sim 1$ Field Galaxies

David C. Koo

UCO/Lick Observatory and Board of Studies in Astronomy and Astrophysics
University of California, Santa Cruz, Calif. 95064 USA

Abstract. The first of the two 10m Keck Telescopes has been operational for over two years and is providing realistic previews of how well field galaxies at very high redshifts $z \sim 1$ or greater can be directly studied. Although detection and recognition of such galaxies are not difficult, gathering the wealth of information needed to decipher the physical mechanisms and their role in evolution will remain a challenge for future observers. To illustrate the value of adding such information as well as the potential of the VLT, we present highlights from three pilot projects from the DEEP program: redshifts of faint field galaxies; line widths and masses of some field galaxies with strong emission lines; and rotation curves of several high redshift spirals. The scientific results from all three surveys have relied on combining spectroscopy from Keck with high spatial resolution imaging from the Hubble Space Telescope.

1 Introduction

As the title of this workshop emphasizes, one of the major motivations for constructing large optical telescopes is to understand the early universe, especially in the context of the formation and evolution of galaxies. The emphasis of this review will be to use actual observations of field galaxies from the Keck I telescope to illustrate empirically the capabilities of VLT.

1.1 Overview of Keck Instruments for Studies of High Redshift Galaxies

As listed in Table 1, the Keck I and II (due for operation in fall 1996) telescopes have a number of instruments that allow observers to tackle high redshift field galaxies. Keck also has two mid-IR instruments (Long Wavelength Spectrometer and Long Wavelength Infrared Camera) that might also be used, but their potential for studying galaxies at high redshift has yet to be tested.

Among the first generation instruments, the workhorse for gathering the redshifts of the faintest galaxies is LRIS (Oke et al. 1995). This instrument has already yielded results that range from very faint number counts based on deep images (e.g., Smail et al 1995); redshifts of clusters of galaxies beyond $z \sim 1$ (see Dickinson in these proceedings); various redshift surveys of distant field galaxies by, e.g., Cowie et al. 1996, Cohen et al. 1996, and Koo et al. 1996; to the spectacular set of redshifts $z \sim 3$ of very faint galaxies selected by their UV-faint colors (Steidel et al. 1996; also see Giavalisco in these proceedings). This

Table 1. Keck Instruments for Studies of High Redshift Galaxies

Instrument (a)		Gen (b)	FOV (c)	λ (d)	R (e)	No. (f)	Eff. (g)
NIRC	Near InfraRed Camera	1	38"	1-5	600	1	
HIRES	HIgh Resolution Echelle Spectrograph	1	10"	0.3-0.7	3	1	7%
LRIS	Low Resolution Imaging Spectrograph	1	7'	0.4-1.0	30	30	20%
LRIS-B	LRIS-Blue	2	7'	0.3-0.5	60	30	
ESI	Echellete Spectrograph Imager	2	20"	0.4-1.1	11	1	20%
DEIMOS	Deep Imaging Multi-Object Spectrograph	2	16'	0.4-1.0	25	150	25%
NIRC-2	NIRC-two	2	40"	1-5	25	1	
NIRSPEC	NearInfraRed SPECtrograph	2	30"	1-5	5-60	1	

a Keck I or II Instrument
b First or second generation
c Field of view in arcsecs(") or arcmin(')
d Wavelength range in μ
e Spectral resolution in km-s^{-1}
f Number of targets per exposure
g Peak efficiency in %

work by Steidel et al. (1996) is a major breakthrough for high redshift observers, because the Keck demonstrates that such high redshifts can be recognized via *absorption lines* at rest-frame UV but redshifted to the optical, bypassing the need for strong Lyman α or other emission lines to secure the redshift. Prior to these observations, the next generation of near-infrared spectrographs (such as NIRSPEC on Keck – see Table 1; or ISAAC on VLT as described by Moorwood in these proceedings) appeared to be the most promising instruments with which to detect the usual set of strong optical emission lines in faint galaxies (e.g., OII(3727) or H_α) after they redshift to the near-infrared.

The other two first-generation Keck instruments have also been used for high redshift galaxy observations, but somewhat more indirectly. For example, extragalactic HIRES targets are almost exclusively bright QSO's, which do yield information about distant galaxies, but via their gas that produce absorption lines. As shown later, however, Keck gathers enough photons to allow direct and powerful observations of the gas kinematics of very low mass galaxies. NIRC, on the other hand, with its access to the near-IR, has been extensively used to study various classes of high redshift galaxies. In imaging mode, for example, Soifer et al. (1994) used broadband filters to study a QSO at redshift $z \sim 4.9$ and deep K band counts, while Pahre and Djorgovski (1995) used the narrow-band mode to search for primeval galaxies.

The second-generation Keck instruments will be even more powerful for the study of high redshift galaxies. The benefits of accessing the near-IR for spectroscopy (NIRSPEC) and of having higher spatial resolution imaging (NIRC-2) are clear. LRIS-B, with its access to the UV, will allow the commonly strong emission lines of either OII[3727] or Lyα to be observable over the full redshift range of redshifts $z \lesssim 7$, with no gaps. ESI, with its high throughput, will enable high spectral resolution work to be routine for very faint galaxies, but only through one slit. DEIMOS, in contrast, will allow ~ 150 slitlets, and thus provide a practical means to undertake massive surveys of 1000's of distant galaxies, but with only moderate spectral resolution (see discussion below).

1.2   Detection, Recognition, and Understanding of High Redshift Galaxies

What types of high redshift galaxy work is feasible with large telescopes? Detection of $z \sim 1$ galaxies is easy! A typical (L*) galaxy is visible on CCD images ($I \sim 22.6$) or infrared images ($K \sim 20$) taken with 4-m class telescopes. Recognizing that a particular object is indeed at high redshift takes more work, but is also not difficult. For example, astronomers have been working on samples of high redshift galaxies for several years via clusterings of objects found around distant radio galaxies (see Dickinson in these proceedings) or QSOs or via the study of the host galaxies around high-redshift QSOs (Heckman et al 1991). Another technique that is relatively efficient is to use multicolors to estimate redshifts (e.g., Connolly et al 1995); the UV drop method that has been adopted by Steidel et al (1996) or Guhathakurta et al (1990) are examples of using multicolor photometry to search for candidates at $z \sim 3$. Spectroscopic redshifts of any *candidates* remain, of course, the most trusted confirmation. Finally, note that many galaxies with $z \sim 1$ have already been acquired with 4-m class telescopes (e.g., Lilly et al 1995, or see contribution by F. Hammer in these proceedings). These examples show that detection or recognition of $z \sim 1$ field galaxies is not the exclusive domain of 8-10m telescopes.

The true challenge for astronomers is to gather the wealth of data needed to unravel the nature of high redshift galaxies. Knowing just the redshift, and presumably luminosity and perhaps colors from the photometry, is not enough. For interpreting the variety of physical processes likely to be affecting the wide diversity of galaxies that exist in a wide range of enviroments, astronomers also benefit from morphologies and spatial structure (from HST, lensing events, excellent ground-based seeing, or adaptive optics); the kinematics (velocity widths of spectral lines or rotation curves) and mass (when galaxy sizes and disk inclinations are also available); the age and metallicity of the underlying populations (from spectral indices measured from high S/N spectra); and the ionization, dust content, temperature, density, etc. of the gas (requiring observations in the IR, radio HI and CO, X-ray, sub-mm, etc.). Moreover, given the probable need to divide samples into related subclasses (e.g., by luminosity, mass, size, environment, redshift, etc.), large samples are generally needed for comprehensive surveys. Despite Keck's early entry into the large telescope area, there will remain an enormous need for new VLT observations of high redshift galaxies to unravel what is really happening.

2   Examples of Keck Surveys that Combine HST with Keck

DEEP is a multi-institution collaboration now underway to exploit the Keck 10-m Telescope for a spectroscopic survey of faint field galaxies (Mould 1993, Koo 1995). The scientific goals are broad and include explorations of evolution in the spectral, dynamical, and spatial distribution of distant galaxies. These results will be used to probe dark matter on the scale of galaxies and clusters of galaxies in the past and to provide new estimates of the geometry of the universe via tests of correlations among galaxy properties (e.g, fundamental plane) as well as classical tests such as the volume density versus redshift. The ultimate goal is to acquire spectra for 10,000 galaxies or more to B $\sim$ 25. Such spectra will be used not only to measure redshifts for the full sample but also to measure kinematics (via rotation curves or velocity dispersions) as well as age and chemical abundances for a brighter subset of galaxies. A new imaging spectrograph for Keck called DEIMOS (see Table 1) is being constructed to complete such a massive survey. By exploiting the higher reflectivity of the silver-coated Keck II mirrors, 8Kx8K CCD mosaics in *each* of the two cameras, and a substantial field of view of 16'x5' for each of two multi-slit masks, DEIMOS will achieve a throughput gain of over 10x that of the current Keck I and LRIS system, at similar spectral resolution. Until DEIMOS is operational, the interim DEEP science programs will be limited in scope and will rely on LRIS and HIRES that are currently operational. The following three examples highlight some pilot DEEP programs that show the potential of combining HST with VLT to study distant galaxies.

2.1   Redshift Survey of $z \sim 1$ Field Galaxies

A strong rationale for large telescopes is to push redshifts to fainter limits. This was the goal of our first DEEP pilot program in which we combined these redshifts from the Keck Telescope with photometry, colors, and morphologies from refurbished HST images (Koo et al. 1996). Though this early Keck redshift sample has only 33 galaxies (due to the loss of 80% of the run to weather), the magnitudes are so faint (22 galaxies with $22 \lesssim I \lesssim 24$) and the redshifts so high (median $z \sim 0.8$), that this survey provides a unique probe of the nature of faint, distant field galaxies of *typical luminosities (L*)* at an epoch beyond half the Hubble age.

Some examples of the spectra are shown in Fig. 1. Figure 2 summarizes our results in a V-I color versus redshift diagram for the Keck sample, including 18 galaxies from Forbes et al. (1996). The curves provide guides to the intrinsic colors of the galaxies. Figure 2 shows how well the galaxies fall within the bounds seen in the colors of normal, local galaxies. We find neither unusually blue nor unusually red galaxies. Of perhaps more interest, we have detected the presence of several bulges in spiral-like galaxies which appear much redder, and thus presumably older, than their surrounding disks. Similarly, we have confirmed that at high redshifts $z \geq 0.7$, some *field* galaxies do exist with intrinsic colors comparable to that found for local ellipticals. Furthermore, although HST images show strong and frequent hints of mergers, interactions, other peculiar patterns, and infall of minor galaxies into larger hosts, galaxies with normal morphologies are also visible. The morphologies of $z \sim 1$ galaxies are thus not confined to late-type, peculiar systems and, conversely, the late-type galaxies seen in deep HST images are not predominantly at low redshifts (i.e., of low-luminosity).

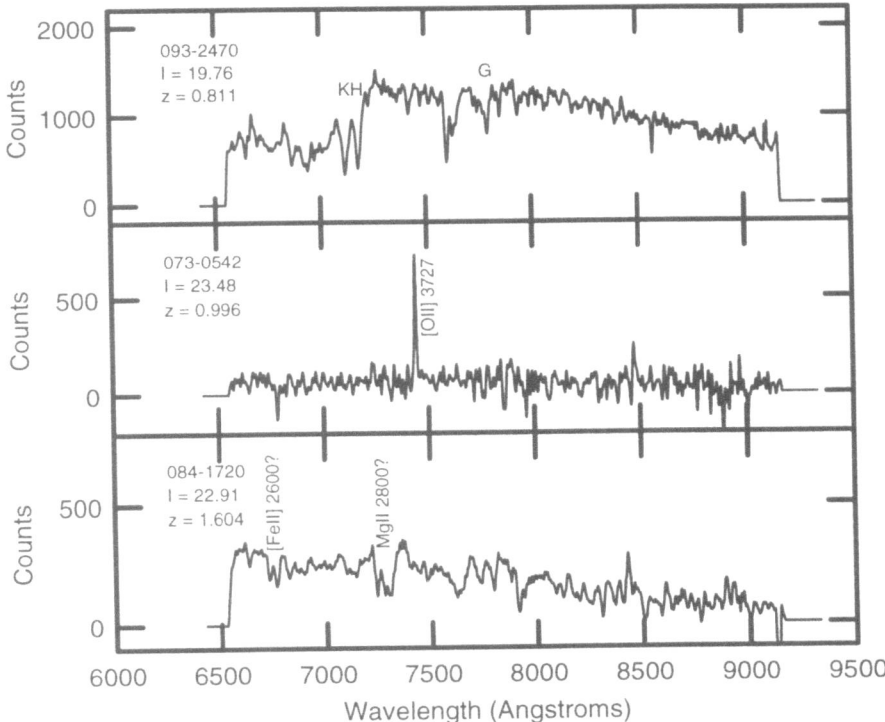

Fig. 1. Counts vs. wavelength for a sample of Keck spectra, each labeled with the source name, total I mag, and redshift. The redshift is uncertain for the bottom object.

This glimpse of very faint $z \sim 1$ field galaxies strongly suggests that we need to invoke an unknown, possibly complex, mixture of physical processes to account for the faint blue field galaxies, rather than to rely on a single dominant mechanism, such as mergers or bursting dwarfs.

2.2 Low Mass Progenitors of Dwarf Spheroidals

One of our more pleasant surprises in the early phase of Keck I operations was the serendipitous discovery that HIRES is able to yield important new information about faint field galaxies. Due to the failure of LRIS a few days before

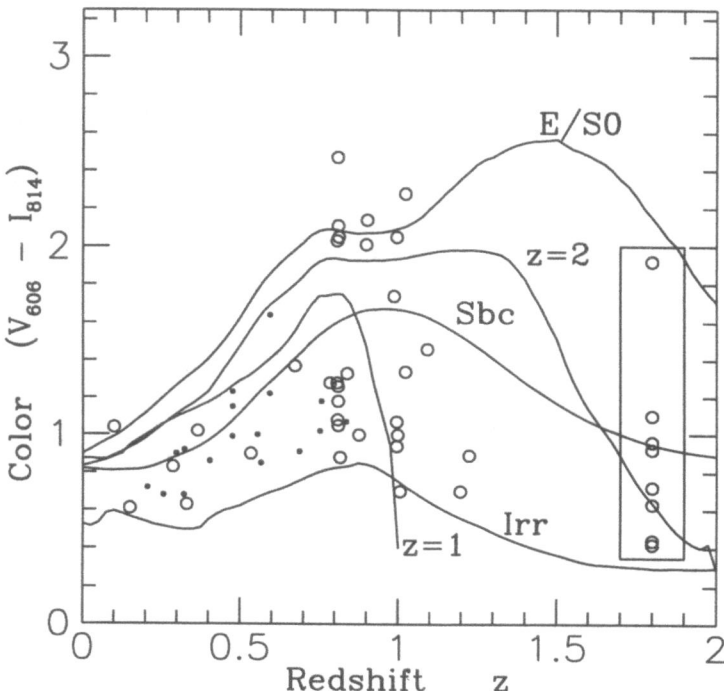

Fig. 2. $V - I$ color vs redshift plot of Keck targets from Koo et al. (1996 – open circles) and Forbes et al. (1996 – points). Objects without redshifts are placed in the separate box to the right. Several labeled lines show the expected colors for various spectral energy distributions, including one resulting from an instantaneous burst of star formation at redshift $z = 2$ that becomes almost as red as a non-evolving local elliptical or S0 (E/S0) by $z < 1$; another resulting from a model burst at $z = 1$ that might be compared to the bursting dwarfs in the model of Babul and Ferguson (1996); another to the colors of a local Sbc galaxy; and the bluest one for N4449, a very actively star-forming Irr galaxy.

our run, HIRES was instead used to study a sample of $B \sim 22$ galaxies. Based on prior KPNO 4m spectra of the HIRES targets, these galaxies had measured redshifts z from 0.1 to 0.7, and thus inferred luminosities close to typical (L^*) galaxies. These galaxies were considered good HIRES targets because they were also known to possess unusually strong emission lines and very small sizes. The resulting line profiles were found to be well fit by Gaussians, with velocity widths ranging from 28-157 km-s^{-1} (see Fig. 3 for some examples). Though such narrow widths were easily resolved given the HIRES resolution of ~ 3 km-s^{-1}, these observations emphasize the importance of high spectral resolution for faint galaxy work.

Furthermore, we have been able to obtain HST images for a subsample of our HIRES targets. The addition of an accurate size enables us to then make

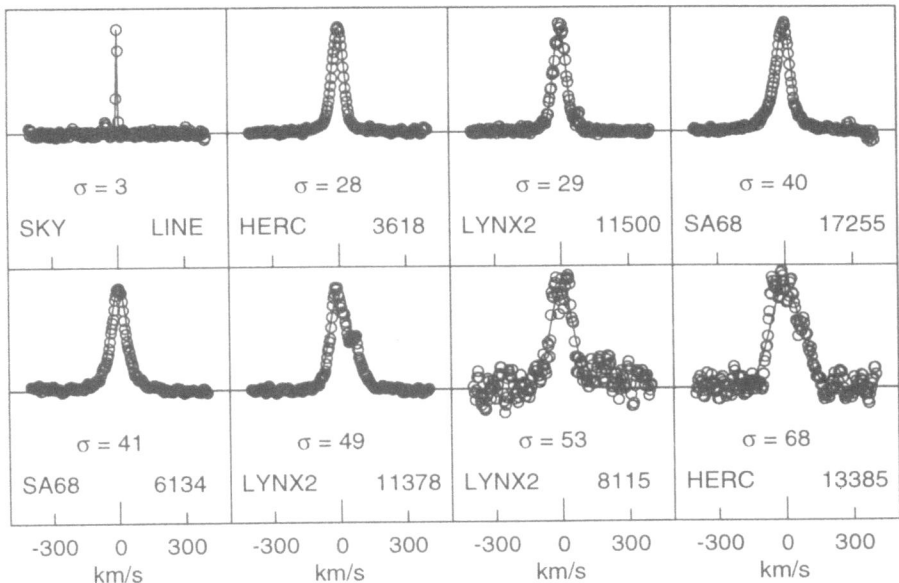

Fig. 3. Panel of HIRES emission line profiles from a sample of CNELG's that also have HST images (see Guzman et al. 1996 for more details). The HIRES instrumental profile is shown in the upper-left panel.

mass estimates (see Fig. 4) and we find CNELG's to have masses between 10^9 to $10^{10} M_\odot$. Such low masses are much more typical of dwarfs than that of typical (L^*) galaxies. As discussed by Koo et al. (1995) and Guzman et al. (1996), the properties of these these compact, narrow-emission line galaxies (CNELGs) are all consistent with the view that we might be watching a sample of dwarf galaxies undergoing their final burst of star formation, strong enough to expel their remaining gas and to fade in luminosity to become some of today's brighter dwarf spheroidals (e.g., NGC 205).

2.3 Rotation Curves and Optical Tully-Fisher Studies of Distant Spirals

Instead of velocity widths, another DEEP program that probes masses of distant galaxies uses the rotation curves from optical emission lines, similar to the measurements undertaken for nearby galaxies (e.g., Rubin et al. 1985). The feasibility of this approach was tested in a pilot study of 9 faint field galaxies with redshifts $0.1 \lesssim z \lesssim 1$ (see Fig. 5) as discussed in more detail by Vogt et al. (1996). HST images provided the complementary photometry, colors, galaxy morphology, and, most importantly, the orientation and inclination. In contrast to local measurements in which the slit-width is small compared to the galaxy, significant corrections are needed to derive a maximum circular velocity for distant galaxies, observed even under good seeing conditions. Except for one very

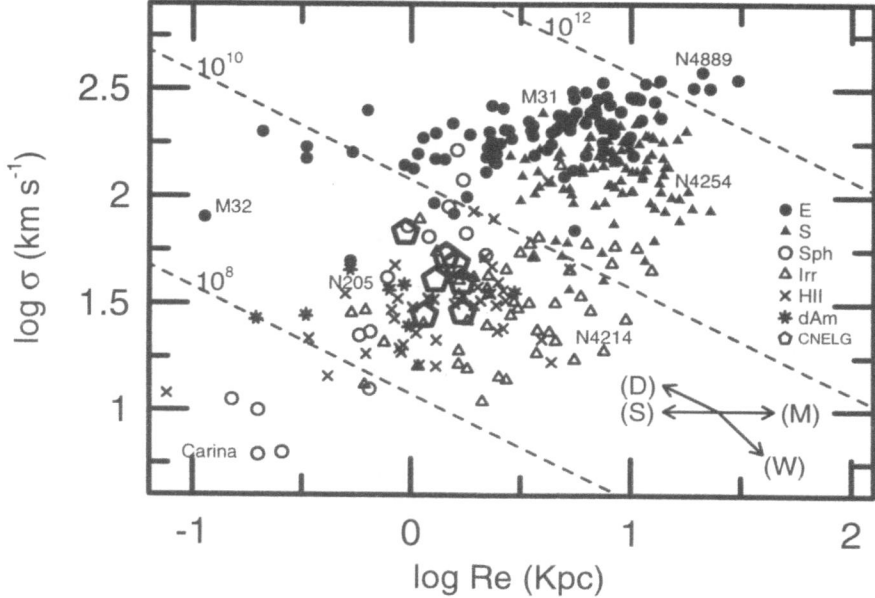

Fig. 4. Half-light size versus velocity-width relation that compares the CNELGs against local samples of galaxies. The dashed lines indicate the masses of 10^8, 10^9, and $10^{10}\,M_\odot$. As discussed in more detail in Guzman et al. (1996), the arrows show the direction of change due to a variety of physical processes: dissipation (D), tidal stripping (S), merging (M), and galactic winds (W).

elongated galaxy (with an unusually low-amplitude rotation curve that might suggest a merger seen perpendicular to the collision path), the remaining rotation curves appear similar to those of local galaxies in both form and amplitude. When the equivalent of a Tully-Fisher relation is plotted and compared to that of local galaxies, the high redshift sample reveals an upper-limit of 0.6 mag to the brightening in the past of spiral galaxies (see Fig. 6). Though this result is tentative due to the small sample size, these observations, nevertheless, demonstrate that optical rotation curves can be measured to high redshifts.

I would like to thank the Keck Telescope staff for the success of our DEEP projects. These results are contributions from a team effort of DEEP and especially its many younger astronomers who do most of the work: R. Brunner, A. Connolly, D. Forbes, C. Gronwall, R. Guzman, A. Phillips, N. Vogt, and K. Wu. DEEP is a project sponsored by the Center for Particle Astrophysics at UC Berkeley. Funding for this work was provided by NSF grants AST91-20005 and AST-8858203 and NASA grants AR-5801.01-94A and GO-2684.04-87A.

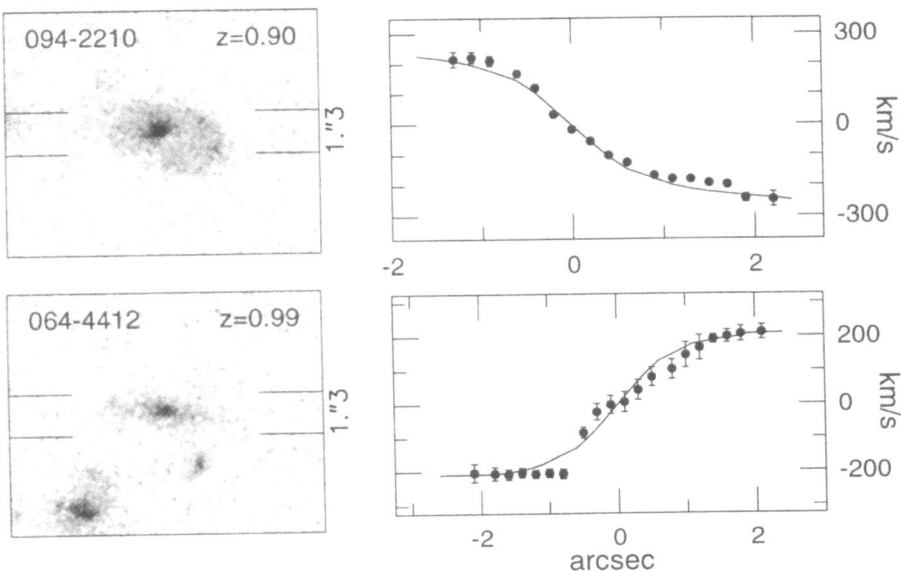

Fig. 5. Panel showing the HST image and Keck rotation curve of two high-redshift galaxies. Besides the ID and redshift at the top, the images show the orientation and width of the slit.

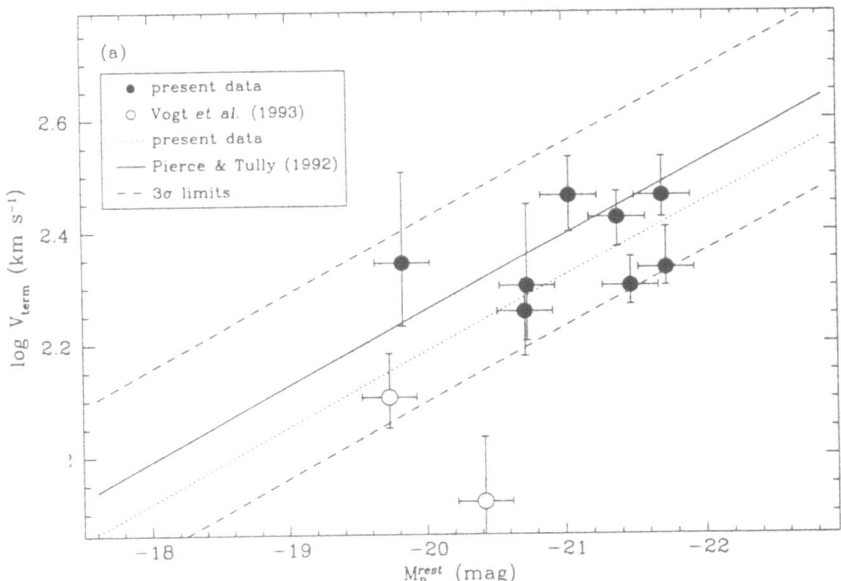

Fig. 6. Terminal velocity versus absolute magnitude in rest-frame blue (Tully-Fisher diagram) of the Keck and Vogt et al. (1993) sample compared to the best-fit B band relation (solid line) and 3σ limits (dashed lines) from Pierce and Tully (1992). The best fit to the Keck data (dotted line) has an offset of 0.55 mag.

References

Babul, A., and Ferguson, H. C. (1996): ApJ, **458**, 100–119

Cohen, J. G., Hogg, D. W., Pahre, M. A., and Blandford, R. (1996): ApJ, **462**, L9-L12

Connolly, A. J., Csabai, I., Szalay, A. S., *et al.* (1995): AJ, **110**, 2655-2564

Cowie, L. L., Songaila, A., Hu, E. M., and Cohen, J. G. (1996): preprint

Forbes, D. A., Phillips, A. C., Koo, D. C., & Illingworth, G. D. (1996): ApJ ,**462**, 89–103

Guhathakurta, P., Tyson, J. A., and Majewski, S. R. (1990): ApJ, L9-L12

Guzmán, R., *et al.* (1996): ApJ, **460**, L5–L9

Heckman, T. M., Miley, G. K., Lehnert, M. D., and van Breugel, W. (1991): ApJ, **370**, 78-101

Koo, D. C. (1995): in proceedings of *Wide Field Spectroscopy and the Distant Universe*, eds. S. Maddox and A. Aragón-Salamanca (World Scientific: Singapore), 55-62

Koo, D. C., Guzman, R., Faber, S. M., *et al.* (1995): ApJ, **440**, L49-L52

Koo, D. C., Vogt, N. P., Phillips, A. C., *et al.* (1996): ApJ, in press for Oct

Lilly, S. J., Le Fèvre, O., Crampton, D., Hammer, F., *et al.* (1995): ApJ, **455**, 50-59

Mould, J. (1993): in *Sky Surveys: Protostars to Protogalaxies*, ed. B. T. Soifer, PASPC, **43**, 281-282

Oke, J. B., Cohen, J. G., Carr, M., *et al.* (1995): PASP, **107**, 375-385

Pahre, M. A., and Djorgovski, S. G. (1995): ApJ, **449**, L1-L4

Pierce, M. J., and Tully, R. B. (1992): ApJ, **387**, 47-55

Rubin, V. C., Burstein, D., Ford, W. K., Jr., and Thonnard, N. (1985): ApJ, **289**, 81-98

Smail, I., Hogg, D. W., Yan, L., and Cohen, J. G. (1995): ApJ, 449, L105-L108

Soifer, B. T., *et al.* (1994): ApJ, **420**, L1-L4

Steidel, C. C. *et al.* (1996): ApJ, **462**, L17-L20

Vogt, N. P., Herter, T., Haynes, M. P., and Courteau, S. (1993): ApJ, **415**, L95-L98

Vogt, N. P., Phillips, A. C., *et al.* (1996): ApJ, **465**, L15-L18

VIRMOS: Deep Redshift Surveys with the VLT

O. Le Fèvre[1], P. Vettolani[2], J.G. Cuby[1,8], D. Maccagni[3], D. Mancini[4],
A. Mazure[5], Y. Mellier[6], J.P. Picat[7]

[1] DAEC, Observatoire de Paris-Meudon, France
[2] Istituto di Radioastronomia CNR, Bologna, Italy
[3] Istituto di Fisica Cosmica CNR, Milano, Italy
[4] Osservatorio di Capodimonte, Napoli, Italy
[5] Laboratoire d'Astronomie Spatiale, Marseille, France
[6] Institut d'Astrophysique, Paris, France
[7] Observatoire Midi-Pyrénées, Toulouse, France
[8] European Southern Observatory, Garching, Germany

Abstract. We present the status of the feasibility study of the Visible InfraRed Multi Object Spectrograph, conducted under contract with ESO for the VLT. The aim of this study is to establish the scientific objectives, identify the opto-mechanical concepts, and predict the performances of a wide-field multi-object spectrograph dedicated to the deepest galaxy and AGN redshift surveys that will be possible with the VLT. We have identified that a concept based on a four quadrant imaging–spectrograph, using multi-slit aperture plates, is the most efficient concept in this context. The maximum number of slits 10 arcseconds in length, that can be observed simultaneously, is in excess of 800, offering a spectacular multiplexing gain in a field of $4 \times 7 \times 8$ arcmin^2, for a wavelength coverage from 0.37 to 1.8μm. We are proposing a dedicated wide-field deep redshift survey, to collect $\sim 10^5$ galaxies brighter than $I = 24$, in the redshift range $0.3 < z < 4$ or more, in less than 200 nights at one Nasmyth focus of the VLT. The timescale of the instrument development is expected to be 4 years, with first light in early 2001, provided that the VIRMOS concept is selected by ESO early in 1997.

1 Introduction

With the VLT we will be in a position to greatly expand the exploration of the universe, at epochs only a small fraction of its present age, in much the same way as is being done in our local environment. Among the main questions that observational cosmology strives to address are the evolution of galaxies and AGNs, and the evolution of large scale structures. These two main topics contain a wealth of questions which will be adequately answered only if we gather large samples of galaxies in large volumes, at epochs up to a small fraction of the age of the universe, to produce results with a high *statistical* significance. Only a dedicated instrument, aimed at high efficiency in collecting large numbers of redshifts from simultaneous measurements, can address these goals.

In early 1994, a working group created at the initiative of the ESO Science and Technical Committee, reviewed the needs for a 2nd generation VLT instrumentation, and identified the need for a dedicated spectrograph to conduct the deepest possible galaxy and AGN redshift surveys. Two groups of largely identical people proposed two concepts: NIRMOS, a Near InfraRed Multi Object

Spectrograph, and WFIS, a Wide Field Imaging Spectrograh. The combination of visible and near-infrared wavelength coverage was felt to be a necessity to investigate evolution in a continuous manner out to z~4 or more, and in particular sample the redshift range $1 < z < 2$, a major epoch in the star formation history. Both optical concepts were based onto the original optical design by Delabre et al. (1994), a four quadrant imaging spectrograph, allowing a large field of view at one of the VLT Nasmyth focii. The two groups joined their forces to form a French-Italian consortium: VIRMOS for Visible InfraRed Multi Object Spectrograph, which was awarded a contract by ESO for a one year feasibility study. We present here the mid-term results of this study.

2   The Science Case

We concentrate here on two main topics: *the evolution of galaxies* and *the evolution of large scale structures*

2.1   Evolution of Galaxies

For many years, astronomers have aimed to provide an explanation for the large excess in the number of galaxies over the expectations of simple no-evolution models, as observed in the deep counts, the excess being larger for counts in the bluer bandpasses (see e.g. Lilly et al., 1991). Several physical processes have been invoked, including a dramatic increase of the luminosity of all or part of the galaxy population, wide spread mergers, gradual fading of low-mass galaxies, some or all of which are possibly at work together (Babul & Rees, 1992; Broadhurst et al., 1992; see Lilly et al., 1993 and Koo&Kron, 1992, for recent reviews).

Considerable progress has been made in recent years in constraining the evolution of normal galaxies from epochs ~40% the age of the universe (redshifts ~1). For the first time, several hundred redshifts of normal galaxies have been measured out to large look-back times, $z \sim 1$ (Canada-France Redshift Survey: CFRS, Lilly et al., 1995, Le Fèvre et al., 1995; LDSS, Colless et al., 1993; Cowie et al., 1995). These large samples allow to study e.g. the time evolution of the luminosity function of galaxies, the morphological mix, or the star formation rate with good statistical significance.

To significantly expand upon the current knowledge, there are two avenues that will require a major effort, and on which the VLT with VIRMOS can make an essential contribution. There remains a number of galaxy evolution issues to investigate at redshifts $z < 1$. We now have proofs of the existence of a substantial population of massive galaxies lying at $z > 1$ (Cowie et al., 1995), with the deepest surveys now reaching $z > 3$ (Steidel et al., 1996), and the epoch $1 < z < 2$ has been identified as an important period in the evolution of massive galaxies (Lilly et al., 1996, Cowie et al., 1996). As the spectral features of interest are redshifted in the near-IR for $1.5 < z < 2.5$, a multi-object capability in the visible, J and H bands, is required.

The main questions related to the evolution of galaxies, out to look-back times $\sim 10\%$ the current age of the universe ($z \sim 3$) or less, can be summarized as follows:

1. What is the behavior of the faint end of the luminosity function vs. time ?
2. What is the history of star formation ?
3. What is the evolution of the different galaxy types ?
4. What is the evolution of the fundamental plane, or projections of it ?
5. What is the relationship galaxy evolution $\leftrightarrow$ local matter density ?
6. What is the relationship between field galaxies and AGNs ?
7. What is the influence of interactions and mergers ?

The number of galaxies to be studied can be evaluated as follows: taking a luminosity range of 8 magnitudes to probe from $-24.5 \leq M_B \leq -16.5$, 3 galaxy types, and 3 types of galaxy density environments, and measuring the luminosity function in 7 redshift bins, with 50 galaxies in each elemental bin for proper statistics, requires $50 \times 8 \times 3 \times 3 \times 7$ galaxies per field. Ideally several (~ 5) independant fields should be explored. A total of $\sim 10^5$ $I < 22$ galaxies, and $\sim 5 \times 10^4$ $I < 24$ galaxies, is thus required.

2.2   Evolution of Large Scale Structures

The comparison of the most recent numerical SPH simulations with the total mass and the galaxy distributions provide important constraints on theoretical models of cosmic scenarii. Redshift surveys already show that galaxies are located in sheet-like structures, suggesting that the mass could be strongly clustered in filaments, superclusters and clusters, at the periphery of large voids (Vettolani et al., 1996, de Lapparent et al., 1995), even out to $z \sim 1$ (Le Fèvre et al., 1996).

The deep redshift surveys only recently allowed to trace the evolution of the clustering of galaxies from $z \sim 1$. The evolution of $\xi(r)$ has been shown to proceed somewhat faster than stable clustering (Le Fèvre et al., 1996). At redshifts beyond $z > 1$, very little is known about the evolution of large scale structures: only a few large structures have been identified so far (Giavalisco et al., 1994; Le Fèvre et al., 1994; Pascarelle et al., 1996; Francis et al., 1996). At these high redshifts, the prevalence of clusters, proto-clusters, or other large scale structures in the galaxy distribution is as yet unknown, and the evolution of large scale structures may well be at a critical stage (Peebles et al., 1989).

The observation of LSS at $z > 1$ will set strong constraints on the theories of large scale structure formation and evolution, in particular it will investigate:

1. Evolution of the spatial correlation function $\xi(r)$ and of the power spectrum
2. Evolution of clustering for star forming vs. quiescent galaxies
3. Redshift-space distorsion of the clustering pattern $\Omega^{0.6}/b$
4. Distribution of galaxies on the largest scales ~ 100 Mpc or more
5. Clustering topology
6. Evolution of the space density of clusters

One of the most promising new tool to probe the distribution of matter (dark+visible) is to use the small anisotropies in the shapes of galaxies as a result of the shear produced by gravitational lensing in the weak regime (Fort & Mellier, 1994). Combining the information provided by the deep redshift surveys with the information provided by the weak shear analysis of the faintest galaxy images should allow to solve the ambiguity in associating the lensing shear pattern, observed in projection, to one single 3D structure, as the line of sight is crossing many overdensities (Le Fèvre et al., 1996). Deep imaging capability is thus required, to allow a shear measurement at the 2% level.

To compute $\xi(r)$ at $z > 1$ requires samples on order 500 galaxies per redshift bin studied. From z=1 to z=4, 6 redshift bins would require to observe ~3000 galaxies to have an accurate estimate of $\xi(r)$. To do so for 3 galaxy types requires ~ 10^4 I=24 galaxies for each field, and one would need ~ 5 fields to smooth out any field-to-field variation, therefore a total of 5×10^4 galaxies is required.

3   Optical Design

The optical designs have been inpired by the original work of Delabre et al. (1994). A team of opticians (Conconi, Delabre, Molinari, Pares), studied several optical designs based on a 4 quadrant imaging spectrograph, with 4 separate, and identical, optical channels, each with its own detector. The advantages are numerous: *(i)* it allows for a wide field of view, $4 \times 8 \times 7$ arcmin^2, the practical maximum permitted by the unvigneted field of view at the VLT Nasmyth focus, *(ii)* the design is of the "focal reducer" type, which has proved its efficiency at several observatories (see e.g. Le Fèvre et al., 1994), and which allows to switch from imaging to spectroscopy, by simply inserting or removing a dispersing element, *(iii)* each detector can be expanded in the dispersion direction to accomodate more spectral elements and hence allow higher spectral resolution, *(iv)* the mechanical placement of multi-slit aperture masks at the entrance focal plane, and positioning on the sky, is straightforward, and *(v)* efficiency is high over the 0.4 to 1.8 μm domain, provided efficient multi-layer coatings are applied on the optical elements.

Two design concepts have been retained. One combines visible and near-IR capabilities into one single instrument (fig.1), the second one calls for two separate instruments, one for the visible, one for the near-IR. In the first case, the difficulty is to identify highly efficient multi-layer coatings to allow for high throughput at the two extremes of the wavelength range, and to manage an increased instrument complexity. In the second case, the throughput is improved, and each instrument has reduced complexity, while overall cost and operational constraints might be higher. In addition, a design based on the OH-suppressor concept (Maihara et al., 1993), projecting a high resolution R~5000 intermediate spectrum onto a mirror with zero reflectivity at the location of strong OH sky emission lines, has also been studied. As described in section 5, such a concept has been shown to be less efficient than the optical concept described above.

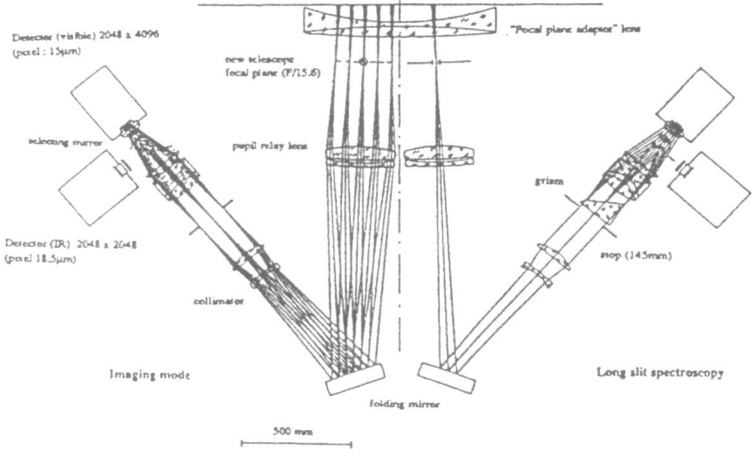

Fig. 1. VIRMOS optical design: combined visible and IR design

4   Mechanical Design

The main structure of the instrument is similar to a classical telescope Serrurier structure in order to increase the stiffness-to-weight ratio, and minimize the flexures at the level of the detectors. Gratings are installed in a carroussel and filters in a wheel, for an easy and fast configuration of the system. Each of the 4 channels has a supply of 15 pre-loaded aperture masks. Multi-slit masks, with slits in any number and shape, can be cut with an auxiliary facility based on a milling machine with a high accuracy X-Y stage.

5   Performances

The performances of the 4 quadrant imaging spectrograph have been compared with the performances of an OHS-suppressor spectrograph. It has been found that in the wavelength domain where strong OH lines are present ($\lambda > 0.8\mu$m), the most efficient is to work at spectral resolution $R \geq 2000$, and subtract the sky emission numericaly, after data acquisition on the detector. Hardware OH emission subtraction is about 3 times less efficient in this context. This is shown in the simulations described in figure 2, comparing low/high spectral resolution, and numerical/hardware OH suppresion. Table I summarize the performances expected from the VIRMOS design.

6   The VLT-VIRMOS Deep Survey

To address the observational goals, as identified in section 2, the following is needed:

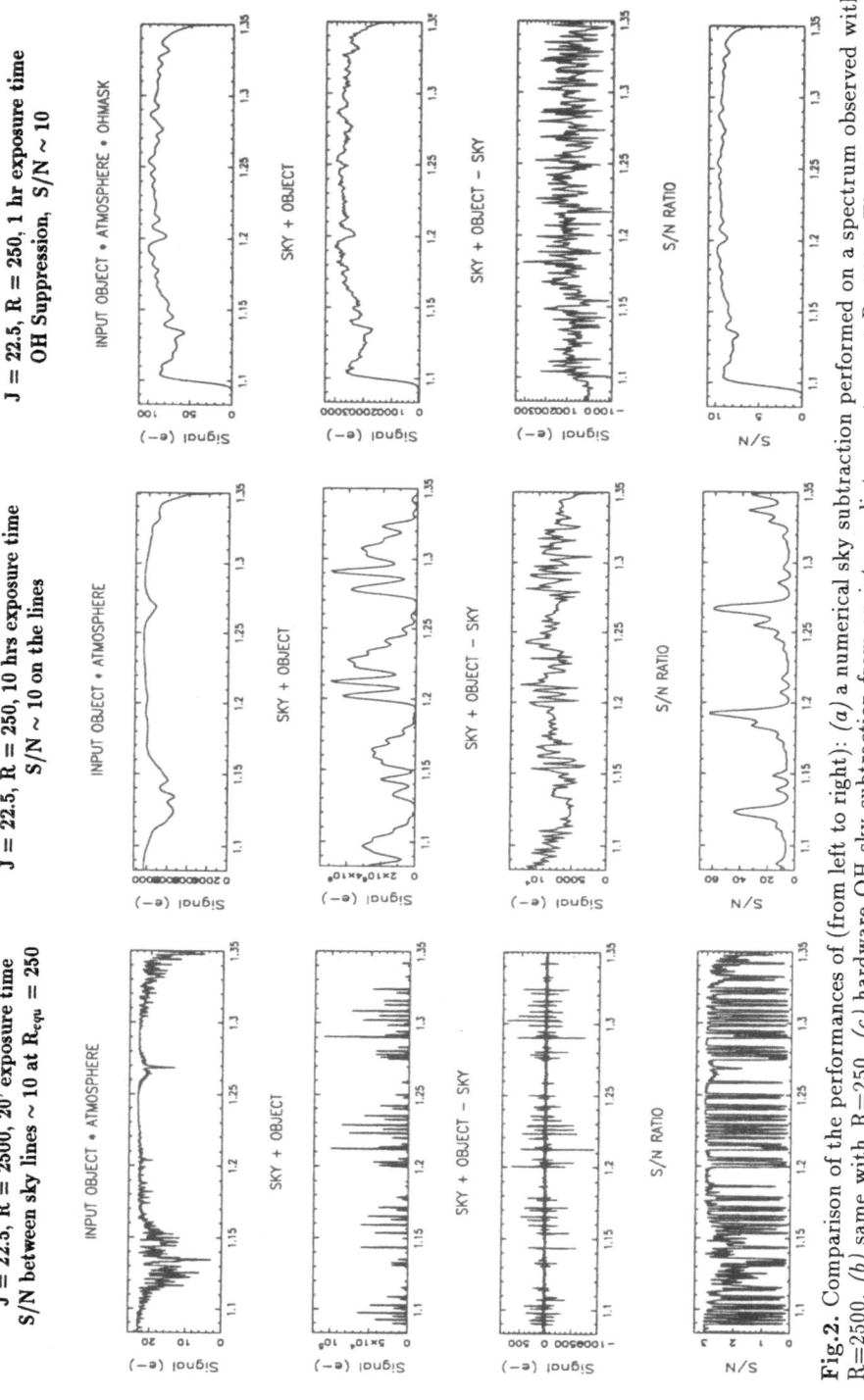

Fig. 2. Comparison of the performances of (from left to right): (*a*) a numerical sky subtraction performed on a spectrum observed with R=2500, (*b*) same with R=250, (*c*) hardware OH sky subtraction from an intermediate spectrum at R=5000. The final S/N quoted have been scaled to the same R=250 resolution. The case (*a*) outperforms the case (*c*) by a factor 3 in efficiency.

Table 1. Main characteristics and performances of the VIRMOS design

Imaging Spectrograph	Visible	IR
Field of view	$7' \times 8'$	$8' \times 6'$
Multiplex capability	~800 (R=250)	~600 (R=250)
	~170 (R=2000)	~150 (R=2000)
Image quality in the visible	0.4 arcsec	
(spectroscopy and imagery)	at 80% encircled energy	
Gravitational shear accuracy	0.2%, S/N=5	
	50 × 50 arcsec²	
Limiting magnitude	I=24, R=250	J=23, R=2000, 2hrs
S/N=10	6 hours	H=21, R=2000, 1h

- Gather a sample of $\sim 10^5$, $I \leq 22$, galaxies out to $z = 1$
- Cover $\sim 100h^{-1}$ Mpc at $z \sim 1$
- Observe $\sim 5 \times 10^4 L^*$, $I \leq 24$, galaxies out to $z \sim 3.5$ or beyond
- Observe a few tens of clusters of galaxies with $0.5 < z < 2$ or beyond
- Couple weak shear measurements at the 2% level from imaging to redshift catalogs
- Observe $> 10^3$ AGNs out to $z \sim 3.5$ or beyond
- Explore a volume of space $\sim$ comparable to the SLOAN or 2dF local redshift surveys
- Obtain the deepest possible spectra in selected areas with integral field spectroscopy ($I \sim 26$)

The VIRMOS design could allow to address these points by observing the following fields (indicative), for a total number of nights of around 200 on 1 focus of one VLT 8m unit: 5 strips of $16' \times 3$ deg to $I = 22$, 3 strips of $16' \times 1$ deg to $I = 24$, a small area 80×80 arcsec² to $I = 26$

At a redshift $z = 1$, 100 comoving Megaparsecs projects to ~ 2 degrees on the sky ($H_0 = 50$, $q_0 = 0.5$). The geometric layout of the survey could cover an area at least 4 square degrees, possibly separated in several strips, each e.g. $16' \times 3°$. At I=22, there are 25000 galaxies per square degree, hence a total of 10^5 galaxies would be observed. From the recent deep redshift surveys, we know that most of the $I < 22$ galaxies will be at $z < 1$.

To reach out to $z = 3.5$ or more, one needs to select galaxies with $I = 24$, as shown by Steidel at al. (1996). At these magnitudes, the number density of galaxies projected on the sky is $\sim 6 \times 10^4$ per square degree. A strip $16' \times 1°$ would contain 16000 galaxies. Predictions on the redshift distribution show that $\sim 1/2$ of the $I < 24$ galaxies will be at $z > 1$.

A small area 80×80 arcseconds can be picked up in the surveyed fields. Integral field spectroscopy can be performed to yield spectroscopic information on more than 400 galaxies down to $I = 26$ for the deepest probe into the far universe ever attempted.

If we extrapolate the number counts of AGNs with a slope similar to what is observed for $20 < m_B < 22.5$, we can estimate to observe ~ 600 AGNs in a $16' \times 3°$ strip.

From the expected performances of the VIRMOS instrument concept, we expect to complete such a survey in less than 200 nights at the VLT.

7   Conclusion

The VIRMOS concept combines a wide field imaging spectrograph with high multiplex capabilities in the wavelength range $0.37 - 1.8\mu$m, to allow ultra deep surveys on large numbers of galaxies.

This instrument would allow to conduct a survey of 10^5 $I \leq 24$ galaxies and AGNs in less than 200 nights at one Nasmyth focus of the VLT. This should provide the VLT a place of choice in the field of observational cosmology.

References

Babul, A., & Rees, M.J., 1992, MNRAS, 255, 346

Broadhurst, T.J., Ellis, R.S., Glazebrook, K., 1992, Nature, 355, 55

Colless, M., Ellis, R.S., Broadhurst, T.J., Taylor, K., Peterson, B., 1993, MNRAS, 261, 19

Cowie, L.L., Hu, E., Songaila, A., 1995, Nature, 377, 603

Cowie, L.L., et al., 1996, in press

Delabre, B., et al., 1994, SPIE symposium "Instrumentation for the 21st century", p.2198

de Lapparent et al., 1995, Ap.J., 455, L103

Fort and Mellier, 1994. AetAR 5, 329

Francis et al., 1996, Ap.J., 457, 490

Giavalisco, M., Steidel, C., Macchetto, F.D., 1996, Ap.J., in press

Koo, D.C., & Kron, R.G., 1992, ARA&A, 30, 613

Le Fèvre, O., Crampton, D., Hammer, F., Lilly, S.J., Tresse, L., 1994, Ap.J., 424, L14

Le Fèvre, O., Crampton, D., Felenbok, P., Monnet, G., 1994, A&A, 282, 325

Le Fèvre, O., Crampton, D., Hammer, F., Lilly, S.J., Tresse, L., 1995, Ap.J., 455, 60

Le Fèvre, O., Hudon, D., Lilly, S.J., Crampton, D., Hammer, F., Tresse, L., 1996, Ap.J., 461, 534

Lilly, S.J., Cowie, L.L., Gardner, J.P., 1991, Ap.J., 369, 79

Lilly, S.J., 1993, Ap.J., 411, 501

Lilly, S.J., Le Fèvre, O., Crampton, D., Hammer, F., Tresse, L., 1995a, Ap.J., 455, 50

Lilly, S.J., Le Fèvre, O., Hammer, F., Crampton, D., Ap.J., 1996, 460, L1

Maihara et al., PASP, 105, 940

Pascarelle et al., 1996, Ap.J., 456, L21

Peebles et al., 1989, ApJ, 347, 563

Steidel, C.C., Giavalisco, M., Pettini, M., Dickinson, M., Adelberger, K., Ap.J., 462, L17

Vettolani et al., 1996, A&A, preprint

The AUSTRALIS Instrument Concept – an Interim Report

Keith Taylor[1] and Matthew Colless[2]

[1] Anglo-Australian Observatory, PO Box 296, Epping, NSW 2121, Australia
[2] Mount Stromlo and Siding Spring Observatories, Weston Creek, ACT 2611, Australia

Abstract. Since September 1995 an Australian consortium of astronomers and engineers (from the Anglo-Australian and Mount Stromlo Observatories and the University of New South Wales) have been contracted by ESO to perform a one-year concept design study for a near-infrared multi-object spectrograph whose primary scientific motivation is the detection and study of high-redshift galaxies. The scope of the study includes the elucidation of the main science drivers and the development of a VLT instrumentation strategy best suited to those goals.

Our underlying instrumental philosophy has been to supply a significant object multiplex at a high enough spectral resolution to resolve the internal kinematics of galaxies. This science-driven goal also permits digital OH sky-suppression, yielding better S/N and spectral coverage than at lower resolutions. A full contiguous wavelength coverage (0.9μm to 1.8μm) is achieved through the use of multiple HgCdTe-based spectrograph cameras. A preliminary optical design for the spectrographs is already in place, while a 400-fibre positioner and integral field unit is presently being designed by the AUSTRALIS technical team. This will allow the proposed instrument to carry out both statistical studies of large numbers of objects and detailed studies of individual objects.

We detail here the basic conclusions to the scientific study together with their impact on instrumentation philosophy.

1   The Main Science Drivers

The main science drivers for a near-infrared multi-object spectrograph on the VLT are undoubtedly studies of galaxies at redshifts $z>1$. Scientific highlights that would emerge from such studies are:

- A census of the evolving galaxy population at redshifts from $z=0.5$ to $z=3.5$.
- Physical studies of the processes underlying galaxy evolution.
- A direct picture of hierarchical structure formation.
- The detection and study of primeval galaxies.

Although the instrument is therefore to be optimised for high-redshift studies, important science can also be carried out in a number of other fields. Some notable examples are:

- Follow-up spectroscopy for sky surveys.
- Spectral mapping of nearby AGN.
- Protostars and T-Tauri stars.
- Kinematics and abundances in globular clusters and Local Group galaxies.

2   Implications for Instrument Design

The broad requirements for the instrument capabilities which emerge from consideration of the science drivers are as follows:

2.1   Imaging Capability

Although there is a clear need for wide-field optical/infrared imaging to provide targets for spectroscopic follow-up (as well as imaging for its own sake), there is no strong reason why this should take the form of an imaging spectrograph.

- Wide-field broad-band imaging of sufficient depth for selecting spectroscopic samples for an 8m-class telescope can be achieved on a 4m-class telescope (although wide-field imaging programmes in general would of course prefer an 8m if the field size was effectively comparable).
- Broad-band imaging in the near-infrared (I through K bands) is optimal for selecting targets for most of the main science programmes, although a visible imaging capability is also desirable for studies of QSOs and very high redshift galaxies.
- Narrow-band imaging, such as could be carried out with a tunable filter imager, is extremely desirable in searching for primeval galaxies, QSOs and other strong-lined objects, and for selecting line-flux limited samples for studies of galaxy star-formation rates or internal motions.

A two-instrument approach to imaging and spectroscopy is advocated as the most efficient way to meet these goals: a dedicated wide-field spectrograph in conjunction with a broadband/tunable-filter, wide-field, optical/near-infrared imager on either a 4m or 8m telescope.

2.2   Efficiency and Multiplex

The science that can be performed depends critically on the limiting magnitude/flux achieved by the spectrograph and on the multiplex gain that can be realised. Almost all the main science programmes envisioned for the instrument will work at or close to the limiting magnitude, and will benefit from any faintward extension of this limit that can be achieved. Assuming that the field of view is a significant fraction of that available at VLT Nasmyth, most of the main science drivers have high enough target density that any multiplex up to at least 1000 can be fully utilised. However a number of programmes, notably spectroscopic follow-ups to sky surveys and stringently-selected samples of various kinds (e.g. candidate QSOs or primeval galaxies) will have only ~100 targets per field.

2.3   Spectral Range

The spectral range is clearly set by the main science drivers, which essentially involve the study of galaxies at $z>1$. Objects in this redshift range have all the familiar strong features of the rest-frame optical shifted out into the near-infrared, whereas there are no common strong emission lines at these redshifts in the optical (i.e. rest-frame UV). By covering the range 0.9 to 1.8μm, a near-infrared spectrograph can see Hα from $z=0.4$ to 1.7 and [OII] 3727Å and the 4000Å break from from $z=1.4$ to 3.5. Thus galaxies can readily be identified and studied over the whole range from $z=0.5$ to 3.5. Extending the spectral range into the optical is therefore mainly of interest to programmes involving QSOs and very high redshift primeval galaxies or low-redshift galaxies and stellar studies, which are given lower scientific priority (and which are, in part at least, catered for by MFAS). The red end of the spectral range is less clear-cut scientifically, as higher redshift objects or redder spectral features can be followed out into the K band (which is also of interest for some $z\sim0$ programmes). However the numbers and magnitudes of very high redshift objects are highly uncertain, and the technical difficulties attached to covering the K-band are considerable. Certainly if one had to pick a single octave of wavelengths to cover, the range 0.9 to 1.8μm would be the prime choice.

An associated issue is the *useful* spectral range. Those parts of the spectrum severely contaminated with OH line flux will not achieve sufficient S/N to be useful. Performance model calculations show that significant gains in useful spectral range are made with increasing resolving power up to about $\mathcal{R}\sim5000$. This is a very strong argument for employing some form of OH sky suppression (or sky avoidance) technique.

2.4   Spectral Resolution

It is clear that there are very significant qualitative and quantitative gains to be made from having a resolution sufficient to resolve the internal motions of galaxies. In order to reach any way down the luminosity function (especially in light of recent evidence of evolution in the Tully-Fisher relation) this will in practice mean at least $\mathcal{R}\sim3000$, with further gains up to $\mathcal{R}\sim5000$ being valuable.

The instrumental gains are that $\mathcal{R}>3000$ offers the possibility of very effective digital OH sky suppression, with resulting improvements in the average S/N for continuum spectra (i.e. in useful spectral range) and in the S/N in a line, especially where the line is just resolved (as is the case for most galaxies at $\mathcal{R}\sim3000$). It is important to note that although digital OH sky suppression can be applied at lower resolutions, considerable S/N gains continue to be made as the resolution is increased up to at least $\mathcal{R}\sim5000$.

2.5   Field of View

Scientifically there is always some advantage in a larger FoV. However given the limit to object multiplex of ~1000 set by detector area and other technical difficulties, many of the main science drivers could be efficiently carried out with FoV~$10'$. Nonetheless *all* the science programmes making use of the multiplex capability could benefit from fields of view as large as the VLT Nasmyth limit, and those with low sky densities of targets (such as follow-up spectroscopy of sky-survey samples or stringently-selected samples of candidate QSOs, primeval galaxies or other rarer objects) would benefit in proportion to the field area.

2.6   Integral Field Capability

An integral field mode (or modes) would greatly increase the flexibility of the instrument (probably at relatively little cost) and significantly extend the range of science programmes which it could undertake. Specific examples of such programmes include mapping the dusty central regions of AGN, studying the detailed kinematics of z~0.5–1 galaxies or the dynamics in the crowded cores of globular clusters, and studies of young stellar objects.

2.7   Spatial Resolution

In multiplex mode, where typically the targets are very faint and are only slightly extended, the issue is not spatial resolution but the optimal aperture size (with S/N being the relevant figure of merit). Detailed modelling based on the light profiles of faint galaxies, and previous experience with faint-object spectrographs, suggests that the optimal aperture is about $1.5''$ across (though the broad 'optimal' range covers at least 1–$2''$).

In integral field mode there is a trade-off to be made between field area and resolution. The optimal arrangement would be a resolution that is adjustable between about $0.25''$ and $0.5''$. In the AUSTRALIS concept this would yield integral fields of $9''$–$18''$ diameter, which would be suitable for most of the proposed integral field spectroscopy programmes.

3   Instrument Strategy

There is clearly a very strong case for an instrument which has a resolution $\mathcal{R}>3000$. Given that, in the general case, the time to study n objects is proportional to ne, where e is the instrument efficiency, it is clearly the case that digital OH sky suppression outperforms optical OH sky suppression, since $e_{digital} \gg e_{optical}$ and $n_{digital} \gg n_{optical}$. However general instrumental considerations mean that, unless one is building an instrument dedicated entirely to a single observational topic, n reaches a natural limit of ~1000. The science team considered that, given an instrument which could achieve $\mathcal{R}>3000$, a perfectly acceptable object multiplex would be in the range 100 to 500 (preferably, of course, towards the upper end of that range).

Given (i) the high dispersion ($\mathcal{R}>3000$) requirement, (ii) the optimal matching of aperture to object size, (iii) the maximum spectral packing density, and

(iv) the above object-multiplex goal, we arrive at a 400-fibre (1.5" diameter effective aperture) robotic positioner concept (similar to the AAO's 2dF project, but in the NIR) feeding a dual white-pupil spectrograph configuration, each spectrograph having a set of 4 very fast transmission cameras with HgCdTe detectors (8 cameras in all). This configuration permits the choice of either full coverage of the 0.9 to 1.8μm spectral window with an object multiplex of 200 or two wavelength bites (e.g. all of J *or* all of H) with a 400 object multiplex—whichever is considered optimal for the particular observational science goals.

A natural addition to such a flexible instrument concept is one and/or several fibre integral field units (IFUs) which can facilitate 3D spectroscopy of spatially resolved sources. Given the strength of the science case in this area, this was one of the several drivers favouring the use of fibres rather than slits.

4   Conclusions

The AUSTRALIS instrument concept is summarised as follows:

- *Spectral range:* 0.9 to 1.8μm (I, J and H passbands).
- *Field of view:* 20' diameter (unvignetted portion of VLT Nasmyth field).
- *Spectrograph feeds:* the spectrograph will have at least two feed modes: (i) up to 400 separately positioned microlens/fibre units, each subtending 1.5"; (ii) an integral field unit (IFU) consisting of a hexagonal microlens array feeding 1387 fibres, each subtending 0.25–0.5", with a total FoV of 9–18" diameter. (A loose-packed IFU and multiple smaller IFUs are also options.)
- *Multiplex factor:* in multiobject mode, either 200 simultaneous spectra over the whole wavelength range or 400 simultaneous spectra over half the wavelength range; in integral field mode, 1387 simultaneous spectra over the whole wavelength range.
- *Resolving power:* effective resolution of $\mathcal{R}\sim4500$ (allowing digital OH sky-suppression); lower resolutions by binning.
- *Detectors:* 8 Rockwell 1024^2 HgCdTe devices (18.5μm pixels).

5   Acknowledgements

Thanks are due to:

- **The AUSTRALIS science team:** Jeremy Bailey, Joss Bland-Hawthorn, Russell Cannon, Karl Glazebrook and Stuart Lumsden (AAO); Charlene Heisler, Peter McGregor, Jeremy Mould, Bruce Peterson and Peter Wood (MSSSO); Warrick Couch and John Webb (UNSW).
- **The AUSTRALIS technical team:** Peter Conroy and Heath Jones (MSSSO); John Storey and Michael Ashley (UNSW); Fred Watson (AAO); Ian Parry (IoA, Cambridge) and Damien Jones (Prime Optics; Qld).

Numerical Simulations of Galaxy Formation

Matthias Steinmetz

Department of Astronomy, University of California, Berkeley, CA 94720, USA

Abstract. An overview over the current status of modeling galaxies by means of numerical simulations is given. After a short description of how galaxies form in hierarchically clustering scenarios, success and failures of current simulations are demonstrated using three different applications: the morphology of present day galaxies; the appearance of high redshift galaxies; and the nature of the Ly-α forest and metal absorption lines. It is shown that current simulations can qualitatively account for many observed features of galaxies. However, the objects which form in these simulations suffer from a strong overcooling problem. Star formation and feedback processes are likely to be indispensable ingredients for a realistic description even of the most basic parameters of a galaxy. The progenitors of todays galaxies are expected to be highly irregular and concentrated, as supported by recent observations. Though they exhibit a velocity dispersion similar to present day $L \gtrsim L^*$ galaxies, they may be much less massive. The filamentary distribution of the gas provides a natural explanation for Ly-α and metal absorption systems. Furthermore, numerical simulations can be used to avoid misinterpretations of observed data and are able to alleviate some apparent contradictions in the size estimates of Ly-α absorption systems.

1 Introduction

Within the last two or three years, the advent of 10m class telescopes like the KECK and the superb imaging quality of the refurbished Hubble-Space-Telescope (HST) have revolutionized the way we can look at the formation of galaxies. While a few years ago observations were restricted to galaxies which were at best a few billion years younger than the Milkyway, we can now routinely study galaxy formation at redshift $z > 1$, when galaxies were only 20% of their present age. And even the first billion years of the life of a galaxy seem to be reachable (Steidel et al. 1996). Though these observations are still in their childhood, we can expect that with the advent of several 8m class telescopes (including the VLT) we will construct a dense coverage of the morphologies of galaxies from today back to redshifts of $z > 3$. These observations are not only likely to provide important clues on the formation process of galaxies themselves, they will also constrain cosmological background models.

Confronted with these detail-rich observations, however, theoretical models of galaxy formation, which mainly classify galaxies according to their Hubble type and therefore according to their global star formation history, seem to be outdated. Also so-called quasianalytic models which were successfully applied to the evolution of different populations of galaxies, e.g. to the Butcher-Oemler effect (Kauffmann 1995) and to the large scale distribution of galaxies (Kauffmann, Nusser & Steinmetz 1996)), say little about the spatial distribution of light within individual galaxies at different redshifts. Only highly resolved numerical simulation appear to be capable to provide the adequate theoretical framework for these new observations. In the following, I try to give a brief overview on the current status of studying galaxy formation by numerical simulation. Besides describing progress and failure in modeling current galaxy populations, I will also draw connections to the appearance of high redshift galaxies and the origin of Ly-α absorption lines.

2   Galaxy Formation in Hierarchically Clustering Universes

Hierarchical clustering is at present the most successful theory of structure formation. In this scenario, structure grows as systems of progressively higher masses merge and collapse to form newly virialized systems. Over the last two decades, the build-up of the mass hierarchy has been investigated in detail by means of N-body simulations (for a review see, e.g., Davis et al. 1992). However in comparing the results of N-body simulations with observed galaxies, it seems unlikely that there is a very close correspondence between galaxies on the one hand side and the dark matter halos on the other. The circular velocity of a galaxy is not directly correlated to the circular velocity of a dark matter halo (Navarro, Frenk & White 1996), nor does each dark matter halo necessarily contain one and only one galaxy (see, for example, Kauffmann et al. 1996).

Within the last few years, we have began to simulate the dynamical evolution of the baryonic component by including the effects of gas dynamics, shock heating and radiative cooling (for an overview, see e.g. Steinmetz, 1996a and references therein). First attempts to mimic the effects of star formations have been performed. Though many of the physical and numerical issues of these advanced simulations are still matter of debate, some qualitative features describing galaxy formation seem common in all of these simulations.

The simulations presented below have been performed using the smoothed particle hydrodynamics code GRAPESPH (Steinmetz 1996b). They have a mass resolution of several $10^6 \, M_\odot$ and a spatial resolution of a few kpc. Using a multi-mass technique, the tidal field exerted by surrounding matter up to radii of 30 Mpc is included. In spite of their high resolution, the evolution of these galaxies is nevertheless followed up to the present epoch. The achieved numerical resolution therefore surpasses that of any other cosmological galaxy formation simulations performed so far by a factor of five or more. Details of the simulations themselves can be found in Navarro & Steinmetz (1996) and in Haehnelt, Steinmetz & Rauch (1996). Though the simulations were performed using the standard cold dark matter (CDM) cosmogony (i.e., $\Omega_0 = 1$, $h = 0.5$, $\Lambda = 0$, $\sigma_8 = 0.63$), many of the results are quite generic for any hierarchically clustering scenario. Furthermore, differences between different cosmological scenarios are partially eliminated due to the different normalizations: the normalization is usually chosen so to match the observed abundance of rich clusters ($\sigma_8 \approx 0.63 \, \Omega^{-0.6}$).

3   Cooling, Mergers and the Morphology of Galaxies

While the build-up process of dark matter halos in hierarchically clustering scenarios is fairly well understood, the dynamics of the gaseous component is much less clear. It is usually assumed that most of the gas within a dark matter halo is able to cool radiatively. Due to the high cooling capabilities, no substantial gas pressure can be established and gas collapses until it settles in a rotationally supported disk. According to the tidal torque theory the gas can (marginally) acquire enough angular momentum to explain the size of todays disk galaxies (Fall & Efstathiou 1980). Mergers of disk galaxies (or their progenitors at higher redshift) provide a plausible explanation for the formation of ellipticals. This picture is indeed qualitatively supported by the outcome of numerical simulations as those presented at this conference: Halos which form in the field or in a filament typically experience a relatively quiescent merging history and no major merger is involved at redshifts smaller than about one. Within such a halo, the cooling gas settles and forms a rotationally supported disk. In a denser environment like a group of galaxies, a more violent merging history arises: During a major merger of two dark halos, the fusion of the two gaseous disks at the center of the dark matter halos ends up in a very compact, slowly rotating gas concentration. Simulations which include star formation (Katz 1992, Steinmetz & Müller 1995) show that during such a merging event gas is efficiently transformed into stars and an ellipsoidal distribution of stars arises – a scenario for the formation of bulges and/or elliptical galaxies.

However, though the general picture seem to be pretty promising, a closer look at these galaxy-like objects exhibit quite substantial differences to observed galaxies: First of all it has still to be shown that the right fraction of elliptical to spiral (as a function of the environment) can be achieved. This also may critically depend on the cosmological background model. Furthermore, though the arising objects visually represent spiral galaxies, they are far too concentrated. In contrast to the assumption of Fall & Efstathiou (1980) the gas has not been accreted axisymmetricly but by a series of merging events. Consequently angular momentum is efficiently transported from the gas to the dark matter halo (Navarro, Frenk & White, 1995) and the specific angular momentum of the gaseous object is only 10 to 20 per cent of that of the dark halo. A more detailed analysis (Navarro & Steinmetz 1996) demonstrates that gas which falls in diffusely (and also less bound gas lumps which are tidally disrupted early on) settles down to form a disk. However, most of the gas lumps are sufficiently tightly bound to resist the tidal field of the halo. They spiral down to the center due to dynamical friction and deliver most of their angular momentum to the dark halo. The numerical models predict the formation of disks, as observed, but far too much mass is acquired by the central "bulge". Note that this is an immediate consequence of the cooling catastrophe (White & Rees 1978, Blanchard, Valls-Gabaud & Mamon 1992) and the neglect of feedback processes. Since cooling times scale inversely with density, the dissipative collapse of gas is more efficient at high redshift because the dark matter halos present at that time (and the universe as a whole) were denser.

Up to the present, there has been no self–consistent high–resolution simulation which avoids the angular momentum problem. In order to solve it, one has to take care that more gas is diffusely accreted. A variety of different physical processes which might be able to solve this problem are currently under discussion:

1. One possibility is to change the merging history of the forming galaxy, e.g. by using a fluctuation spectrum with less power on small scales as predicted, for example, by a cosmogony with hot and cold dark matter. Changing the cosmological parameters Ω_0, Λ_0 has probably rather little influence: Although the merging rate in the near past can be changed a lot, the merging history, expressed in terms of number of mergers and mass distribution of progenitors, is quite similar (Lacey & Cole 1993). However, it still has to be investigated to what extend the non-linearity of the merger process itself affects the result. For example, low Ω models typically have a higher baryon fraction and, therefore, self gravitation of the gas component may be more important and may reduce the efficiency of the angular momentum transport.

2. A photoionizing UV background with a strength as required to explain the Ly-α forest (see below) may suppress the formation of small structures ($v_c \lesssim 50$ km/sec, Efstathiou 1992) and gas might fall in more diffusely. Therefore, the angular momentum transport to the halo might be reduced. However, the simulations presented here are only slightly influenced by the UV background: Most of the central gas clump has formed at sufficiently high redshifts (i.e. at sufficiently high densities) that recombination ($\propto \varrho^2$) dominates over photoionization ($\propto \varrho$).

3. A realistic galaxy formation model has to include the effects of star formation, and the solution of the angular momentum problem is most likely related to feedback due to supernova and stellar winds. First simulations which account for feedback due to star formation (Katz 1992) assume that supernovae increase the thermal energy of the surrounding gas. Most of this energy, however, is immediately radiated away due to the high cooling capability of the gas. As a result, the formation of small lumps of gas cannot be prevented. The main influence of star formation is to transform a dense knot of gas into a slightly more diffuse lump of stars, but the extensive transport of angular momentum to the dark halo has not been overcome. It is conceivable that momentum input due to supernovae might have a much stronger effect (Navarro & White 1993).

4   The Appearance of High Redshift Galaxies

Recently, Steidel and coworkers (1996, see also the contributions of Giavalisco and Macchetto to this conference) have detected and spatially resolved galaxies at redshifts $z > 3$. These galaxies appear to be very compact and often show substantial substructure. In case of a high Ω_0 universe, these objects seem to be at least half as abundant as L^* galaxies today. It has been argued that

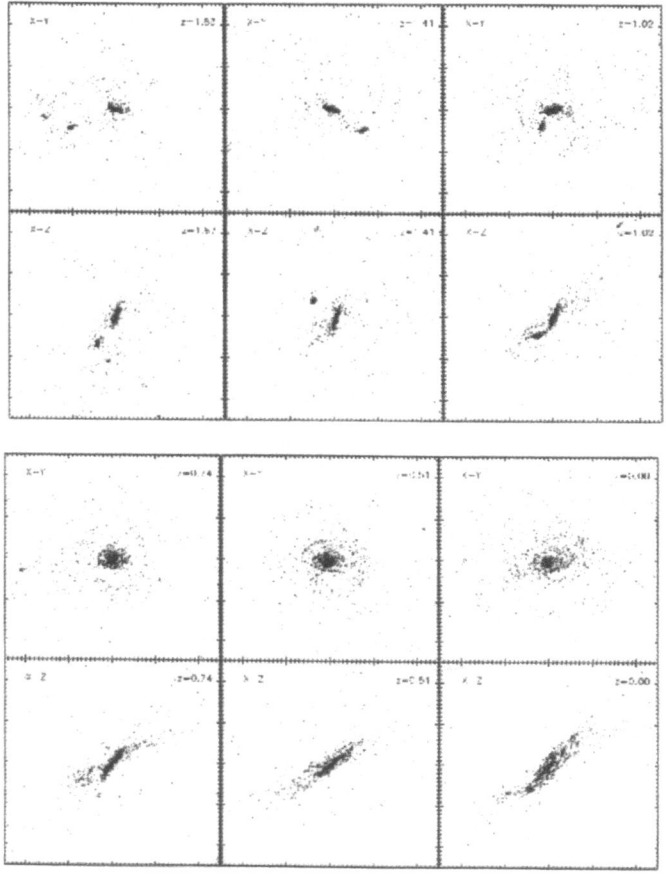

Fig. 1. The distribution of gas projected in the X-Y and Y-Z plane shown for 6 different redshifts. The gas infall is mainly lumpy. Diffusely infalling gas settles down to form a rotationally supported disk.

the equivalent width of saturated absorption lines implies velocities dispersions for these galaxies of the order of 180-320 km/sec, though it is still matter of debate whether these velocity dispersions are gravitational. By comparing with the expected number densities of dark matter halos with velocity dispersions higher than 180 km/sec, Mo & Fukujita (1996) argued that this observation can be used to rule out some cosmological scenarios.

Contrary to some recent claims, size, abundance and substructure of these objects are consistent with most structure formation scenarios, and by parts even predicted (see, e.g., Katz 1992). To demonstrate that, figure 2a shows an artificial I-band CCD image of a galaxy at redshift 3.1 as it forms in the numerical simulations presented above. The picture has been created by translating the ages of the formed star particles into I band luminosities using spectrophotometric models (Contardo, Steinmetz & Fritze-von Alvensleben, in preparation). A point spread function similar to that of the HST has been assumed. No assumptions on noise and absorption due to intervening dust have been done. The object exhibits a total apparent luminosity of $m_I = 22.5$, the central surface brightness is $\mu_I = 20$. Analyzing the velocity dispersion of the system, typical values of 200 km/sec can be found, while the halo circular velocity is about 30 per cent smaller. The total mass of the object is less than about 20 per cent of that of the corresponding object at the present epoch, though the circular velocity is similar. Comparing these objects with the corresponding galaxies at the present epoch, the early formed stars can be dominantly found close to the center and may correspond to the formation of (or parts of) the bulge component. However, most of the mass of the galaxy at $z = 0$ is not yet collapsed but is dispersed over a few hundred kpc, partially concentrated in several less massive subclumps.

The high velocity dispersion of the progenitors may be surprising, but it is easily understandable in the context of hierarchical clustering scenarios. In figure 2b, the circular velocity and in figure 2c the mass of the most massive progenitor of halos with circular velocities between 80 and 200 km/sec at $z = 0$ is shown, normalized to the mass and circular velocity at the present epoch, respectively. Going to higher redshifts, one can see that the circular velocity is rising up to redshifts of about 2, though the mass is decreasing by more than a factor of 3. Only at redshift close to 4, mass and circular velocity are rapidly dropping. The rise of circular velocity can be understood as following (for the sake of simplicity of the argument, $\Omega = 1$ and $h = 0.5$ has been assumed): for a given circular velocity v_c, the mass within the virial radius (i.e. the radius within which the average overdensity is 178 times the critical density) is given by

$$M(v_c, z) = 4\ 10^{12}\left(\frac{v_c}{200\mathrm{km/sec}}\right)^3 (1+z)^{-1.5}, \tag{1}$$

i.e. for a constant circular velocity, an object at higher redshift has a lower mass, but also a correspondingly smaller virial radius. According to the Press-Schechter algorithm, an object of mass M has on average acquired half of its mass at a "formation" redshift of

$$z_f = \sqrt{2^{(n+3)/3} - 1}\left(\frac{M}{M_*}\right)^{-(n+3)/6}, \tag{2}$$

n being the spectral index of the power spectrum (for CDM, it is $n \approx -1 \ldots -2$ on scales of galaxies). M_* is the typical non-linear mass scale. Normalizing a CDM like power spectrum to cluster abundances, we obtain

$$M^* = 4\ 10^{13}\,(1+z)^{-6/(3+n)}\,\mathrm{M}_\odot \tag{3}$$

(White 1996). By comparing equations 1 and 2, the circular velocity at the formation redshift is larger as long as $M(v_c, z) > C_n M^*(z)$, with $C_n = (\sqrt{2^{(3+n)/3} - 1}/(2^{2/3} - 1))^{6/(3+n)}$. For typical values of the spectral index n on scales of galaxies, $C_n \approx 0.5 \ldots 2$. The circular velocities of high redshift galaxies, are expected to drop only if $M(v_c, z) \gg M^*(z)$. Their inferred abundance being similar to todays galaxies (namely L^*) therefore implies that these galaxies (or at least their halos) would have roughly the same circular velocities as today galaxies (namely about 200 km/sec), though their mass is much smaller (a factor of 10). This also reflects another feature seen in N-body simulations, namely that as long a $M(v_c, z) \lesssim M^*(z)$, every dark matter also has a well defined progenitor at higher redshifts, while if $M(v_c, z) \gg M^*(z)$, the object has been formed by merging several objects of comparable mass. Concerning the abundances of these objects, one also has to keep in mind the weak correlation between the actual circular velocity of a galaxy and the circular velocity of a dark matter halo, as mentioned above: First of all, the maximum of the circular velocity can be up to 40 per cent higher (Navarro, Frenk & White, 1996) and it is achieved at radii much smaller than the virial radius. Secondly, as the baryons accumulate near the center of the galaxies the depth of the potential well is further increased. This implies that the circular velocity of dark matter halos is likely to be (substantially) smaller than that of the observed galaxy. Since their circular velocity is smaller, they are more abundant. Constraints derived via the Press-Schechter algorithm are likely less severe than they seem to be on the first view. Similar caveats also hold using the abundance of damped Ly-α systems to constrain cosmological models.

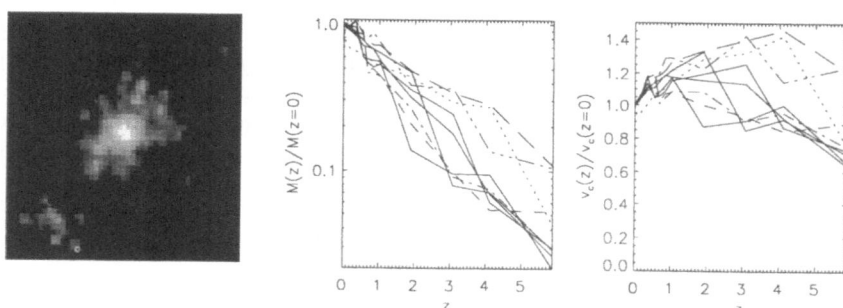

Fig. 2. Left: Artificial I-band CCD image of a z=3 galaxy as formed in a CDM simulation. The image is 7.5 arcsec across; Middle: mass of the most massive progenitor at redshift z normalized to the halo mass at $z = 0$; Right: circular velocity of the most massive progenitor at redshift z normalized to the circular velocity at $z = 0$.

5   Ly-α Absorbers and Metal Line Systems

Considering the difficulties if implementing models of star formation, and, therefore, of modeling the light emitted by galaxies, it is an intriguing alternative to study the absorption characteristics of the gas. This exercise has been successfully performed within the last two years (Cen et al. 1994, Petitjean et al. 1995, Hernquist et al. 1996, Haehnelt et al. 1996). It seems that the gas distribution as seen in large scale structure simulations can account for the large dynamic range of column densities seen in Ly-α absorption systems, from column densities of a few $10^{13}\,cm^{-2}$ up to $10^{20}\,cm^{-2}$ or higher. Very low column density systems

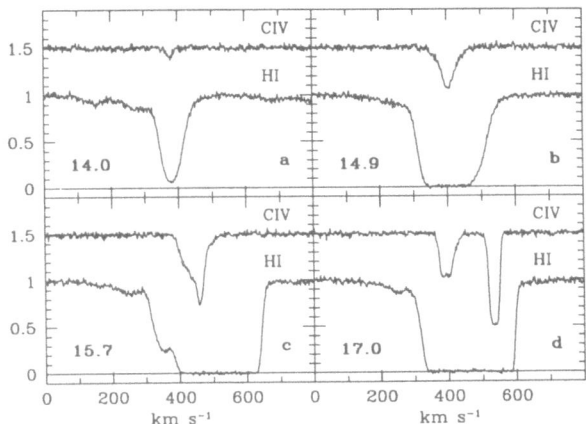

Fig. 3. HI and CIV absorption as produced by the numerical simulations for lines-of-sight of different column densities. HI and CIV are shifted relative to each other by 0.5. Only $\lambda 1538$ line of the CIV doublet is shown.

($10^{13}\,\mathrm{cm}^{-2}$) typically arise for light rays penetrating voids, higher column densities arise if filaments ($10^{15}\,\mathrm{cm}^{-2}$) or even individual gaseous halos ($> 10^{17}\,\mathrm{cm}^{-2}$) are penetrated. Typical absorption lines as they are produced in numerical simulations are shown in figure 3. Sometimes, these systems do not even represent physically connected systems, but are caustics in the velocity space. Matching the column density distribution to observations, however, requires adjustment of the ratio J_{21}/n_b by about a factor of $\approx 1/3$ from standard values.

The low physical density ($n \lesssim 10^{-3}\,\mathrm{cm}^{-3}$) in systems with column densities less than about $10^{16}\,\mathrm{cm}^{-2}$ also implies that the cooling time scales for the gas can be similar or even larger than the local dynamical time scales. Therefore, substantial deviation of the actual gas temperature from the photoionization equilibrium temperature can arise. Due to the rather strong temperature dependence of the ionization state of hydrogen, carbon and other relevant metals, the assumption of photoionization temperatures leads in many cases to errors in the determination of total density and ionization fractions. For example a modest change in the temperature by a factor of two can change the CII/CIV ratio by an order of magnitude and the line-of-sight extent of CIV absorption systems inferred from photoionization models depends very strongly on the actual temperature (Haehnelt, Rauch & Steinmetz 1996). For more details on metal absorption systems see the contribution of Haehnelt in this volume.

6   Summary and Conclusions

Numerical simulations are likely the only way to model galaxy formation which can account for the detail-rich appearance of galaxies at low and high redshifts as it is expected (and partially already seen) by observations using the new generations of 8m class telescopes. From this point of view, the outcome of current numerical simulations on the formation of galaxies in hierarchically clustering scenarios has been presented, specifically the morphology of present galaxies, the appearance of high redshift galaxies and the origin of the Ly-α forest and metal absorption lines. It is encouraging that the same simulation, which was designed to study the properties of present galaxies, allows at least qualitatively to understand a large variety of features seen in objects at different redshifts. However, the simulations only follow the evolution of individual galaxies and demonstrate that the investigated scenario is potentially able to explain different morphological types at different redshifts. It is still to be shown that also the statistical behavior of a galaxy population as a whole can be reproduced. Furthermore, the simulations exhibit problems in the hierarchically clustering picture, which require further investigation. The most prominent problem is the overcooling and the related overly small angular momentum of disk galaxies. The solution of these problems is likely related to feedback processes due to stellar evolution, like, e.g. supernovae and stellar winds.

References

Blanchard, A., Valls-Gabaud, D., Mamon, G.A. (1992): *A&A*, **264**, 365

Cen, R., Miralda-Escude, J., Ostriker, J.P., Rauch, M. (1994): *ApJ*, **437**, L9

Davis M., Efstathiou, G., Frenk, C.S., White, S.D.M (1992): *Nature*, **356**, 489

Efstathiou, G. (1992): *MNRAS*, **256**, 43p

Fall, S.M., Efstathiou, G. (1980): *MNRAS*, **193**, 189

Hernquist, L., Katz, N., Weinberg, D.H., Miralda-Escude, J. (1996): *ApJ*, **457**, L1

Lacey, C., Cole, S. (1993): *MNRAS*, **262**, 627

Katz, N. (1992): *ApJ*, **391**, 502

Kauffmann, G. (1995): *MNRAS*, **274**, 153

Kauffmann, G., Nusser, A., Steinmetz, M. (1996): *MNRAS*, in press

Navarro, J.F., Frenk, C.S., White, S.D.M. (1995): *MNRAS*, **275**, 56

Navarro, J.F., Frenk, C.S., White, S.D.M. (1996): *ApJ*, **462**, 563

Navarro, J.F., Steinmetz, M. (1996), *ApJ*, submitted

Navarro, J.F., White, S.D.M. (1993): *MNRAS*, **265**, 271

Haehnelt, M., Steinmetz, M., Rauch M. (1996): *ApJ*, **465**, L95

Haehnelt, M., Rauch M., Steinmetz, M., (1996): *MNRAS*, in press

Steidel, C.C., Giavalisco, M., Pettini, M., Dickinson, M., Adelberger, K.L. (1996): *ApJ*, **462**, L17

Steinmetz, M., Müller E. (1995): *MNRAS*, **276**, 549

Steinmetz, M. (1996a): Proc. Int. School of Physics "Enrico Fermi" – *Dark Matter in the Universe*, Varenna, Italy, July 24 - August 4 1995, IOP, Bristol in press

Steinmetz, M. (1996b): *MNRAS*, **278**, 1005

White, S.D.M. (1996) *Lecture Series of the Les Houches Summer School on 'Dark Matter and Cosmology'* (North Holland ed R Schaeffer) in press

White, S.D.M., Rees, M., (1978): *MNRAS*, **183**, 341

The Calar Alto Deep Imaging Survey for Primeval Galaxies

K. Meisenheimer, S. Beckwith, R. Fockenbrock, J. Fried, H. Hippelein, U. Hopp, Ch. Leinert, H.-J. Röser, E. Thommes, C. Wolf

Max-Planck-Institut für Astronomie, Königstuhl 17, D-69117 Heidelberg, Germany

The Calar Alto Deep Imaging Survey (CADIS) employs a combination of deep exposures in three broad bands (B,R,K) and up to 13 medium bands ($R = \lambda/\Delta\lambda = 25\ldots50$) with spectroscopic scans through an imaging Fabry-Pérot-Interferometer ($R \simeq 450$). The survey will cover 10 fields of $> 100\square'$, i.e. a total area of $0.3\ \square°$. Although primarily designed to detect faint emission line galaxies at various intermediate ($z = 0.2\ldots1.4$) and very high redshifts ($z = 4.7,\ 5.7, 6.5$) its multi-color strategy will also allow the detection and classification of hundreds of early type galaxies, faint QSOs, and extremely faint stars in our Galaxy. This contribution outlines the survey concept and its principal goals and then focuses on the expectations and first results of the search for "primeval" Ly-α emitting galaxies at $z \gtrsim 5$. More details about the data analysis and first results on faint field galaxies at intermediate redshift are presented in the contributions by Fockenbrock et al. and Thommes et al. to these proceedings.

1 The CADIS Concept

The principal idea of CADIS is to search a significant area of the sky for faint emission line galaxies, by means of imaging spectroscopy using a Fabry-Pérot-Interferometer (FPI). Its wavelengths resolution ($R \simeq 450$) is tailored to an optimum detection of massive galaxies with a velocity dispersion of $\gtrsim 500\,\mathrm{kms}^{-1}$. With a very deep flux limit of $3 \times 10^{-20}\ \mathrm{Wm}^{-2}$ it should be able to detect bright spirals out to a redshift $z > 1$, and the first burst of star formation in primeval galaxies at $z \gtrsim 5$ down to star formation rates (SFR) of 10 to 50 $M_\odot/\mathrm{yr}$. The wavelength range scanned with the FPI is 12 to 15 nm, corresponding to $\Delta z \simeq 0.018,\ 0.03$ and 0.1 at fiducial redshifts $z = 0.25$, 1 and 6, respectively. A completely new feature of the CADIS strategy are very deep images in a set of narrow band filters ($R \simeq 50$, typical exposure times: 10 h at a 2.2 m-telescope), which are selected in such a way that for every prominent emission line falling into the FPI scan (e.g. H$_\alpha$) a second or third line (e.g. [OII]372.7 nm or [OIII]500.7 nm) should show up in them (see example in Fig. 1). We call these narrow band filters veto-filters since a signal in one of them excludes that the line detected in the FPI is Ly-α (cf. Fig. 1). Combining the veto-filter information with the accurate spectroscopy through the FPI, we expect to determine the redshift of the majority of emission line galaxies ($0.25 < z < 1.0$) to an accuracy of $75\ldots250\,\mathrm{kms}^{-1}$ (depending on the S/N-ratio).

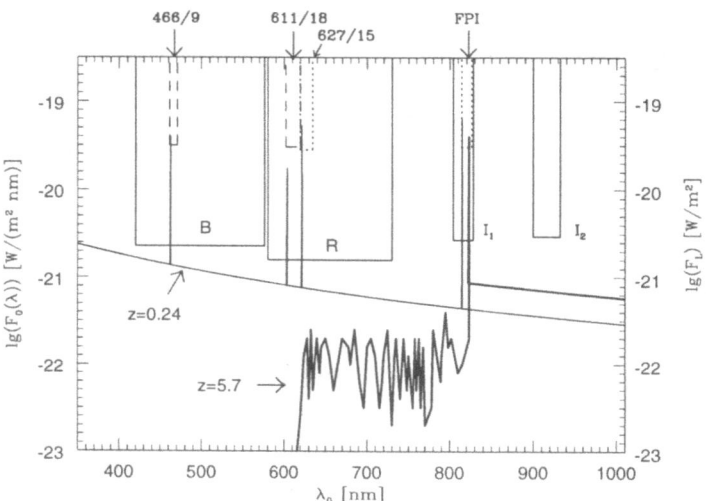

Fig. 1. Demonstration of the *veto*-filter strategy in CADIS. Solid boxes indicate the band width and 5σ limit of the deepest broad and medium band filters of CADIS (B,R,I_1,I_2, refer to scale on the left). The dotted box represents the 5σ limit for line detection in the FPI scan, and dashed boxes refer to 3σ limits in the *veto*-filters 466/9, 611/18 and 627/15 (scale on the right). Note that a typical Ly-α galaxy at $z = 5.7$ will only be detected in the FPI and (perhaps) the I-bands, whereas a foreground galaxy at $z = 0.24$ with very strong emission lines should show up in both 466/9 and 627/15.

Since the detection of emission line objects requires several "off-line" continuum exposures anyway, we decided to add in a complete set of broad and medium band images, which are optimized both for the continuum determination and the identification of one of the most severe contaminants – faint M-stars in our Galaxy. This optical multi-color survey (flux limit $F_\lambda \lesssim 3 \times 10^{-21}\,\mathrm{Wm^{-2}nm^{-1}}$) is supplemented by an equally deep K' image with the new OMEGA camera at the prime focus of the Calar Alto 3.5 m-telescope ($6.6 \times 6.6\,\square'$ field). This gives us a global view of the spectral energy distribution (SED) of every object in the field. Due to the long-term nature of the CADIS project (we need > 10 *clear* nights per field, which will be collected over ~ 3 years) it will be possible to repeat the R-band images every half year or so, in order to accumulate a rather deep variability survey (down to R = 23.5 for detecting 10% variability). CADIS is the extragalactic key-project of the Max-Planck-Institut für Astronomie.

The size of one CADIS field is limited by the field-of-view of the focal reducers at the Calar Alto 2.2 m- and 3.5 m-telescopes ($120\,\square'$). Altogether CADIS will survey 10 fields distributed over the northern sky ($\delta \gtrsim -5°$), which were selected for their absence of bright stars (R $\lesssim 16.0$) and an extra-ordinary low flux on the IRAS 100 μm maps ($\lesssim 2\,\mathrm{MJy/sr}$). Thus the total survey area will be at least $0.3\square°$. The FPI scans will cover the following ranges: **(A)** $\lambda \in [694, 706\mathrm{nm}]$, **(B)** $\lambda \in [814, 826\mathrm{nm}]$, and **(C)** $\lambda \in [910, 926\mathrm{nm}]$ with 9 wavelength settings each. They fit into local minima of the night-sky emission.

2 The Major Goals of CADIS

CADIS mainly aims for the detection, classification and redshift determination of five classes of objects:

- Primeval galaxies at redshift $z \gtrsim 5$. Since the audience expressed some resentment about the – often misused – term "primeval galaxy" here is our definition: A *primeval galaxy* is a gravitationally bound system in which the first burst of star formation (from primordial material) occurs. CADIS searches for massive systems at high redshift (*i.e.* $M_b > 10^{10} \, M_\odot$ at $z \gtrsim 5$), that are the progenitors of current day massive spheroids (ellipticals and bulges of early-type spirals). Since there seems to be virtually no way to avoid intense Ly-α from the formation of the first O-stars, CADIS is tailored to detect the Ly-α line which should be the ultimate sign of a violent starburst in primordial material. (Some details about the expected number of primeval galaxies will be given in the next section.)
- Faint emission line galaxies at intermediate redshift. Since $\gtrsim 98\%$ of the faint emission line galaxies detected in the FPI scans will lie at intermediate redshifts $z \leq 1.4$, CADIS will provide a unique list of up to 10^4 emission line galaxies located in various redshift bins between $z = 0.07$ and $z = 1.4$ (see Table 1). This sample will probe the faint end of the galaxy luminosity function to unprecedented depth and will allow, for the first time, a determination of the spatial 2-point correlation function out to $z \simeq 1$ (based on the accurate redshift measurement in the FPI scans).
- The well-sampled multi-color data of CADIS will detect and determine the redshift (to $\sigma_z \simeq 0.02$) of several hundred early type galaxies in the range $0.5 < z < 1.2$. They will complete the luminosity function at the bright end.
- We expect to constrain the stellar luminosity function at the very faint end, since the medium band exposures in the range $680 < \lambda < 920 \, \mathrm{nm}$ are optimized to identify faint M-stars.
- Both the multi-color survey (which is fine-tuned to identify the most important contaminants: M-stars) and the variability survey in CADIS are ideally suited to detect faint QSOs beyond $z = 3$. We expect to find several hundred QSOs which will confine the faint end of the QSO luminosity function and could even contain the first QSOs beyond $z = 5$ (if they exist).

Table 1. Redshift intervals for the emission line search

$\langle z \rangle$	Δz	line detected in the FPI	$\langle z \rangle$	Δz	line detected in the FPI
0.07	0.018	H$_\alpha$ in (A)	1.20	0.032	[OII]372.7 in (B)
0.25	0.018	H$_\alpha$ in (B)	1.46	0.036	[OII]372.7 in (C)
0.40/0.44	0.024	H$_\alpha$ in (C),	4.76	0.10	Ly-α in (A)
		[OIII]500.7 / H$_\beta$ in (A)	5.74	0.10	Ly-α in (B)
0.64/0.69	0.024	[OIII]500.7 / H$_\beta$ in (B)	6.55	0.13	Ly-α in (C)
0.83/0.89	0.032	[OIII]500.7 / H$_\beta$ in (C)			
		[OII]372.7 in (A)			

3   The Search for Primeval Galaxies: How Many Do We Expect?

The search for primeval galaxies (PGs, refer to definition above) is a story of permanent frustration. Starting from the first predictions of Partridge & Peebles (1967), over the first Ly-α prediction by Meier (1976), to the more advanced models by *e.g.* Baron & White (1987) or Pritchet (1994), one fact has remained constant: The observations failed to detect any PG at the predicted flux level.

We think that this is no reason for despair. By recognizing that the Ly-α bright phase of primeval star formation is likely to be rather short-lived, Thommes & Meisenheimer (1995) could demonstrate that all previous predictions have been far too optimistic. Their refined model follows the standard approach to estimate the mass function of PGs from the locally observed mass function of spheroids. The scaling law between mass and early $SFR \propto L_{\text{Ly}-\alpha}$ is taken from numerical simulations by Steinmetz & Müller (1995). However, based on the assumption that the onset of galaxy formation occurs early (between $z = 30$ and 10) and taking into account that the Ly-α bright phase could be as short as 10^8 years (*e.g.* due to rapid dust formation), while the entire epoch of galaxy formation is likely to span several 10^9 years, the expected number of PGs turns out to be at least 10× lower than previous models had predicted. Therefore, even the deepest survey so far (the Palomar FPI survey which searched out to $z = 4.7$, Thompson *et al.* 1995) had no realistic chance to find a single PG.

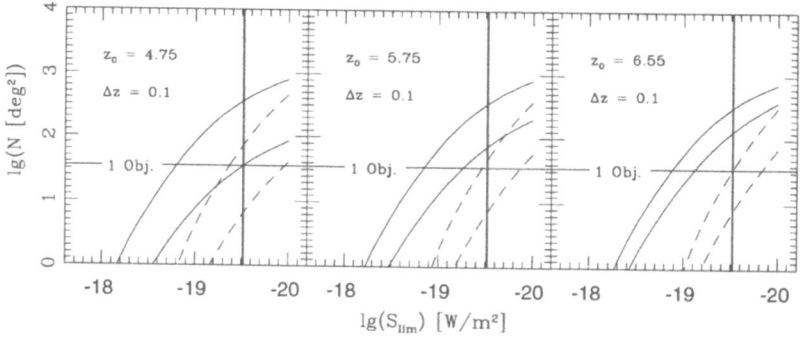

Fig. 2. Predicted number of primeval galaxies above the flux limit S_{lim} for the three wavelength regions searched by CADIS (solid curves: $q_0 = 0.5$, dashed: $q_0 = 0.1$, from Thommes 1996). The thick vertical line refers to the CADIS limit. The horizontal line labelled "1 Obj." indicates the number density which is necessary to find on average 1 object per CADIS field (0.03□°).

Fig. 2 shows the refined predictions for the three FPI ranges (A), (B), (C) to be observed in CADIS. Obviously CADIS should find between 1 ($q_0 = 0.1$) and 20 ($q_0 = 0.5$) PGs per field if all three bands are observed. More significant than these absolute numbers is the fact that the detection probability will be

$100\times$ larger than in the best previous surveys – due to the much larger area and deeper flux limit probed by CADIS. The enhanced probability depends only on dN/dS_{lim} and is rather independent of the choice of model parameters.

4 The Search for Primeval Galaxies: First Results

Due to an unforeseen delay in the delivery of science grade $2k \times 2k$ CCDs and the refurbishment of the 2.2 m telescope on Calar Alto, CADIS has not yet reached the state of routine data collection. However, we managed to get meaningful data (several FPI settings, B,R exposures plus $\sim 1/4$ of the attempted integration time in the veto-filters) for two fields using a smaller CCD (60$\square'$ field).

The data of one field have been fully analysed for emission line objects (Thommes 1996), according to the following criteria:

- image detection at ≥ 1 FPI wavelength (using FOCAS),
- Photometric detection[1] above 5σ at ≥ 1 FPI wavelength,
- presence of an emission line at a level of 3.5σ[1] above the continuum.

On this very limited data set ($\lesssim 2\%$ of the CADIS project) we find 193 "emission line objects" (here we included 26 objects which do not obey the strict criteria, but show a clear emission line in the FPI pre-filter and ≥ 1 veto-filter). Inspection of their overall spectra reveals 46 of them to be artifacts (reflections in the FP etalon), or late M-stars which happen to show a pronounced continuum peak around 815 nm. Thus we are left with **147 true emission line galaxies**. 104 of these show at least a marginal detection in our blue band ($\lambda_0/\Delta\lambda = 490/160$ nm) and are therefore classified as Faint Blue Galaxies. Since the blue band probes rest wavelengths $\lambda < 85$ nm for objects at $z = 5.7$, they can readily be excluded from our list of Ly-α candidates and have to be at intermediate redshifts (details are given by Thommes *et al.* these proceedings). Of the remaining 43 emission line galaxies, 35 show at least a marginal detection in one of the veto-filters and can therefore be identified as being either at $z = 0.24$ or $z = 0.63/0.68$ (see Thommes *et al.*). The remaining 8 galaxies have to be regarded as candidate primeval galaxies at $z = 5.7$. However, one of them lies close to a large galaxy at $z = 0.24$ and is presumably a HII region at that redshift. This leaves us with **7 good Ly-α candidates** at $z = 5.7$. They fall into two groups:

(i) Three candidates without any continuum flux shortward of the emission line at $\lambda \simeq 816$ nm, and

(ii) four candidates with a marginal detection in our red filter (*i.e.* R$\lesssim 26.0$).

The average spectra of both groups are displayed in Fig. 3. Before we discuss these Ly-α candidates and their possible contamination with foreground objects, we would like to point out several robust results:

---

[1] Typically the total integration at one FPI setting, or narrow band filter, consists of $n = 5 \ldots 20$ individual exposures. The photometric error ($= 1\sigma$) is determined from the scatter around the mean of the n flux values measured on the individual frames.

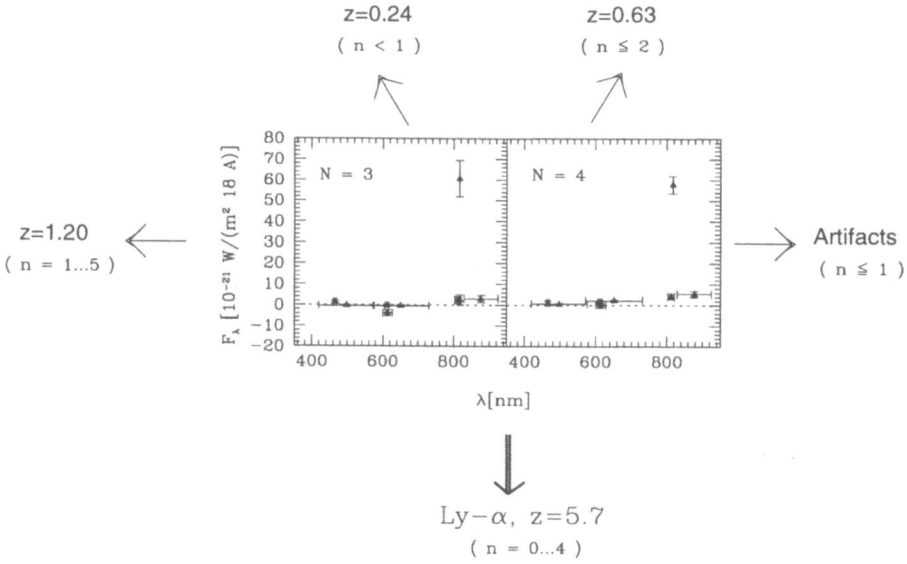

Fig. 3. Averaged spectra of the 3 Ly-α candidates without any continuum blue-wards of the emission line (left panel), and the 4 candidates with marginal detection in the R-band (right panel). This sample may contain 5 classes of objects: From the observed properties of artifacts and galaxies at $z \simeq 0.24$ we conclude that their contribution is negligible. Due to the under-exposed *veto*-filter images which aim to reject galaxies at $z \simeq 0.63$, we expect that one or two of these galaxies escaped identification. The most important contamination should be caused by galaxies at $z \simeq 1.2$ which show a strong [OII] line but no continuum above B$\simeq$ 26. They have to be separated from the true Ly-α galaxies by accurate slit spectroscopy.

+ Even in the averaged spectra there is no indication of any flux in either the blue band or one of the *veto*-filters. So the "average candidate" exactly follows the spectral behaviour expected for a primeval Ly-α galaxy at $z = 5.7$ (*cf.* Fig. 1).

+ There is a significant continuum step in going from the blue to the red side of the emission line – as expected for Ly-α galaxies due to foreground absorption by the Lyman-forest.

+ The average line flux $F_L = 6 \times 10^{-20}$ Wm^{-2} is well above the detection limit $(3.5\sigma \simeq 4 \times 10^{-20}$ Wm$^{-2})$, excluding that spurious objects at the noise limit play an important role.

+ Despite of the under-exposed *veto*-filter images the reliability of our *veto*-strategy seems already very promising: only $7/147 \simeq 5\%$ of the true emission line objects pass the "high redshift test". This should be compared with the expected rate of Ly-α galaxies $(0.2 \ldots 2\%$, *cf.* Fig. 2) and our detection rate of galaxies without any blue continuum ($43/147 \simeq 30\%$). Thus our *veto*-filter exposures already reject 83% of the Ly-α candidates, which would otherwise have to be sorted out by time-consuming spectroscopy.

Nevertheless, it would be premature to claim the unanimous detection of the first primeval Ly-α galaxies at $z = 5.7$, since we expect that our sample of 7 candidates is still contaminated by several types of foreground objects. The most likely contaminants are summarized in Fig. 3. Statistical considerations indicate that artifacts and emission line galaxies at $z \simeq 0.24$ and $z \simeq 0.63$ could hardly account for more than half of the candidates. The unknown fraction of galaxies at $z = 1.20$ with a very strong [OII] line but undetectable blue continuum could, however, well make up for the other half of the candidates. Therefore, high S/N-ratio ($\gtrsim 10$ at R$\simeq 25$) slit-spectroscopy at medium resolution ($\Delta\lambda \simeq 0.5\,\text{nm}$) is required to identify the true Ly-α galaxies among our candidates. This is clearly the domain of telescopes in the 8-10 m class !

On the other hand, since we are pretty optimistic that some objects will, in fact, turn out to be at $z = 5.7$ it is worthwhile to inspect the morphology of the primeval galaxy candidates. Exactly as expected from standard scenarios of galaxy formation, most of our candidates are clearly resolved ($\gtrsim 2''$ FWHM, see typical examples in Fig. 4). This opens up very exciting perspectives for detailed investigations of their kinematics with very large ground-based telescopes.

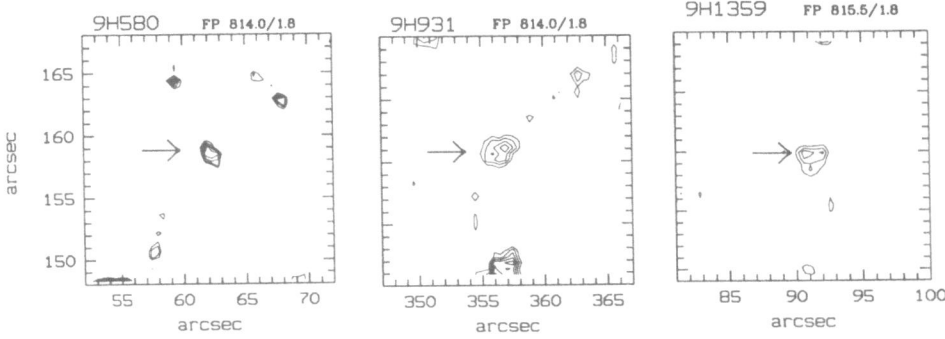

Fig. 4. Emission line morphology of three typical Ly-α candidates. Candidate 9H580 represents the 2 objects which are essentially unresolved. Candidates 9H931 (no continuum) and 9H1359 (marginal detection in R) demonstrate that well extended morphologies are frequently found in both groups.

5   Summary and Future Perspective

Based on a minute subset of the entire CADIS data, we have established the feasibility of detecting the Ly-α emission of primeval galaxies at very high redshifts $z \gtrsim 5$. Although the first seven candidates presented here still have to pass the final test – slit-spectroscopy at about 0.5 nm resolution – it is already quite likely that CADIS will eventually produce a statistically meaningful sample of Ly-α bright galaxies at $4.7 < z < 6.6$. After the first true Ly-α galaxies have been confirmed, we hope to be in the position to design a stricter criterion based on the CADIS multi-color data to separate [OII] galaxies (at $0.9 < z < 1.4$) from the most promising Ly-α candidates.

Thus, until mid-1998 we expect to produce large and rather clean samples of Ly-α galaxies (with $\lesssim 50\%$ contamination by foreground emission line galaxies). At about this time ISAAC and FORS should become fully operational. Sensitivity, spectral resolution, and (combined) wavelength coverage of both first generation VLT instruments are ideally suited to attack the most important issues in primeval galaxy research: The stellar spectral energy distribution and the kinematics of the emission line gas. The latter should give a first insight into the amount of matter aggregated in the object and the 3-dimensional structure of the newly forming galaxies. We think that the CADIS schedule is perfectly in phase with the VLT project. This should allow us to capitalize, from the very first day, on the unique opportunities of the VLT for one of the most fascinating projects of modern astronomy: The direct observational study of galaxy formation in the young universe !

References

Baron, E. and White, S.: 1987, *Astrophys. J.* **322**, 585.

Fockenbrock, R. *et al.*: 1996, *these proceedings*, p. ??

Meier, D.: 1976, *Astrophys. J.* **207**, 343.

Partridge, R.B. and Peebles, P.: 1967, *Astrophys. J.* **147**, 868.

Pritchet, C.: 1994, *Publ. Astron. Soc. Pac.* **106**, 1052.

Steinmetz, M. and Müller E.: 1995, *Mon. Not. R. Astron. Soc.* **276**, 549

Thommes, E.: 1996, *PhD Thesis*, Universität Heidelberg.

Thommes, E. and Meisenheimer, K.: 1995, in "Galaxies in the Young Universe", Hippelein *et al.* (eds.), Springer Lecture Notes, No. 463, p. 242.

Thommes, E. *et al.*: 1996, *these proceedings*, p. ??

Thompson, D., Djorgovski, S. and Trauger, J.: 1995, *Astron. J.* **110**, 963.

Faint Emission Line Galaxies Detected in CADIS

E. Thommes, R. Fockenbrock, H. Hippelein, K. Meisenheimer, H.-J. Röser

Max-Planck-Institut für Astronomie, Königstuhl 17, D-69117 Heidelberg, Germany

1   Introduction

The **C**alar **A**lto **D**eep **I**maging **S**urvey (CADIS) is a very deep emission line survey (using a Fabry-Pérot (FP)) combined with deep broad- and medium-band photometry (for details see Meisenheimer et. al., these proceedings). Though this survey project is specifically designed to detect primeval galaxies it will in addition produce a large data base for investigations of faint galaxies at intermediate redshifts ($0.2 < z < 1.2$). In this contribution we present some first results concerning these forground objects which we got from the first data recorded with the CADIS strategy. These data were taken with the 2.2m telescope at Calar Alto in the CADIS field 9H. Due to delays in getting the 2k$\times$ 2k CCDs, we employed an a 1k$\times$1k CCD (field of view 8'$\times$ 8'). We got four FP settings in the wavelength region 814nm to 818.5nm (resolution=1.8nm). Every setting consists of 7 individual exposures of 1500 s integration. We reached a 5σ detection limit of $S_{lim}(5\sigma) \approx 5 \times 10^{-20}\mathrm{W/m^2}$. To get an estimate of the continuum near the emissions lines detected in the FP, we did exposures with a filter $\lambda/\Delta\lambda$=812/17 nm ($F_{lim}(5\sigma) \approx 5.8 \times 10^{-21}\mathrm{W/(m^2nm)}$). The FP exposures were supplemented by broad band exposures with the filters BV (centered at 500 nm, 5σ limit $\approx 25.^m8$), R$_c$ (5σ limit $\approx 25.^m0$) and I (5σ limit $\approx 23.^m1$). Further narrow band exposures with the filters 466/9 ($F_{lim}(5\sigma) \approx 10.7 \times 10^{-21}\mathrm{W/(m^2nm)}$), 612/10 ($F_{lim}(5\sigma) \approx 10.8 \times 10^{-21}\mathrm{W/(m^2nm)}$) and 614/28 ($F_{lim}(5\sigma) \approx 6.6 \times 10^{-21}\mathrm{W/(m^2nm)}$) enable to detect further emission lines of foreground objects. The data reduction technique is outlined in Fockenbrock et. al. (these proceedings). Emission line objects are selected by the requirement, that they have a 5 σ detection in at least one FP wavelength setting and that the line flux exceeds the contiuum by more than 3.5σ.

2   Results of Galaxy Number Count Statistics

As a first test for the photometry and calibration of the data we did galaxy number counts on the BV and R$_c$ exposures. For the number density of galaxies per $\square^\circ$ and 0.5 mag-bin as a function of the BV magnitude we found
$$\lg(N[\square^\circ 0.5 mag]) = (0.49 \pm 0.02) \times BV[mag] - 7.89 \pm 0.09.$$
Number counts on the R$_c$ exposure gave
$$\lg(N[\square^\circ 0.5 mag]) = (0.36 \pm 0.04) \times R_c[mag] - 4.56 \pm 0.11.$$ These results are in excellent agreement with previous results by Metcalfe et.al. (1991, *Mon. Not. R. Astron. Soc.* **249**, 498-522. We used the same magnitude intervals for our fits as they used.)

3   Redshift-Classification of the Emission-Line Objects

3.1   Classification Procedure

With the FP information and the additional filter information it was possible to assign a redshift to most of the objects with an significant emission line in the FP. We did this in the following way. First we inspected the photometric spectra bye eye and made an estimate for the line identification in the FP. Due to the narrow velocity resolution of the FP the estimate for the line identification gives an accurate estimate for the redshift z. With this redshift estimate, we calculated the exact wavelength positions of other lines, which could be detected in the narrow band filters. The location of any emission line in the transmission characteristic of the filters was taken into account to deduce the true line fluxes. The plausibility of an identification was checked by comparing the observed line ratios (or their limits) with those of nearby emission line galaxies and HII regions. Fig. 2 and 3 show the photometric spectra of representative emission line objects. The most likely line identification according to the described classification procedure are indicated.

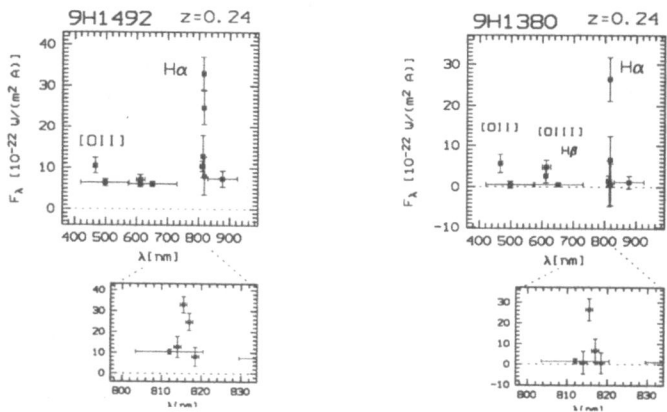

Fig. 1. *Photometric spectra of representative emission line objects. The most likely line identification and the corresponding redshifts are indicated.*

3.2   Results of Classification

To 62 emission-line objects we assigned a redshift in the intervall [0.239,0.249] (like the examples in the Fig. 1). It is very interesting, that 22 of these only have bearly detectable continuum in the BV and R_c filter. Without the additional narrow band filter information, they would be mistaken for primeval galaxy candidates (see Meisenheimer, these proceedings). 47 objects are classified as

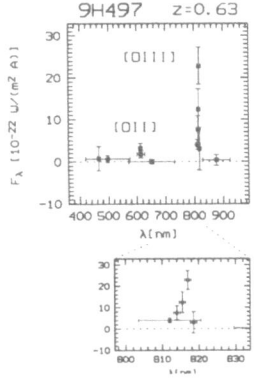
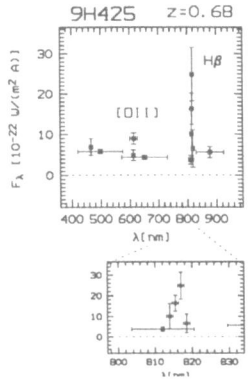

Fig. 2. *Photometric spectra of representative emission line objects. The most likely line identification and the corresponding redshifts are indicated. Notice that the [OII] line in the spectrum of the object 9H425 (right diagramm) is detected in the 614/28 filter but not in the filter 612/10. For that reason the line in the FP is identified as the Hβ line (and not as [OIII]).*

objects with a redshift in the range [0.624,0.687] (like the objects in Fig. 2). Again, 13 of these objects have no detectable continuum in the BV filter. 38 emission line objects show only an emission line in the FP-bands. For these objects it is very difficult to assign a redshift. It is quite possible that many of them are galaxies at a redshift of $z \approx 1.2$ from which we detected their [OII] line in the FP. In addition, this class of objects with only an emission line in the FP contains 8 primeval galaxy candidates (see Meisenheimer, these proceedings). A nice way to illustrate the main features of the spectra of a certain class of objects is to average the spectra of this class. Fig. 3 shows the averaged spectra of objects without blue continuum in the redshift bins $z \approx 0.24$ and $z \approx 0.64$. The spectra are already transformed to the rest-frame of the objects. The line ratios are very reasonable. This proofs, that our classification procedure is correct on average.

4   Discussion

The data presented here are only a very small part of what is planed for every CADIS field. Specifically the additional narrow band exposures are underexposed by more than a factor 4. Nevertheless, the data show that the concept of CADIS works (see also Meisenheimer et.al., these proceedings). In addition, it seems that we have detected a new class of objects: Many of the detected emission line galaxies (35 % in the z-bin around 0.24 und 28% in the z-bin arround 0.64) have no detectable continuum in the BV and R band ($W_{[OII]} > 40$nm in the rest frame). These objects are overseen in broadband selected redshift surveys. In Fig. 4 we compare the luminosity function as derived from our data for the redshift bins 0.24 and 0.64 with the results of the Autofib redshift survey (see Colles et.al., 1996 these proceedings). It is very astonishing that the lower limits

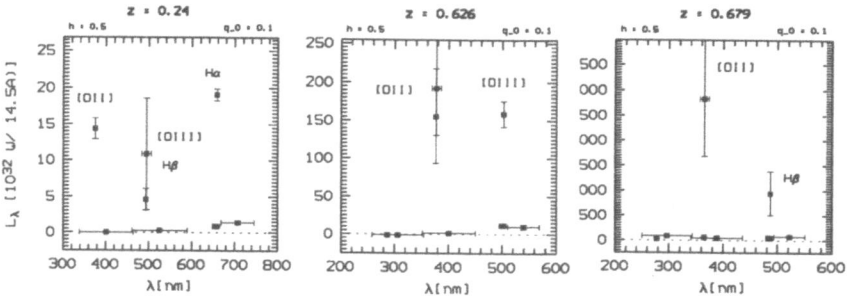

Fig. 3. Left: *Averaged spectrum of objects in the redshiftbin [0.239,0.249] (Hα in FP) without blue continuum.* **Middle:** *Averaged spectrum of objects in the redshiftbin [0.642,0.637] ([OIII]500.7 in FP) without blue continuum.* **Right:** *Averaged spectrum of objects in the redshiftbin [0.673,0.687] (Hβ in FP) without blue continuum. Notice that the [OII] line is detected in the 614/28 filter but not in the filter 612/10. For that reason the line in the FP is identified as the Hβ line (and not as [OIII]).*

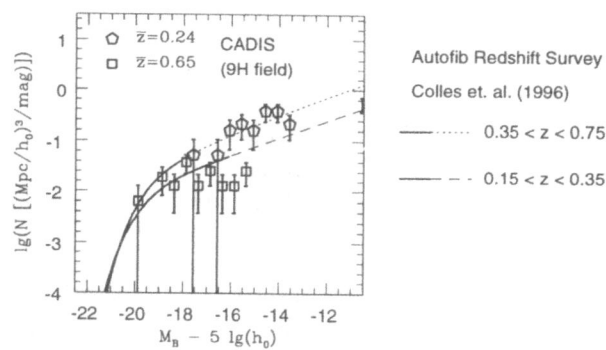

Fig. 4. *Comparison of the luminosity function determined from the first CADIS data in the field 9H for the redshift bins $z \approx 0.24$ and $z \approx 0.64$ ($q_0 = 0.5$) with the results of the Autofib redshift survey (see Colles et.al., 1996, these proceedings). The solid lines are Schechter function fits to the data of the Autofib survey. The dotted and dashed extensions show the extrapolation of these Schechter functions to magnitudes below the limits of the Autofib survey.*

to the luminosity function deduced from this pure emission line CADIS data are in such good agreement with the results of the Autofib redshift survey. But there are already some hints for differences. Although we included only emission line galaxies, our datapoints for $z \approx 0.24$ show significantly higher densities at faint absolute magnitudes than the Autofib redshift survey. The covered spatial angle of our data is yet too small to draw definite conclusions, but they demonstrate the power of the CADIS method to attack the problem of redshift-evolution of the luminosity function especially at very faint absolute magnitudes.

A Global Approach to Star and Galaxy Formation

S. Michael Fall

Space Telescope Science Institute, 3700 San Martin Drive, Baltimore, MD 21218, USA

Abstract. We present a new method to compute the cosmic emissivity $\mathcal{E}_\nu$ and background intensity J_ν. Our method is based entirely on data from quasar absorption-line studies, namely, the comoving density of HI and the mean metallicity and dust-to-gas ratio in damped Lyα galaxies. These observations, when combined with models of cosmic chemical evolution, are sufficient to determine the comoving rate of star formation as a function of redshift. From this, we compute $\mathcal{E}_\nu$ and J_ν using stellar population synthesis models. Our method includes a self-consistent treatment of the absorption and reradiation of starlight by dust. In all of our calculations, the near-UV emissivity declines rapidly between $z \approx 1$ and $z = 0$, in agreement with estimates from the Canada-France Redshift Survey. The background intensity is consistent with a wide variety of observational limits and with a tentative detection at far-IR wavelengths.

1 Introduction

The mean emissivity of the universe $\mathcal{E}_\nu$ and the mean intensity of background radiation J_ν are important cosmological probes. The former is defined here as the power radiated per unit frequency per unit comoving volume, while the latter is defined as the power received per unit frequency per unit area of detector per unit solid angle of sky. These quantities, of course, are not independent; J_ν is given by an integral of $\mathcal{E}_\nu$ over redshift. It is likely that $\mathcal{E}_\nu$ and J_ν are dominated at near-UV, optical, and near-IR wavelengths by the direct radiation from stars and at far-IR wavelengths by reradiated starlight from the dust within galaxies. Thus, they contain potentially valuable information about the global history of star formation. Observationally, $\mathcal{E}_\nu$ and J_ν have proven to be elusive; until recently, it was only possible to estimate $\mathcal{E}_\nu$ at $z \ll 1$ and to place weak constraints on J_ν. Theoretically, they have also proven to be elusive. Most calculations of $\mathcal{E}_\nu$ and J_ν are based on the properties of present-day galaxies and some assumed evolution in the past, often specified by several free parameters (see Lonsdale 1995 for a review).

Here, we describe a new method to compute $\mathcal{E}_\nu$ and J_ν (Fall, Charlot, & Pei 1996); in essence, we predict the "emission history" of the universe from its "absorption history." From the absorption lines in the spectra of distant quasars, it is possible in principle to determine the global rates of gas consumption, metal production, and hence star formation in galaxies at redshifts up to $z \approx 4$. The focus here is on the damped Lyα galaxies, which contain most of the cool, neutral gas in the universe and appear to be the progenitors of present-day galaxies

(Lanzetta, Wolfe, & Turnshek 1995). We have already combined absorption-line data with models of cosmic chemical evolution to compute the comoving densities of stars and dust in damped Lyα galaxies (Pei & Fall 1995). Here, we employ stellar population synthesis models to compute $\mathcal{E}_\nu$ and J_ν, including the absorption and reradiation of starlight by dust. In this first illustration of the method, we have deliberately kept the analysis as simple as possible in order to highlight the main ideas and assumptions. Moreover, the existing data on damped Lyα galaxies are too sparse to warrant a more elaborate analysis.

2   Models

Our first task is to relate the cosmic emissivity and background intensity, $\mathcal{E}_\nu$ and J_ν, to the comoving densities of stars and dust, Ω_s and Ω_d (expressed here in units of the present critical density, $\rho_c = 3H_0^2/8\pi G$). We ignore other sources of radiation (primarily active galactic nuclei) and exclude photons that are absorbed in the galaxies in which they were produced. Thus, for the stellar part of the emissivity, we write

$$\mathcal{E}_{s\nu} = (1 - A_\nu)\mathcal{E}_{s\nu}^0, \tag{1}$$

where A_ν is the mean fraction of absorbed photons, and $\mathcal{E}_{s\nu}^0$ is the emissivity before absorption. The latter is given by

$$\mathcal{E}_{s\nu}^0(t) = \rho_c \int_0^t dt' S_\nu(t - t') \frac{\dot{\Omega}_s(t')}{1 - R(t')}, \tag{2}$$

where $S_\nu(\Delta t)$ is the power radiated per unit frequency per unit initial mass by a generation of stars with an age Δt, R is the returned fraction, and the dot denotes differentiation with respect to cosmic time. We compute S_ν and R using the latest version of the Bruzual-Charlot models (described by Charlot, Worthey, & Bressan 1996). In most of the calculations reported here, the stellar initial mass function (IMF) is assumed to be a power law, $\phi(m) \propto m^{-(1+x)}$, with $x = 1.5$ and upper and lower cutoffs at $100 M_\odot$ and $0.1 M_\odot$. We further assume that all H-ionizing photons are absorbed in the local interstellar medium, that 68% of them are converted to Lyα photons (the fraction appropriate for case B recombination in gas at 10^4 K), that all of these are absorbed by dust (as a consequence of resonant scattering by HI), and that the remaining energy is radiated uniformly in wavelength between 3000 and 7000 Å (a range that includes most of the relevant emission lines). This treatment of ionizing radiation is consistent with far-UV observations of starburst galaxies (Leitherer et al. 1995).

We relate the mean fraction of photons absorbed by dust to the other properties of galaxies as follows. For simplicity, we assume that the stars and dust have the same spatial distributions within galaxies (so that the source function is constant along any ray), and we ignore the influence of scattering on absorption (which should be a good approximation for average quantities such as A_ν). In this case, the fraction of photons absorbed along a ray that intercepts a galaxy with an optical depth τ_ν is given by

$$a(\tau_\nu) = 1 - \tau_\nu^{-1}[1 - \exp(-\tau_\nu)]. \tag{3}$$

We now define $\phi(\tau_\nu)d\tau_\nu$ to be the fraction of such rays with optical depths between τ_ν and $\tau_\nu + d\tau_\nu$ when all galaxies and all positions and directions within them are considered. This enables us to express the mean fraction of photons absorbed by dust in the form

$$A_\nu = \frac{\int_0^\infty d\tau_\nu \tau_\nu \phi(\tau_\nu) a(\tau_\nu)}{\int_0^\infty d\tau_\nu \tau_\nu \phi(\tau_\nu)}. \tag{4}$$

The optical depth and HI column density N along a ray are related by $\tau_\nu = \kappa_\nu k_m m_H N$, where κ_ν is the opacity (i.e., mass absorption coefficient), k_m is the dust-to-HI mass ratio, and m_H is the mass of an H atom. Thus, for a single value of k_m, the (true) distributions of τ_ν and N are related by $\phi(\tau_\nu) \propto f(N)$. For consistency with the models of chemical evolution, we adopt the gamma distribution, $f(N) = (f_*/N) \exp(-N/N_*)$ (Pei & Fall 1995). This implies $\phi(\tau_\nu) \propto \tau_\nu^{-1} \exp(-\tau_\nu/\tau_{*\nu})$ with $\tau_{*\nu} = \kappa_\nu k_m m_H N_*$. Equations (3) and (4) and the relations $\Omega_d = k_m \Omega_{HI}$ and $\Omega_{HI} = (8\pi G m_H/3cH_0) f_* N_*$ then give

$$A_\nu = 1 - \tau_{*\nu}^{-1} \ln(1 + \tau_{*\nu}), \tag{5}$$

$$\tau_{*\nu} = (3cH_0/8\pi G f_*)\kappa_\nu \Omega_d. \tag{6}$$

If the stars and dust have different spatial distributions or there is a dispersion in the dust-to-gas ratio, these expressions should be regarded as approximations. We have checked that our results are not sensitive to the particular form of A_ν.

The starlight absorbed by dust is reradiated thermally. We approximate the corresponding emissivity by that of a single blackbody:

$$\mathcal{E}_{d\nu} = 4\pi \rho_c \Omega_d \kappa_\nu B_\nu(T_d). \tag{7}$$

The effective temperature of the dust T_d is then determined by the condition of energy balance

$$4\pi \rho_c \Omega_d \int_0^\infty d\nu \kappa_\nu [B_\nu(T_d) - B_\nu(T_{CMB})] = \int_0^\infty d\nu A_\nu \mathcal{E}_{s\nu}^0, \tag{8}$$

where $T_{CMB} = 2.73(1 + z)$K is the temperature of the cosmic microwave background radiation. Equation (7) should be a good approximation at wavelengths beyond the peak $[\lambda_{max} \approx 150(T_d/20 \text{ K})^{-1} \mu\text{m}]$, where the emission is likely to be dominated by large amounts of cool dust, but a poor approximation at shorter wavelengths, where the emission is likely to be dominated by small amounts of warm dust (e.g., small, transiently heated grains). We return to this point later. For the optical properties of the dust, we adopt the Draine & Lee (1984) model but with the proportions of graphite and silicates adjusted so as to reproduce the mean UV extinction curve in the Large Magellanic Cloud (Pei 1992). In this model, the opacity at long wavelengths has the form $\kappa_\nu \propto \nu^2$, and the left-hand side of equation (8) is proportional to $\Omega_d(T_d^6 - T_{\mathrm{CMB}}^6)$. Given Ω_s and Ω_d, we can now compute the total emissivity $\mathcal{E}_\nu = \mathcal{E}_{s\nu} + \mathcal{E}_{d\nu}$ from the equations above, and the corresponding background intensity at $z = 0$ from

$$ J_\nu = \frac{c}{4\pi} \int_0^\infty dz \mathcal{E}_{(1+z)\nu} \left| \frac{dt}{dz} \right|. \tag{9} $$

Here, we have neglected absorption between the sources of radiation and the observer. This is appropriate because ionizing photons are assumed to be absorbed locally and because relatively few non-ionizing photons are absorbed by the dust in intervening galaxies.

The comoving densities of stars and gas in galaxies, Ω_s and Ω_g, and the mean metallicity in the interstellar medium Z, including dust, are governed by the equations of cosmic chemical evolution:

$$ \dot{\Omega}_g + \dot{\Omega}_s = \dot{\Omega}_f, \tag{10} $$

$$ \Omega_g \dot{Z} - y\dot{\Omega}_s = (Z_f - Z)\dot{\Omega}_f. \tag{11} $$

Here, y is the IMF-averaged yield, and the source terms on the right represent the inflow or outflow of gas with metallicity Z_f at a rate $\dot{\Omega}_f$. For purposes of illustration, we assume that just over half of the heavy elements are locked up in dust grains and that the ionized and molecular components of the interstellar medium are negligible, i.e., $\Omega_d = 0.55Z\Omega_g$ and $\Omega_g = 1.3\Omega_{\mathrm{HI}}$. The constancy of the dust-to-metals ratio can also be expressed in the form $k(z)/Z(z) = k(0)/Z(0) = 0.8/Z_\odot$, where k is the ratio of τ_B, the extinction optical depth in the rest-frame B band, to $N/10^{21}$ cm^{-2} (see Table 2 of Pei 1992). This agrees to within a factor of two with the dust and metal content of present-day galaxies and damped Lyα galaxies at $\bar{z} = 2.2$ (Pei, Fall, & Bechtold 1991; Pettini et al. 1994). Given Ω_{HI} and some assumptions about $\dot{\Omega}_f$ and Z_f, it is now possible to solve equations (10) and (11) for Ω_s and Ω_d. We adopt the solutions presented by Pei & Fall (1995). These are of three types: a closed-box model ($\dot{\Omega}_f = 0$), a model with inflow of metal-free gas ($\dot{\Omega}_f = +\nu\dot{\Omega}_s$, $Z_f = 0$), and a model with outflow of metal-enriched gas ($\dot{\Omega}_f = -\nu\dot{\Omega}_s$, $Z_f = Z$). The adjustable parameters are the "initial" comoving density of gas in galaxies $\Omega_{g\infty}$ and the relative inflow or outflow rate ν. The models include self-consistent corrections for the damped Lyα galaxies that are missing from optically selected samples as a result of the obscuration of background quasars (Fall & Pei 1993). The models were designed to reproduce the observed comoving density of HI in damped Lyα galaxies at $0 < z \lesssim 4$ (Lanzetta et al. 1995; Storrie-Lombardi, McMahon, & Irwin 1996). They also reproduce the observed mean metallicity $Z \approx 0.1Z_\odot$ at $\bar{z} = 2.2$ (Pettini et al. 1994) and are consistent with the average properties of present-day galaxies. The results presented here, unless otherwise noted, are based on models with $\Omega_{g\infty} = 4 \times 10^{-3}h^{-1}$, $\nu = 0.5$, and $h = 0.5$, $q_0 = 0.5$, $\Lambda = 0$ (with $h \equiv H_0/100$ km s^{-1} Mpc^{-1}). For comparison, we also display results from the analogous models without dust.

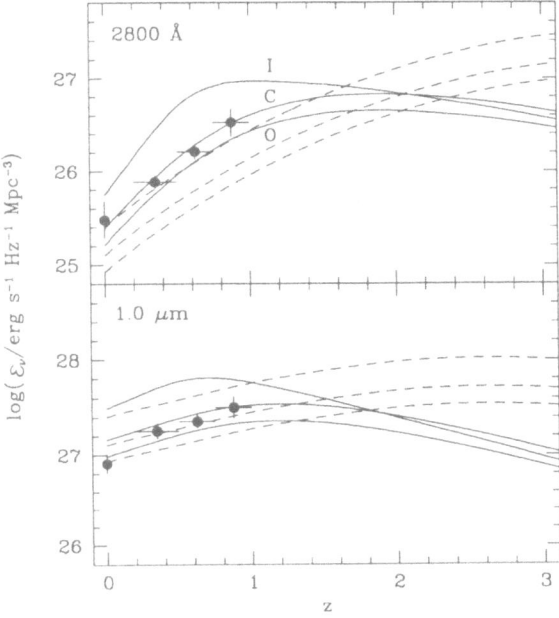

Fig. 1. Cosmic emissivity $\mathcal{E}_\nu$ as a function of redshift z at rest-frame wavelengths of 2800 Å (*top*) and 1.0 μm (*bottom*). The solid curves are from the closed-box (C), inflow (I), and outflow (O) models with dust, while dashed curves are from the analogous models without dust. The data points with error bars are estimates from local samples and the Canada-France Redshift Survey (Lilly et al. 1996).

3 Results

Figure 1 shows the evolution of $\mathcal{E}_\nu$ at rest-frame wavelengths of 2800 Å and 1.0 μm. The former is dominated by the light from young, massive stars and is thus nearly a direct indicator of current star formation, while the latter includes the light from a wide range of stellar types and thus reflects a combination of past and current star formation. The solid curves in Figure 1 represent the closed-box (C), inflow (I), and outflow (O) models with dust, while the dashed curves represent the analogous models without dust. The data points with error bars are estimates by Lilly et al. (1996) from local samples of galaxies ($z = 0$) and from the Canada-France Redshift Survey ($z = 0.3, 0.6,$ and 0.9). These include corrections for incompleteness at faint magnitudes. The observed emissivity at 2800 Å decreases by an order of magnitude between $z = 0.9$ and $z = 0$, indicating a similar decrease in the comoving rate of star formation. Evidently, the slopes of the predicted and observed $\mathcal{E}_\nu(z)$ relations agree to within the uncertainties, both at 2800 Å and 1.0 μm. The amplitudes match for the closed-box and outflow models with dust but not for the inflow model with dust. However, the predicted amplitudes are somewhat sensitive to the adopted IMF in

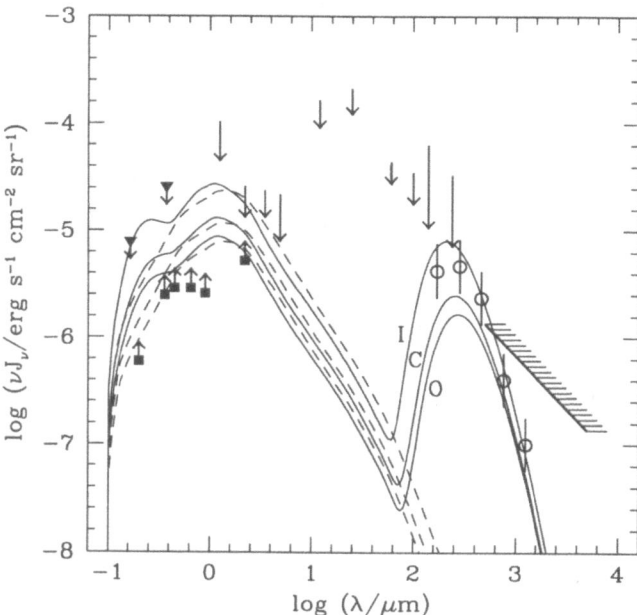

Fig. 2. Background intensity J_ν at $z = 0$ times frequency ν as a function of wavelength λ. The solid curves are from the closed-box (C), inflow (I), and outflow (O) models with dust, while the dashed curves are from the analogous models without dust. The symbols with arrows and the hatched line are observational limits. The open circles with vertical bars are a tentative detection derived from $COBE$/FIRAS data (Puget et al. 1996).

the population synthesis models. If the IMF slope were increased to $x = 2.0$ or the lower cutoff were decreased to $0.01\,M_\odot$, the predicted amplitudes would be reduced by factors of $2 - 3$. With this freedom to adjust the IMF, the emissivities in the inflow model with dust can be brought into rough agreement with the observed emissivities. We count this as a success because models with very different histories of star formation cannot be made consistent with the observations for any choice of the IMF. For example, models with $\Omega_{g\infty} \lesssim 1 \times 10^{-3} h^{-1}$ or $\Omega_{g\infty} \gtrsim 8 \times 10^{-3} h^{-1}$ have emissivities that decline too slowly or too rapidly. Our predicted emissivities at $z > 2$ are consistent with lower limits derived from recent HST and Keck observations (Madau et al. 1996; Steidel et al. 1996).

Figure 2 shows νJ_ν as a function of wavelength. Again, the solid curves represent the closed-box (C), inflow (I), and outflow (O) models with dust, while the dashed curves represent the analogous models without dust. The symbols with arrows and the hatched line represent various observational limits (see Fall et al. 1996 for references). Evidently, our models are consistent with these constraints, over four and a half decades in wavelength. The near-UV background is higher in the models with dust because they have more star formation at low redshifts than the models without dust (see Figure 1 here and Pei & Fall 1995). The open circles with vertical bars in Figure 2 represent the tentative detection of an extragalactic far-IR background by Puget et al. (1996). This result, derived from *COBE*/FIRAS data, is uncertain because it depends critically on the removal of foreground emission by interplanetary and interstellar dust. Our models with dust are consistent with this detection. The effective temperature of the dust remains near $T_d \approx 20$ K at $z \gtrsim 1$ and then decreases to $T_d \approx 15$ K at $z = 0$. This produces a peak in νJ_ν at $\lambda \approx 240$ μm. The valley at $\lambda \approx 60$ μm is an artifact of our assumption that the spectrum of the dust emission is that of single blackbody [see equation (7)]. We have experimented with more realistic, two-temperature models and find that the valley can easily be filled in by emission from small amounts of warm dust with little effect on the emission at longer wavelengths. This indicates that the far-IR background is a robust feature of our models. The amplitude of J_ν depends on the IMF in the same way as the amplitude of $\mathcal{E}_\nu$; in particular, the background would be weaker at all wavelengths if the proportion of massive stars were reduced. Moreover, the values of $\Omega_{g\infty}$ that are consistent with the data on J_ν are essentially the same as those that are consistent with the data on $\mathcal{E}_\nu$.

4   Discussion

We have computed the cosmic emissivity $\mathcal{E}_\nu$ and background intensity J_ν with input only from absorption-line studies of damped Lyα galaxies. These objects have $N \gtrsim 10^{20}$ cm^{-2} (by definition) and are the probable sites of most star formation in the universe. They appear to be the progenitors of present-day galaxies, although many of their properties remain to be determined. In particular, we do not yet know the sizes and morphologies of the damped Lyα galaxies. It is possible that most of those at low redshifts are disks, while most of those at high redshifts are spheroids. We emphasize that the results presented here are not affected by such issues. The reason for this is that all of the quantities required in our analysis can be computed directly from statistics of the absorption along random lines of sight [e.g., $\Omega_{\rm HI}$ is given by the integral of $Nf(N)$ over N]. The primary restriction on our results is that they do not include any galaxies that consumed their HI before $z \approx 4$, the highest redshift probed systematically by quasar absorption-line studies. Since our calculations include several approximations and idealizations, we regard them as illustrative rather than definitive. Nevertheless, we find reasonable agreement with all of the available data on $\mathcal{E}_\nu$ and J_ν without any fine-tuning of parameters. In particular, our models reproduce the rapid decline in the near-UV emissivity between $z \approx 1$ and $z = 0$ reported by Lilly et al. (1996). They are also consistent with a wide variety of observational limits on the extragalactic background and a tentative detection at far-IR wavelengths reported by Puget et al. (1996). It is therefore possible that this background is produced mainly or entirely by galaxies at $z \lesssim 4$.

The material presented in this article has recently been published elsewhere (Fall et al. 1996). I am grateful to my coauthors, Stéphane Charlot and Yichuan Pei, for permission to quote liberally from the original article.

References

Charlot, S., Worthey, G., & Bressan, A. 1996, ApJ, 457, 625

Draine, B.T., & Lee, H.M. 1984, ApJ, 285, 89

Fall, S.M., Charlot, S., & Pei, Y.C. 1996, ApJ, 464, L43

Fall, S.M., & Pei, Y.C. 1993, ApJ, 402, 479

Lanzetta, K.M., Wolfe, A.M., & Turnshek, D.A. 1995, ApJ, 440, 435

Leitherer, C., Ferguson, H.C., Heckman, T.M., & Lowenthal, J.D. 1995, ApJ, 454, L19

Lilly, S.J., Le Fevre, O., Hammer, F., & Crampton, D. 1996, ApJ, 460, L1

Lonsdale, C.J. 1995, in Extragalactic Background Radiation, ed. D. Calzetti, M. Livio, & P. Madau (Cambridge University Press), 145

Madau, P., Ferguson, H.C., Dickinson, M.E., Giavalisco, M., Steidel, C.C., & Fruchter, A. 1996, MNRAS, in press

Pei, Y.C. 1992, ApJ, 395, 130

Pei, Y.C., & Fall, S.M. 1995, ApJ, 454, 69

Pei, Y.C., Fall, S.M., & Bechtold, J. 1991, ApJ, 378, 6

Pettini, M., Smith, L.J., Hunstead, R.W., & King, D.L. 1994, ApJ, 426, 79

Puget, J.-L., Abergel, A., Bernard, J.-P., Boulanger, F., Burton, W.B., Désert, F.-X., & Hartmann, D. 1996, A&A, 308, L5

Steidel, C.C., Giavalisco, M., Pettini, M., Dickinson, M., & Adelberger, K.L. 1996, ApJ, 462, L17

Storrie-Lombardi, L.J., McMahon, R.G., & Irwin, M.J. 1996, MNRAS, in press

Metal Absorption from Galaxies in the Process of Formation

Martin G. Haehnelt

Max-Planck-Institut für Astrophysik, Karl-Schwarzschild-Straße 1, 85740 Garching, Germany

Abstract. In a hierarchical cosmogony present-day galaxies build up by continuous merging of smaller structures. At a redshift of three the matter content of a typical present-day galaxy is split into about ten individual protogalactic clumps. Numerical simulations show that these protogalactic clumps have a typical distance of about 100 kpc, are embedded in a sheet-like structure and are often aligned along filaments. Artificial QSO spectra were generated from hydrodynamical simulations of such regions of ongoing galaxy formation. The metal and hydrogen absorption features in the artificial spectra closely resemble observed systems over a wide range in HI column density. Detailed predictions of the column density as a function of impact parameter to protogalactic clumps are presented for HI,CII,CIV,SiIV,NV and OVI. The expected correlations between column densities of different species and their role in understanding the physical properties of the gas from which galaxies form are discussed. The model is able to explain both high-ionization multi-component heavy-element absorbers and damped Lyman alpha systems as groups of small protogalactic clumps.

1 Introduction

While at low redshift metal absorption systems have been convincingly demonstrated to arise in the haloes of rather normal galaxies (Boisse & Bergeron 1991, Steidel 1995), much less is known about the nature of metal absorption system at redshifts $z \gtrsim 2$ (Sargent, Boksenberg & Steidel 1988; Petitjean & Bergeron 1994; Aragon-Salamanca et al. 1994). Recently, it was shown that the prominent complex CIV absorption features observed at high redshift can be well reproduced by the absorbing properties of regions in which galaxies form by hierarchical merging (Haehnelt, Steinmetz & Rauch 1996). Here we present further results concerning the metal absorption properties of such regions.

2 Numerical Simulations

The simulations were performed using GRAPESPH (Steinmetz 1996). The cosmological background model is a $\Omega = 1$, $H_0 = 50$ km s^{-1} Mpc^{-1} cold dark matter (CDM) cosmogony with a normalization of $\sigma_8 = 0.63$. The baryon fraction is $\Omega_b = 0.05$. The gas particle masses is 5×10^6 M$_\odot$. The high resolution region of the simulation box is about 5.5 comoving Mpc across and contains three galaxies with circular velocities between 100 and 200 km s^{-1} at redshift zero. For the UV

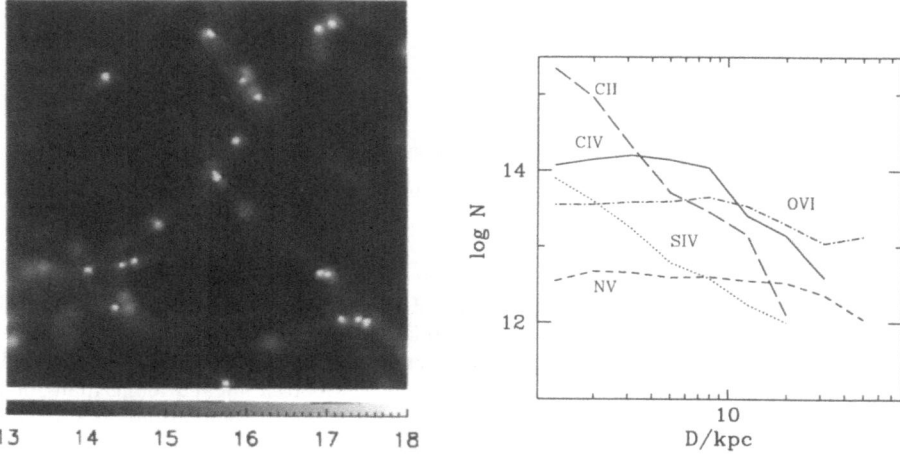

Fig. 1. The left panel shows the HI column density (log N(HI)) distribution of the inner 700 kpc (proper length) of the simulation box projected along one of the axis at $z = 3.07$. In the right panel the column densities of a set of ions is plotted as a function of the impact parameter to a typical protogalactic clump in the simulation. A homogeneous metallicity of 0.01 solar is assumed.

background a power-law spectrum with spectral index $\alpha = -1.5$, a normalization of $J_{21}(z=3) = 0.3$ and a redshift dependence as given by Vedel, Hellsten & Sommer–Larsen (1994) was assumed. The photoionization code CLOUDY (Ferland 1993) was used to calculate the ionization state of the gas (see the contribution by Steinmetz in this volume for a more detailed description of the simulation).

3   The Gas Distribution

There are about 20 collapsed protogalactic clumps (PGC's) aligned in a filamentary matrix which itself is embedded in a sheet-like structure (Figure 1a). Figure 2a shows the observed column density distribution of HI and a set of other ionic species observable in absorption systems of intermediate HI column density at high redshift. The crosses show the observed HI distribution $(f(N)N)$ obtained by Petitjean et al. (1993). Apart from the high column density end the shape of the observed and simulated HI distribution correspond rather well. There self-shielding becomes important which is not taken into account in the simulation. The normalization was freely adjusted as the simulation box is too small to be a fair sample of the universe. In Figure 2b the typical column density (mean log N) is shown as function of absorption-weighted overdensity $\delta = |\rho - \bar{\rho}|/\bar{\rho}$. There is a tight correlation between column density and density confirming the

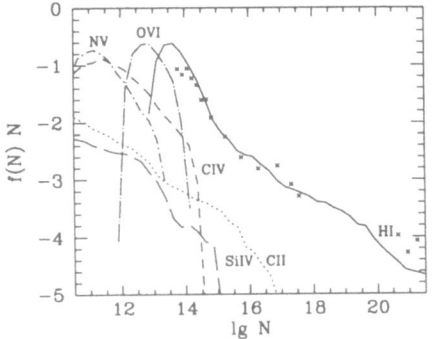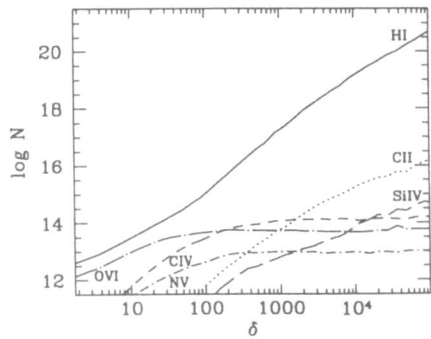

Fig. 2. The left panel shows the fraction of the projected simulation box which has column density N (per ln N). In the right panel the mean log N is plotted against the absorption weighted mean overdensity along the line-of-sight. A homogeneous metallicity of 0.01 solar is assumed.

visual impression from Figure 1a that higher column densities probe the centre of the collapsed PGC's. Low column densities arise from the more diffuse gas in sheets and filaments between the collpased objects. Figure 1b and 2b show that the spatial distribution of different ionic species differs. Species like CII and SIV are only strong in the dense inner regions of the PGC's and the column density falls off rather rapidly on scales of 10 kpc or less. Higher ionization species probe the gas further away from the PGC's. The high CIV column density extends to scales of typically 30 kpc. A closer inspection of the simulation shows that CIV is a good tracer of the filamentary matrix connecting the PGC's. NV and OVI probe even lower densities. Especially OVI is a good tracer of the diffuse intergalactic medium in the sheets between the PGC's.

4   Column Density Ratios of Different Ionic Species

In Figure 3 we show a selection of column density ratios as function of HI and CIV column density assuming a homogeneous metallicity of 0.01 solar. These ratios can be used as diagnostic of the UV radiation field and the metallicity of the gas. As consequence of the tight density-column density correlation the ratio of the species [CII/HI] and [SiIV/HI] drops rather fast towards lower HI column density. [CIV/HI] has a pronounced peak around $\log N(HI) \sim 14$ to 15 while [OVI/HI] rises rapidly towards small column densities. Recently, there have been measurements of CII, SIV, CIV and NV column densities down to $\log N(HI) =$ 14.5 by Soingaila & Cowie (1996). Even so the number of detected lines are small the measured [SiIV/CIV] seems to be larger than those shown in Figure 3b by a factor of 3 at the relevant CIV column density while [NV/CIV] is smaller by about the same factor. A preliminary analysis suggests that this indicates

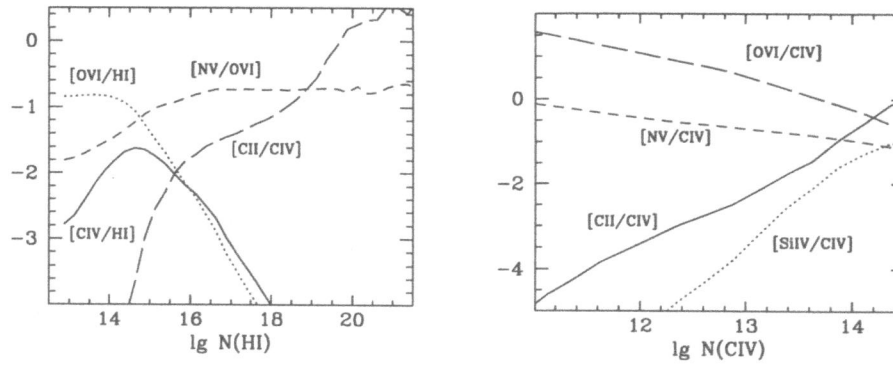

Fig. 3. A set of column density ratios is shown as function of HI and CIV column density. A homogeneous metallicity of 0.01 solar is assumed.

an ionizing spectrum softer than the assumed power law at high energies (see Rauch, Haehnelt & Steinmetz 1996 for a further discussion). OVI has not yet been searched for at small HI column densities. If OVI should turn out to be weak below $\log N(HI) = 14.5$ this would strongly argue for a metallicity gradient towards low-density regions.

References

Aragon-Salamanca A., Ellis R.S., Schwartzenberg J.-M., Bergeron J.A., 1994, ApJ, 421, 27

Bergeron J., Boissé P., 1991, A&A, 243, 344

Ferland G.J., 1993, University of Kentucky Department of Physics and Astronomy Internal Report

Haehnelt M.G., Steinmetz M., Rauch M. 1996, ApJ, 465, L95

Petitjean P., Webb J.K., Rauch M., Carswell R.F., Lanzetta K., 1993, MNRAS, 262, 499

Petitjean P., Bergeron J.A., 1994, A&A, 283, 759

Rauch M., Haehnelt M.G., Steinmetz M., 1996, submitted

Sargent W.L.W., Boksenberg A., Steidel C.C., 1988, ApJS, 68, 539

Songaila A., Cowie L.L., 1996, Astronomical Journal, in press

Steidel C.C., 1995, in *QSO Absorption Lines*, Proc. ESO Workshop, ed. G.Meylan (Heidelberg: Springer), p. 139.

Steinmetz M., 1996, MNRAS, 278, 1005

Vedel H., Hellsten U., Sommer–Larsen J., 1994, MNRAS, 271, 743

Weak Lensing at the Limit of the Sky Background Noise

Yannick Mellier[1,2] and Bernard Fort[2]

[1] Institut d'Astrophysique de Paris CNRS, 98 bis Boulevard Arago, 75014 Paris, France
[2] Observatoire de Paris, DEMIRM, 61 Avenue de l'Observatoire, 75014 Paris, France

Abstract. Recent weak lensing observations have pushed the use of 4 meter-class telescopes to the limits of their capabilities with exposure times exceeding several hours. The leading idea is that the surface density of faint galaxies up to very faint magnitude ($B > 28 - 30$) raises continuously thus potentially offering a dense template of distant sources whose intensity contrast is at the sky noise level. In complement to the Peter Schneider's presentation on dark matter search from weak lensing (this conference), we review some of these recent advances in weak lensing analysis based on this extreme faint population of galaxies in order to explore:

1. the dark matter distribution on large scales,
2. the redshift ditribution of lensed sources at very large distance,
3. and eventually the values of cosmological parameters.

For each observational topic we will briefly discuss these new methods as compare to more classical lensing studies as well as the possible VLT scientific impact in the domain.

1 Cosmology with Gravitational Lenses

Gravitational lenses effects on distant galaxies can potentially probe either the dark matter distribution from low mass compact objects to large-scale structures, or the angular distances of sources that depend on their redshift distribution and to a smaller extend on the cosmological parameters. This has motivated a lot of observational efforts devoted to large multiple arcs and arclets in rich clusters of galaxies (see Fort & Mellier 1994 and Narayan & Bartlemann 1996 for reviews). Indeed, we still expect important results from the strong lensing regime with the HST and the VLTs since recent improvements in instrumentation or data analysis provide unprecedented observational capabilities with these facilities. But except in the core of compact clusters of galaxies, arc(let)s are relatively rare events which cannot neither easily represent a large and fair sample of faint distant galaxies at large redshifts (luminosity bias) nor probe the mass distribution outside high condensation of masses. In this domain the greatest hope could be that HST observe an exceptional event with several multiple images systems at different redshifts that will allow a unique test on the geometry of the Universe.

So, after the pioneer analysis of Tyson et al (1990) on the shear field around A1689 and the extensive work of theoreticians (see Schneider, this conference), it appears clear that weak lensing can in principle be observed everywhere in the sky if we have very deep CCD images of the galaxy population at large redshift: the densest the population of background galaxies, the highest the visibility of the shear effect (better angular resolution and/or higher signal to noise ratio). Simultaneously crucial questions raised to observers: how deep is it possible to go in order to detect a large number of sources that can be statistically used to measure a coherent lensing signal on a given sky aperture with the highest signal-to-noise ratio? Is it actually possible to correct all the atmospheric and instrumental distortions of astronomical images whose amplitudes can be 5 to 10 times larger than the predicted amplitude of the gravitational lensing signal?

A technical answer to the second question is not trivial and far beyond the scope of this paper because all the ground based telescopes are affected by seeing effect and have been so far constructed without such an ultimate image quality in mind. Without a strong and dedicated effort on image correction of non axi-symmetric geometrical Point Spread Function, the measurement of weak shear below a few percent may be only possible with the HST on a few limited field of view for a while. This important remark has to be kept in mind all along the reading of this paper because we implicitly suppose that the problem has been solved.

For the first question, we know that for long exposure time with excellent seeing[1], the photometry of the faintest galaxies is limited to about $B = 26.5$ or $I = 25$. Beyond these magnitudes, the sources are still there but hidden in the photon noise of the sky background. The good linearity of CCD gives in fact a unique opportunity to detect this underground population and to measure their number density or some global coherent geometrical feature, like the weak shear induced by the deviation of light by condensation of mass. In the following we show first how the pixel to pixel autocorrelation function of the sky background reveals the existence of such a large population of extremely faint galaxies and how it is possible to use them to map the shear with an unprecedented signal to noise ratio. Then we discuss the possibility claimed by Fort et al (1996a) to detect the possible positions of sources down to 0.5σ of the sky level ($B = 28$) and to use the number counts to study the magnification bias effect around cluster of galaxies. Such an approach open a new way to explore the redshift distribution of galaxies to limiting magnitudes far beyond the other methods (spectroscopy and lensing inversion), and to search for new constraints on the value of the cosmological constant.

In this review, we focus on these recent developements which use faint sources at the noise level. Though still in their infancy, they will demonstrate their full scientific interest when used with the outstanding capabilities of the VLTs.

2   The Autocorrelation Function of Pixels

All standard methods which are used to determine the projected mass density from a lensing inversion of the shear map proceed in the same basic way. Very

---

[1] The meaning of this is not clearly defined. Let say that it corresponds basically to 3-5 hours on a 4 meter telescope with a median seeing below 0.7 arcsec.

faint objects are detected down to a threshold limit that strongly depends on the seeing. Then, the center, shape, size and magnitude of every sources are calculated and averaged within a given solid angle (scanning aperture on the sky) that defines the angular resolution of the shear map. The averaged ellipicity is finally linked to the potential and projected mass distribution on the sky. Though pratical and easy to implement, the method depends on the detection threshold, the convolution mask used for the measurements and the local statistical properties of the noise (Bonnet & Mellier 1995, Kaiser et al. 1995). Several critical issues of the method, related to the identification and delimitation of individual objects have led Van Waerbeke et al. (1996) to consider faint sources down to the noise level as a global density field and to measure the weak lensing effects from the analysis of the autocorrelation function of pixels (ACF) in CCD images. The concept is simple and can be easily understood: the Fourier Transform of an elliptical distribution of light is a conjugate elliptical distribution of density in the u, v plane with the same ellipticity (rotated by $\pi/2$). Thus the ACF on the angular scale of faint distant galaxies is a new mathematical object that sum up all the u, v sources down to the noise and immediately reveals the shear of all the sources in the scanning aperture. An important point is that ACF avoids measurement of centroids and shape parameters of individual galaxies which considerably reduces the uncertainties coming from sources of errors on the geometry of small and noisy objects.

Following Van Waerbeke et al. (1996) and Van Waerbeke & Mellier (1996) we can express the relation between the surface brightness, $I(\boldsymbol{\theta})$, in the image plane at the position $\boldsymbol{\theta}$ and the surface brightness in the source plane $I^{(s)}$ by

$$I(\boldsymbol{\theta}) = I^{(s)}(\mathcal{A}\boldsymbol{\theta}), \tag{1}$$

which can be extended to the ACF (e.g. the two-point autocorrelation function of the light distribution in a given area),

$$\xi(\boldsymbol{\theta}) = \xi^{(s)}(\mathcal{A}\boldsymbol{\theta}) , \tag{2}$$

and the thin lens equation can be re-written as

$$\xi(\boldsymbol{\theta}) = \xi^{(s)}(\boldsymbol{\theta}) - \theta \ \partial_{\theta}\xi^{(s)}(\boldsymbol{\theta})[1 - \mathcal{A}] . \tag{3}$$

It proves that the local ACF behaves like a new object. In the image plane $\xi(\boldsymbol{\theta})$ can be understood as the sum of an isotropic unlensed term, $\xi^{(s)}(\boldsymbol{\theta})$, an isotropic lens term which depends on κ, and an anisotropic term which depends on γ_i.

The weak lensing information is now given by the shape matrix $\mathcal{M}$ of the ACF,

$$\mathcal{M}_{ij} = \frac{\int \mathrm{d}^2\theta \ \xi(\boldsymbol{\theta}) \ \theta_i \ \theta_j}{\int \mathrm{d}^2\theta \ \xi(\boldsymbol{\theta})} . \tag{4}$$

and the shape matrix in the image plane is simply related to the shape matrix in the source plane $\mathcal{M}^{(s)}$ by $\mathcal{M}_{ij} = \mathcal{A}_{ik}^{-1}\mathcal{A}_{jl}^{-1}\mathcal{M}_{kl}^{(s)}$. If the galaxies are isotropically distributed in the source plane, $\xi^{(s)}$ is isotropic, and $\mathcal{M}_{ij}^{(s)} = M\delta_{ij}$, where δ_{ij} is

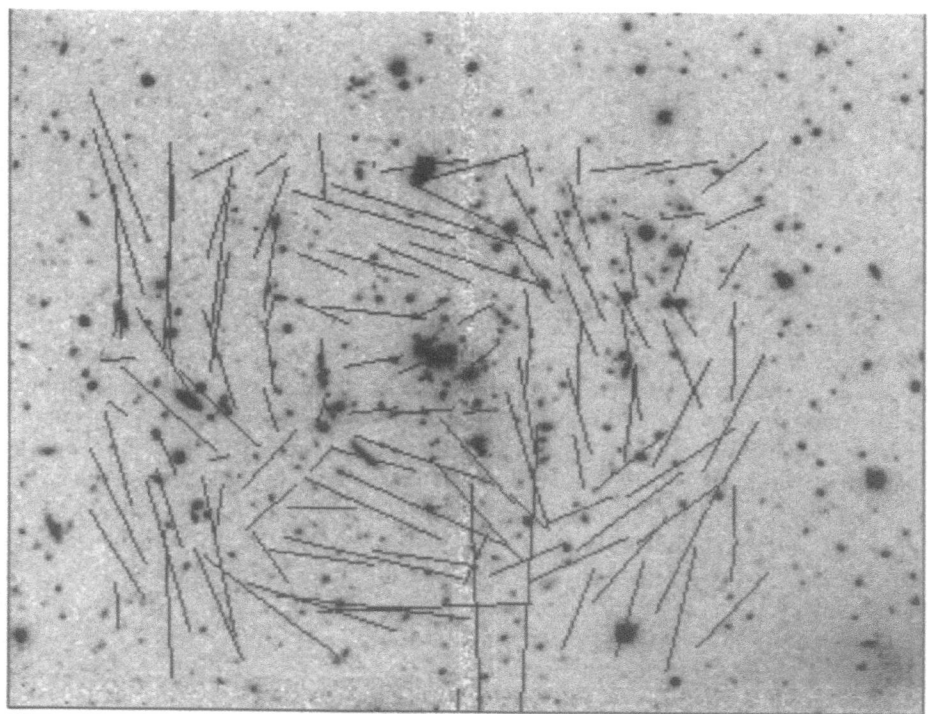

Fig. 1. Shear map around Cl0024 from the ACF method obtained from a mosaic of two CCD images. The map can be compared with the first one obtained by Bonnet et al. (1993). The resolution is better and in particular the innermost pattern shows now that the central mass distribution Cl0024 is bimodal.

the identity matrix. Using the expression of the amplification matrix $\mathcal{A}$ we can formally write $\mathcal{M}$ as follows:

$$\mathcal{M} = \frac{M(a + |g|^2)}{(1 - \kappa)^2 (1 - |g|^2)} \begin{pmatrix} 1 + \delta_1 & \delta_2 \\ \delta_2 & 1 - \delta_1 \end{pmatrix} . \tag{5}$$

Finally, the observable quantities (distortion δ_i and magnification μ) are given in terms of the components of the ACF shape matrix,

$$\delta_1 = \frac{\mathcal{M}_{11} - \mathcal{M}_{22}}{\mathrm{tr}(\mathcal{M})} \;\; ; \;\; \delta_2 = \frac{2\mathcal{M}_{12}}{\mathrm{tr}(\mathcal{M})} \;\; ; \;\; \mu = \sqrt{\frac{\det(\mathcal{M})}{M}}, \tag{6}$$

where $\mathrm{tr}(\mathcal{M})$ is the trace of $\mathcal{M}$ and $\det(\mathcal{M})$ is the determinant of $\mathcal{M}$.

As for classical methods, we see that the distortion is available from a direct measurement in the image plane while the magnification measurement requires to know the value of M which is related to the light distribution in the source plane or in any unlensed reference plane. The main advantage of the ACF method is that it provides a new way to measure δ_i and μ which does not depend upon the geometry of the scanning aperture on the CCD image. Furthermore, its signal to noise ratio is proportional to the number density n of background galaxies: $S/N \propto n$ instead of $S/N \propto \sqrt{n}$ for the standard methods (see figure 1).

A full description of the practical implementation of the ACF and first results are given in Van Waerbeke et al. (1996), Van Waerbeke & Mellier (1996) and Mellier et al. (1996). Clearly, the ACF is potentially the most powerful and promising technique to map very weak shears as those already detected at the outer periphery of clusters, those around bright quasars that seems magnifed by (unseen) condensations of mass (Fort et al. 1996b) or those predicted by large scale structures.

As the signal to noise of the ACF increases as the galaxy number density, the method will be very attractive on deep VLT images with good image qualities (seeing < 0.7"). FORS should be a unique instrument for some of these programmes.

3   The Distance of Background Sources with $B > 25$

3.1   Present Status of the Arc(et)s Redshift Distribution

Spectroscopic redshifts of luminous giant arcs allow to calculate the angular distances $D_{\rm d}, D_{\rm ds}$ and $D_{\rm s}$ (see Eq. (1) of Schneider in this proceedings) and to get the absolute scaling of the lens potential. If arclets are also observed in regions where the potential are properly probed by the modelling of giant arcs it is then possible to infer the most probable redshift of each individual arclet from their position and their ellipticity. The method has been successfully developped by Kneib et al. (1994) using some cluster lenses as *very low resolution gravitational spectrograph*. The most probable redshift for each arclet is obtained by assuming it corresponds to the source plane where the distortion is minimum. Potentially, it allows to provide redshift ($\delta z \pm 0.1$) of equivalent un-magnified galaxies down to $B = 27$ if the lens modelling is reliable and the shape of each arclet is measurable with a good accuracy. Recently, Kneib et al. (1996) used this technique on deep HST images of A2218 and show that, as for A370 (Kneib et al. 1994), the colour-redshift diagram seems consistent with the deepest redshift surveys. However, because it is a probabilistic approach, its efficiency must be also checked from spectroscopy. This long term programme started one year ago and spectroscopic surveys of the "brightest" arclets in A2218, A2390 and others are underway. Ebbels et al. (1996) and Bézecourt & Soucail (1996) purposely selected arclets showing bright spots of stars forming regions on HST images in order to detect an emission line and get secure redshift (see figure 2). The rather good agreement between spectroscopic redshifts and lensing inversions demonstrate that gravitational redshift are reliable. However, the observations require a large amount of nights on 4 meter telescopes and cannot be extended to the faintest arclets.

The spectroscopic observation of arc(et)s for the strong lensing modelling, for future cosmological tests, as well as for the study of the distribution and

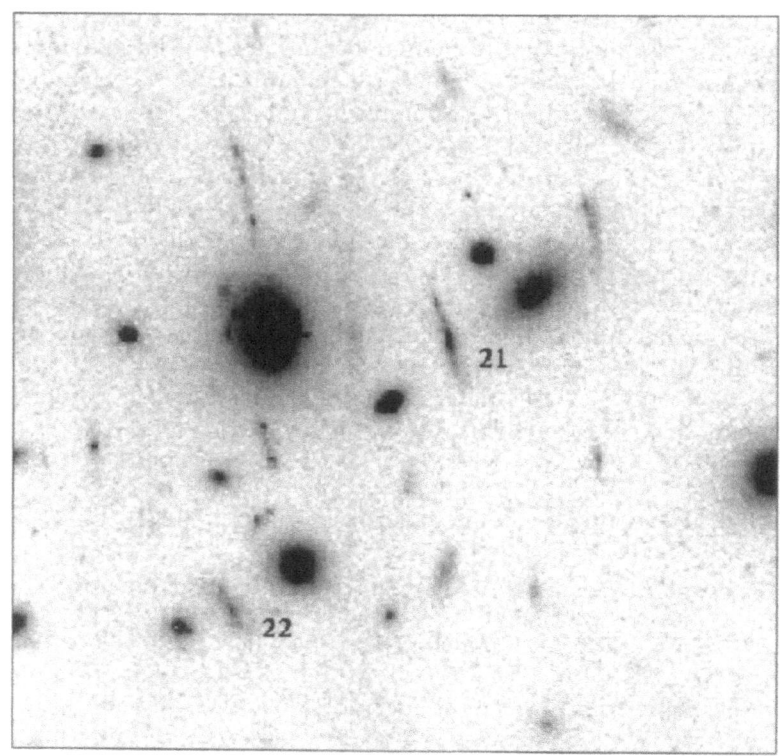

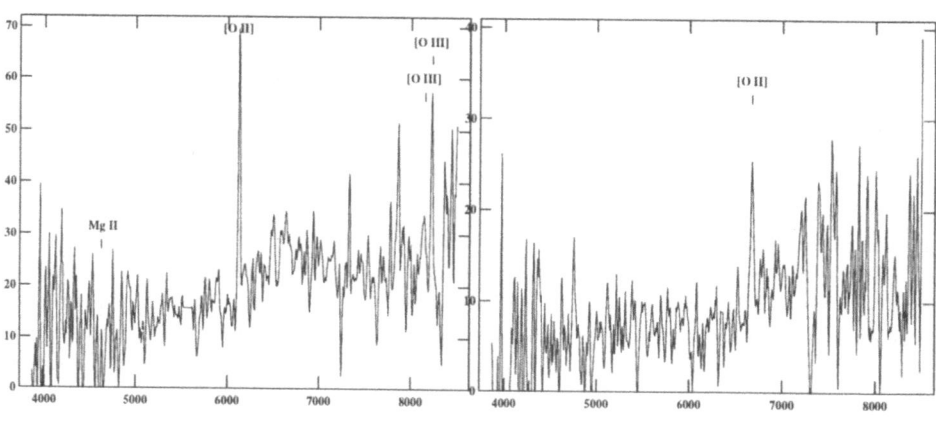

Fig. 2. Spectroscopy of faint arclets in A2390. The top panel is a deep HST image of its central region. The arclets are clearly visible with some of them showing image multiplicity with image parity changes. The arclets 21 and 22 have been observed by Bézecourt and Soucail (1996) with the MOS multiobject spectrograph at CFHT. As expected, they show the [OII]λ3727 emission line from which the redshifts are easily measured ($z = 0.643$ and $z = 0.790$ respectively. Courtesy J. Bézecourt).

evolution of faint distant galaxies will be continued with the VLT because one can reasonnably go at least 0.5 magnitude beyond 4 meter telescopes. In fact, one may be able to go a bit deeper by using FORS with the Va-et-Vient spectroscopic mode proposed by Cuillandre et al. (1994). This technique works well in low resolution spectroscopy in particular to derive the spectral energy distribution of faint distant galaxies without emission lines or QSOs with broad emision lines (for galaxies with thin emission line higher spectral resolution may be more suited). According to the gain estimated by Cuillandre et al., the flat field residuals are removed with a much better accuracy than in standard spectroscopic observations and it is possible to go 1 magnitude deeper. In that case, FORS with the Va-et-Vient spectroscopic mode could observe arclets as faint as 26.5 which corresponds to 20-30 arclets per cluster in three hours.

Since the lensing magnification is large, redshift surveys of arc(let)s could probe the redshift distribution of galaxies with $B > 24$. But unfortunately, the galaxy sample is biased in three ways. First, only arclets with star forming emisson lines are selected. Second, sky features and redshift effect conspire to offer peculiar windows of redshift visibility. Finally, as it is discussed in the next section, the magnification bias also favours observations of blue galaxies rather than red. So, at present the spectroscopy of arclets is crucial for the lens modelling but the redshift distributions obtained from these methods are still questionnable.

3.2   Probing the Redshift of Sources up to $B = 28$ with the Magnification Bias

Broadhurst (1995) first demonstrated that the radial surface density of background galaxies around a cluster lens like Abell 1689 varies according to the predicted magnification bias of the lens and can be used to measure κ directly and to map the projected mass density of the cluster. More attractive, this (anti) magnification bias effect provides a good way to break the intrinsic degeneracy of the inversion methods based on gravitational weak shears: an additionnal plan of constant mass density on the line of sight does not change the shear pattern but just the convergence of light beams.

The radial behavior of the so-called Broadhurst's effect results from the competition between the gravitational magnification that increases the detection of individual objects above the limit of detection and the deviation of light beam that decreases the apparent number counts. Therefore the amplitude of the magnification bias depends explicitly on the slope of the galaxy counts as a function of magnitude and on the magnification factor of the lens:

$$N(< m, r) = N_0(< m) \; \mu(r)^{2.5\gamma - 1} \;, \tag{7}$$

where $\mu(r)$ is the magnification factor of the lens, $N_0(< m)$ the intrinsic number density in a nearby empty field and γ is the intrinsic count slope:

$$\gamma = \frac{d log N(< m)}{dm} \;. \tag{8}$$

The radial magnification bias $N(< m, r)$ shows up only when the slope γ is different from the value 0.4; otherwise, the increasing number of magnified sources is exactly cancelled by the apparent field dilatation and there is no effect on $N(< m, r)$. As noticed by Broadhurst, a radial amplification bias cannot be observed for $B(< 26)$ since the slope is almost this critical value (Tyson 1988, Shanks 1996) but it can be detected in the R or I bands when the slopes are close to 0.3 (Smail et al. 1995).

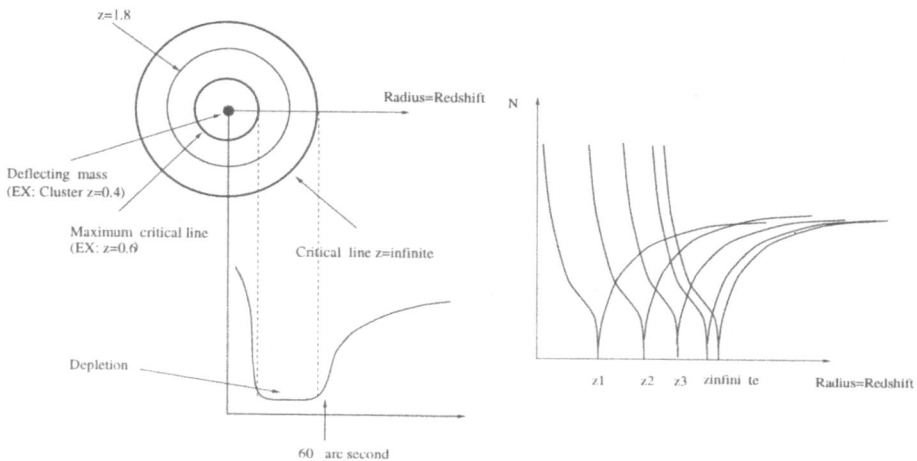

Fig. 3. Measurement of redshifts from depletion curves. The left panel shows the depletion by a singular isothermal sphere as it would be observed on the sky (top left) and the radial projected density of galaxies (bottom left). For a single source redshift, when the lens is perfectly known, the minimum of a depletion curve is sharp and its radial position is formally equivalent to a redshift (for example the position $z1$ of the right panel). The radial position of the miminum increases with the redshift of sources but the depletion curves tighten and converge towards the curve corresponding to sources at infinity. In a realistic case, the redshift distribution is broad and the individual curves must be added. The bottom left panel shows the depletion as it would be observed: instead of the single peaked depletion we expect a more pronounced minimum between two radii (i.e. two redshifts) whose angular positions depend on the cosmological constant for high redshift sources. Thus, if the mass distribution of the lens is well known as in the rich lensing cluster Cl0024+1654, the distribution of sources and λ can be inferred from the shape of the depletion curve.

When $\gamma < 0.3$ a sharp decrease of the number of galaxies is expected in regions of strong magnification close to the critical radius of the lens corresponding to the redshift of the background sources. Since the critical radius increases with redshift, every population of galaxies at different redshifts will display a distinct peaked annular depletion at its own critical line. It can result a shallower depletion between the smallest and the largest critical line which depends on the redshift distribution of the galaxies. (Figure 3 and 4). In short, the lens gives us a new way to sort out different class of galaxies versus their redshift distribution. It was first used by Fort et al (1996a) with the cluster Cl0024+1654 to study the faint distant galaxies population in the extreme range of magnitude $B = 26.5 - 28$ and $I = 25 - 26.5$ after a detection of the sources in the sky background noise. For this selected bins of magnitude they found on their CFHT blank fields that the counts slope was near 0.2, well suited for the study of the Broadhurst's effect. After analysis of the shape of the depletion curve (figure 4), $60\% \pm 10\%$ of the B-selected galaxies were found between $z = 0.9$ and $z = 1.1$ while most of the remaining 40% galaxies appears to be broadly distributed around a redshift of $z = 3$. The I selected population present a similar distribution with two maxima, but spread up to a larger redshift range with about 20% above $z > 4$. In fact, many of the I selected galaxies were not detected in B as if their Lyman α discontinuity has already crossed the B filter band (redshift range $z > 3.5$).

The main characteristic of this observation is the long exposure time with an excellent seeing condition. Otherwise the detection of such faint objects with an accuracy almost comparable to the HDF would be impossible. In that sense, it is quite clear that such observations are perfectly suited for FORS during the best seeing period on Paranal. So far it seems that the magnification bias is probably a good technique to study the abundance of galaxies at very large distance. Just one limitation comes from the fact that the critical lines of a gravitational lens tend to rapidly merge with the critical line at infinity for a redshift larger than 3. So we can just estimate the relative population of faint galaxies above this redshift. Conversely, this limitation was in turn used by Fort et al (1996a) for a first attempt of a direct measurement of the value of the cosmological constant.

4   Measuring the Cosmological Constant from Depletion Curves

The qualitative profile of the depletion curve in a direction perpendicular to the critical line is given on figure 3. When a distribution of detectable galaxies spread up to large redshifts all the most distant ones beyond $z = 3$ will form their sharp depletion at almost the same radial distance that, for simplicity, we identify here to rc_∞. The most abundant the very distant galaxies are, the stronger is the rising discontinuity at the end of the depletion (figure 4). Basically, Fort et al (1996a) have used the modelling of the lens inferred from the giant arc and the location of the last critical line R_I detected in the I bandpass to implement a cosmological test on the value of the cosmological constant. For a flat universe they found rc_∞ at a too large distance from the center of the cluster for a non zero cosmological constant[2]. With a thin lens hypothesis and despite large uncertainties their observations seems to favour an Ω_Λ-dominated flat universe with a cosmological constant ranging from 0.6 to 0.9. On the other hand, statistics of gravitational lens events on QSOs seems to rule out models

---

[2] except if there is an additional, and so far unseen, deflector hidden behind Cl0024+1654.

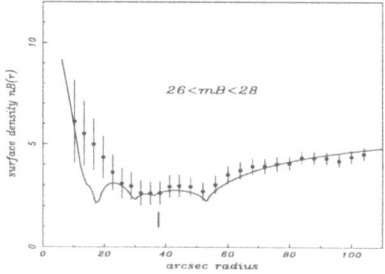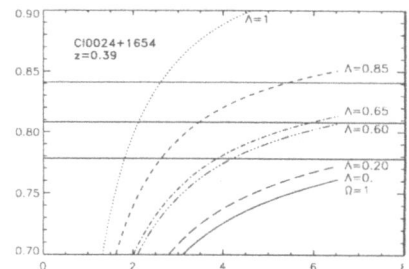

Fig. 4. A measurement of the cosmological constant from a depletion curve. The left panel shows the depletion curve measured in Cl0024+1654. By using the redshift corresponding to the two limits of the plateau observed on the curve, one can constrain the cosmological constant from the position of the second minimum. Whatever the redshift of the most distant sources visible on the images, we see that the angular position where the depletion curve raises again imposes that $\lambda > 0.65$

with $\Omega_\Lambda > 0.6$ (Kochanek 1995). Thus, if the conclusions of the magnification bias and of the QSO statistics are correct, few room is left for a possible value of the cosmological constant.

The Fort et al. (1996a) preliminary result is still questionable but at least it demonstrates that the magnification bias is worth to be explored further with many clusters, both to study the distribution and evolution of faint distant galaxies and to provide cosmological tests in clusters of galaxies with large multiple arcs. A reasonnable sample of 20 rich lensing clusters with simple geometry observed with FORS on the VLT would provide very deep exposures (to $B \approx 28 - 29$) rapidly with a very low flat field residuals which is crucial for detection in the noise. It will be a challenging programme with FORS on the VLT.

5   Conclusion

It is now established from the weak lensing analyses we have described here as well as from the HDF data and recent deep spectroscopic surveys (this conference) that it is possible to detect and study the population of galaxies at redshift beyond $z = 4$. Most of them, despite probably strong star forming activities, will be close to or within the sky background noise. It is already demonstrated with 4 meter class telescopes that their detectability is not a big challenge for VLTs during period of excellent seeing. Their observations on a regular basis is a tremendous chance for weak lensing studies and their large cohort of exciting applications. However, for the measurement of very faint shear ($\approx 1\%$), it is crucial to control the stability of the image quality to a level which was never thought of before. A dedicated and a large comprehensive effort must be done in this domain by instrument builders and observers in order to reach the ultimate imaging capabilities of the VLTs.

Acknowledgments

We thank J.-C. Cuillandre, R. Pelló, P. Schneider, C. and S. Seitz, and L. Van Waerbeke for stimulating discussions about lensing and prospective aspects. J. Bézecourt kindly provided the spectra shown in this review.

References

Bézecourt, J., Soucail, G. 1996, SISSA preprint astro-ph/96006064.
Bonnet, H., Mellier, Y., Fort, B. 1994, ApJ 427, L83.
Bonnet, H., Mellier, Y. 1995, A&A 303, 331.
Broadhurst, T., Taylor, A.N., Peacock, J. 1995, ApJ 438, 49.
Broadhurst, T. 1995, SISSA preprint astro-ph/9511150.
Cuillandre, J.-C., Fort, B., Picat, J.-P., Soucail, G., Altieri, B., Beigbeder, F., Dupin, J.-P., Pourthié, T., Ratier, G. 1994, A&A 281, 603.
Ebbels, T. M. D., Le Borgne, J.-F., Pelló, R., Ellis, R. S., Kneib, J.-P., Smail, I., Sanahuja, B. 1996, SISSA astro-ph/9606015.
Fort, B., Mellier, Y. 1994, A&A Review 5, 239, 292.
Fort, B., Mellier, Y., Dantel-Fort, M. 1996a, SISSA preprint astro-ph/9606039.
Fort, B., Mellier, Y., Dantel-Fort, M., Bonnet, H., Kneib, J.-P. 1996b, A&A 310, 705.
Kneib, J.-P., Mathez, G., Fort, B., Mellier, Y., Soucail, G., Longaretti, P.-Y. 1994, A&A 286, 701.
Kneib, J.-P., Ellis, R. S., Smail, I., Couch, W. J., Sharples, R. M. 1996, SISSA preprint astro-ph/9511015.
Kochanek, C. S. 1995, SISSA preprint astro-ph/9510077.
Mellier, Y., Van Waerbeke, L., Bernardeau, F., Fort, B. 1996, Proceedings of the VIIIth Rencontres de Blois "Neutrinos, Dark Matter and the Universe". Blois, France 1996.
Narayan, R., Bartelmann, M. 1996, SISSA preprint astro-ph/9606001
Shanks, T. 1996, Proceedings of the 37th Herstmonceux Conference "HST and the High Redshift Universe". Cambridge 1996. N. Tanvir, A. Aragón-Salamanca, J. V. Wall eds.
Smail, I., Hogg, D., Yan, L., Cohen, J. G. 1995, ApJ 449, L105.
Tyson, A. J. 1988, AJ 96, 1.
Tyson, J. A., Valdes, F., Wenk, R. 1990, ApJ 349, L1.
Van Waerbeke, L., Mellier, Y., Schneider, P., Fort, B., Mathez, G. 1996, A&A in press. SISSA preprint astro-ph/9604137.
Van Waerbeke, L., Mellier, Y. 1996, Proceedings of the XXXIst Rencontres de Moriond "Dark Matter in Cosmology. Quantum Measurement, Experimental Gravitation". Les Arcs, France 1996. SISSA preprint astro-ph/9606100.

Morphology of High Redshift Galaxies with HST

F. D. Macchetto

Space Telescope Science Institute, 3700 San Martin Drive,
Baltimore, Maryland 21218.
On assignment from the Space Science Department of ESA

1 Introduction

The identification of the population of primeval galaxies, namely the first galaxies to form stars in the Universe, is essential to provide the much needed observational underpinning to modern theories of galaxy formation and evolution. Despite decades of intensive search, primeval galaxies have not been identified and, thus, the epoch and the early physics of galaxy formation are still substantially unknown.

The combination of ground-based photometry and spectroscopy with HST's high-angular resolution, is rapidly changing this situation.

We are now obtaining empirical evidence that a population of galaxies of relatively normal luminosity was already in place at redshifts $z > 3$ and in an evolutionary state characterized by active star formation.

The evolutionary state of galaxies at intermediate redshifts ($z < 1.0$), corresponding to < 55 % of the life of the Universe ($H_o = 50 km^{-1} Mpc^{-1}$ and $q_o = 0.5$), has been probed by several surveys, during the last few years.

Deep redshift surveys have been carried out by Lilly et al. 1995a and 1995b; Steidel et al. 1995; Glazebrook et al. 1995a; Cowie et al. 1994; Cowie, Hu & Songaila 1995a.

At higher redshifts, deep (post-refurbishment) HST imaging surveys have been carried out by Driver et al. 1995a and 1995b; Glazebrook et al. 1995b; Schade et al. 1995; Cowie Hu & Songaila 1995b and more recently by the HDF Team (Williams et al 1996).

A common conclusion of these authors is that the evolution of the luminosity function has followed rather diverse tracks for galaxies of different luminosity and morphological type.

The luminous galaxies, i.e., massive ellipticals and spirals, are characterized by at most a modest amount of evolution in luminosity and/or number density since $z \sim 1$. Typical luminosity evolution is a factor of ~ 3 to 10.

This suggests that the most massive and oldest systems must have formed at much earlier epochs. Therefore, the population of galaxies possessing a substantial spheroidal component (i.e., ellipticals and early-type spirals) has been remarkably quiescent over the redshift range probed by the redshift surveys, suggesting that the important epoch for their formation lies far beyond $z \sim 1$.

Our ground-based and HST observatories have allowed us to discover a population of normal galaxies at $z > 3$ which appear to be in the process of forming a substantial fraction of their stellar population. The high spatial resolution HST images allow us to probe the rest-frame ultraviolet spectrum and to carry-out a quantitative analysis of their morphological properties. We can, therefore, characterize for the first time the evolutionary status of galaxies at a time when the Universe was $\simeq$ 20% or less of its current age.

2   Early EFOSC Observations

Our early observations were made with the ESO 3.6m and NTT telescopes and EFOSC. We obtained narrow-and broad-band imaging observations of the field containing the quasar QSO 0000 + 2619. The imaging observations were designed to be sensitive to Lyman alpha emission in the redshift range [3.363, 3.436]. The observations were carried out at the 3.6 m telescope and EFOSC in November 1991 with seeing in the range 2" - 1.5" FWHM. Follow-up spectroscopy was also obtained for candidates identified in the initial search. We determined the redshift of a Lya-emitting galaxy at z = 3.428 (galaxy G2). (Macchetto et al. 1993, Giavalisco et al 1994, 1995). This discovery of a radio-quiet Lya-emitting galaxy at redshift z = 3.428 was extremely important.

We measured the Lyα equivalent width to be 720Å. The Lyα emission-line flux is $7.2 \times 10^{-16} ergs^{-1} cm^{-2}$, which corresponds to a luminosity of $1.8 \times 10^{43} ergs^{-1}$. Assuming that all the Lyα luminosity is due to stars and in the absence of dust obscuration, the inferred star formation rate is about 18 M$\odot$ yr^{-1}.

3   Follow-Up Imaging

Following that initial (but limited) success, we adopted a different strategy. To detect galaxies at similar distances, we used a custom made photometric system, UnGR, designed to reach very faint flux leves and provide accurate color photometry (Steidel & Hamilton 1992).

The technique relies on the presence of the 912 A Lyman discontinuity in the very blue and flat rest-frame UV spectral energy distribution. The method takes advantage of the fact that the observable Lyman discontinuity from high-redshift sources is entirely dominated by the intervining QSO absorption systems, although in addition any realistic IMF will produce a pronounced intrinsic Lyman discontinuity (Madau 1994). (Fig. 1.)

With the UnGR system, we observed a number of fields around QSOs with known optically-thick absorption systems, and general sky fields, building a sample of more than 120 galaxy candidates with redshifts $z > 3$.

Star forming galaxy candidates with redshifts $z > 3$ are selected from a colour-colour plane to have very red U_n - G colours and blue (G - R) colours, respectively. A subset of these galaxies have now been observed with Keck (see paper by Giavalisco at this conference). We have also embarked on a systematic program of HST imaging of the $z > 3$ candidates to study their morphological properties.

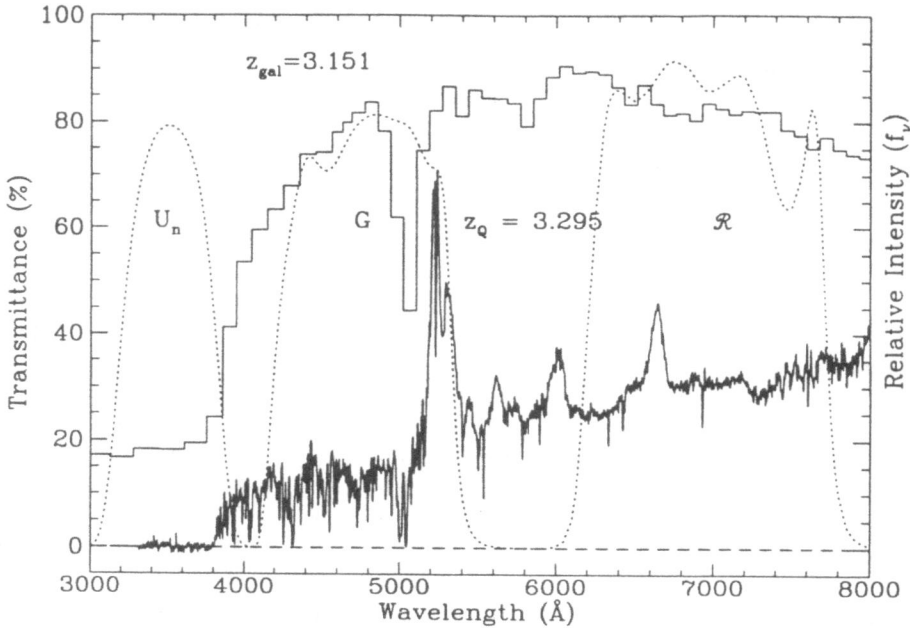

Fig. 1. Transmittance curves of the UnGR photometric system. Also shown in the synthetic spectrum of our unreddened star-forming galaxy and the spectrum of Q2233+1310.

4 HST Imaging of the $z > 3$ Galaxies

We have studied the morphological properties of the $z > 3$ galaxy candidates with HST and WFPC2. To date we have obtained images for approximately 30 galaxies in several fields (0000 - 263, 0347 - 383 and 2219 - 003).

The HST data consists of images taken with filters F606W, F702W or F814W and typically probes the rest-frame UV in the range 1400-1900 A. These observations have high enough resolution to determine the galaxy's size and light distribution. (Fig. 2.)

The majority of the $z > 3$ galaxies have compact morphologies characterized by a bright "core" surrounded by more diffuse nebulosities with significantly lower surface brightness.

We have studied the morphology of the galaxies in two ways. First, we have

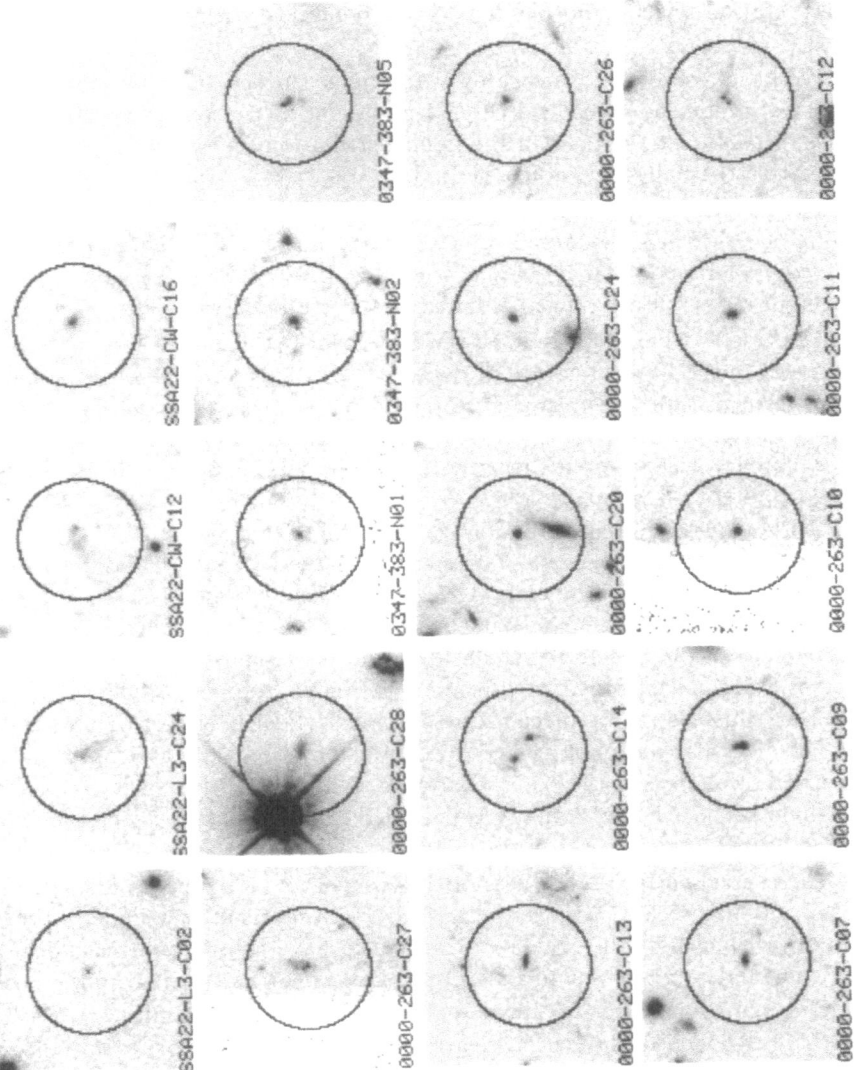

Fig. 2. A mosaic of the HST images of candidate $z > 3$ galaxies. Note compact cores surrounded by diffuse nebulosities.

analyzed their radial light profile and have found that many can be described in terms of the $r^{1/4}$ or exponential profiles. These fits are intended to broadly classify the light profiles and are not meant to imply that a particular galaxy rigorously follows a given model. Note that we are fitting models of the light distribution of present-day galaxies at 4500 A rest-frame to galaxies at $z > 3$ observed at 1600 A rest-frame with a star-formation rate one order of magnitude (or more) higher. Clearly, we do not yet know whether the same physical interpretation relating morphology to the dynamical state of the galaxies would hold.

We have also derived measures of the light concentration by measuring the isophotal magnitudes (using FOCAS) and a set of concentric aperture magnitudes, centered at the peak of the light distribution to produce a growth curve and derive the half-light radius. (Fig. 3.)

Most of the $z > 3$ galaxies are characterized by a compact morphology, with diameters less than 1.5 arcsec, which corresponds to 10.5 h_{50}^{-1} kpc ($q_o = 0.5$). The core contains about 90-95% of the total luminosity of the galaxy, and has half-light radius in the range 0.2-0.3 arcsec, corresponding to 1.4-2.1 h_{50}^{-1} kpc.

At the observed rest-frame far-UV wavelengths the emission is directly proportional to the formation rate of massive stars. Therefore, 90-95% of the stars being formed are concentrated in a region whose size is that of a present-day luminous galaxy. If the morphology of the massive stars is a good tracer of the overall stellar distribution, then the light distribution of the core is consistent with a dynamically relaxed structure.

The core is very often surrounded by low surface brightness nebulosities which may extend for a few arcsecs. These nebulosities are seen as deviations from the more regular profile of the core. The presence of such halos is consistent with the intense star-formation activity observed in the $z > 3$ galaxies, even if the morphology of the underlying stellar distribution is relatively regular. The presence of extended gaseous components is expected from the intense supernovae rate that must be taking place in these systems (Ikeuchi & Norman, 1991). Halos with irregular morphology and lower star formation rate are consistent with dissipative collapse in the cores (Baron & White 1987). In such a scenario, a core-halo segregation is actually expected due to the increased cloud collision rate and SFR in the denser, more collapsed regions.

The central surface brightness of the compact $z > 3$ galaxies is consistently close to 23 mag arcsec^{-2}. Since the surface brightness is proportional to the star-formation efficiency, these galaxies must also have comparable star-formation efficiency, indicative of a similarity of physical processes in the core regions.

There are four cases of diffuse galaxies, whose light profiles are very well fitted by exponential laws. The central surface brightness of these galaxies is significantly lower than that of their more compact counterparts, showing an overall reduced star-formation efficiency. One is candidate to be one of the two damped Lyman α systems in the spectrum of Q0000- 263, either at $z = 3.052$ or $z = 3.390$.

In 3 cases out of 19, we observe galaxies with multiple morphology. They have two major light concentrations separated by about 1 arcsec or less, corresponding to $\leq 7h_{50}^{-1}$ kpc. The individual sub-components are spatially resolved in the first case, barely resolved in the second case, and too small to conclude anything in the third case.

In all cases, the galaxies show extended diffuse nebulosity around or extending from them, suggestive of systems in interaction. This could be evidence of

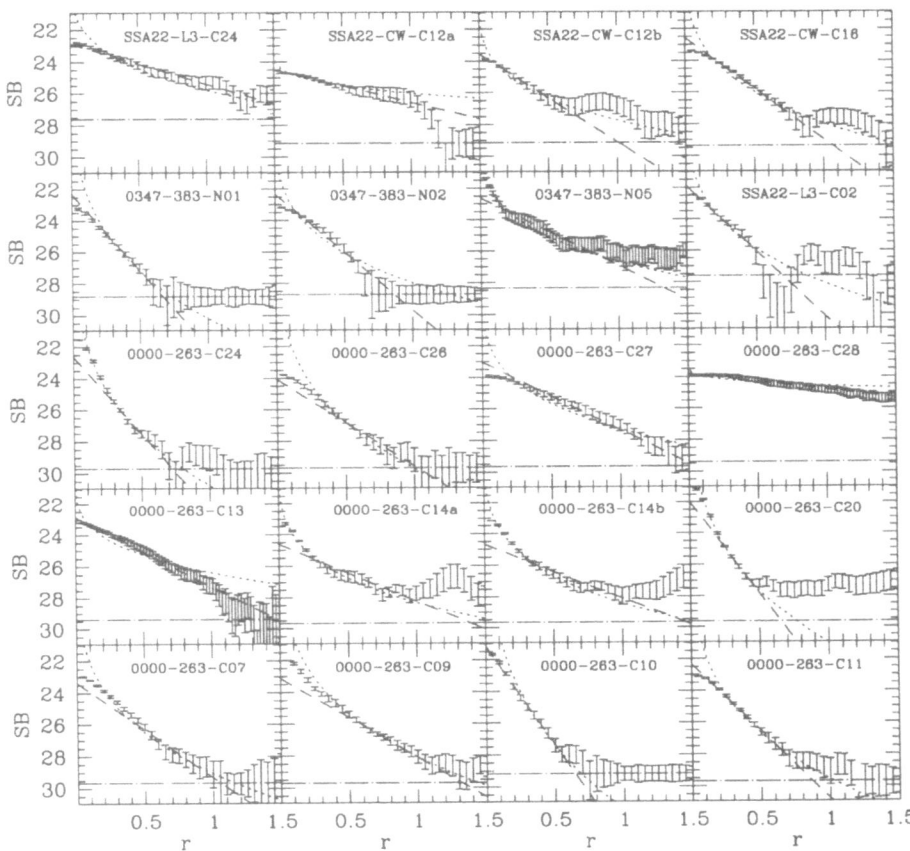

Fig. 3. Radial profiles for the $z > 3$ galaxies. The vertical axis is the surface-brightness and the horizontal axis is the radius in arcsec. The dotted lines are fitted $r^{1/4}$ profiles and the dashed lines are exponential fits.

hierarchical merging of sub-units into more massive systems with time scales about an order of magnitude shorter than the time stretch of the cosmic epoch proved, namely $t_{merg} \approx 5 \times 10^7$ yr.

The "merging" units have smaller luminosity than the other systems, but comparable central surface brightness. This is fully consistent with the interpretation of these galaxies as young spheroidal systems.

The geometry of the cores seems characterized by a high degree of spherical symmetry. There are no cases of highly elongated structures, such as the "chain galaxies" discussed by Cowie et al. (1995b). Those chain galaxies are apparently formed by strings of star-forming regions of similar surface brightness.

Given their knotty structure and the fact that they seem to have comparable star-formation rates to those of the $z > 3$ galaxies if placed at $z > 3$, we would expect that the "morphological k-correction" would make them appear even more elongated, because of $(1 + z)^4$ surface- brightness dimming of the more diffuse regions that surround the knots. None among the 19 galaxies observed so far has, even approximately, such a morphology. We have measured the axial ratios of the isophotes, which provide an upper limit to the axial ratios of the galaxy core regions. These values have a mean of 1.7. They are larger than $\approx$ 1.5-2 only for markedly exponential galaxies and are never larger than $\approx$ 3. In comparison, Cowie et al. report unconvolved axial ratios as high as 9.5, with a mean value of 4.7 for the chain galaxies.

5   Conclusions

We have discovered a population of star- forming galaxies with redshifts $3.0 \le z \le 3.5$ using a colour-selection technique whose efficiency is very high (Giavalisco, Steidel & Macchetto, 1996).

This allows a systematic study of the nature of galaxies at such large redshifts. Since we use a flat-UV spectrum selection criterion, then dust obscuration may not be an important limiting factor in searches for high- redshift galaxies.

The space density, star formation rates, morphologies and physical sizes, masses, and early epoch of the galaxy population support the conclusion that the progenitors of the present-day bright galaxies have been observed during a phase characterized by intense star formation. Rough dynamical estimates suggest that these galaxies are massive systems.

Since spheroids must have formed relative early to attain a state of quiescent evolution by $z \sim 1$, the centrally concentrated star formation that characterizes the $z > 3$ population is compelling evidence that we are observing directly the ongoing formation of the spheroid components of what would become the luminous galaxies of the present epoch.

At $3 \le z \le 3.5$, we are probably seeing an epoch where the star formation was concentrated primarily in the central regions of massive galaxies (the "spheroid epoch").

This star formation appears to have "migrate" over time to more morphologically peculiar objects, and finally to the present where the bulk of the star formation is distributed in spiral disks.

Our results demonstrate that by $z > 3$ massive galaxy formation was well underway, and the galaxies that we have identified are the sites of the most active star formation at that epoch.

The properties of these objects must be reproduced by any theory attempting to explain the formation of normal galaxies.

What should the VLT do in this field of research? It clearly should invest large amounts of time to: search for truly primeval galaxies, find many new fields with high-redshift galaxy candidates, compare the morphological and physical properties of the high - redshift ($z > 3$) population to the lower - redshift ($z \approx$

1-2) population, study their SEDs and emission lines properties, investigate evolutionary effects, and determine the clustering properties at different redshifts.

A comprehensive multiyear project in this area will be fundamental to our understanding of the processes of galaxy formation and evolution.

References

Baron, E., White, S.D.M. (1987): ApJ. **322**, 585

Cowie, L.L., Gardner, J.P., Hu, H.M., Songaila, A., Hodapp, K.W., Wainscoat, R.J. (1994): Ap. J. Suppl. **94**, 461

Cowie, L.L., Hu, E.M., Songaila, A., (1995a): Nature **377**, 603

Cowie, L.L., Hu, E.M., Songaila, A., (1995b): AJ. **110**, 1576

Driver, S., Windhorst, R., Griffith, R., (1995a): ApJ. **453**, 48

Driver, S., Windhorst, R., Ostrander, E., Keel, W., Griffith, R., Ratnatunga, K., (1995b): ApJ. **449**, L23

Giavalisco, M., Macchetto, F.D., Sparks, W.S., (1994): A&A. **288**, 103

Giavalisco, M., Macchetto, F.D., Madau, P., Sparks, W.B., (1995): ApJ **441**, L13

Giavalisco, M., Steidel, C.C., Macchetto, F., (1996): ApJ in press.

Glazebrook, K., Ellis, R., Colless, M.M., Broadhurst, T.J., Allington-Smith, J.R., Tanvir, N.R., (1995a): MNRAS. **273**, 157

Glazebrook, K., Ellis, R., Santiago, B., Griffith, R., (1995b): MNRAS. **275**, L19

Ikeuchi, S., Norman, C.A., (1991): ApJ. **375**, 479

Lilly, S., Le Fevre, O., Crampton, D., Hammer, F., Tresse, L., (1995a): ApJ. **455**, 50

Lilly, S., Tresse, L., Hammer, F., Crampton, D., Le Fevre, O., (1995b): ApJ. **455**, 108

Macchetto, F.D., Lipari, S., Giavalisco, M., Turnshek, D.A., Sparks, W.B., (1993): ApJ. **404**, 511

Madau, P., (1995): ApJ. **441**, 18

Schade, D., Lilly, S.J., Crampton, D., Hammer, F., Le Fevre, O., Tresse, L., (1995): ApJ. **451**, L1

Steidel, C.C., Hamilton, D., (1992): AJ. **104**, 941

Steidel, C.C., Hamilton, D., (1993): AJ. **105**, 2017, SH93

Steidel, C.C., Dickinson, M., Persson, S.E., (1994): ApJL. **437**, L75

Steidel, C.C., Pettini, M., Hamilton, D., (1995): (Paper III), AJ. **110**, 2519

Williams, R., et al., (1996): ApJ. in press

Search for Galaxies at $z > 4$ from a Deep Multicolor Survey

E. Giallongo[1], S. Charlot[2], S. Cristiani[3], S. D'Odorico[4], A. Fontana[1]

[1] Osservatorio Astronomico di Roma, I-00040 Monteporzio, Italy
giallo@coma.mporzio.astro.it
[2] Institut d'Astrophysique du CNRS, 98 bis Boulevard Arago, 75014 Paris, France
[3] Dipartimento di Astronomia, Università di Padova, Vicolo dell'Osservatorio 5,
I-35122 Padova, Italy
[4] European Southern Observatory, K. Schwarzschild Strasse 2, D-85748 Garching,
Germany

Abstract. We present deep BVrI multicolor photometry in the field of the quasar BR1202-07 ($z_{em} = 4.694$) aimed at selecting field galaxies at $z > 4$. We compare the observed colors of the galaxies in the field with those predicted by spectral synthesis models including UV absorption by the intergalactic medium and we define a robust multicolor selection of galaxies at $z > 4$. We provide spectroscopic confirmation of the high redshift QSO-companion galaxy ($z = 4.702$) selected by our method. The first estimate of the surface density of galaxies in the redshift interval $4 < z < 4.5$ is obtained for the same field, corresponding to a comoving volume density of $\sim 10^{-3}$ Mpc^{-3}. This provides a lower limit to the average star formation rate of the order of 10^{-2} M$_\odot$ yr^{-1} Mpc^{-3} at $z \sim 4.25$.

1 Introduction

Deep images from the Hubble Space Telescope and ground based telescopes (Keck, NTT, CFHT) are providing new exciting information about abundance and morphology of galaxies in a wide redshift interval up to $z \sim 4.5$.

In particular the search of high redshift galaxies is relevant not only to extract information about the physical processes which control the formation of individual objects, but also to probe the cosmological evolution of the formation of galactic structures in the Universe. Indeed cosmological scenarios are attempting to follow the evolution in cosmic time of the galaxy formation, describing in detail the history of the star formation. In the standard CDM cosmology, for example, most of the stars are formed at intermediate redshifts ($z \sim 1$; e.g. Cole et al. 1994).

A useful parameter which allows a more direct comparison between theoretical predictions and data interpretations is the star formation rate per unit comoving volume. A reference value at the present epoch of $\sim 5 \times 10^{-3}$ M$_\odot$ yr^{-1} Mpc^{-3} (for a Salpeter IMF) has been recently given by Gallego et al. (1995) on the basis of an Hα galaxy survey.

Recent estimates of the galaxy luminosity function of faint galaxies up to $z = 1$ are providing the first evidence of strong evolution by a factor of ten of the cosmological star formation rate in the redshift interval $z = 0 - 1$ (Lilly et al. 1996; Cowie et al. 1996). This of course implies that more than half of the stars formed at intermediate redshifts, in good agreement with theoretical expectations.

Nevertheless, the same models predict a fraction $< 2\%$ of the present mass density in stars at $z > 4$ (Cole et al. 1994). It is therefore at these very high redshifts that cosmological scenarios for galaxy formation are more vulnerable to observational constraints.

Efficient selection criteria are needed to find out galactic structures at these very high redshifts. Well known examples of high z sources are luminous Active Galactic Nuclei like quasars. Their absorption spectra provide unique information on the abundance and ionization state of the intergalactic medium at very high z. The presence of the IGM has a twofold cosmological relevance. First, Lyman absorption by the IGM along any line-of-sight produces strong depression of the UV spectrum of high redshift sources. Moreover, its high ionization level requires a large background of UV ionizing photons up to $z \sim 4 - 5$ (Giallongo et al. 1994,1996). This large UV background is only marginally consistent with that produced by the observed quasars (Haardt & Madau 1996), leaving room for a possible UV contribution by a large number of star-forming galaxies at $z > 4$.

2   The Multicolor Selection

In selecting galaxies which are actively forming stars at very high redshifts, two different approachs can be followed. It is possible to exploit the intrinsic spectral properties expected from star formation activity, or, in case we want to select galaxies at $z > 4$, it is better to exploit the complex but universal opacity to the UV photons of the intergalactic medium.

The selection criteria based on the intrinsic spectral properties exploit the main UV features of the star-forming galaxies like the possible presence of strong emission lines and/or the Lyman absorption break of the flat UV continuum due to the stellar evolutionary properties plus Lyman continuum absorption by the interstellar medium present inside the same galaxy.

Surveys based on the detection of intense Lyman alpha emission by means of narrow band imaging in the optical/IR band in the redshift interval $1.8 < z < 6$ have provided no systematic detections of high z galaxies with only few exceptions (e.g. Macchetto et al. 1993).

A very efficient method based on the detection of the Lyman break present in a flat rest-frame UV continuum has been proposed by Steidel & Hamilton (1992). An appropriate choice of a set of broad band colors can allow the detection of the Lyman break in a given, relatively narrow redshift interval. Steidel used in particular a set of U G R filters adopted to select Lyman break galaxies in the redshift interval $2.8 < z < 3.4$. Since the spectrum of an actively star-forming galaxy is flat longward of the Lyman break, an average color of G–R~ 0.5 is expected in the selected redshift interval. At the same time strong reddening is expected in the U–G color which samples the drop of the emission shortward of the Lyman break (U–G> 1.5) for $z \sim 3$ galaxies.

Given the faintness of the galaxies ($R \sim 24 - 25$), low resolution spectra with good s/n are beeing produced only in the last period from observations with the Keck LRIS instrument. Steidel et al. (1996) confirmed with low resolution spectra the identification of 15 galaxy candidates in the expected redshift interval showing the high success rate ($> 70\%$) of this multicolor selection. Extrapolating the success rate obtained for the subsample of their candidates, Steidel et al. (1996) provide a first estimate of the surface density of galaxies at $z \sim 3$ of the order of 0.4 arcmin^{-2} corresponding to a comoving volume density of 3.6×10^{-4} h$_{50}^{3}$ Mpc^{-3}. The average rest frame UV luminosity of these galaxies would imply a cosmological star formation rate SFR$\sim 3 \times 10^{-3}$ M$_{\odot}$ yr^{-1} Mpc^{-3}.

However, the selection of galaxies at redshift $z > 4$ becomes considerably more efficient if we take into account the complex absorption produced by the intergalactic medium in the UV spectrum of high z sources. The reddening produced by the IGM in the colors of high z quasars was investigated by Giallongo & Trevese (1990) and a considerable number of very high z quasars has been discovered by means of this multicolor technique (e.g. Warren et al. 1991; Irwin, McMahon & Hazard 1991). Recently, Madau (1995) has refined and applied this multicolor method to the selection of high z star forming galaxies.

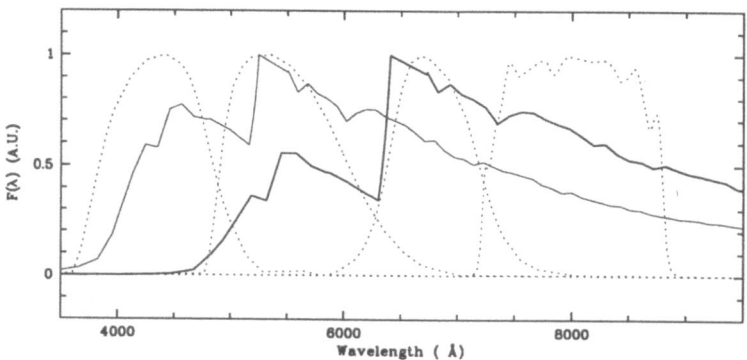

Fig. 1. Spectra of constant star-forming galaxy emitting at $z = 3.25$ (thin line) and $z = 4.25$ (thick line) depressed by Lyman absorption by the intergalactic medium. A BVrI filter set (from left to right) has been superimposed (dashed lines).

We have plotted in Fig.1 the average IGM absorption affecting the spectral properties of a constant star-forming galaxy emitting at $z = 3.25$ or $z = 4.25$. We have adopted the Madau (1995) absorption model and the galaxy spectrum by Bruzual & Charlot (1993) with a Salpeter IMF. First, it is to notice that, at given redshift, the absorption by IGM is characterized by the average Lyman alpha forest absorption present just shortward of the galaxy Lyman alpha wavelength and by the absorption of the overall Lyman series down to the Lyman continuum absorption, where the IGM is fully opaque to the UV radiation. While at $z \sim 3$ the Lyα forest absorption produces a fractional decrement of only $\sim 30\%$, at $z \sim 4.5$, 60–70% of the galaxy emission is lost causing a strong and easily detectable reddening in the broad band colors which sample the relevant wavelength interval.

An efficient sampling of this complex absorption requires at least 4 broad band filters. We have chosen the BVI Johnson and Gunn r filters to extend the multicolor selection up to $z \sim 4.5$ (Fontana et al. 1996). These filters are plotted in Fig.1 superimposed to the galaxy spectra.

The r-I color can select the intrinsic flat spectrum of any star-forming galaxy up to $z \sim 4.5$, while the V-r and B-r colors provide evidence of the strong reddening expected because of the Lyα and Lyman continuum IGM absorption, respectively.

To examine how robust is the color selection of $z > 2$ galaxies, we have computed the expected colors as a function of redshift in our photometric system (Fontana et al. 1996) adopting the spectral synthesis Bruzual & Charlot (1993) model. Models of these kind have a number of parameters whose uncertainties can be large in some cases. However, the resulting color changes of a few tenths do not alter the robustness of our color selection, as shown in the following.

To explore how the colors of different galaxy spectral populations are distributed in redshift, we have considered the *e-folding* star-formation timescale τ as the main interesting parameter. Different τ values reproduce different spectral types. For example, a star-formation timescale of $\tau \sim 1$ Gyr is more appropriate for an early type galaxy, while $\tau > 3$ Gyr represent the spectral properties of different late type galaxies. At each "observed" redshift, different ages (i.e. different formation redshifts ranging from 1 to 7) have been considered for galaxies with a given τ. A Salpeter IMF and a solar metallicity have been adopted.

Our relevant colors B-r, V-r, r-I are reproduced as a function of z in Fig.2 only for the case $\tau = 1$ Gyr.

The first remark that should be done is that the r-I color is sampling the intrinsic spectrum of galaxies in a wide redshift interval from $z = 0$ to $z \sim 4.5$. At $z > 4.5$ IGM absorption in the r band produces appreciable reddening in the r-I colors. In selecting galaxies in the redshift range $2.5 < z < 4.2$ the fundamental property of galaxies of all spectral types is the flatness of their rest-frame UV spectra revealed in their r-I colors (see Fig.2). Indeed, in the relevant z interval is always r-I<0.2 due to the intense star formation activity. At $z < 2.5$ the r-I colors are sampling progressively longer rest-frame wavelengths where the galaxy spectra are in general steeper, always resulting in r-I>0.2. Thus, it appears that the r-I color selection is very useful to discriminate high z galaxies in the field. Of course the presence of non-negligible photometric errors suggests the use of bluer colors to select high z galaxies with high confidence.

From Fig.2 it can be seen that the IGM absorption produces strong reddening first in the B-r colors with B-r$\sim$1 at $z \sim 3$ then in the V-r color with V-r$\sim$1 at $z \sim 4$. Thus, the simultaneous presence of the three colors at the average expected values can select galaxies at $\langle z \rangle \sim 3$ and at $\langle z \rangle \sim 4$ or even more. Any possible contamination by an old population with steep blue spectra (a pronounced 4000 Å break) producing red B-r and V-r colors at $z = 0.5 - 1$ can be avoided just requiring a "flat" r-I color.

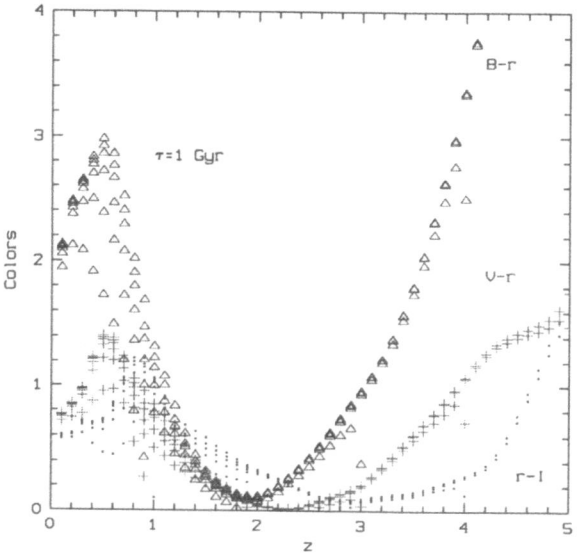

Fig. 2. Colors as a function of redshift for galaxies with star-formation timescale $\tau = 1$ Gyr. Different formation redshifts have been adopted in the interval $z = 1 - 7$.

3   A QSO Companion Galaxy at $z = 4.702$

We have applied this multicolor technique to the field around one of the brightest high z QSO BR1202-07 at $z = 4.694$ (McMahon et al. 1994, Storrie-Lombardi et al. 1996) where at least one very high z galaxy is close to the line of sight to the QSO as shown by the detection of a damped absorption system at $z \sim 4.4$ (Giallongo et al. 1994).

Deep BVrI images were obtained during the 1994 at the NTT with the SUSI direct imaging CCD camera in very good seeing conditions (FWHM$\sim$ 0.5-0.6 for the stellar objects in the r and I images). A diffuse object clearly stands out 2.2 arcsec NW of the QSO with an r magnitude of r=24.3. The companion galaxy has the unusual colors expected for star-forming galaxies at $z > 4$, i.e. r-I=0.2, V-r=1.9, B-r>3 (Fontana et al. 1996). On the basis of our multicolor selection criterion we estimated a probable redshift range $4.4 < z < 4.7$ depending on the intensity of the galaxy Lyman alpha emission. On the basis of the 1500 Å continuum flux measured in the I band we derived a star formation rate ~ 16 $M_\odot$ yr^{-1} (for a Salpeter IMF). This galaxy has also been detected in the K band by Djorgovski who estimated a magnitude K$\sim$ 23. The r-K$\sim$1 color so derived implies a very young age $< 10^8$ yr independently of details on the assumed metallicity and star-formation timescale.

This galaxy has also been observed in narrow band imaging centered at the Lyman α QSO redshift by Hu et al. (1996) and in imaging spectroscopy by Petitjean et al. (1996). Both authors discovered a Lyman α emission in the galaxy spectrum at $z \sim 4.7$. The strong Lyα emission increases the r flux ($\Delta r \sim -0.8$ mag) keeping a flat r-I color up to $z \simeq 4.7$ despite the strong attenuation in the r band due to the presence of the Lyα forest.

We have recently obtained a low resolution (15 Å) spectrum of this galaxy at the NTT with EMMI (D'Odorico et al. 1996, in preparation) which extends well in the red up to 9000 Å. The spectrum is shown in Fig.3 where the strong Lyα emission is detected at $z = 4.702$ corresponding to a proper distance from the QSO $\sim$ 600 kpc or equivalently to a velocity difference $\Delta v \sim 400$ km s^{-1}. The line flux $f \simeq 2 \times 10^{-16}$ ergs s^{-1} cm^{-2} corresponds to a luminosity $L_{Ly\alpha} \simeq 3.8 \times 10^{43}$ ergs s^{-1}. Although this line luminosity could be formally converted into a star-formation rate, some contamination by reprocessing of the QSO UV continuum can not be excluded even at distances $\sim$ 100 kpc. More important is the absence of any CIV emission within the flux measured in the I band. This implies that the redshifted I flux can be converted in the star-formation rate of ~ 16 $M_\odot$ yr^{-1} previously mentioned.

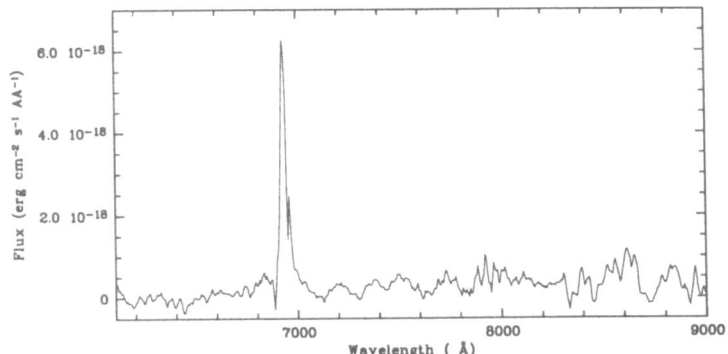

Fig. 3. NTT 15 Å spectrum of the QSO-galaxy companion showing a Lyα emission at $z = 4.702$.

These observations provide the first evidence of strong star formation activity at $z \gtrsim 4.5$.

4   A Sample of High Redshift Galaxies

In the 2.2×2.2 arcmin^2 field centered on the QSO position we have detected and counted galaxies in the r band down to r$\simeq$ 26 mag by means of the SExtractor software package (Bertin 1994). Reliable colors have been obtained for galaxies with r$\leq$ 25 mag. We have selected galaxies in two different redshift ranges. First, galaxies satisfying the criterion r-I<0.2 and B-r>1 are expected to lie in the redshift interval $3 \lesssim z \lesssim 4$. We found 11 galaxies at r$\leq$ 25 mag in this z interval corresponding to a surface density of 2.3 arcmin^{-2}. The derived average comoving volume density at $\langle z \rangle \simeq 3.5$ is $\phi \sim 10^{-3}$ Mpc^{-3}. The redshift interval $4 \lesssim z \lesssim 4.5$ has been selected imposing r-I<0.4, V-r>1 and B-r>2. We found 5 galaxies in the field corresponding to a surface density of 1 arcmin^{-2} and to a comoving volume density $\phi \sim 10^{-3}$ Mpc^{-3} at $\langle z \rangle \simeq 4.25$ (see Fig.4).

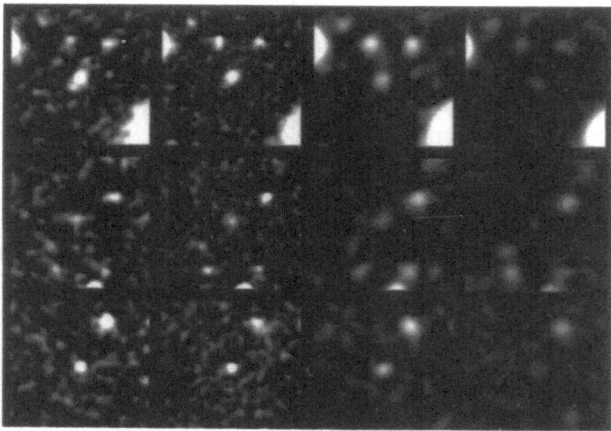

Fig. 4. IrVB (from left to right) images of three $z \gtrsim 4$ galaxy candidates.

Of course these estimates have to be considered as lower limits since galaxies at fainter r magnitudes will contribute somewhat to the volume density. Moreover, the selected galaxies have colors consistent with dust free spectral models. Although an intrinsic reddening E(B-V)$\lesssim$ 0.1 does not alter appreciably the r-I color selection, some high z dusty galaxies could be lost by our multicolor selection. The average $\langle I \rangle \sim 24.5$ mag of the galaxies at $3 \lesssim z \lesssim 4.5$ corresponds to an average star-formation rate ~ 8 M$_\odot$ yr^{-1}. The corresponding cosmological SFR per unit comoving volume is $\sim 10^{-2}$ M$_\odot$ yr^{-1} Mpc^{-3} in agreement with the value found by Steidel et al. (1996) at $\langle z \rangle \simeq 3.25$. This limit is about 2 times higher than the present value derived by Gallego et al. (1995) assuming a Salpeter IMF and 5 times lower that at $z \sim 1$ (Lilly et al. 1996). Thus the cosmological SFR increases by a factor of 10 from $z = 0$ to $z = 1$ then it seems to decline by a factor 5 or less up to $z = 4.5$. Assuming a fiducial local stellar mass density $\sim 3 \times 10^8$ M$_\odot$ Mpc^{-3} (Cowie et al. 1995) and an age for the $z > 4$ galaxies of a few 10^8 yr, a lower limit to the luminous matter density at $z \sim 4.25$

could be of the order of 1% of the local value. Of course our estimates are derived in a small field of 4.8 arcmin^2 centered on a high z QSO. Larger areas are needed to reduce density fluctuations.

5 Prospects for the VLT

The large collecting area of the VLT can be exploited to confirm and study high z galaxy candidates selected by multicolor photometry. The first and most obvious follow-up is the spectroscopic observation of galaxies down to r$\sim$ 25 mag by means of the mos capability present in FORS. For objects fainter than r$\sim$ 25.5, a different approach has to be pursued. Intermediate band filters (200–300 Å) can be used to extend the redshift identification to r$\sim$ 26.5 $-$ 27 mag in a reasonable observing time (Fontana et al. 1996, this volume).

References

Bertin, E. (1994), SExtractor Manual (Paris: IAP)

Bruzual, A. G., Charlot, S., (1993): ApJ, 405, 538

Cole, E., Aragón-Salamanca, A., Frenk, C. S., Navarro, J. F., Zepf, S. E. (1994): MNRAS, 271, 781

Cowie, L. L., Hu, E. M., Songaila, A. (1995): Nature, 377, 603

Cowie, AL. L.,Songaila, A., Hu, E. M., Cohen, J. G. (1996) AJ in press

Fontana, A., Cristiani, S., D'Odorico, S., Giallongo, E., Savaglio, S. (1996): MNRAS, 279, L27

Gallego, J., Zamorano, J., Aragón-Salamanca, A., Rego, M. (1995): ApJ, 455, L1

Giallongo, E., Cristiani, S., D'Odorico, S., Fontana, A., Savaglio, S. (1996): ApJ, in press

Giallongo, E., D'Odorico, S., Fontana, A., Savaglio, S., McMahon, R. G., Cristiani, S., Molaro, P., Trevese, D. (1994): ApJ, 425, L1

Giallongo, E., Trevese, D., (1990): ApJ, 353, 24

Haardt, F., Madau, P., (1996): ApJ, 461, 470

Hu, E. M., McMahon, R. G., Egami, E. (1996): ApJ, 459, L53

Irvin, M. J., McMahon, R. G., Hazard, C. (1991): *The Space Distribution of Quasars* (San Francisco: ASP), 21, 117

Lilly, S. J., Le Févre, O., Hammer, F., Crampton, D. (1996): ApJ, 460, L1

Macchetto, F. D., Lipari, S., Giavalisco, M., Turnshek, D. A., Sparks, W. B. (1993): ApJ, 404, 511

Madau, P. (1995): ApJ, 441, 18

McMahon, R. G., Omont, A., Bergeron, J., Kreysa, E., Haslam, C. G. T. (1994): MNRAS, 267, L9

Petitjean, P., Pécontal, E. Valls-Gabaud, D., Charlot, S. (1996): Nature, 380, 411

Steidel, C. C., Giavalisco, M., Pettini, M., Dickinson, M., Adelberger, K. L. (1996): ApJ, 462, L17

Steidel, C. C., Hamilton, D., (1992): AJ, 104, 941

Storrie-Lombardi, L. J., McMahon, R. G., Irwin, M. J., Hazard, C. (1996): ApJS, in press

Warren, S. J., Hewett, P. C., Osmer, P. S. (1991): ApJS, 76, 23

Part 5

Clusters and
Large-Scale Structure

Large-Scale Structure at High Redshift

Simon D.M. White

Max-Planck-Institut für Astrophysik, Karl-Schwarzschild-Straße 1,
D-85740 Garching bei München, Germany

Abstract. I discuss and illustrate the development of large-scale structure in the Universe, emphasising in particular the physical processes and cosmological parameters that most influence the observationally accessible aspects of structure at large redshift. Statistical properties of this structure can be measured from the apparent positions of faint galaxies and quasars; the structure can be mapped in three dimensions by obtaining redshifts for large samples of such objects; it can be studied using foreground absorption in the spectra of quasars; finally the mass distribution can be constrained by measuring the gravitationally induced distortion of background galaxy images. The first and last of these techniques require deep imaging of large areas of the sky with the best possible image quality. The second and third will require 8m-class telescopes with efficient multiobject spectrographs. For QSO absorption line spectroscopy high spectral resolution is also important.

1 Introduction

The term "large-scale structure" normally refers to the distribution of matter on scales larger than those of individual galaxies or galaxy clusters. On these scales objects have not yet had time to collapse fully and to come to equilibrium, and so their observed morphology is determined principally by the properties of the initial fluctuations from which they grew, and by the physical processes which amplified those fluctuations. As a result, there is hope that by studying large-scale structure we may learn directly about the mechanisms which imprinted irregularities on our otherwise almost homogeneous Universe.

Recent observational work on large-scale structure has focussed primarily on galaxy redshift surveys, the most recent example, and also the largest so far, being the Las Campanas Redshift Survey of more than 25,000 galaxies (Shectman et al 1996). The planned 2dF and Sloan surveys will increase this already impressive number by more than an order of magnitude. Surveys of this kind can be analysed in many ways but two broad approaches can be distinguished, quantitative analysis of low-order statistics and determinations of the morphology and topology of structure through detailed maps. The first approach typically aims to discriminate between specific models such as the many variants of the cold dark matter (CDM) model, while the second is more empirical and might, for example, provide a test for the broad class of theories which assume that structure grew through gravitational amplification of an initially gaussian density fluctuation field. In the nearby Universe independent distances to galaxies can be measured sufficiently well for studies of peculiar velocities and large-scale flows to provide powerful additional constraints on the large-scale mass distribution.

Theoretical discussions of large-scale structure vary from the development of purely descriptive statistics for the present distribution of galaxies (e.g. power spectra, position and velocity correlations of all orders, counts-in-cells, void probabilities and the relations between these quantities) through dynamical treatments of structure growth based on perturbation theory, to massive attempts to simulate the development of structure from the linear into the fully nonlinear regime. In this contribution I will focus mainly on the latter since simulations produce results that are easy to appreciate visually and can be compared directly with observational data. The descriptive statistics do, of course, provide the main quantitative comparison between simulation and observation, while the perturbation theory gives a powerful means for checking that the numerical simulations are, in fact, correct. The main difficulty when comparing theory and observation is that most current simulations predict the distribution of dark matter whereas almost all the observational data refer to the distributions of gas or of galaxies. The first attempts to simulate both gas dynamics and galaxy formation have now been carried out, but much of the physics cannot be treated properly and the results must be regarded as very uncertain.

In the next section I discuss simulations of structure formation in somewhat more detail in order to to show how the evolution of large-scale structure depends on the cosmological model in which it is occurring. I then summarise our current understanding of the relation between visible structure and that in the mass in order to indicate how the predictions of N-body simulations may be related to observation. Section 4 discusses the techniques available for measuring large-scale structure at high redshift and assesses what may be achievable with the next generation of telescopes if current ideas about structure formation are correct.

2   Evolution of Structure in the Dark Matter

Standard structure formation theories suppose the dark matter to be pregalactic, collisionless, and gravitationally dominant. The only significant agent affecting the recent development of its spatial structure is then gravity, and on large scales the distribution of the visible material is also structured primarily by the gravity of the dark matter. The currently most popular, and certainly the most thoroughly investigated structure formation models suppose that the dark matter is non-baryonic and that the initial deviations from uniformity were produced by quantum fluctuations during an early inflationary period. The latter assumption implies that density fluctuations at early times are a gaussian random field and so are fully specified by their power spectrum alone. The form of this power spectrum depends on the details of the inflation model and on the nature of the dark matter. Since early work showed that hot dark matter, specifically neutrinos with a mass of a few tens of eV, cannot produce the kind of large-scale structure we see (they produce structure too late and on scales which are too large) inflationary models now all assume the dominant matter constituent to be some form of cold dark matter (CDM). Topological relics of an early phase transition are another possibility for imposing structure on an otherwise uniform universe (e.g. Brandenberger 1994); I will not discuss them further here.

Within the general family of CDM models, several cures have been proposed for the inability of the original "standard" CDM model simultaneously to fit data on large-scale structure and the fluctuation amplitude measured by COBE. All involve adding an additional complexity to the model. Thus "tilted" CDM (or TCDM) supposes that a non-standard inflation model produces fluctuations with a slightly different scaling of amplitude with wavelength; hot plus cold, or mixed dark matter models (H+CDM or MDM) suppose that that a small fraction of the dark matter is in the form of stable massive neutrinos; τCDM supposes that the decay of an unstable massive neutrino at early times has left a relativistic neutrino background of higher density than in the standard model; ΛCDM supposes that a cosmological constant makes a significant contribution to the present energy density; open CDM (OCDM) supposes the curvature radius of the Universe to be comparable to its observable extent, rather than much larger as predicted by standard inflation. With a suitable choice of the additional free parameter each of these models can be made to give a rough fit both to the COBE amplitude and to observed large-scale structure.

To break the degeneracy between these models, one must appeal to other data. Large-scale flows may exclude low density models (Dekel 1995); combining measures of the Hubble constant and of globular cluster ages may exclude high density models (Freedman et al 1994); in a high density universe the observed baryon fraction in galaxy clusters may be inconsistent with big bang nucleosynthesis (White at al. 1993); the Hubble diagram for distant SNIa or the frequency of gravitational lensing of quasars may rule out models with a substantial cosmological constant (Perlmutter et al 1996; Kochanek 1995). For the purposes of this talk, however, the major difference between the high and low density models lies in the predicted evolution of large-scale structure with redshift.

This difference is illustrated in figures 1 to 3. These plots show thin slices through some large N-body simulations of a "standard" CDM model (SCDM) and of some variants that are generally thought to be consistent both with COBE and with present-day large-scale structure. These pictures were made by Joerg Colberg from simulations carried out on the Garching T3D parallel supercomputer as part of the programme of the Virgo Consortium (Jenkins et al, in preparation). The simulations used 17 million particles to follow the evolution of the matter distribution within comoving cubic regions of present size $240h^{-1}$Mpc; they are able to resolve structures down to a linear scale of $25h^{-1}$kpc and a mass scale corresponding to the halo of a Milky Way-like galaxy. The thickness of each slice is about 10% of its width. The dark matter distribution is smoothed adaptively to give an overdensity which is represented on the same logarithmic colour scale in all plots. Objects containing fewer than 20 particles are not visible.

Structure is much more prominent at $z = 0$ in the low density models (OCDM and ΛCDM) than in the Einstein-de Sitter models (τCDM and SCDM). This is a reflection of the well-known "bias' needed to make high density models consistent with the observed galaxy distribution. Although all four models have

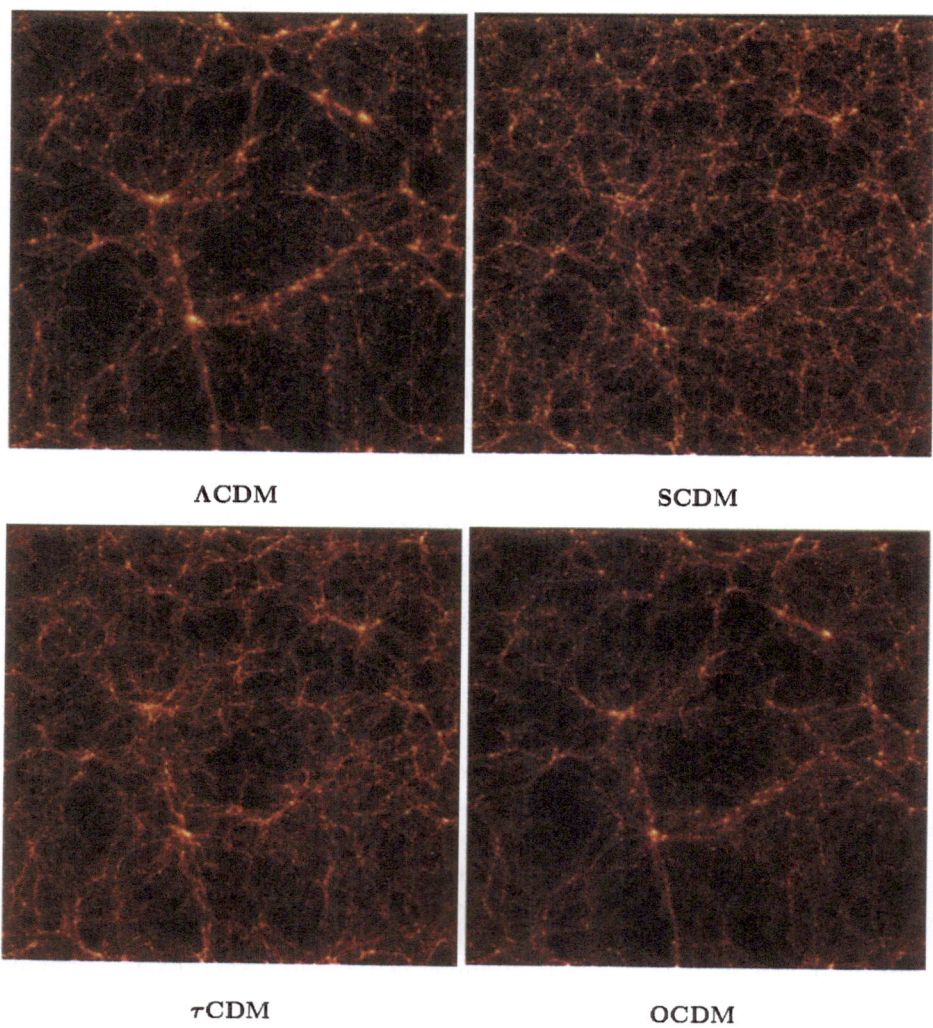

ΛCDM

SCDM

τCDM

OCDM

Fig. 1. Slices through simulations of four cosmological models at $z = 0$.

about the same abundance of massive quasi-equilibrium objects – rich galaxy clusters – models with a high total matter content achieve this with relatively low fluctuation amplitudes, lower, in fact, than those measured for the galaxy distribution. For low density models the required amplitude is a better match to the observed strength of galaxy correlations on large scales. Biasing must therefore enhance the contrast of structure in the galaxy distribution for the $τ$CDM and SCDM models (see, for example, fig. 16 of Davis et al 1985) whereas

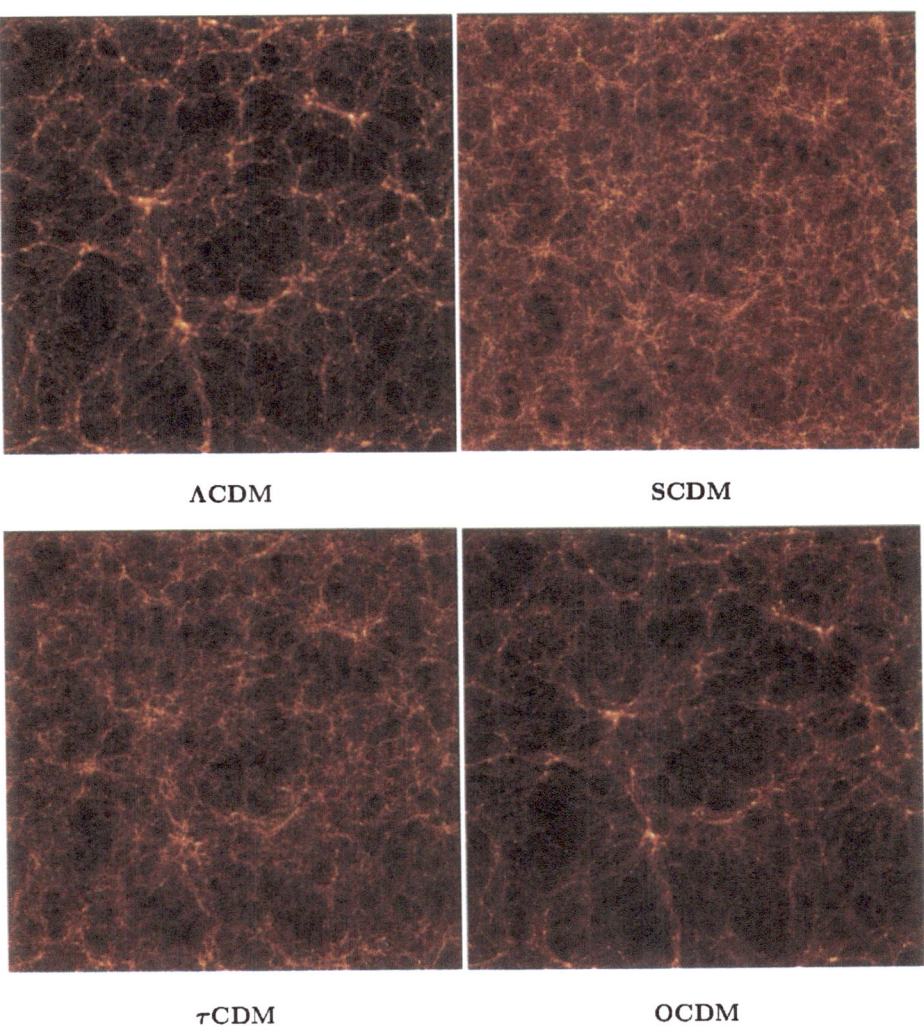

ΛCDM

SCDM

τCDM

OCDM

Fig. 2. Slices through simulations of four cosmological models at $z = 1$.

for the other models it is not required. Fig. 1 suggests, independent of this, that more fine-scale structure is to be expected in high density models. The difference between the two high density models gives a visual impression of the "lack of large-scale power" which has often been cited as ruling out standard CDM; large-scale correlations are consistent with those measured for galaxies and galaxy clusters in the τCDM model, but are too weak in SCDM.

The differences in evolution between the various models are quite striking at

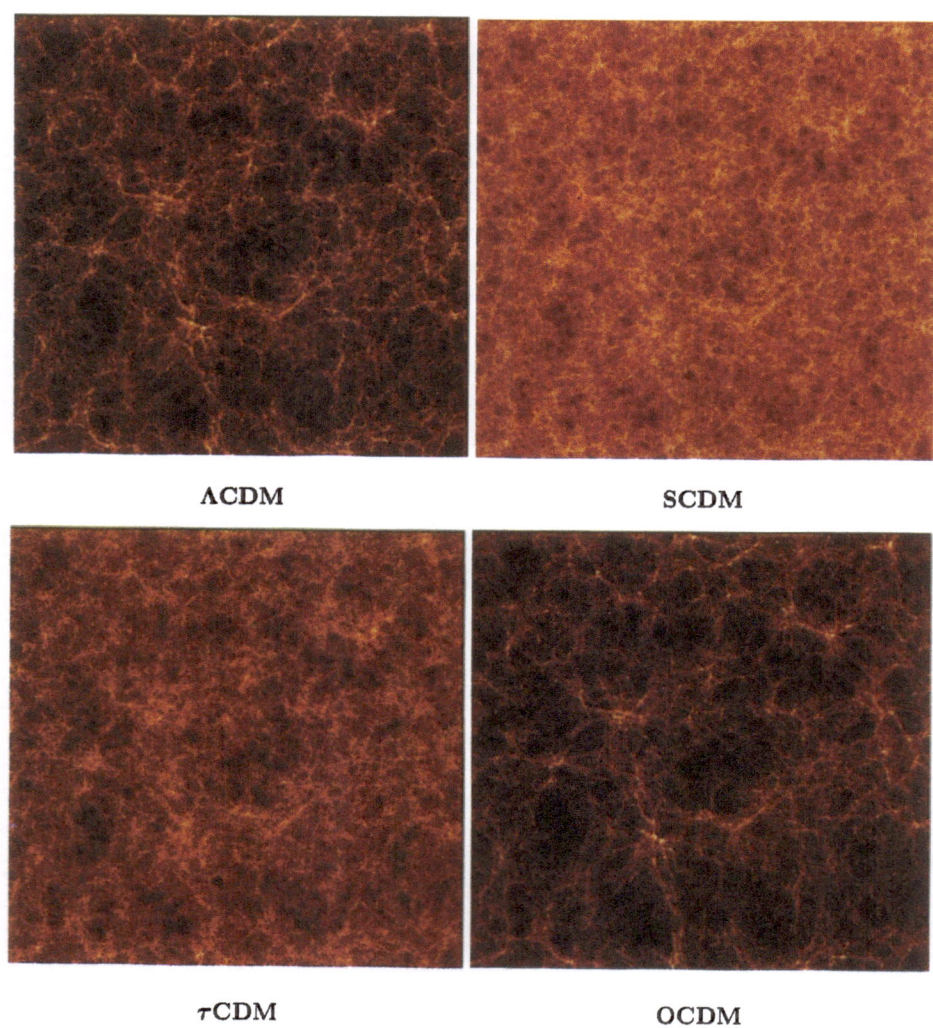

ΛCDM   SCDM

τCDM   OCDM

Fig. 3. Slices through simulations of four cosmological models at $z = 3$.

higher redshift. By a redshift of three the $\Omega = 1$ models look much more uniform, the ΛCDM model has changed rather little, and the OCDM model has hardly changed at all. These differences reflect, of course, the different behaviours of the linear growth factor. In the open case growth effectively "switches off" at $1+z \sim \Omega_0^{-1} \sim 5$. In the ΛCDM case this switch-off occurs at $1+z \sim \Omega_0^{-1/3} \sim 1.5$, while for an Einstein-de Sitter universe growth continues until $z = 0$. Thus in low density models we expect much more large-scale structure in the high redshift mass distribution than if $\Omega = 1$. Because of the bias galaxy clustering is expected to evolve less rapidly, but, as I discuss next, the galaxies themselves evolve more rapidly in this case. It is interesting to note that the pattern of the final large-scale structure is still visible at $z = 3$ in the high density models and would be more prominent in the biased but observable galaxy distribution.

3   Bias and its Evolution with Redshift

Except in the few situations where its gravitational lensing effects can be measured directly, the large-scale structure seen in figs 1 – 3 must be investigated using "tracers" like galaxies, galaxy clusters, quasars, or the gas seen as quasar absorption lines. The simplest method is to map out the spatial distribution of the tracer, to characterise its properties by some appropriate statistics, and then to use a model to relate the statistics to those of the dark matter. Although direct measurements of peculiar velocities for nearby objects are good enough to map the local mass distribution (Dekel 1995; Strauss and Willick 1996), this is impossible at higher redshift (except that peculiar velocities for galaxy clusters may be measurable using the kinematic Sunyaev-Zel'dovich effect, e.g. Haehnelt and Tegmark 1996). Peculiar velocities can be measured statistically at high redshift through the anisotropies they induce in the apparent spatial clustering of galaxies. Here, however, as with all clustering statistics, the interpretation hinges critically on the relation between the tracer and the mass distributions, in other words on the "bias".

In hierarchical clustering theories a good model for the bias of galaxy clusters can derived from the gaussian initial conditions (Kaiser 1984). The current amplitude of superclustering depends on cluster abundance and on the shape and amplitude of the linear power spectrum of mass fluctuations; it has no direct dependence on Ω and Λ (Mo et al 1996). The evolution of superclustering *does* depend strongly on the cosmological parameters because they change the history of the linear growth factor and so the amplitude of linear fluctuations at high z. To apply this test to a sample of distant clusters one would need to know only: (a) the abundance of the sample; (b) that it is effectively complete for all clusters more massive than some (possibly unknown) threshold; and (c) the amplitude of cluster-cluster correlations.

For more abundant tracers like galaxies or absorbing gas, much more physics must be included to get a realistic model for bias. In hierarchical theories residual gas is supposed to collapse dissipatively within the halos provided by the dark matter, settling to form centrifugally supported star-forming disks at their centres (White and Rees 1978; Fall and Efstathiou 1980). Recent work has shown that such a model, supplemented by Toomre's (1976) idea that ellipticals and bulges form by the merging of early stellar disks, can account qualitatively (and often quantitatively) for most of the systematics of the observed galaxy population; e.g. the present distributions of luminosity, colour, and morphology, and their correlation with environment (Kauffmann et al 1993), the counts, redshift distributions and morphologies of distant galaxies (Cole et al 1994; Kauffmann et al 1994; Heyl et al 1995; Baugh et al 1996a), the observed evolution of the population in rich clusters (Kauffmann 1995, 1996a; Baugh et al 1996b), and the star formation history of disk galaxies as inferred from nearby spirals and from the damped Lyα aborbers in quasar spectra (Kauffmann 1996b). This "semi-analytic" approach uses simplified but physically based models to treat each of the important processes (cooling, star formation, feedback of energy and of metals, evolution of the stellar populations, rates for galaxy merging). Its results can be compared with a much broader range of data than any feasible simulation. The main current difficulty, visible in most of the papers cited above, is a substantial overprediction of the number of faint galaxies in the local Universe. Much of the assembly and star-formation of galaxies is predicted to take place late (at or below $z \sim 1$) if $\Omega = 1$, suggesting rapid evolution of the tracers of large-scale structure; earlier formation is possible in low density universes.

Direct simulations which include a dissipative gas component have confirmed (or inspired) several aspects of the above work, showing that gas does cool off to make centrifugally supported disks at the centres of dark matter halos, and that these can plausibly be identified as the progenitors of galaxies and galaxy clusters (Cen and Ostriker 1992; Katz et al 1992; Evrard et al 1994). It is currently impossible to simulate the formation of individual galaxies in regions large enough to study large-scale structure, and the differing compromises with numerical limitations made by different groups show up as substantial discrepancies in their predictions for galaxy masses and sizes, for the fraction of gas turned into galaxies, etc. Although the results of these simulations are encouraging, none of their quantitative predictions for "galaxy" clustering can yet be considered reliable. The situation may be better in the case of quasar absorbers. Simulations which include both the dark matter and a dissipative gas component subject to a photoionising UV background seem to give considerable insight into the nature of the absorbers and into their spatial distribution. It appears relatively easy to explain both the observed abundance as a function of HI column density and the observed coincidence rate between neighboring lines-of-sight (Cen et al 1994; Hernquist et al 1996; Katz et al 1996). It is nevertheless still too soon to conclude that the simulations have converged to the physically correct answer.

The most promising approach to understanding galaxy bias and its evolution may be a combination of the semi-analytic galaxy formation models either with similar semi-analytic models for clustering (e.g. Mo and White 1996) or with N-body simulations which do not explicitly follow the gas. A first attempt at each of these routes was made by Kauffmann et al (1996). This paper shows how the bias in the present galaxy distribution can be calculated as a function of galaxy luminosity, colour, or morphological type, as well as how the consequences of bias for any particular statistic can be evaluated using N-body simulations. Extensions of this work to larger simulations such as those shown in figs 1 – 3 should allow much more detailed predictions for the evolution of large-scale structure in the galaxy distribution. These can then be compared directly with the kinds of data reviewed in the next section.

4   Measuring Large-Scale Structure at High Redshift

4.1   Angular Correlations

As shown most recently by the Hubble Deep Field, at faint magnitude limits the sky is covered with galaxies. With ground-based telescopes it is possible to get magnitudes, positions and colours for objects as faint as $B \sim 27$ whereas spectroscopy, even on 10m telescopes is limited to $B \sim 24$. The faintest galaxies seen are plausibly (although not necessarily!) the most distant, although the colours of those found in the HDF suggest, somewhat surprisingly, that only a few percent are at redshifts beyond 2.5 (Madau et al 1996; Lanzetta et al 1996; Steidel et al 1996). This appears to require most of the star formation and assembly of present-day galaxies to take place *after* $z = 2.5$. For these faint samples almost the only available clustering information is the angular two-point correlation function. This may be written as:

$$w(\theta) = A_\gamma \int_0^\infty dz \left(\frac{1}{N} \frac{dN}{dz} \right)^2 \xi(\theta d_A, z) \frac{H(z) d_A(z)(1+z)}{c} \tag{1}$$

where A_γ is a numerical constant which depends weakly on the slope of the correlation function, $N(z)$ is the number of objects per unit redshift, $\xi(r, z)$ is their spatial two-point correlation, $d_A(z)$ is the usual angular size distance, and $H(z)$ is the Hubble ratio. It is clear that three factors contribute to the observed $w(\theta)$: the evolution of the galaxies themselves affects $N(z)$; the evolution of their clustering affects $\xi(r, z)$; and the background cosmological model affects H and d_A. Notice that $N(z)$ and $\xi(r, z)$ depend on the precise magnitude and colour criteria used to define the sample since galaxy abundances and clustering amplitudes depend sensitively on luminosity and colour. Notice also that intrinisically different galaxy populations will contribute to the integral at different redshifts.

Current data show a steady weakening of $w(\theta)$ as fainter and fainter samples are considered (Brainerd et al 1995; Villumsen et al 1996). For $R > 25$ significant correlations have so far only been detected for $\theta \leq 1$ arcmin. This corresponds to scales well below 1 Mpc and so is not really what is normally thought of as large-scale structure. Comparison with data at $z = 0$ requires consideration of the evolution of clustering in the strongly *nonlinear* regime. The current results could be substantially improved by constructing good deep photometric samples over large fields. A particularly interesting possibility would be the use of colour criteria to isolate high redshift subsamples. This should be possible with the newest wide-field imagers on big telescopes. Preliminary studies of the dependence of $w(\theta)$ on colour selection criteria already show a strong, and as yet poorly understood effect (Landy et al 1996).

4.2   Deep Redshift Surveys

Recent deep redshift surveys include the Canada-France Redshift Survey discussed in this volume by F. Hammer and O. LeFèvre, the Anglo-Australian B-limited surveys discussed here by R. Ellis and M. Colless, and the Hawaii deep survey (Cowie et al 1996). The CFRS, for example, contains redshifts for almost 600 galaxies and is about 85% complete to a magnitude of $I = 22.5$. Its median redshift is greater than 0.5. As samples get deeper it becomes *much* harder to analyse them in an analogous way to local surveys. This is not merely because it is more difficult to get redshifts for individual galaxies, but also because the sampling volume has a very large extent in the redshift direction ($\sim 10^3 h^{-1}$Mpc) so that many redshifts have to be obtained before there are enough galaxies within any given structure (of size 20 to 50 h^{-1}Mpc) for it to be mapped clearly. The situation is made worse by the expected weakening of large-scale structure with increasing redshift, and, at very high redshift, by the fact that only a few percent of faint galaxies are at $z > 2$.

As can be seen from O. LeFèvre's contribution, considerations of this kind led the VIRMOS project to conclude that they need redshifts for 10^5 objects. The use of colour criteria could provide a well defined sample of preselected high redshift objects to give this kind of project a better lever-arm for studying the evolution of clustering. In practice, studies of large-scale structure at high redshift are likely to be restricted, at least initially, to measuring two-point correlation functions for galaxies on scales of a few Mpc. The considerations of previous sections suggest that the major difficulty in interpreting the results will lie in understanding the "bias" of the particular galaxy population observed. Indeed, this is already true for the CFRS where the major uncertainty in interpreting the measured correlations at $z = 0.5$ is in knowing which population of galaxies they should be compared with at $z = 0$ (LeFèvre et al 1996). It is unlikely that the problem of understanding the development of large-scale structure can be decoupled from that of understanding galaxy evolution.

Other approaches to structure at high redshift could involve samples of galaxy clusters or of quasars. Selecting clusters from optical data is much more complex than, say, measuring a correlation function, and it is particularly hard at high redshift because of the large number of foreground and background galaxies. Multiband colours ("photometric redshifts") can undoubtedly play a major role in enhancing the apparent contrast of clusters, and distant cluster selected by X-ray luminosity or Sunyaev-Zel'dovich decrement may eventually be available. Critical points when analysing such samples will be the influence of the observational selection criteria and the relation of the distant objects to nearer clusters. For quasars, of course, similar considerations apply. In both cases the samples are sparse and so the clustering signal is difficult to measure.

4.3   Large-Scale Structure in Absorption

There a number of major advantages to using quasar absorption lines to probe large-scale structure: the probability of detecting any particular absorber depends only weakly on its position along the line-of-sight; absorbing systems are abundant along each line-of-sight; usable lines-of-sight are quite common – quasars with $B < 20$ have a typical separation of about 15 arcmin; the lines appear due to relatively unevolved material so that their relation to other components, for example the dark matter, may be relatively easy to understand; the absorbing gas is plausibly the raw material for galaxy formation and studies of its distribution and metallicity should therefore clarify how galaxies form. The best strategy for carrying out a substantial survey of large-scale structure at $z \sim 2$ to 2.5 is probably to use a multiband, wide-angle photometric sky survey to identify quasar candidates; intermediate resolution spectroscopy on a 4m-class telescope can then yield a confirmed quasar sample with a suitable redshift distribution; finally, high resolution multi-object spectroscopy on an 8m-class telescope would provide good absorber samples along each line-of-sight.

The Sloan and 2dF surveys will be able to carry out the first two of these functions, but they will not have the resolution or sensitivity to see the abundant, low column density absorbers along lines-of-sight to quasars with $B \sim 20$. As a result their ability to study clustering of the absorbers, although very useful, will be limited by the sparseness of their absorber sample. (For example, the comoving abundance of detected CIV systems will be comparable to the local comoving abundance of Abell clusters.) High resolution spectroscopy on a large telescope is critical to being able to measure structure reliably on the relatively small scales where it is expected to be significant at $z \sim 2$.

4.4   Large-Scale Structure from Gravitational Lensing

Gravitational lensing can be used in at least two different ways to detect large-scale clustering in the mass distribution. The first employs the fact that coherent gravitational shearing of the images of background galaxies induces polarisation, that is to say, images which are near each other on the sky have a weak tendency to line up. This effect can be detected by correlating the orientations of galaxies as a function of their angular separation using large, deep, and high quality photometric images taken during excellent seeing. This is a difficult measurement because the gravitationally induced excess ellipticities are of the order of one per cent. So far only upper limits (Brainerd et al 1995) or tentative detections (Villumsen 1996) of the effect have been published. With better cameras on large telescopes firm detections are quite feasible. The measured quantity, the polarisation correlation function, depends on the redshifts of the background galaxies, on the amplitude, shape and evolution of the power spectrum of mass fluctuations, and on the cosmic geometry (see Blandford et al 1991). For geometric reasons most of the effect is induced at $z \sim 0.5$.

A second effect of lensing is produced by its magnification and demagnification of background galaxies. This can result in an apparent clustering even if the background objects are, in fact, unclustered. The strength of the effect depends on whether the increased abundance of galaxies in magnified regions, caused by the lifting of faint systems above the sample magnitude limit, is outweighed by the increased separation, which magnification also produces (Broadhurst et al 1995). In practice the combined effect is quite weak and must be considered in combination with the intrinsic clustering of the faint galaxies. A first theoretical analysis is given by Villumsen et al (1996). Since image orientations are not used, there are no additional requirements on image quality beyond those normally needed to measure angular correlations to faint magnitude limits.

References

Baugh C.M., Cole S., Frenk C.S., 1996a, MNRAS, in press
Baugh C.M., Cole S., Frenk C.S., 1996b, MNRAS, in press
Blandford R.D., Saust A.B., Brainerd T.D., Villumsen J.V., 1991, MNRAS, 251, 600
Brainerd T.G., Smail I., Mould J.R., 1995, MNRAS, 275, 781
Brandenberger R., 1994, Int.J.Mod.Phys., A9, 2117
Broadhurst T.B., Taylor A.N., Peacock J., 1995, ApJ, 438, 49
Cen R., Ostriker J.P., 1992, ApJ, 393, 22
Cen R., Miralda-Escude J., Ostriker, J.P., Rauch, M. 1994, ApJ, 437, L9
Cole S. et al 1994, MNRAS, 271, 781
Cowie L.L., Songaila A., Hu E.M., Cohen J.G., 1996, AJ, in press
Davis M., Efstathiou G., Frenk C.S., White S.D.M., 1985, ApJ, 292, 371
Dekel A, 1994, Ann.Rev.A.Ap., 32, 371
Evrard A.E., Summers F.J., Davis M., 1994, ApJ, 422, 11
Fall S.M., Efstathiou G., 1980, MNRAS, 193, 189
Freedman W.L. et al, 1994, Nature, 427, 628
Haehnelt M.G., Tegmark M., 1996, MNRAS, 279, 545
Hernquist L., Katz N., Weinberg D.H., Miralda-Escude J., 1996, ApJ, in press
Heyl J.S., Cole S., Frenk C.S., Navarro J.F., 1995, MNRAS, 274, 755
Kaiser N., 1984, ApJ, 284, L9
Katz N., Hernquist L., Weinberg D.H., 1992, ApJ, 399, L109
Katz N., Weinberg D.H., Hernquist L., Miralda-Escude J., 1996, ApJ, in press
Kauffmann G., 1995, MNRAS, 274, 161
Kauffmann G., 1996a, MNRAS, 281, 487
Kauffmann G., 1996b, MNRAS, 281, 475
Kauffmann G., Guiderdoni B., White S.D.M, 1994, MNRAS, 267, 981
Kauffmann G., Nusser A., Steinmetz M., 1996, MNRAS, in press
Kauffmann G., White S.D.M., Guiderdoni B., 1993, MNRAS, 264, 201
Kochanek C.S., 1996, ApJ, in press
Landy S.D., Szalay A.S., Koo D.C., 1996, ApJ, in press
Lanzetta k., Yahil A., Fernandez-Soto A., 1996, Nature, 381, 759
LeFèvre O. et al, 1996, ApJ, in press
Madau P. et al, 1996, ApJ, submitted
Mo H.J., Jing Y.P., White S.D.M., 1996, MNRAS, in press
Mo H.J., White S.D.M., 1996, MNRAS, in press
Mould J. et al, 1994, MNRAS, 271, 31
Perlmutter S. et al 1996, in Canal R., Ruiz-LaPuente P., Isern J. eds, Thermonuclear
    Supernovae, in press
Shectman S.A. et al, 1996, ApJ, in press
Steidel C.C. et al, 1996, ApJ, in press
Strauss M.A, Willick J.A., 1996 Phys.Rep., in press
Toomre A., 1976, in Tinsley B.M., Larson R.B., eds, Evolution of Galaxies and Stellar
    Populations, Yale Univ. Obs, p401
Villumsen J.V., 1996, MNRAS, submitted
Villumsen J.V., Freudling W., DaCosta L.N., 1996, ApJ, submitted
White S.D.M., Rees M.J., 1978, MNRAS, 183, 341
White S.D.M., Navarro J.F., Evrard A.E., Frenk C.S., 1993, Nature, 366, 429

The CNOC Cluster Survey

R. G. Carlberg[1,2], H. K. C. Yee[1,2], E. Ellingson[1,3], S. Morris[1,4],
R. Abraham[1,4,5], P. Gravel[1,2], F. D. A. Hartwick[6] J. E. Hesser[4],
J. B. Hutchings[4], J. B. Oke[4], C. J. Pritchet[1,6], & T. Smecker-Hane[1,4,7]

[1] Visiting Astronomer, Canada–France–Hawaii Telescope, which is operated by the
National Research Council of Canada, le Centre National de Recherche Scientifique,
and the University of Hawaii.
[2] Department of Astronomy, University of Toronto, Toronto ON, M5S 3H8 Canada
[3] Center for Astrophysics & Space Astronomy, University of Colorado, CO 80309, USA
[4] Dominion Astrophysical Observatory, Herzberg Institute of Astrophysics, National
Research Council of Canada, 5071 West Saanich Road, Victoria, BC, V8X 4M6,
Canada
[5] Institute of Astronomy, Madingley Road, Cambridge CB3 OHA, UK
[6] Department of Physics & Astronomy, University of Victoria, Victoria, BC, V8W 3P6,
Canada
[7] Department of Physics & Astronomy, University of California, Irvine, CA 92717,
USA

Abstract. The CNOC cluster survey was designed to measure the cluster mass to
light ratio and the luminosity density of the Universe in the same range of redshifts
for the purpose of estimating the parameter Ω. We find that the clusters at $z \simeq$
1/3 have a mean ratio of virial mass to k-corrected Gunn r luminosity of $M_v/L =$
$288 \pm 49h\,\mathrm{M}_\odot/\mathrm{L}_\odot$, where quantities are estimated with $H_0 = 100h\,\mathrm{km\,s^{-1}\,Mpc^{-1}}$ and
$q_0 = 0.1$. The radially resolved profiles of galaxy density and velocity dispersion support
the hypothesis that the clusters are effectively in equilibrium and that the galaxies are
distributed similarly to the total mass. However, the virial mass needs to be reduced
by a factor of 0.70 ± 0.06, on the average, to account for background contamination,
which increases the virial radius, and the oft-neglected "surface term" of the virial mass
estimate. In the same photometric system the mean co-moving luminosity density has
a closure mass-to-light ratio of $1017 \pm 144h\,\mathrm{M}_\odot/\mathrm{L}_\odot$. After correcting for the virial mass
overestimate and the small differential evolution of galaxy luminosities in and out of
clusters, we find that $\Omega_0 = 0.18 \pm 0.05$ for those components of the mass field that fall
into clusters.

1 Galaxy Clusters and Ω

The mass-to-light ratio of rich galaxy clusters multiplied with the field luminosity
density estimates the mean mass density of the universe, ρ_0 (Oort 1958). The
cosmological density parameter, $\Omega_0 \equiv \rho_0/\rho_c$, is therefore the ratio of the cluster
M/L value to the $(M/L)_c \equiv \rho_c/j$ for closure (*e.g.* Gunn 1978). The resulting
Ω estimate has no dependence on H_0 for dynamically measured cluster masses
(Gott & Turner 1976). Rich galaxy clusters are unique because any internal
biases in the M/L value and population variations with respect to field galaxies
can be explicitly measured.

The CNOC sample was designed to create a dataset that allows complete internal control of most aspects of the cluster Ω estimate. It is essential that the sample be able to make a useful test of the equilibrium hypothesis and whether the virial mass is biased in some way. Furthermore, the sample should allow differential evolution of cluster and field galaxies to be measured. On the basis of some n-body simulation data (Carlberg et al. 1994) it was argued that these criteria could be best met within the constraints of available observational resources with a set of a dozen or so clusters at $z \simeq 1/3$ with a total of 3000 or so accurate redshifts. The cluster sample was chosen from the Einstein Medium Sensitivity Survey Catalogue of X-ray clusters (Gioia et al. 1990, Henry et al. 1992, Gioia & Luppino 1994) to have a high X-ray luminosity, $L_x > 4 \times 10^{44}$ erg s^{-1}, which helps guarantee that the clusters are reasonably virialized and have relatively high masses, makes them easier to study. The clusters are sufficiently distant that they have a significant redshift interval over which the surrounding field galaxies are nearly uniformly sampled in redshift. Observations were made at CFHT in 24 assigned nights in 1993 and 1994, of which 22 were usefully clear. The primary catalogues contain Gunn r magnitudes and $g - r$ colors for 25,000 objects, of which 2592 currently have velocities, with an average accuracy of 100 km s^{-1}. About one-third of the sample is cluster galaxies with the other two-thirds being field galaxies.

Table 1. Dynamical Parameters of the CNOC Clusters

Name	z	r_v h^{-1} Mpc	σ_1 km s^{-1}	r_{200} h^{-1} Mpc	M_v/L_r^k $h\,\mathrm{M_\odot/L_\odot}$	$\epsilon_{M/L}$
A2390	0.2280	3.154	1095	1.51	337	54
MS0016+16	0.5465	1.639	1243	1.32	260	79
MS0302+16	0.4246	0.874	646	0.75	262	105
MS0440+02	0.1965	1.843	611	0.87	383	110
MS0451+02	0.2008	2.261	927	1.31	410	76
MS0451−03	0.5391	1.269	1353	1.45	360	105
MS0839+29	0.1930	1.030	860	1.23	418	157
MS0906+11	0.1706	0.790	1834	2.67	1041	238
MS1006+12	0.2604	0.890	912	1.22	338	117
MS1008−12	0.3061	0.899	1066	1.37	316	88
MS1224+20	0.3255	0.994	798	1.01	330	135
MS1231+15	0.2351	1.467	645	0.88	226	47
MS1358+62	0.3290	2.393	910	1.15	229	30
MS1455+22	0.2568	1.027	1168	1.57	808	249
MS1512+36	0.3727	1.803	697	0.85	413	185
MS1621+26	0.4275	2.244	841	0.98	210	41

The virial mass is

$$M_v = \frac{3}{G}\sigma_1^2 r_v, \tag{1}$$

where σ_1 and r_v are defined below in Equations 2 and 3. Deciding which galaxies in redshift space are cluster members is fundamentally ambiguous. That is, a cluster galaxy with a completely reasonable line-of-sight velocity of 1000 km s^{-1} appears in redshift space at $10h^{-1}\,\text{Mpc}$ from the cluster's center of mass, far outside the virialized cluster and overlaying field galaxies. This complication increases in severity at larger distances from the cluster center, which our sample is specifically designed to probe. One straightforward solution to this problem is to subtract explicitly a mean density of field galaxies from the cluster.

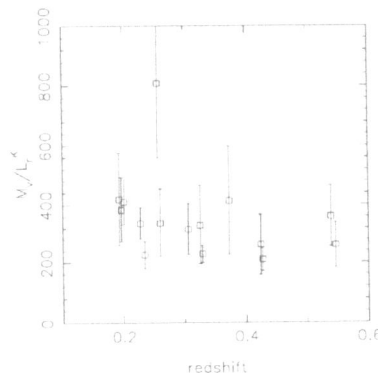

Figure 1: The virial mass to light ratios as a function of redshift.

The calculation of the characteristic internal energy of each cluster, σ_1, is based on iterating the classical estimator,

$$\sigma_1^2 = \left(\sum_i w_i\right)^{-1} \sum_i w_i (\Delta v_i)^2,\qquad(2)$$

where the $\Delta v_i = c(z_i - \bar{z})/(1 + \bar{z})$ are the peculiar velocities in the frame of the cluster and $\bar{z}$ is the weighted mean redshift of the cluster. The weights are calculated from the redshift and photometric catalogues to allow for statistical incompleteness of sampling (Yee, Ellingson & Carlberg 1996). The key to the use of this is to have an objective choice of the galaxies that are cluster members. The adopted method (Carlberg et al. 1996) is as follows. First a choice of the cluster redshift range is made which gives a set of galaxy velocities from which a trial σ_1 is calculated. Then all galaxies between 5 and 15 σ_1 are used to make a background estimate, which is subtracted from the weights of the galaxies selected to be in the cluster. If the σ_1 calculated from the background subtracted weights is within the 68% bootstrap confidence range (Efron & Tibshirani 1986), then the procedure is stopped. Otherwise the redshift limits are increased or decreased to find a convergence.

For each cluster an RMS "ringwise" estimate of the virial radius of the observed distribution is calculated (see Carlberg et al. 1996 for details)

$$r_v^{-1} = \frac{2}{\pi} \left(\sum_i w_i\right)^{-2} \sum_{i<j} w_i w_j \frac{2}{\pi(r_i + r_j)} K(k_{ij}),\qquad(3)$$

where $k_{ij}^2 = 4R_iR_j/(R_i + R_j)^2$ and $K(k)$ is the complete elliptic integral of the first kind in Legendre's notation. The angular extent of the sample is set by the observed field size. The ringwise estimate is an overestimate of the virial radius of a flattened cluster, for which we correct by a factor of 1.28 on the basis of comparison with the r_{ij} estimator. The resulting ratios of the cluster virial masses to the k-corrected cluster light measured in the Gunn r band are shown in Figure 1 and given in Table 1.

Our data are designed to include the entire virialized mass of the cluster. Analytic models (Gott & Gunn 1972) and simulations (Cole & Lacey 1996) find that the virialized mass is generally contained inside the surface where the mean interior density is $200\rho_c$. The calculation of r_{200}, the radius where the mean interior density is $200\rho_c(z)$, needs a crude assumption as to how the mass is distributed. Simplicity, and a great deal of work on clusters of galaxies and the properties of dark halos in general, suggests that the first approximation for the density-radius relation is the singular isothermal sphere, $\rho \propto r^{-2}$, in which case we find,

$$r_{200} = \frac{\sqrt{3}}{10} \frac{\sigma_1}{H_0\sqrt{\Omega_0(1 + z)^3 + (1 - \Omega_0)(1 + z)^2}}. \tag{4}$$

Note that $r_{200} \propto \sigma_1$ and is independent of the angular extent of the sampling, in so far as the cluster's σ_1 value has no radial dependence, which for the shallow gradients of the velocity dispersion profile found below is effectively true. At $z = 0.3$ the low density flat universe gives a value of r_{200} about 13% larger than in a open universe.

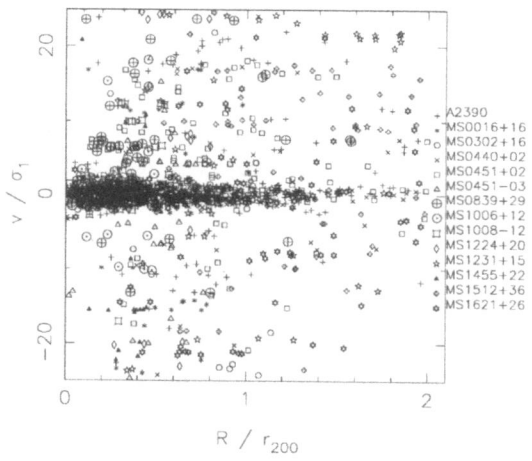

Figure 2: 14 clusters combined in velocity from the mean versus projected separation from the BCG.

One of the clusters selected, MS0906+11, is a redshift space binary and has an ill determined velocity dispersion and was dropped from all analyses. The cluster MS1358+62 having a substantial substructure on one side moving at 1000 km s^{-1} relative to the cluster center and is dropped from the profile analyses below. The scaled velocity-radius data for the remaining clusters are displayed in Figure 2. Note that the sampling in both radius and velocity is not uniform.

2   The Number Density and Velocity Dispersion Profiles

The surface density of the cluster is calculated as the sum of the statistical weights (which correct for nonuniform sampling: Yee, Ellingson & Carlberg 1996) for objects within $n_c = 3$ velocity units of zero velocity, divided by the area, minus the expected background over the same velocity interval. The position of the brightest cluster galaxy (BCG) is used as the center of each cluster.

The surface number density of cluster galaxies, $\Sigma_N(R)$, assuming symmetry about the BCGs is displayed in Figure 3. The volume galaxy density as a function of radius, $\nu(r)$, is needed for the tests of whether the galaxies are distributed like the mass. The relation between the volume density and the surface density is,

$$\Sigma_N(R) = 2 \int_R^\infty \nu(r) \frac{r}{\sqrt{r^2 - R^2}} \, dr. \tag{5}$$

A statistically adequate, analytically convenient model that describes these data is the Hernquist (alternatively designated as an $\eta = 2$) model (Hernquist 1990, Tremaine $et\ al.$ 1994). We fit the data centered on the BCG of Figure 3 to the projection of the volume density,

$$\nu(r) = \frac{A}{r(r + a)^3}. \tag{6}$$

The form $Ar^{-1}(r+a)^{-2}$ (Navarro, Frenk & White 1996) is an equally acceptable fit.

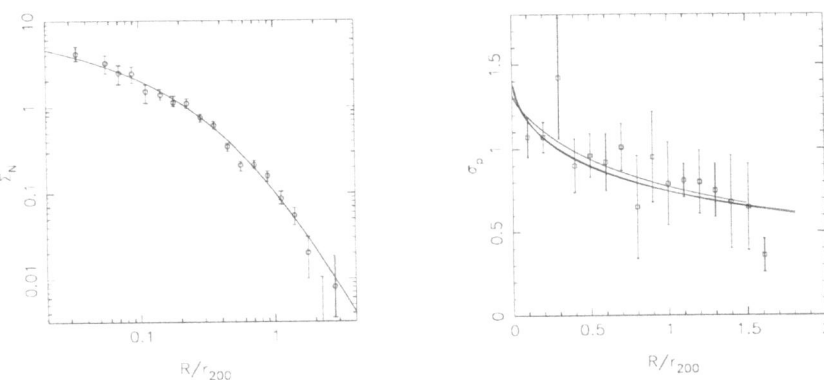

Figure 3: The projected number density profile (left) and velocity dispersion (right).

To derive a $\sigma_r(r)$ which is consistent with the observed $\sigma_p(R)$ we use the same approach as we used for the density profile. That is, we choose a reasonable functional form, then adjust two parameters, a scale length and a normalization, until its calculated projection, Eq. 8, minimizes the χ^2. The projected velocity dispersion data are shown in Figure 3. The adopted $\sigma_r(r)$ function should be finite at the origin, at large r it should tend to a Keplerian $r^{-1/2}$ for a convergent mass distribution, and it should be a simple smooth function in between. It is important that this function not assume that the galaxy populations are self-consistent with the mass density of the potential. We adopt the simple form,

$$\sigma_r^2(r) = \frac{B}{b+r},\tag{7}$$

for the radial velocity dispersion. This form, with B=1/4 and $b = a$, is the solution of the Jeans Equation (Equation 9 below) for $\beta = 1/2$ and $A = 1/(2\pi)$ in our assumed $\nu(r)$. The parameters of $\sigma_r(r)$ are adjusted until its projection,

$$\sigma_p^{\,2}(R)\Sigma_N(R) = \int_R^\infty \nu(r)\sigma_r^2(1 - \beta\frac{R^2}{r^2})\frac{r}{\sqrt{r^2 - R^2}}\,dr,\tag{8}$$

minimizes the χ^2 with the observed projected velocity dispersion.

3   Mass-to-Light Profiles

To test the relative distribution of mass and light we integrate our fitted $\nu(r)$ to give $L(r)$, the "light profile" (actually a number density profile), which is converted to a mass profile using a global M_v/L, calculated in the same manner as for the individual clusters. Then we compare the light-traces-mass predicted mass to the $M(r)$ derived from the fitted $\sigma_r(r)$ and $\nu(r)$ using Jeans' Equation,

$$\frac{\sigma_r^2}{r}\left[\frac{d\ln\sigma_r^2}{d\ln r} + \frac{d\ln\nu}{d\ln r} + 2\beta\right] = -\frac{GM(r)}{r^2},\tag{9}$$

where $\beta = 1 - \sigma_\theta^2/\sigma_r^2$ is the velocity anisotropy parameter. For the purposes of this paper β will be taken to be a constant, with a value suggested by n-body simulations. Our observational data have an RMS velocity dispersion over the observed radial range normalized to unity. Therefore the constant relating $\nu(r)$ and the mass profile here is fixed via the same constraint.

For constant β Equation 9 has a formal solution,

$$\sigma_r^2(r) = \frac{\int_r^\infty GM(x)\nu(x)x^{(2\beta-2)}\,dx}{\nu(r)r^{2\beta}}.\tag{10}$$

Satisfying the Jeans equation is a necessary condition for an equilibrium to exist, but it is not sufficient. The galaxy density profile, which we will call the light profile, is normalized to a mass using a mass-to-light ratio calculated in the same manner as done for our clusters individually. The $M(r)/L(r)$ ratio, in units where the virial mass-to-light ratio is unity, is displayed in Figure 4 for $\beta = 0., 0.5$ and 1.0. The errors are estimated using the variances of the observed $\sigma_p(R)$. Note that if $\beta > 2/3$ the implied mass is negative at small radius. The ratio $M(r)/L(r)$ is not a strong function of radius, or β, for $r > 0.5r_{200}$. We conclude from Figure 4 that the light-traces-mass assumption is consistent with the data at all radii. An average over the radial range $0.8 \leq r/r_{200} \leq 1.2$ for $\beta = 0.5$ gives $M(r_{200})/L(r_{200}) = 0.70 \pm 0.06\overline{M_v/L}$. That is, the virial mass-to-light ratio overestimates the true value of the mass. Doing the same fit for other β gives values that are either the same or smaller, with essentially identical errors.

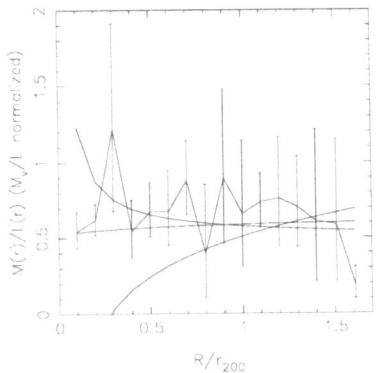

Figure 4: The normalized $M(r)/L(r)$ for $\beta = 0, 1/2$ and 1. The value would be unity if the virial mass predicted the Jeans mass correctly. At $R = r_{200}$ the result has essentially no β sensitivity and reasonable β indicate the light follows mass.

4   Corrections and Error Analysis

There is no significant difference in the galaxy luminosities or colours at any radius outside the cluster. That is, the field galaxies near the cluster appear to be indistinguishable from those far away. The color gradient appears inside $2r_{200}$, which is approximately where one expects cluster X-ray gas to be encountered. Compared to the field the average cluster galaxy (excluding the BCG) is between 0.05 ± 0.04 and 0.11 ± 0.07 magnitude fainter, depending on precisely how the measurement is made. We apply a correction of 10% to lower the field M/L estimate to $183 \pm 49 \, M_\odot / L_\odot$.

Splitting the cluster sample into independent blue and red subsamples shows the perils of the virial mass estimator. The blue galaxies are much more extended and have a 20% higher velocity dispersion than the red galaxies, which leads to a virial mass about four times larger than the red galaxies give. However each of these subsamples independently gives the same mass profile from the Jeans equation as found from the full sample, which we take as strong support for the assumption that both subsamples are effectively in equilibrium with the cluster potential.

5   Discussion and Conclusions

The overall goal of the CNOC survey is to use clusters of galaxies to derive a value of Ω_0 with a well determined error budget with particular emphasis on systematic errors. The major innovation of our analysis is that it is completely self-contained, with the key assumptions being testable, and that the error estimates are derived from the data themselves. We have shown that at r_{200} $M_{Jeans}/L = 0.70 \pm 0.06\overline{M_v/L}$, with essentially no dependence on assumptions about velocity anisotropy. The dominant source of error is random cluster to cluster variations, rather than the internal error from individual clusters. The global mass-to-light ratio (in our photometric system) of our sample clusters of galaxies is constant within our typical errors of 25% at a value of $288 \pm 49h \, M_\odot/L_\odot$ wich we have correct to a field value of $183 \pm 49 \, M_\odot/L_\odot$. Over the same redshift range we measure the closure value, ρ_c/j to be $1017 \pm 144h \, M_\odot/L_\odot$ (Carlberg et $al.$ 1996). After allowing for the slighly lower luminosities of cluster galaxies and the systematic overestimate of $\overline{M_v}$, we find that $\Omega_0 = 0.18 \pm 0.05$.

References

Carlberg, R. G., Yee, H. K. C., Ellingson, E., Pritchet C., Abraham, R., Smecker-Hane, T., Bond, J. R., Couchman, H. M. P., Crabtree, D., Crampton, D., Davidge, T., Durand, D., Eales, S., Hartwick, F. D. A., Hesser, J. E., Hutchings, J. B., Kaiser, N., Mendes de Oliveira, C., Myers, S. T., Oke, J. B., Rigler, M. A., Schade, D., & West, M. 1994, JRASC, 88, 39

Carlberg, R. G., Yee, H. K. C., Ellingson, E., Abraham, R., Gravel, P., Morris, S. M, & Pritchet, C. J. 1996, ApJ, 462, 32

Cole, S. & Lacey, C. 1996, MNRAS, submitted

Efron, B. & Tibshirani, R. 1986, $Statistical$ $Science$, 1, 54

Gioia, I. M. & Luppino, G. A. 1994, ApJS, 94, 583

Gioia, I. M., Maccacaro, T., Schild, R. E., Wolter, Stocke, J. T., Morris, S. L., & Henry, J. P. 1990, ApJS, 72, 567

Henry, J. P., Gioia, I. M., Maccacaro, T., Morris, S. L., Stocke, J. T., & Wolter, A. 1992, ApJ, 386, 408

Gott, J. R. & Gunn, J. 1972, ApJ, 176, 1

Gott, J. R. & Turner, E. L 1976, ApJ, 209, 1

Gunn, J. E. 1978, in $Observational$ $Cosmology$, eds. Maeder, A., Martinet, L, & Tammann, G. 1978 (Geneva Observatory: Sauverny)

Hernquist, L. 1990, ApJ, 356, 359

Navarro, J. F., Frenk, C. S., & White, S. D. M. 1996, ApJ submitted

Oort, J. H. 1958, in La $Structure$ et $L'Évolution$ de $L'Univers$, Onzième Conseil de Physique, ed. R. Stoops (Solvay: Bruxelles) p. 163

Tremaine, S., Richstone, D. O., Byun, Y-I., Dressler, A., Faber, S. M., Grillmair, C., Kormendy, J., & Lauer, T. R. 1994, AJ, 107, 634

Yee, H. K. C., Ellingson, E. & Carlberg, R. G. 1996, ApJS, 102, 269

Clusters of Galaxies and the VLT

Guido Chincarini[1,2], Luigi Guzzo[2], Davide Lazzati[1,2], and Roberto Scaramella[3]

[1] Università degli Studi di Milano, via Celoria 16, I-20100 Milano, Italy
[2] Osservatorio Astronomico di Brera, via E. Bianchi 46, I-22055 Merate (LC), Italy
[3] Osservatorio Astronomico di Roma, via dell'Osservatorio 2, I-00040 Monteporzio Catone (Rm), Italy

1 Introduction

In recent years we have witnessed a tremendous progress in the study of clusters of galaxies. This has been largely due to the detection of evolutionary phenomena as the Butcher–Oemler effect, the advent of the HST which allows to gain morphological information at any redshift, the detection of arcs and arclets as probes of the gravitational potential and of population of galaxies at larger redshifts and the surveys in the X–ray band. Among the most important goals to be achieved are the estimate of the mass and the temperature functions and the measure of the fraction of clusters with substructures, (Richstone et al. 1992), both as a function of z. In addition a large effort should go in determining at various redshifts the X–ray, and possibly the optical, Luminosity function.

The making of 8-10 meter-class telescopes will allow detailed studies of clusters at cosmological redshifts aimed to a full understanding of evolution, distribution of physical parameters in relation to the cosmological models and determination of cluster cosmography as a function of redshift. To fully exploit the capabilities of very large telescopes, however, we must know the physics and statistics of galaxy clusters at the present epoch, have and comprehend catalogues of clusters at high redshifts, and focus on key problems.

2 Catalogues of Clusters of Galaxies

The completeness and bias of the Abell (1958) and ACO (1989) catalogues have been discussed in the literature by many authors with different conclusions (see for instance Scaramella et al. (1991) and references therein). While it is certainly true that some clusters are missing at redshifts larger than $z = 0.1$ and others are affected by projection effects, the catalogues are reasonably complete for $R > 0$ and for redshifts smaller than $z = 0.12$, Scaramella et al. (1991); indeed most Abell clusters emit also in the X-ray band.

Local cluster densities for $R > 0$ are of about 8.7 (Abell catalog) and 12.5 (ACO catalog) in units of $10^6 (h^{-1} Mpc)^{-3}$ (Scaramella et al. (1991) among many others). At larger z ($z > 0.2$) the catalogues are rather incomplete. The main problem in the optical survey of distant clusters being that the contrast between the cluster and the background galaxies decreases rapidly with redshift, see for instance Postman et al. (1995). Gunn et al. (1986) in their pioneering search find at $z < 0.5$ a density of about 11 clusters/sqdegree and expect 45 to 63 at $z \sim 1$ without taking into account evolution. Later, Couch et al. (1991) compiled a catalogue of distant galaxy clusters lying mainly in the redshift range 0.2–0.6. Postman et al. (1995) in their extensive search find that clusters with $R \geq 1$ have a comoving density which is about 5 times larger than the Abell/ACO cluster density and it remains constant for $z < 0.6$. This catalogue, which consist of 79 clusters and reach redshifts beyond $z = 1$ urgently need a spectroscopic follow up. In spite of the large observational effort required, it is fundamental to carry out new complete and deep optical surveys.

In principle the best way to detect galaxy clusters is that of surveying the sky in the X–rays. Indeed the X–ray background is very low and deep observations would allow detection at any redshift. Available surveys, however, suffer mainly of the poor resolution off axis of the Wolter I optics and, in part, from the soft passband and small collecting area. In spite of these limitations, which could be cured in the future, with the Einstein Observatory and especially with the ROSAT satellite we gained, and are gaining, fundamental information on clusters.

The EMSS, Henry et al. (1992), is yet one of the most complete sample and consists of a total of 94 X–ray detected galaxy clusters. Rapid evolution has been claimed from this sample, based on an apparent lack of clusters at $z > 0.4$. The effect is, however, only marginally significant, as stressed by the authors themselves (see Rosati and Della Ceca, 1995, for a discussion of problems related to sample completeness). The EMSS counts ($f_X > 2 \times 10^{-13} erg/cm2/s$) are in good agreement with a Shechter luminosity function ($\Phi^* = 1.20 \times 10^{-6}$, $\alpha = 1.11$, $L^* = 1.34 \times 10^{44} erg/s$) and do not give any sign of evolution. Counts at these fluxes, on the other hand, are not very sensitive to evolution unless this is drastic and at very low redshifts. That clusters in the redshift range $0.4 < z < 0.7$ are missing, see figure 1, might be due to fluctuations due to statistics and incompleteness.

Cluster identification in the ROSAT All–Sky Survey (RASS) is under way. In particular, an ESO key–programme of redshift survey of ROSAT clusters in the Southern hemisphere , Guzzo et al. (1995), is progressing rapidly and we have recently completed the analysis of a subsample of bright clusters, De Grandi (1996). The sample, 111 clusters with flux $f_X > 3 \times 10^{-12}$, is estimated to be 97% complete and the distribution in redshift is fitted quite well by the preliminary Schechter Luminosity Function, figure 2, which has been derived by De Grandi in her dissertation work ($\Phi^* = 4.51 \times 10^{-7}$, $\alpha = 1.32$, $L^* = 2.63 \times 10^{44}$ erg/s). In this case the main identification procedure is based on the search for galaxy surface density enhancements corresponding to the X-ray source, called C-search. In general the need of looking at the optical counterpart to identify a cluster is a strong limitation especially at high z. Our strategy obviously introduce also a redshift cut–off in the identification, due to the requirement of actually seeing enough galaxies. This cut–off lies above $z = 0.2$ and does not affect the bulk of this sample even at lower fluxes.

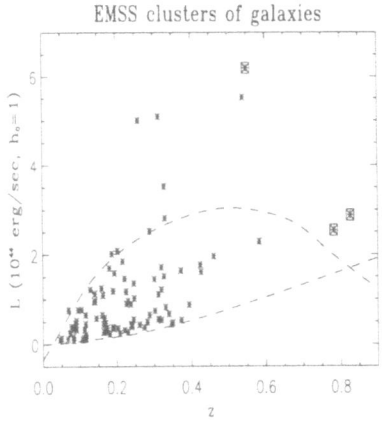

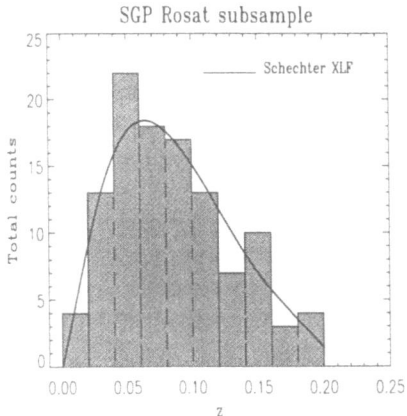

Fig. 1. Distribution of the EMSS Clusters in the Luminosity–redshift plane. The bottom line marks the sensitivity limit of the survey at various redshifts while the top line is given by the equation $N_{(>L,z)}$ =const (Chincarini (1996)).

Fig. 2. Redshift distribution of the 111 X–ray clusters in the sample. The continuous line is the expected distribution assuming a luminosity function as described in the text.

Table 1. Surveys of clusters: OP=optical selected; XRC=X–ray confirmed; HS=hybrid samples (X–ray + opt. cat.); PXRS=pure X–ray selected; SD=survey data; PD=pointed data.

OP & XRC	HS	PXRS
Ebeling '95 - SD	ESO KP - SD	NRASS - SD
Bower '94 - PD	XBCS - SD	NEP - SD
Castander '94 - PD	SGP - SD	RIXOS - PD
	Crawford '95 - SD	RDCS - PD

3 Surveys Underway

Except for the search in the I band mentioned by Postman et al. (1996) and some work progressing in the States (a search for galaxy clusters is forseen also with the Sloan Digital Sky Survey), too little is going on searching for clusters optically. In the X–ray band, on the other hand, we have a lot of activity in progress and that is summarized in table 1.

The Rosat Deep Cluster Survey (RDCS, Rosati 1995), in particular, is the deepest serendipitous search of X–ray clusters currently underway. It uses deep ($t > 15$ ksec) high galactic latitude ROSAT-PSPC pointed observations drawn from the public archive. The survey covers 26 sqdgrees with 130 PSPC fields, has 100 cluster candidates with a Flux $F_{[0.5-2.0]}$KeV $> 1 \times 10^{-14}$erg cm^{-2} s^{-1}, an off axis angle less than 15 arcmin and an extension larger than the PSF at a 3σ level. The CCD imaging and spectroscopic follow up show a very high success rate of identifications.

4 Conclusions

It is clear that in the last years we gained a lot of knowledge on galaxy clusters. That is not enough, however, and we really do not know yet whether or not we have a strongly biased picture. To improve our knowledge especially regarding statistics and distribution in the Universe at high redshifts we must plan:

- NTT surveys: preparatory work is progressing at ESO in collaboration with the european community.
- VLT spectroscopic followup & deep surveys on reasonably large area. To this end we could use of the second generation VLT instrumentation.
- X–ray Survey using high resolution and large field of view X–ray optics as it seems feasible using the technology of the Astronomical Observatory of Brera, Chincarini & Citterio (1995).

We thank Piero Rosati of Johns Hopkins University and Sabrina De Grandi of MPE for very usefull discussions and the ROSAT-KP team for the use of unpublished data.

References

Abell, G.O., 1958, ApJS,3,211

Abell,G.O., Corwin, H.G., Olowin, R.P., 1989,ApJ, 70,1

Bahcall, N.A., 1979. ApJ, 232, 689

Chincarini, G., 1996, in preparation

Chincarini, G. and Citterio, O., 1994, Milano preprint N. 140.

Couch, W.J., Ellis, R.S., Malin, D.F. and MacLaren, I., 1991, MNRAS, 249, 606

De Grandi, S., 1996, Ph. D. Dissertation at the Universita' degli Studi di Milano.

Gunn, J.E., Hoessel, J.G., Oke, J.B., 1986, ApJ, 306, 30

Guzzo,L. et al., in "Wide Field spectroscopy and the distant Universe", Ed. S.J. Maddox and A. Aragon- Salamanca, World Scientific Publisher

Henry, J.P., Gioia, I.M., Maccacaro, T., Morris, S.L., Stocke, J.T. and Wolter, A., 1992, ApJ, 386, 408

Postman, M., Huchra, J.P., & Geller, M.J., 1992, ApJ, 384, 404

Postman,M., Lubin, L.M., Gunn, J.E., Oke, J.B., Hoessel, J.G., Schneider, D.P., Christensen, J.A., 1996, Preprint, Sissa 9511011, To appear in AJ

Richstone, D., Loeb, A., Turner, E.L., 1992, AJ, 393, 477

Rosati, 1995, Sesto "Workshop on Observational Cosmology: from galaxies to galaxy systems", Proceedings

Rosati, P. and Della Ceca, R., 1995, proceedings of "Roentgenstrahlung from the Universe", Wuerzburg, Germany, Sept. 1995

Scaramella, R., Zamoramin, G., Vettolani, G., Chincarini, G., 1991, AJ, 101, 342

Visible Multi-Object Spectroscopy: Prospects and Requirements

Giampaolo Vettolani[1], Dario Maccagni[2]

[1] Istituto di Radioastronomia CNR, via Gobetti 101, I-40129 Bologna, Italy
[2] Istituto di Fisica Cosmica CNR, via Bassini 15, I-20133 Milano, Italy

Abstract. We present some considerations about the future role of redshifts surveys leading to constraints on the next generation of instruments on 8 meter class telescopes. We stress the fact that surveying galaxies at faint magnitudes and high redshift will require multislit spectrographs with high multiplex gain and working both in the visible and infrared spectral domains.

1 Introduction

Our present knowledge of the structure of the universe relies on the extension of the local redshift surveys to medium distances, and on snapshots of fainter field galaxies on which a considerable amount of data begins to accumulate. With respect to these faint objects, we are now in a situation similar to when surveys over small areas firstly showed the existence of large scale walls and voids later confirmed as general structures by the first CfA redshift survey. As it was the case then, the questions raised are more numerous than the answers found.

Shortly after the faint blue galaxy counts became available, it has been realized that starting from the local galaxy luminosity function and mix of morphological types, it was not possible to fit the data with a simple non evolving model, i.e. a model in which the luminosity and the spectral energy distribution of galaxies do not change in time. The excess of the observed counts over the prediction is of the order of 5 at $b_J \simeq 24$ (see Koo and Kron (1992) for a recent review).

Models allowing some luminosity evolution and changes in the spectral energy distributions (blueing due to star formation) give a better fit to the blue counts, especially with universes with low q_0. However, these models have been ruled out by the fact that they could not contemporaneously fit the redshift distribution of faint galaxies, since they predicted too many high z galaxies and a peak in the redshift distribution of magnitude limited samples at redshifts systematically higher than observed.

To solve this contradiction many recipes have been proposed, from an increase of merging activity with z, to the fading of dwarf galaxies in recent times, (which, in both cases, means changing the galaxy number density with cosmic time), to the introduction of a cosmological constant.

Much work is however still required both on the theoretical modelling and on the observational side. For instance the Hubble Deep Field images show that the morphology mix of galaxies is totally different at magnitudes around $I \simeq 25$, thus indicating a remarkable increase of irregular galaxies at high z (Abraham et al. 1996). Which is their redshift distribution? Are these objects undergoing a (merging induced?) starburst at some point of their life or are they galaxies forming their stars for the first time? Furthermore the density of the environment is a parameter that is known to influence the evolution of the stellar populations in galaxies, but up to now no detailed study of galaxy properties as a function of the environment is available, even at moderate lookback times.

2   Medium-Deep Surveys

Table 1 presents a summary of the medium and deep redshifts surveys currently available. The decrease in the number of observed objects (and area) with magnitude and mean redshift is obvious.

Table 1. Galaxy Surveys: Slices and Deep Fields

Name	mag	mean z	N galaxies	Instrument	Reference
Las Campanas	$17.7r$	0.08	23500	multifiber	Lin et al. 1996
ESP	$19.4b_J$	0.11	3500	multifiber	Vettolani et al. 1995
BES	$20.0 - 21.5b_J$	0.20	200	multifiber	Broadhurst et al. 1988
Autofib bright	$17.0 - 20.0b_J$	0.15	478	multifiber	Ellis et al. 1996
Autofib faint	$19.5 - 22.0b_J$	0.22	548	multifiber	Ellis et al. 1996
LDSS1	$21.0 - 22.5b_J$	0.25	160	multislit	Colless et al. 1993
ESO Sculptor	$20.5R$	0.30	700	multislit	BdL 1995
LDSS2	$22.5 - 24.0b_J$	0.50	73	multislit	Glazebrook et al. 1995
CFRS	$22.5I_{AB}$	0.60	600	multislit	Lilly et al. 1995a,b
HGS	$20.0K$	0.55	300	multislit	Songaila et al. 1995

Ideally, one would like to survey co-moving volumes larger than the typical size of the structures in the three dimensions at any z. In practice, the line of sight dimension is determined by the limiting magnitude of the sample, and the area on the sky by the galaxy surface density and the available telescope time.

It is evident that the increase in the number surface density of faint galaxies calls for a multislit spectrograph with the highest possible number of slits matching the surface density of galaxies at some faint limit. For example an 8m–class telescope should be equipped with a multislit spectrograph with at least 6 slits per square arcmin if one is going to survey galaxies around $b_j = 24.0$. This means that, if we assume a field of view of 6×6 arcminutes (the larger the better in order to have the most efficient use of telescope time) the ideal number of slits is of the order of 200. Clearly a match between surface density of slits, spectral resolution and detector size must be reached.

3   A Working Example: the ESP

We take as a working example the ESP redshift survey which we have recently completed at ESO (Vettolani et al. 1995). It is located near the South Galactic Pole at a mean declination of $-40°$ (1950). It consists of a one degree wide strip extending $22°$, plus, $5°$ West, a smaller strip extending $4.5°$. We have filled this area with a regular grid of circular fields with a diameter of 32 arcmin. This size corresponds to the field of view of the multifiber spectrograph (OPTOPUS) we have used at the ESO 3.6m telescope. We have obtained 3344 redshifts for galaxies brighter than $b_j = 19.4$ with a median velocity error of ~ 60 km s^{-1}.

3.1   Large-Scale Distribution

Figure 1 shows the observed large scale distribution of the 3344 galaxies of the ESP. Bearing in mind that this magnitude limited survey is $\sim 90\%$ complete and that the peak of the selection function is at $z = 0.12$, we see that voids and wall-like structures have characteristics sizes of the order of 50 and 100 Mpc respectively. Note that this survey has practically only two dimensions: one along the line of sight and only another one perpendicular to it.

3.2   Luminosity Functions

Figure 2 shows the luminosity function of all the ESP galaxies. The dashed line is the best fit Schecter function for magnitudes brighter than $M_{b_j} = -12.4$, and it has parameters $\alpha = -1.20$, $M_{b_j}^* = -19.59$ and $\phi^* = 0.020$. The faint end data points systematically lie above the best fit Schechter function, and we find that overall a better fit can be obtained with a Schechter function with parameters $\alpha = -1.13$, $M_{b_j}^* = -19.54$, $\phi^* = 0.021$, plus a power law with slope $\beta = -1.56$ at magnitudes fainter than $M_{b_j}^* = -16.99$. In any case, there is clear evidence for a steep faint end slope.

This steepening at the faint end of the luminosity function is almost completely due to galaxies with emission lines: in fact dividing galaxies with and without emission lines, using an equivalent width threshold of $5\mathring{A}$ for the [OII] $\lambda 3727$ line, we find significant differences in their luminosity functions. However, the number of intrinsically faint galaxies in the ESP is rather small, and, furthermore they are also faint nearby galaxies.

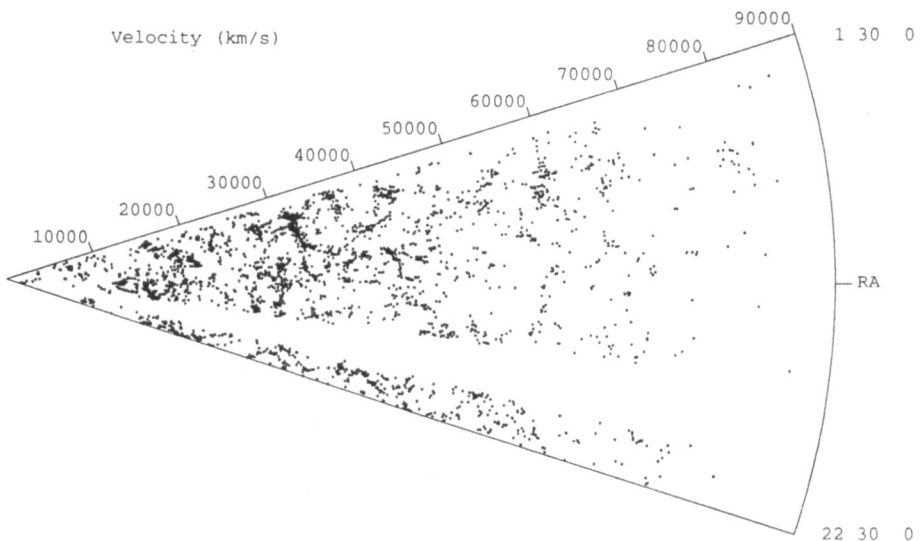

Fig. 1. Wedge Diagram of the ESP galaxies.

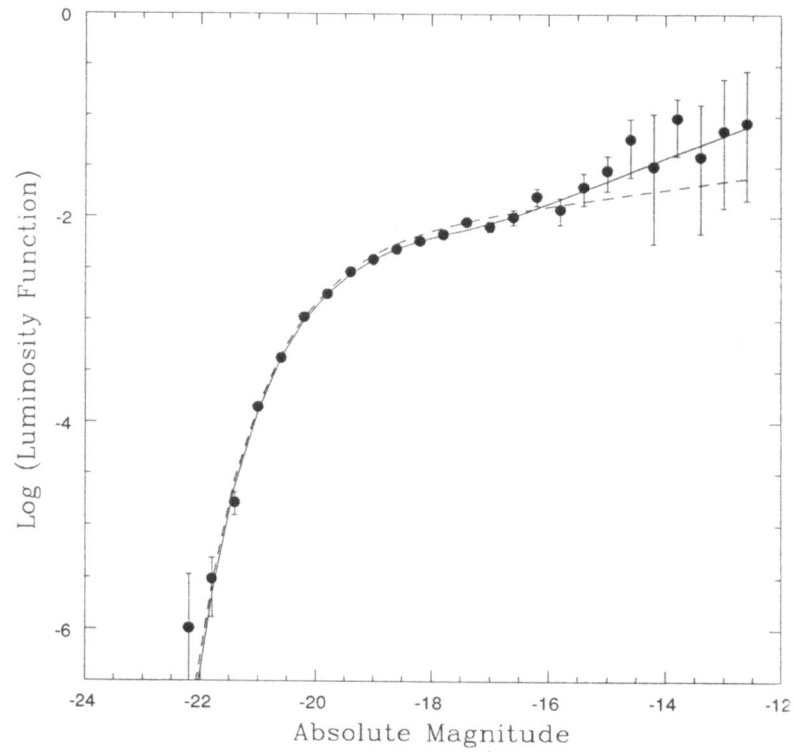

Fig. 2. ESP Luminosity Function

To learn more, and there is more to learn since evidence for evolution in the luminosity function of blue galaxies has been found in the CFRS (Lilly et al. 1995b), we need deeper, well defined and sizable samples of galaxies. A simple recipe which should be used as a guideline is $N = n(p) \times p$ where N is the total number of the objects in the survey, p is the number of parameters one wants to study (for instance, luminosity, density, morphological type) and $n(p)$ is the mean number of objects one needs to observe per bin per parameter for a statistically significant analysis. This easily leads to large numbers for physically meaningful studies (see Le Fèvre 1996, these proceedings).

Note than even the CFRS leaves open the question of the topology of the high-z universe, and only statistical indicators can be obtained (Le Fèvre et al. 1996).

4 Depth and Spectral Range of Future Surveys

Recently the observations by Steidel et al. (1996) of galaxies selected on the basis of the strength of the Lyman break at $912\mathring{A}$ have shown the existence of a population of galaxies at $z \simeq 3.5$ with typical luminosities of L_* galaxies and densities of the order of 1.4×10^3 per square degree (see also Giavalisco 1996, these proceedings). These galaxies represent 2% of the general population in the magnitude range $23.5 \leq R \leq 25.0$.

A sample of faint galaxies, for instance selected at $I \simeq 24$, well within the reach of the new generation of 8m–class telescopes, will span a redshift range between $z = 0$ and $z \sim 4.5$. Figure 3 shows the expected redshift distribution for such a sample. The distribution (Pozzetti 1996, private communication) has been constructed on the basis of the Bruzual and Charlot (1993) galaxy spectral energy evolution library and the pure luminosity evolution model presented in Pozzetti, Bruzual and Zamorani (1996) which fits both the counts and the redshift distributions of faint galaxies for an open universe, assuming the bulk of galaxy formation at $z = 4.5$.

Measuring redshifts requires to be able to identify a sufficient number of spectral features in order to obtain unambiguous determinations. As redshift increases, the most prominent lines move in and out of the visible, J and H range. We can distinguish three different regimes as to the galaxy redshift measurements. The visible domain allows to explore the redshift range $z \leq 1.2$, the J (and H) bands the range $1.2 \leq z \leq 2.5$, and finally in the range $z \geq 2.5$ the visible domain is again preferable.

It is clear that a sample of faint galaxies observed only in the visible or conversely only in the infrared will be quite heavily biased, well dwelling in the nearby low luminosity galaxies and in the high luminosity galaxies at high z, but missing completely the information at the peak of the selection function.

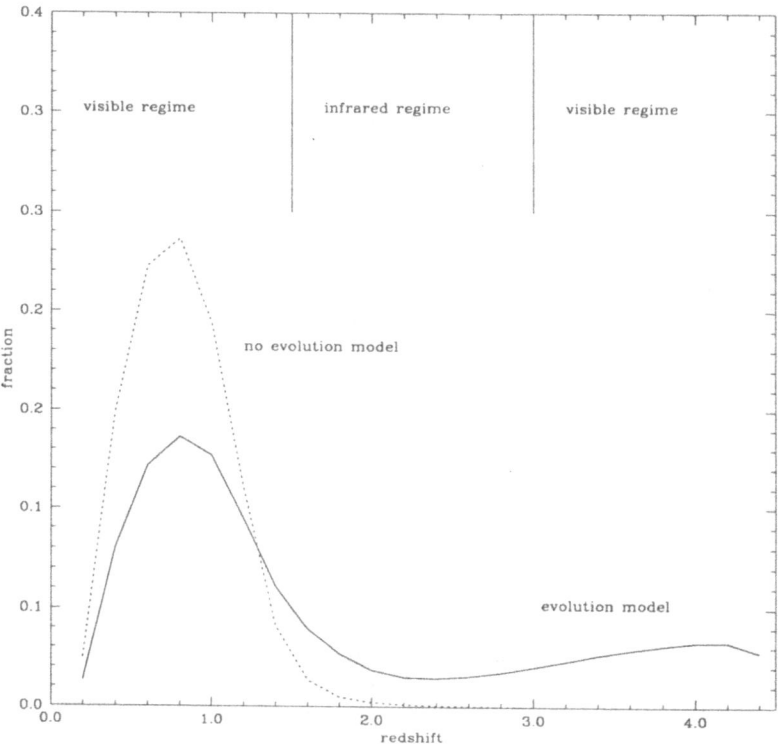

Fig. 3. Expected redshift distribution

5   Conclusions

Visible multi-object spectroscopy at large telescopes is really promising, provided that:

1. The multiplex gain is high, in order to allow obtaining data on large areas (volumes) to construct statistically significant samples. Note that a much larger number of objects must be studied than locally because a new variable (i.e. evolution of the stellar population) enters, plus complications of cosmology.
2. It is coupled with (near)IR multiobject spectroscopy (on the same areas).

If these two conditions were not satisfied, a whole part of the luminosity redshift plane would remain unexplored.

References

Abraham, R.G. et al. (1996): MNRAS, in press
Bellanger, C., de Lapparent, V. (1995): ApJ **455**, L103 (BdL)

Broadhurst, T. et al. (1988): MNRAS **235**, 827

Bruzual, G., Charlot, S. (1993): ApJ **405**, 538

Colless, M. et al. (1993): MNRAS **261**, 19

Ellis, R.S. et al. (1996): MNRAS **280**, 235

Glazebrook, K. et al. (1995): MNRAS **273**, 157

Koo, D., Kron, R. (1992): ARA&A **30**, 613

Le Fèvre, O. et al. (1996): ApJ, in press

Lilly, S.J. et al. (1995a): ApJ **455**, 50

Lilly, S.J. et al. (1995b): ApJ **455**, 108

Lin, H. et al. (1996): ApJ **464**, in press

Pozzetti, L., Bruzual, G., Zamorani, G. (1996): MNRAS, in press

Songaila, A. et al. (1995): ApJS **94**, 461

Steidel, C.C. et al. (1996): ApJ, in press

Vettolani, G. et al. (1995): in 35th Herstmonceaux Conference, p. 115

Dark Matter Searches with Weak Gravitational Lensing

Peter Schneider

Max-Planck-Institut für Astrophysik, Postfach 1523, D-85740 Garching, Germany

Abstract. The main applications of weak gravitational lensing to the investigation of, and the search for, dark matter are briefly reviewed. Particular emphasis is given to the requirements on the observational data quality, and it is stressed that imaging with Very Large Telescopes at excellent sites will play an essential role in this new branch of extragalactic research.

1 Introduction: What Is Weak Lensing, and What Is It Good For?

Light rays from distant sources are deflected if they pass near an intervening matter inhomogeneity. This gravitational lens effect is responsible for the well-established lens systems like multiply-imaged QSOs, (radio) 'Einstein' rings, the giant luminous arcs in clusters of galaxies, and the flux variations of stars in the LMC and the Galactic bulge seen in the searches for compact objects in our Galaxy. These types of lensing events are nowadays called 'strong lensing', to distinguish it from the effects discussed here: light bundles are not only deflected as a whole, but distorted by the tidal gravitational field of the deflector. This image distortion can be quite weak and can then not be detected in individual images. However, since we are lucky to live in a Universe where the sky is full of faint distant galaxies, this distortion effect can be discovered statistically. This immediately implies that weak lensing requires excellent and deep images so that image shapes (and sizes) can be accurately measured and the number density be as high as possible to reduce statistical uncertainties. Weak gravitational lensing can be defined as using the faint galaxy population to measure the mass and/or mass distribution of individual intervening cosmic structures, or the statistical properties of their mass distribution, or to detect them in the first place, independent of the physical state or nature of the matter, or the luminosity of these mass concentrations. In addition, weak lensing can be used to infer the redshift distribution of the faintest galaxies. After introducing the necessary concepts, I will list the main applications of weak lensing as isolated today and discuss some of them in slightly more detail, stressing the need for very deep images of the sky taking with instruments of excellent image quality.

2 Gravitational Lensing of Small Sources

A gravitational lens provides a map from the observer's sky to the undistorted sky, $\beta = \theta - \alpha(\theta)$, where θ is the angular position of a (point) source as seen by the observer, β the angular position of the source in the absence of the deflector, and $\alpha(\theta)$ is the (scaled) deflection angle the light ray undergoes near the lens (see Schneider, Ehlers & Falco 1992 for details). The deflection angle is a linear functional of the dimensionless surface mass density

$$\kappa(\theta) = \frac{4\pi G D_{\mathrm{d}} D_{\mathrm{ds}}}{c^2 D_{\mathrm{s}}} \Sigma(\theta) , \tag{1}$$

where $\Sigma(\theta)$ is the physical surface mass density, and D_{d}, D_{ds} and D_{s} are the angular diameter distances to the lens and source, and from the lens to the source, respectively. The distance factors in (1), and thus in the lens equation, imply that sources at different redshifts are affected differently; this fact will allow to estimate the redshift distribution of faint galaxies. If one considers sources whose angular extent is small compared to the typical angular scale over which the deflection angle changes appreciably, the distortion of such sources can be described by the locally linearized lens equation,

$$\delta\beta = A(\theta)\,\delta\theta , \quad \text{where} \quad A = \begin{pmatrix} 1 - \kappa - \gamma_1 & -\gamma_2 \\ -\gamma_2 & 1 - \kappa + \gamma_1 \end{pmatrix} \tag{2}$$

is the local Jacobian matrix. Its trace depends only on the local surface mass density κ, and its trace-free contribution describes the local tidal field; the shear $\gamma = \gamma_1 + \mathrm{i}\gamma_2$ is again linearly related to κ through

$$\gamma(\theta) = \frac{1}{\pi} \int \mathrm{d}^2\theta'\, \mathcal{D}(\theta - \theta')\,\kappa(\theta') \quad \text{with} \quad \mathcal{D}(\phi) = -\frac{\phi_1^2 - \phi_2^2 + 2\mathrm{i}\phi_1\phi_2}{|\phi|^4} . \tag{3}$$

Together with the fact that the surface brightness is unchanged by gravitational light deflection, the mapping (2) fully specifies the mapping of small sources; a circular source is mapped onto an ellipse, with axis ratio depending on the strength of the shear [or more precisely, on the strength of the reduced shear $g = \gamma/(1-\kappa)$], and the size determined by the magnification $\mu = \left[(1 - \kappa)^2 - |\gamma|^2\right]^{-1}$.

2.1 Shear (or Distortion) Effects

Thus, if one had a sample of circular sources, the matrix elements of A, and thus the local reduced shear, could be determined from the shape of the lensed images. But galaxies are not circular intrinsically, and so individual images do not constrain the local shear. However, making the basic assumption that the intrinsic orientations of an ensemble of galaxies taken from a large cosmic volume are randomly distributed[1] one can statistically infer the local distortion from an ensemble of galaxy images. Thus, any net orientation (or net ellipticity) of this ensemble in excess of Poisson noise is due to light propagation. One component of

---

[1] This assumption is not seriously challenged. The two-point angular correlation function of galaxies decreases rapidly with decreasing flux, implying that galaxies from a small angular region in the sky are typically all at different redshifts and thus not physically associated.

this light propagation occurs in the telescope and camera, so that an anisotropic PSF can cause a net ellipticity of the images. Putting that aside for a moment, the net ellipticity is due to gravitational light deflection. This implies that the tidal gravitational field can be measured locally: if ϵ denotes the (complex) image ellipticity[2], then its expectation value $E(\epsilon) = g(\boldsymbol{\theta}) = \gamma(\boldsymbol{\theta})/[1 - \kappa(\boldsymbol{\theta})]$. Identifying the local mean of ϵ with the expectation value, the reduced shear can be determined locally. Here, 'local' means a region around a point $\boldsymbol{\theta}$ which is small enough so that the reduced shear does not vary strongly over this region, and which is still large enough to contain a sufficiently large number N of images. The accuracy of a local shear measurement is $\sigma_\epsilon/\sqrt{N}$, where σ_ϵ is the dispersion of the intrinsic ellipticity of the sources, implying a limit on the angular resolution of the measurement. As a rough estimate, the magnitude of the reduced shear is several percent in clusters of galaxies, rising to order unity close to their centers, whereas the typical magnitude of the shear caused by the density fluctuations of the large-scale structure is of order one percent, depending on the cosmological model (see, e.g., Villumsen 1996 and references therein). A coherent shear signal in a cluster has first been discovered by Fort et al. (1988) and Tyson, Valdes & Wenk (1990).

2.2   Magnification Effects

In addition to changes of the image shape, the size of an image relative to the size of the source is affected. Surface brightness conservation then implies that the flux of the image is changed by the magnification μ. This latter effect changes the local source counts: if $n_0(> S)$ denotes the number density of sources with flux greater than S in the absence of lensing, the local number density is changed to $n(> S) = n_0(> S/\mu)/\mu$. Thus, if the counts deviate from an S^{-1} form, comparing the local number density to n_0 yields an estimate of the local magnification (Broadhurst, Taylor & Peacock 1995); this effect has first been discovered in the cluster A1689 by Broadhurst (1996). Alternatively, if $\omega_0(s)$ denotes the (appropriately defined) solid angle subtended by an unlensed source with surface brightness s, then in the presence of a lens, $\omega(s) = \mu\omega_0(s)$, and the magnification can locally be determined from $\mu = \langle \omega(s) \rangle / \langle \omega_0(s) \rangle$ (Bartelmann & Narayan 1995).

2.3   Practical Considerations

In order to obtain good angular resolution and/or high accuracy on the local determination of the shear and the magnification, one has to take very deep exposures to be able to work with a high number density of galaxy images. The images are affected by any residual anisotropic PSF which mimics a shear. In

---

[2] defined such that an elliptical image with axis ratio r has $|\epsilon| = (1 - r)/(1 + r)$, and the phase is twice the angle between the positive θ_1-axis and the major axis of the ellipse; for general image shapes, ϵ is defined in terms of the tensor of second brightness moments

order to correct for these instrumental effects, a stable PSF is needed, and good sampling of the PSF is required. It is also obvious that the seeing is crucial in this game: seeing circularizes small elliptical images and thus significantly reduces the shear signal. In order to regain the image ellipticities 'before seeing', correction factors have to be applied, which are determined by simulating images with the same PSF and comparing the input shear with that estimated from the convolved image with pixelization and noise added. These correction factors can be quite large and reduce the accuracy with which the shear can be measured significantly (for a detailed discussion on these methods, see Bonnet & Mellier 1996; Kaiser, Squires & Broadhurst 1995).

2.4 Digging in the Noise

These 'traditional' methods to determine the shear and magnification use the properties of (isolated) galaxy images; for each one has to determine a center and the tensor of second brightness moments, and the rest of the CCD is unused. Alternatively, one can use the two-point auto-correlation function (ACF) of the light distribution on the CCD, $\xi(\boldsymbol{\theta})$. This is related to the unlensed ACF $\xi^s(\boldsymbol{\theta})$ by $\xi(\boldsymbol{\theta}) = \xi^s(A\boldsymbol{\theta})$. Since the unlensed ACF can be assumed to be isotropic, the anisotropy of the observed ACF immediately yields the reduced shear g. One can calculate the ACF locally and determine g locally. In addition, since the ACF is caused by very many faint galaxies per solid angle, one might suppose that it is a universal function (which can be determined from deep HST exposures); in that case, also the magnification can be determined locally. To avoid being dominated by just the brighter objects on the frame, they can be cut out, so that one works on a field with the topology of a Swiss Cheese. Eventually, if all objects are cut out which are significantly detected, one works in the noise limit. If the ACF of the noise is caused by faint high-redshift galaxies, the value of g determined from the noise should agree with that determined from the images, but gives independent information. This method was proposed and successfully tested both on synthetic images as well as on real data; in the latter case, the shear field obtained from individual galaxy images has been reproduced by the ACF of the noise (van Waerbeke et al. 1996). The ACF method is also a sensitive diagnostics for testing image quality; improper data reduction shows up immediately as artificial features in the ACF.

3 Main Applications of Weak Lensing

In this section I will outline the main applications of weak gravitational lensing as currently known.

3.1 Reconstruction of Cluster Mass Profiles

Though historically not the first application of weak lensing, the reconstruction of the two-dimensional mass distribution of clusters has been the major application of weak lensing up to now. Tyson et al. (1990) discovered a shear field in two clusters and determined the radial mass profiles from that. Kochanek (1990) and Miralda-Escudé (1991) investigated how shear data can be used to constrain the mass profiles of clusters. The pioneering paper by Kaiser & Squires (1993) paved the way for a non-parametric two-dimensional mass reconstruction, by obtaining the inversion of (3),

$$\kappa(\boldsymbol{\theta}) = \frac{1}{\pi} \mathcal{R}e \left(\int_{\mathbb{R}^2} d^2\theta' \, \mathcal{D}^*(\boldsymbol{\theta} - \boldsymbol{\theta}') \, \gamma(\boldsymbol{\theta}') \right) + \kappa_0 \,, \qquad (4)$$

where the asterisk denotes complex conjugation, $\mathcal{R}e(z)$ is the real part of the complex number z, and κ_0 is an undetermined additive constant. The shear γ in the weak lensing regime ($\kappa \ll 1$) can be obtained from the local image ellipticities, as described above, and thus from an ensemble of images, the surface mass density can be evaluated by replacing the integral in (4) by a sum over images. This method was first applied by Fahlman et al. (1994) to the cluster MS1224, and they obtained quite a large lower limit for the mass-to-light ratio of this cluster. Since then, several more clusters have been investigated with that method. The method has been modified to allow the inclusion of strong lensing (Schneider & Seitz 1995; Seitz & Schneider 1995; Kaiser 1995) and to replace the integral over the $\mathbb{R}^2$ by one over a finite region represented by the data field (CCD; Seitz & Schneider 1996; see also Squires & Kaiser 1996). Using these generalizations, together with explicitly accounting for a broad redshift distribution of the galaxies, Seitz et al. (1996) have reconstructed the mass profile of the inner part of the cluster Cl 0939+4713 from a deep image taken with the HST. The resulting detailed two-dimensional mass map, when compared with the distribution of bright cluster galaxies, shows that the light traces the mass very well in this cluster. Also, the number density effect caused by the magnification has been discovered in this cluster. The mass-to-light ratio is only moderate (~ 200, depending on the mean redshift of the galaxies), but that should be no surprise: Cl 0939 is the highest-redshift cluster in the Abell catalog (A851) and therefore expected to have a very high optical luminosity. A low-resolution X-ray map (Schindler & Wambsganss 1996) indicates that also the X-ray emission traces the (dark) mass; this will be checked in more detail once a HRI map of this cluster becomes available.

The prospects of this method are simply excellent: deep images taking under good conditions will allow to study the dark mass distribution in clusters (e.g., the radial density profile, detection of substructure and ellipticity), independent of assumptions about symmetries or dynamical or thermal equilibrium of the matter. It therefore provides the least prejudiced mass distributions, and can be used to calibrate other methods, e.g., those using the X-ray profile and temperature (for example, see Squires at al. 1996). As stressed before, the accuracy of this method depends sensitively on the data quality, and on the available number density of galaxy images – thus on the depth of the observations. The combination of distortion and magnification effects, using maximum-likelihood techniques (Bartelmann et al. 1996), will increase the efficiency and accuracy of the reconstructions.

3.2   Statistical Properties of the (Dark) Mass Distribution in Galaxies

Individual galaxies are not massive enough to produce a significant shear signal, but statistically combining the signals from many (foreground) galaxies can yield a detectable 'relative alignment' of background images relative to the direction of the nearest foreground galaxy. First attempted by Tyson et al. (1984), this effect has now been discovered by Brainerd, Blandford & Smail (1996). Fitting a parametrized model to the alignment data, they have shown that the characteristic velocity dispersion (or rotational velocity) of galaxies is in the range expected from other investigations. In addition, they were able to obtain an interesting lower bound on the spatial extent of the dark halos in galaxies. This study was carried out with a relatively small number of galaxies; Schneider & Rix (1996) have shown that even with moderately-sized samples of galaxies, one can obtain very accurate determinations of model parameters such as σ_* or the characteristic size s_* of an L_* galaxy. In addition, the Tully-Fischer exponent can be probed, as well as the evolution of the mean redshift with apparent magnitude. All that is needed is a collection of wide-field images taking in excellent seeing conditions.

3.3   Detection of 'Dark' Mass Concentrations

On wide-field images, one can search for (dark) mass concentrations by looking for statistically significant alignments of faint galaxy images. Based on the aperture densitometry developed by Kaiser (1995), I have investigated the statistical properties of the appropriately-defined aperture mass calculated from the image ellipticities in annular regions (Schneider 1996). The expectation is to detect isothermal halos with velocity dispersion in excess of $\sim 600\,\text{km/s}$, without any reference to the optical or X-ray luminosity of these halos. Depending on the cosmological model, one expects about 10 such halos per square degree for a standard CDM model, increasing by a factor of order 10 in a COBE-normalized CDM model. This method will thus allow for the first time to investigate the statistics of dark halos without any assumption about bias factors, so that these results can be directly compared to numerical LSS simulations. In fact, dark halos have already been discovered by their shear effects: the 'dark' lens in the double QSO 2345+007 was discovered by the shear field it creates (Bonnet et al. 1993), and significant shear fields have been discovered around several high-redshift radio-loud quasars (Fort et al. 1996), supporting the magnification bias hypothesis for the associations of these QSOs with foreground galaxies (e.g., Bartelmann & Schneider 1994).

3.4   Constraints on the Redshift Distribution of Very Faint Galaxies

The dimensionless surface mass density κ depends on the source redshift through the factor $D_{\text{ds}}/D_{\text{s}}$; hence the lensing strength increases with increasing source redshift. This yields the possibility to obtain information about the redshift distribution of the faintest detectable galaxies, as proposed by Smail, Ellis & Fitchett (1994), Bartelmann & Narayan (1995) and others. In particular, the fact that significant shear was observed in the high-redshift ($z_d = 0.83$) cluster MS 1054−03 (Luppino & Kaiser 1996) shows that a large fraction of the galaxies used in this study ($21.5 < I < 25.5$) must have a redshift significantly larger than 1. For a different study of source redshifts from weak lensing, using the magnification effect, see Fort (1996).

3.5   Determination of the Power Spectrum of Cosmic Density Fluctuations

The density fluctuations of the mass inhomogeneities in the Universe distort light bundles from distant sources and can produce an observable effect. It has been shown in several papers (see, e.g., Villumsen 1996 for references) that the statistical properties of the distortion field are directly related to the power spectrum of the density fluctuations. For example, the two-point correlation function of the image ellipticity caused by the LSS is obtained by a convolution of its power spectrum with a known kernel function. Whereas the expected magnitude of the shear is quite small (of order 1%), its detection and quantitative investigation will allow to study the statistical properties of the density field in the Universe, on (comoving) scales much smaller than those achievable with CMB experiments, again without any assumption about bias factors.

4   Weak Lensing and the VLT

4.1   What Can the VLT Do for Weak Lensing?

One can hope that the VLT will provide deep images with superb image quality of clusters, groups and 'blank fields' (e.g., around QSOs) which can be used to carry out the science outlined above. It can also provide spectra (thus redshifts) of the brighter galaxies which can be used to calibrate the statistical redshift determinations from weak lensing. A dedicated, well-designed program on weak gravitational lensing can give the ESO community a substantial lead in the field. To increase the scientific output, a large field-of-view is desired (e.g., to improve the statistics in galaxy-galaxy lensing and LSS studies, to increase the search area for dark matter halos and the accuracy of mass reconstructions – and help to answer the question of 'where a cluster ends'), and queue scheduling is probably needed: One hour of observing at $0\rlap{.}''5$ seeing is likely to produce more useful data for many weak lensing programs than a whole night at $0\rlap{.}''8$ seeing.

4.2   What Can Weak Lensing Do for the VLT?

What sounds like a funny and provocative question is in fact serious: the high requirements on image quality and on the accuracy with which the PSF (and its spatial dependence) needs to be known implies that weak lensing is an ideal scientific test case for the imaging capabilities of the VLT. As mentioned before, the noise ACF provides a sensitive diagnostics on image artefacts. Of course, image quality can be tested without having weak lensing studies in mind. But due to the large scientific interest in this field, the cooperation of many experts from the ESO community in understanding the imaging properties of the VLT is guaranteed, and they most likely will try to push imaging at the VLT to its limits. It is therefore reasonable to expect that the VLT will not only belong to the handful of instruments with which weak lensing observations can reasonably be carried out, but that it will take the lead in this group.

I would like to thank my colleages M. Bartelmann, L. Da Costa, B. Fort, Y. Mellier, H.-W. Rix, C. Seitz, S. Seitz, and J. Villumsen for many enlightening discussions. This work was supported by the "Sonderforschungsbereich 375-95 für Astro–Teilchenphysik" der Deutschen Forschungsgemeinschaft.

References

Bartelmann, M. & Narayan, R. 1995, ApJ 451, 60.

Bartelmann, M., Narayan, R., Seitz, S. & Schneider, P. 1995, ApJL (in press).

Bartelmann, M. & Schneider, P. 1994, A&A 284, 1.

Bonnet, H., Fort, B., Kneib, J.-P., Mellier, Y. & Soucail, G. 1993, A&A 280, L7.

Bonnet, H. & Mellier, Y. 1995, A&A 303, 331.

Brainerd, T.G., Blandford, R.D. & Smail, I. 1996 ApJ, in press.

Broadhurst, T.J. 1996, preprint.

Broadhurst, T.J., Taylor, A.N. & Peacock, J.A. 1995, ApJ 438, 49.

Fahlman, G., Kaiser, N., Squires, G. & Woods, D. 1994, ApJ 437, 56.

Fort, B. 1996, this volume.

Fort, B., Mellier, Y., Dantel-Fort, M., Bonnet, H. & Kneib, J.-P. 1996 A&A, in press.

Fort, B., Prieur, J.L., Mathez, G., Mellier, Y. & Soucail, G. 1988, A&A 200, L17.

Kaiser, N. 1995, ApJ 439, L1.

Kaiser, N. & Squires, G. 1993, ApJ 404, 441.

Kaiser, N., Squires, G. & Broadhurst, T. 1995, ApJ 449, 460.

Kochanek, C.S. 1990, MNRAS 247, 135.

Luppino, G. & Kaiser, N. 1996, preprint.

Miralda-Escudé, J. 1991, ApJ 370, 1.

Schindler, S. & Wambsganss, J. 1996, preprint.

Schneider, P. 1996, MNRAS (submitted).

Schneider, P., Ehlers, J. & Falco, E.E. 1992, *Gravitational lenses*, Springer: New York.

Schneider, P. & Rix, H.-W. 1996, ApJ (submitted).

Schneider, P. & Seitz, C. 1995, A&A 294, 411.

Seitz, C., Kneib, J.-P., Schneider, P. & Seitz, S. 1995, A&A (in press).

Seitz, C. & Schneider, P. 1995, A&A 297, 287.

Seitz, S. & Schneider, P. 1996, A&A 305, 383.

Smail, I., Ellis, R.S. & Fitchett, M.J. 1994, MNRAS 270, 245.

Squires, G. & Kaiser, N. 1996, preprint.

Squires, G. et al. 1996, ApJ (in press).

Tyson, J.A., Valdes, F., Jarvis, J.F. & Mills Jr., A.P. 1984, ApJ 281, L59.

Tyson, J.A., Valdes, F. & Wenk, R.A. 1990, ApJ 349, L1.

Van Waerbeke, L., Mellier, Y., Schneider, P., Fort, B. & Mathez, G. 1996, in press.

Villumsen, J.V. 1996, MNRAS, in press.

Large-Scale Structure at High Redshift

Paul J. Francis

School of Physics, University of Melbourne, Parkville, Victoria 3052, Australia
E-mail: pfrancis@physics.unimelb.edu.au

Abstract. The evidence for large-scale structure at high redshifts is reviewed. I conclude that massive Mpc scale galaxy clusters probably exist at $z > 2$.

The 2139−4430 galaxy cluster at redshift 2.38 is discussed as a case study. This remarkable cluster may consist of ~ 10 Ly-α emitting galaxies, and as many as 100 other, less active galaxies, selected by a novel near-IR photometric technique. Most of these galaxies appear to be quite old ($\gtrsim 1 \mathrm{Gyr}$). Five of these galaxies form a very compact group, which may be a giant elliptical galaxy in the act of bottom-up formation.

1 Background

Does large scale structure exist in the high redshift universe? The answer to this question will be an important cosmological constraint. Theoretically, there should not have been time by $z \sim 2$ for large matter concentrations to have assembled (eg. Brainerd & Villumsen 1994). Galaxies may, however, be much more strongly clustered than the underlying matter; indeed some models predict biasing in the early universe much stronger than that found in the local universe (Fry 1996), especially in low Ω universes.

Low redshift ($z < 1$) observations, such as the weak clustering of faint blue galaxies (Efstathiou et al. 1991), and the rapid evolution in cluster properties out to moderate redshifts (Butcher & Oemler 1978, Edge et al. 1990), suggest that most large scale structure formed at redshifts well below one.

High redshift ($z > 1$) observations are much harder, but present a very different picture. Direct galaxy imaging and redshift surveys will be very difficult until a wide-field near-IR instrument comes on-line on an 8-m telescope, though a few clusters at $z \sim 1$ are being found (eg. Dressler et al. 1993). A more promising approach with existing telescopes is to use QSOs and QSO absorption-line systems.

The small number of existing measurements of clustering at high redshifts are summarised in Fig 1. Metal-line absorption systems, Ly-α emitting galaxies and QSOs all seem to cluster more strongly than galaxies today. This conclusion is reinforced by the discovery of a few spectacular individual absorption-line clusters (eg. Heisler, Hogan & White 1989, Jakobson & Perryman 1992, Foltz et al. 1993). Ly-α forest clouds, however, cluster much less strongly.

The data shown in Fig 1 has many flaws, but if valid, it leads to two important conclusions. Firstly, clustering strength is a strong function of absorption-line column density. The data in Fig 1 can be crudely parameterised as follows:

$$\xi = (r/r_0)^{-1.8} \text{ , where } r_0 = 0.1 + 0.2(\log_{10}(N_H) - 15) \text{ Mpc,} \qquad (1)$$

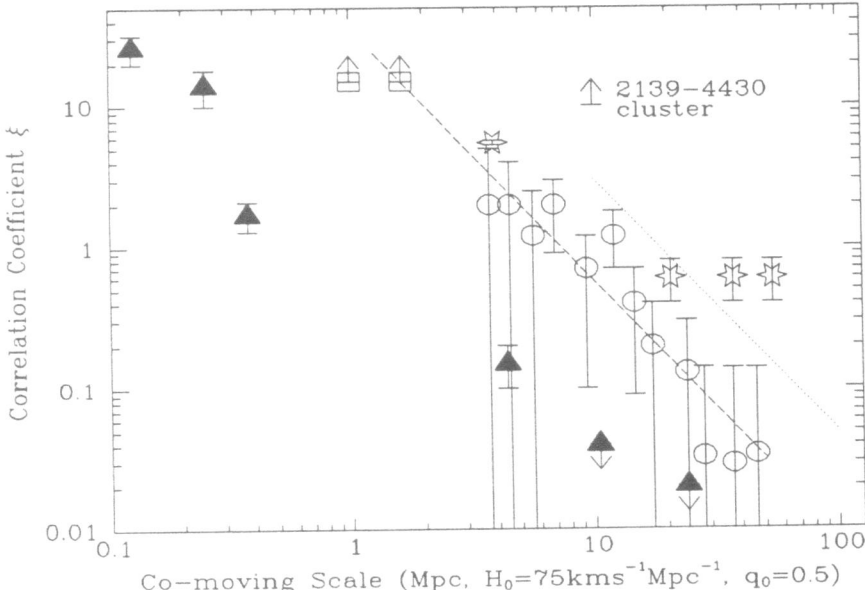

Fig. 1. Measurements of clustering at high redshift. The two-point correlation coeffi-cient ξ is shown as a function of co-moving scale. Triangles are for Ly-α forest systems (Fernández-Soto et al. 1995, Press & Rybicki 1993), circles for QSOs (Shanks & Boyle 1994), squares for the clustering of Ly-α emitting galaxies around damped Ly-α absorp-tion-line systems (Wolfe 1993, Francis et al. 1996) and stars for C IV absorption-line systems (Heisler et al. 1989). For comparison, the dashed line shows bright galaxies today, and the dotted line rich clusters today.

where r is the comoving scale (for $H_0 = 75$km s^{-1}Mpc^{-1} and $q_0 = 0.5$), and N_H the neutral hydrogen column density, per square cm. Secondly, all the classes of objects known to be associated with galaxies seem to cluster strongly, suggesting that large scale structure does exist as high redshifts.

2   The 2139−4430 Cluster

In this paper, I will describe a cluster of galaxies at $z = 2.38$, discovered and studied by my collaborators and I. This is the most convincing case for a cluster of galaxies at $z > 2$, as well as being one of the largest samples of high redshift galaxies discovered to date. As such, this cluster is a case study in the observa-tional techniques and the science that will come from VLT observations of large scale structures at high redshift. I cannot do justice to the scientific richness of this cluster in the space available, but most of the observations described here are published (Francis & Hewett 1993, Francis et al. 1996).

The 2139−4430 cluster was originally identified as a group of Lyman-limit QSO absorption-line systems at matching redshifts in the spectra of two QSOs (LBQS 2138−4427 and LBQS 2139−4434), separated by $8'$ on the sky. In one line of sight, we observed three absorbers, spread over $\sim 800 \mathrm{km\ s^{-1}}$, and with neutral hydrogen column densities of $\sim 10^{18.5} \mathrm{cm^{-2}}$. In the other line of sight, at an almost identical redshift, was a very similar absorption-line (Francis & Hewett 1993).

The odds of seeing this many QSO absorption-line systems in such a small redshift range are tiny ($< 10^{-4}$). This implies that some coherent structure of gas is present, either a Zel'dovich-type pancake, or a rich cluster of neutral hydrogen clouds. This structure, whatever it is, must have a size of at least ~ 10 co-moving Mpc to cause absorption in both lines of sight, and has an absorption-line overdensity of at least a factor of ten, compared to the field.

2.1 Ly-α Imaging

We decided to image this absorption-line structure, to see if we could identify any galaxies associated with it. We chose the unfashionable approach of narrow-band Ly-α imaging. This technique had fallen into disrepute, due to a series of non-detections (eg. Lowenthal et al. 1995, Pritchet & Hartwick 1990). The last three years, however, have seen a renaissance in Ly-α imaging, with a steady trickle of discoveries (eg. Møller & Warren 1993, Pascarelle et al. 1996), and it has the major advantage that it is easy to obtain redshifts for any galaxies identified with this technique, using only 4-m class telescopes.

We imaged a $12' \times 12'$ field centred on the absorption-line cluster, through a narrow-band (60 Å) filter tuned to the wavelength of Ly-α at the absorption-line redshift, $z = 2.38$. The imaging was carried out at prime focus of the AAT and CTIO 4-m telescopes. We detected 12 candidate Ly-α emitting galaxies, mostly extended (~ 2–5 arcsec), with emission-line surface brightnesses $\gtrsim 4.5 \times 10^{-17} \mathrm{erg\ cm^{-2} s^{-1} arcsec^{-2}}$.

Two of these candidates have spectroscopic confirmation (Francis et al. 1996), and are indeed Ly-α emitting galaxies at $z = 2.38$. We consider it probable that most of the remainder are also Ly-α emitting galaxies at this redshift, due to their similarity to the confirmed objects, and their very high observed-frame emission-line equivalent widths (> 100Å). The most likely alternative to Ly-α at $z = 2.38$ as the cause of the narrow-band excess is [O II] at $z = 0.1$, but [O II] emitting galaxies at $z \sim 0.1$ are usually easy to recognise, and typically have much lower equivalent widths. This is one advantage of working at $z \sim 2.5$ rather than $z > 3$; there are fewer rest-frame optical emission-lines to cause confusion, and discriminating between an [O II] emitting galaxy at $z \sim 0.1$ and a Ly-α emitting galaxy at $z \sim 2.5$ is much easier than discriminating between an [O II] emitting galaxy at $z \sim 0.7$ and a Ly-α emitting galaxy at $z \sim 4$.

If the remaining ten candidate Ly-α emitting galaxies are confirmed, this will be the second largest sample of high redshift galaxies found to date (the largest being that of Steidel et al. 1996). The success of our Ly-α imaging is surprising, given that many previous Ly-α searches, covering much larger co-moving volumes with comparable sensitivity, found nothing (eg. Pritchet & Hartwick 1990, Thompson & Djorgovski 1995, Martínez-González et al. 1995).

The most likely reason for our relative success is that the absorption-line structure has an enormous overdensity of Ly-α emitting galaxies. An overdensity factor of at least ten is required, matching the overdensity of absorption-line systems. This overdensity is comparable to that of the 'Great Wall' today, and lies well above the normal clustering amplitude of galaxies (Fig 1). If this is confirmed, it implies that dense, 10 Mpc scale galaxy clusters do exist in the early universe.

These candidate Ly-α emitting galaxies are typically slightly extended, elongated and irregular. Often the continuum cannot be seen in the blue, to a magnitude limit of $B = 26$, but where it is, it tends to be more compact than the line emission, and sometimes slightly offset from it. In one of the confirmed galaxies, for example, the Ly-α emission appears to form a ring around the unresolved blue continuum source. These properties are all quite different from those of the high redshift field galaxies discovered by Steidel et al. (1996), which are very compact. This may reflect the difference between field and cluster galaxies at high redshift, and/or the different selection techniques.

2.2   IR Observations

We have much more detailed observations of a small part ($\sim 10\%$) of the putative cluster; the 10% lying near the background QSO with the three absorption-line components, 2139−4434. The two confirmed Ly-α galaxies lie within this region.

We obtained deep near-IR images of this small region, using the Siding Spring 2.3-m telescope and the ESO/MPI 2.2-m telescope. We also have deep B and I-band images of this small region from the ESO 3.6-m, CTIO 4-m and AAT (Francis et al. 1996).

As Fig 2 shows, this field contains many objects with extremely red $I - K$ colours, objects invisible in the I band to a magnitude limit of 23, but clearly visible in the K band, with $K \sim 20$. We know that at least one of these red objects (B1) is a member of the $z = 2.38$ cluster, as it is one of the two spectrally confirmed Ly-α emitting galaxies.

Why are these objects so red in their $I - K$ colours? The spectral energy distribution of B1 is shown in Fig 3. The dominant feature is a large spectral break at an observed-frame wavelength of $\sim 1.4\mu$m, with a relatively blue spectral energy distribution on both sides of this break. This spectral form is typical of these red objects.

We modelled this spectral break as the redshifted 4000Å and/or Balmer break, seen in the spectra of most stars and galaxies. If correct, this implies that the galaxies are not new-born, as their spectra would be dominated by the light of O and B type stars, which do not show such a break. Instead, we require that most of the stellar population be old; composed of A type stars and later. Such a model can reproduce the wavelength and size of the spectral break (Fig 3). The

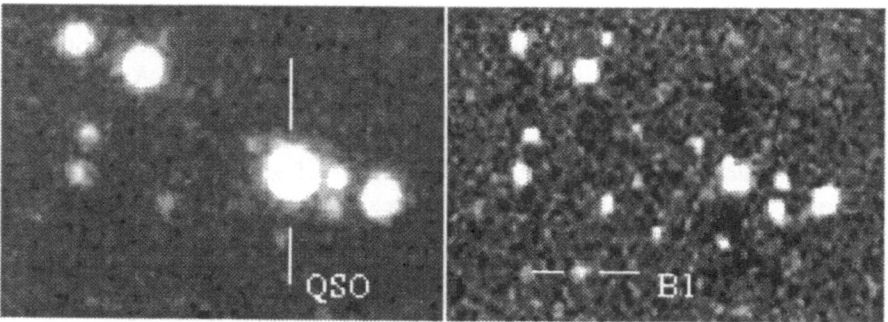

Fig. 2. I (left) and K' (right) images of part of our proposed $z = 2.38$ cluster. The Ly-α emitting galaxy B1 and background QSO 2139$-$4434 are marked. Note the many very red sources, which we hypothesise are other galaxies at $z = 2.38$

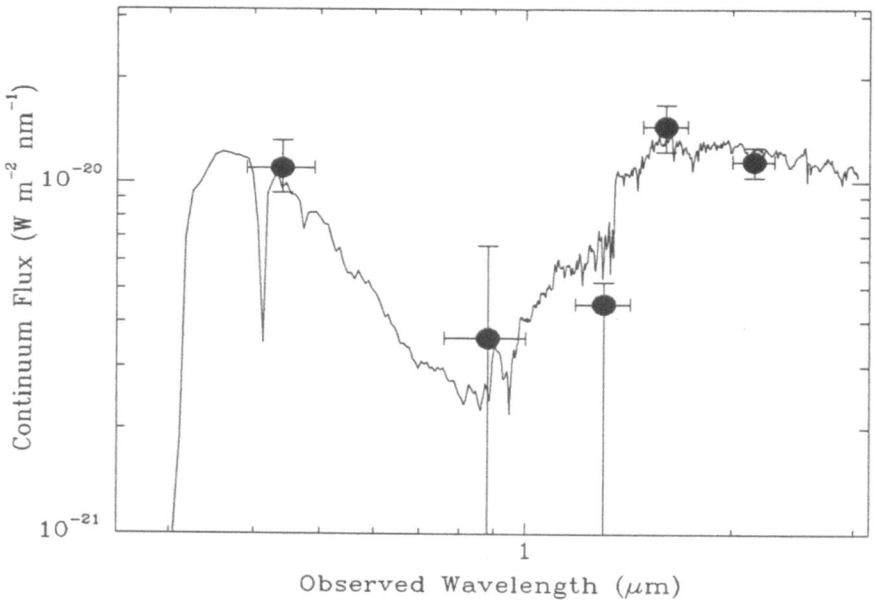

Fig. 3. The observed-frame continuum spectral energy distribution of Ly-α emitting galaxy B1. The fluxes have been corrected for line emission. The line is a synthetic galaxy spectrum, composed 96% of an old (2×10^9 years) stellar population, and $\sim 4\%$ of a young (10^7 years), dust-free starbursting population (Bruzual & Charlot 1996).

galaxies must have an age of at least 10^9 years, and a stellar mass of $\sim 10^{11} M_\odot$. This implies that the galaxies formed before $z = 5$ ($q_0 = 0.5$) or $z = 4$ ($q_0 = 0.1$).

Many radio galaxies show similarly red colours, which are probably caused by Balmer-line emission in the near-IR. For B1, H-α and H-β lie in the K and H filters. We have, however, a narrow-band image of the H-α flux from B1, and it contributes no more than $\sim 5\%$ of the K-band flux; the continuum break is therefore real.

We therefore claim that all objects showing this spectral break are at $z \sim 2.38$. Most do not show Ly-α emission, so we hypothesise that they are non-active cluster members. In this small fields, they outnumber the Ly-α emitting galaxies by a factor of ~ 10; if this ratio holds up throughout the cluster, as many as ~ 100 such galaxies may be present.

We also propose that the presence of a redshifted 4000Å break in the near-IR may be a powerful technique for identifying high redshift galaxies. As Fig 4 shows, when the redshifted 4000Å break lies between two near-IR bands, the colour between those bands becomes very red. We therefore suggest that red colours between any pair of near-IR bands are a signature of high redshift galaxies, and that multi-colour deep wide-field near-IR photometry can be used to identify galaxies at $1.5 < z < 4$, and to constrain their redshifts. Note that this technique is complementary to that of Steidel et al. (1996); their technique is only sensitive to young, star-forming galaxies, whereas our near-IR technique works at lower redshifts, and is only sensitive to more evolved galaxies.

2.3   B1, a Proto-cD?

The region about the Ly-α emitting galaxy B1 is an extraordinary one. B1 itself consists of a compact (< 10 kpc), evolved stellar population of mass $\sim 10^{11.5} M_\odot$. It is surrounded by an elongated emission-line region, seen in Ly-α, C IV and H-α. The line emission extends up to ~ 50kpc beyond the stars. Both the ionisation state and the velocity width (~ 700km s^{-1}) of this extended gas cloud suggest that the ionisation source is a concealed AGN, rather than star formation. Indeed, B1 looks very much like many high redshift radio galaxies, apart from being radio quiet ($F_{2.4GHz} < 0.27$mJy).

Four other IR selected, probable cluster members lie within 100 projected kpc of B1. All have colours suggesting that they are compact evolved galaxies, of age and mass comparable to B1. The presence of so many massive galaxies is such a small region suggests that they will merge on a timescale of $\lesssim 1$Gyr, forming a giant elliptical galaxy; possibly even a cD.

This quintet of massive evolved galaxies is surrounded by a hundred kpc scale lumpy gaseous halo. This is seen in emission where it is photoionised by the AGN within B1. It is also probably responsible for the Ly-α absorption-lines in the spectrum of the background QSO 2139$-$4434, whose line of sight passes 22$''$ from B1 (~ 100 projected kpc).

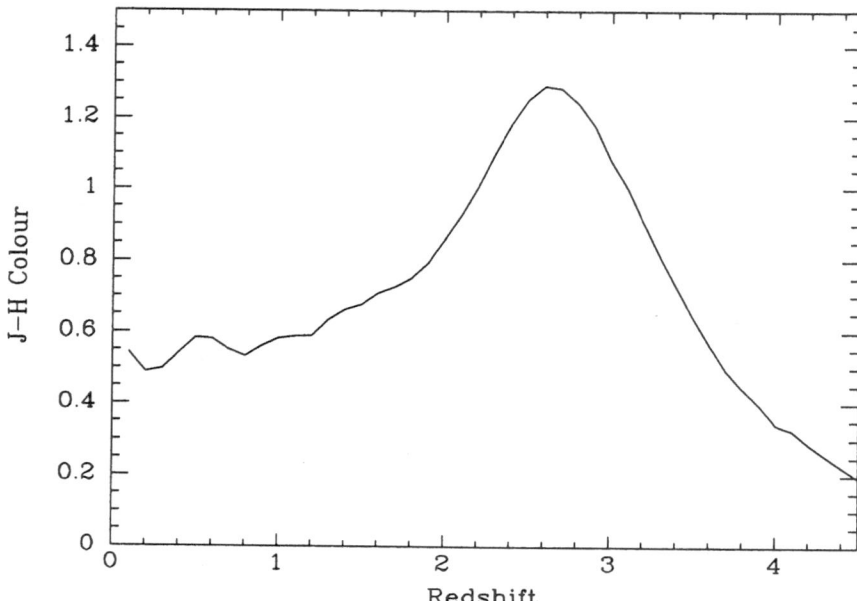

Fig. 4. Predicted $J - H$ colours of an evolved galaxy as a function of redshift (based on the spectral synthesis models of Rocca-Volmerange & Guiderdoni 1988). Note the pronounced reddening as the 4000Å break passes between the two bands. Simulation taken from O'Dowd (1995).

3   Conclusions

1. At least a few massive, large-scale structures exist at $z > 2$.
2. The combination of QSO absorption-line measurements, narrow-band imaging and optical/near-IR photometry is a powerful tool for finding and studying high redshift galaxies.
3. Different selection techniques find different types of high redshift galaxies, so diversity is good!

References

Brainerd, T. G., & Villumsen, J. V. (1994): ApJ, **431**, 477
Bruzual, G. A., & Charlot, S. (1996), in preparation.
Butcher, H., & Oemler, A., Jr. (1978): ApJ, **219**, 18
Dressler, A., Oemler, A., Jr., Gunn, J. E., & Butcher, H. (1993): ApJ, **404**, L45
Edge, A. C., Stewart, G. C.,Fabian, A. C., & Arnaud, K. A. (1990): MNRAS, **245**, 559
Efstathiou, G., Bernstein, G., Katz, N., Tyson, T., & Guhathakurta, P. (1991): ApJ, **380**, L47

Fernández-Soto, A., Lanzetta, K. M., Barcons, X., Carswell, R. F., Webb, J. K., & Yahil, A. (1996): ApJ, **460**, L85

Francis, P. J., & Hewett, P. C. (1993): AJ, **105**, 1633

Francis, P. J., Woodgate, B. E., Warren, S. J., Møller, P., Mazzolini, M., Bunker, A. J., Lowenthal, J. D., Williams, T. B., Minezaki, T., Kobayashi, Y., & Yoshii, Y. (1996): ApJ, **457**, 490

Fry, J. N. (1996): ApJ, **461**, L65

Foltz, C. B., Hewett, P. C., Chaffee, F. H., & Hogan, C. J. (1993): AJ, **105**, 22

Heisler, J., Hogan, C. J., & White, S. D. M. (1989): ApJ, **347**, 52

Jakobson, P., & Perryman, M. A. C., (1992): ApJ, **392**, 432

Lowenthal, J. D., Hogan, C. J., Green, R. F., Woodgate, B. E., Caulet, A., Brown, L. & Bechtold, J. (1995): ApJ, **451**, 484

Martínez-González, E., González-Serrano, J. I., Cayón, L., Sanz, J. L., & Martín-Mirones, J. M. (1995): A & A, **303**, 379

Møller, P., & Warren, S. J. (1993): A & A, **270**, 43

O'Dowd, M. (1995): Honours thesis, University of Melbourne.

Pascarelle, S. M., Windhorst, R. A., Driver, S. P., & Ostrander, E. J. (1996): ApJ, **456**, L21

Press, W. H., & Rybicki, G. B. (1993): ApJ, **418**, 585

Pritchet, C. J., & Hartwick, F. D. A. (1990): ApJ, **355**, L11

Rocca-Volmerange, B., & Guiderdoni, B. (1988): A & AS, **75**, 93

Shanks, T., & Boyle, B. J. (1994): MNRAS, **276**, 33

Steidel, C. C., Giavalisco, M., Pettini, M., Dickinson, M., & Adelberger, K. L. (1996): ApJ, **462**, L17

Thompson, D., & Djorgovski, S. G. (1995): AJ, **110**, 982

Wolfe, A. M. (1993): ApJ, **402**, 411

Clustering of Absorption Line Systems

Patrick Petitjean[1,2]

[1] Institut d'Astrophysique de Paris – CNRS
   98bis, Boulevard Arago
   F-75014 Paris, France
[2] DAEC, URA CNRS 173, Observatoire de Paris–Meudon
   F-92195 Meudon, France

Abstract. Absorption line systems are luminosity unbiased tracers of the spatial distribution of baryons over most of the history of the Universe. I review the importance of studying the clustering properties of the absorbers and the impact of VLT in this subject. The primary aim of the project is to track the evolution of the structures of the Universe back in time.

1 Introduction

Evolution of large scale structures of the Universe is one of the most important issues of modern cosmology. QSO absorption line systems probe material lying on the line of sight to quasars over a large redshift range ($0 < z < 5$). The systems where metal lines are detected have been recognized to be associated with haloes of galaxies (e.g. Bergeron & Boissé 1991, Steidel 1993). Those with very low metal content (the Lyα forest), are generally believed to probe intergalactic gas. Part of this gas could be associated with galaxies however, but how is yet unclear (e.g. Lanzetta et al. 1994, Le Brun et al. 1995, Charlton et al. 1995). In any case, the absorption line systems can be used as luminosity unbiased tracers of the spatial distribution of baryons over most of the history of the Universe. To do so background sources at small projected distances on the sky should be observed to study correlation between absorptions detected along different lines of sight. When the sources are very close (for example two images of the same gravitationally lensed quasar), the lines of sight probe the same clouds and information on the perpendicular sizes and the internal structure of the clouds can be derived. With sources further away from each others, one can study the correlation length of the clouds.

This has been recognized for over a decade (e.g. Shaver & Robertson 1983, Robertson & Shaver 1983). Observations of QSO pairs with projected separations from a few arcseconds to a few arcminutes yield interesting constraints on the size, physical structure and kinematics of galatic haloes, clusters and filaments. Indeed, new constraints have been obtained very recently on the extent of the Lyα complexes perpendicular to the line of sight at high (Smette et al. 1992, 1995, Bechtold et al. 1994, Dinshaw et al. 1994) and intermediate (Dinshaw et al. 1995) redshifts indicating that they could have sizes larger than 300 kpc. Such sizes are more indicative of a correlation length than of real cloud sizes

(Rauch & Haenelt 1996). This is consistent with the picture that the Lyα gas traces the potential wells of dark matter filamentary structures (Cen et al. 1994, Petitjean et al. 1995, Mücket et al. 1996, Hernquist et al. 1996, Miralda-Escudé et al. 1996). Large scale clustering of C IV systems (Heisler et al. 1989; Foltz et al. 1993) or damped systems (Francis & Hewett 1993, Wolfe 1993) have also been detected recently. The advent of 10m-class telescopes will boost this field since observation of faint QSOs in the same field will allow 3-D mapping of the baryonic content of the Universe via absorption line systems (Petitjean 1995).

2 The 1D Correlation Function

2.1 The Lyα Forest

If the spatial distribution of the Lyα gas is related to the mass distribution it is of interest to measure the correlation of the absorption lines as a possible probe of the early stages of the gravitational clustering. Much work has been dedicated to the study of the 1D clustering properties of the Lyα lines along the line of sight to the quasars. Till recently no clustering in velocity space had been detected on scales $300 < \Delta v < 30000$ km s^{-1} (Sargent et al. 1980, Bechtold 1987, Webb & Barcons 1991). Most of the results were obtained using intermediate resolution spectroscopy. The number of lines observed at high resolution and with good S/N ratio has increased dramatically and it has been possible to investigate the clustering of the lines for different column density regimes (Rauch et al. 1992). Cristiani et al. (1995) have found significant clustering, with $\xi \sim 1$, at $\Delta v = 100$ km s^{-1} for lines with log N(H I) > 13.8. For log N(H I) > 13.3, a significant but much weaker signal, $\xi \sim 0.34$, is found. This result has been confirmed by Meiksin & Bouchet (1995) using neighbor probability distribution functions. They find also strong evidence for anticorrelation on the scale of 3–$6h^{-1}$Mpc (see also Hu et al. 1995). This overall is consistent with the idea that strong lines trace the dark matter filaments and weak lines are mostly found in underdense regions (Riediger et al. 1996).

2.2 The Metal Lines

It has been shown convincingly that metal line systems at $z < 1$ are associated with galaxies (Bergeron & Boissé 1991, Steidel 1993). It is thus not surprising to observe that metal line systems do cluster on scales $\Delta v < 600$ km s^{-1} (Sargent et al. 1988). There is however a problem when comparing clustering of absorption lines and clustering of galaxies. Indeed when observed at high spectral resolution, metal lines break up into individual components. The 1D correlation function can be fitted using the sum of two Gaussian distributions with $\sigma = 110$ and 525 km s^{-1} and $\sigma = 80$ and 390 km s^{-1} for the CIV systems at $z \sim 2.6$ (Petitjean & Bergeron 1994) and MgII systems at $z \sim 1$ (Petitjean & Bergeron 1990) respectively. The clustering at small velocities reflects motions of clouds within one individual halo whereas larger velocities indicate clustering of halos.

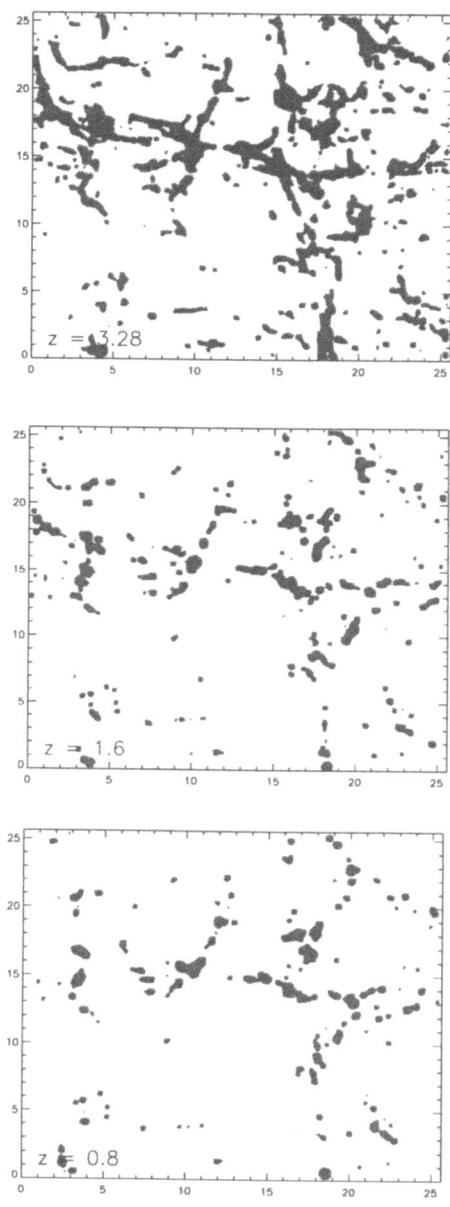

Fig. 1. Spatial distribution of Lyα clouds in a 2 Mpc slice of a $(25 \text{ Mpc})^3$ simulation box at redshifts $z = (3.28, 1.6, 0.8)$ respectively. Gray contours show regions where $N(\text{H I}) > 10^{13} \text{ cm}^{-2}$ through the box.

PKS2000-330

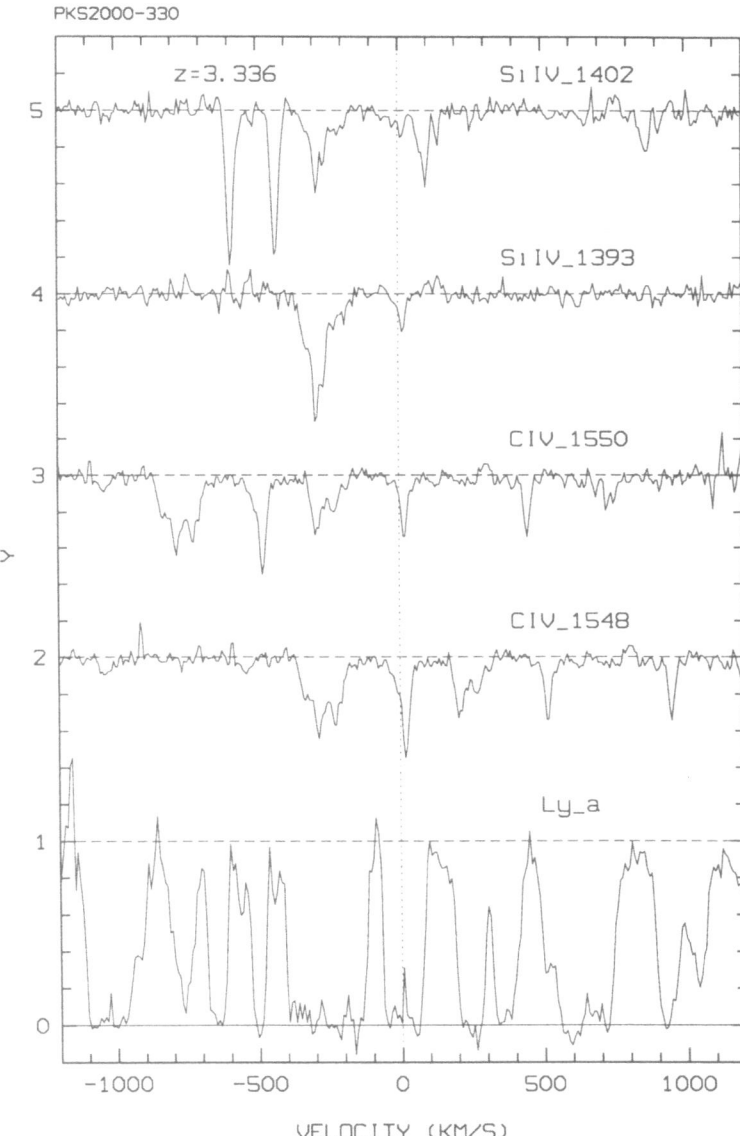

Fig. 2. Cluster of Lyα lines at $z \sim 3.336$ on the line of sight to PKS 2000-33 on a velocity scale. The corresponding C IV and Si IV lines are also shown.

It is thus clear that there is a difficulty in defining what part of the correlation function is related to clustering of haloes and what part is a consequence of motions of clouds within one halo. Only the first part of the function may be relevant for large scale analysis.

2.3 The Weak C IV Systems

Recent observations have shown that, at $z \sim 3$, C IV is found in 90% of the clouds with $N(\text{H I}) > 10^{15}$ cm^{-2} and in about 50% of the clouds with $3 \times 10^{14} < N(\text{H I}) < 10^{15}$ cm^{-2} (Songaila & Cowie 1996, Cowie et al. 1995). Several components are seen in most of these weak systems thus the correlation function shows a signal on scale smaller than 200 km s^{-1}. On this basis, Fernández-Soto et al. (1996) argue that the observed clustering is broadly compatible with that expected for galaxies and that most Lyα absorbers arise in galaxies. We could argue however, on the basis of the discussion in the previous section, that the signal detected in the correlation function has nothing to do with clustering of galaxies. What is seen is just the velocity structure of the Lyα gas inside large complexes.

Indeed a more attractive picture arise from the simulations showing that the Lyα absorption line properties can be understood if the gas traces the development of structures in the Universe (Cen et al. 1994, Petitjean et al. 1995, Mücket et al. 1996, Hernquist et al. 1996, Miralda–Escudé et al. 1996). This is illustrated in Fig. 1 showing the spatial distribution of the Lyα gas in a 2 Mpc slice of a (25 Mpc)^3 simulation box at redshifts $z = (3.28, 1.6, 0.8)$ respectively. In this picture, part of the gas is located inside filaments where star formation can occur very early in small halos that subsequently merge to build-up a so-called galaxy (Haehnelt et al. 1996). This gas contains metals. The remaining part of the gas has no metals and either is loosely associated with the filaments and has $N(\text{H I}) \gtrsim 10^{14}$ cm^{-2} or is located in the underdense regions and has $N(\text{H I}) \lesssim 10^{14}$ cm^{-2}. In this picture it might happen that the line of sight intercepts a filament along its largest dimension. In such a case it is expected to see a cluster of strong Lyα lines with associated C IV lines of very different strengths. Such an observation is shown in Fig. 2. The data have been obtained with EMMI at the ESO NTT. The total integration time is 18 hours (Petitjean et al. 1996).

3   The 3D Clustering

Recent studies indicate large scale clustering of absorbers. Heisler et al. (1989) detected significant correlation signal for C IV systems out to velocities of $\Delta v = 10000$ km s^{-1}. Foltz et al. (1993) found an overdensity of C IV systems in the redshift range $1.57 \lesssim z \lesssim 1.69$ along the lines of sight to 6 QSOs. Francis & Hewett (1995) discovered two damped Lyα systems at similar redshifts in two lines of sight separated by $18h^{-1}$ Mpc comoving. Most of the time however the signal is a consequence of an unusual overdensity of systems along a peculiar line of sight. Large samples of absorption systems have been used to investigate the 3D clustering properties of the absorbers (York et al. 1991, Tytler et al. 1993) with little success mostly due to lack of data.

3.1   An Example: The Field Around Q1037-2704

In this context, the field surrounding the bright ($m_V = 17.4$) high redshift (z_{em} = 2.193) QSO Tol Q1037-2704 is quite promising. Jakobsen et al. (1986) were the first to note the remarkable similarity of the metal-line absorption systems in the spectra of Tol 1037-2704 and Tol 1038-2712 separated by 17'9 on the sky, corresponding to $4.3h_{100}^{-1}$ Mpc for $q_o = 0.5$ at $z \sim 2$. They interpreted this as evidence for the presence of a supercluster along the line of sight to the QSOs. The fact that the number of metal-line systems in both spectra over the range $1.90 \leq z \leq 2.15$ is far in excess of what is usually observed has been considered as the strongest argument supporting this conclusion (Ulrich & Perryman 1986, Sargent & Steidel 1987, Robertson 1987). In a recent paper, Dinshaw & Impey (1996) have presented new data on four quasars in this field. They find that the velocity correlation function of the C IV systems shows strong and significant clustering for velocity separations less than 1000 km s^{-1} and up to 7000 km s^{-1} respectively. The spatial correlation function shows a marginally significant signal on scales of < 18 Mpc. They conclude that the dimensions of the proposed supercluster are at least 30 h^{-1} Mpc on the plane of the sky and approximately 80 h^{-1} Mpc along the line of sight. Moreover, by an analysis of the metal content and ionization state of several C IV complexes in Q1037–2704, Lespine & Petitjean (1996) have shown that the gas lies in intervening systems, supporting the presence of a coherent structure of supercluster dimensions.

3.2   Clustering of Absorbers with VLT

The best approach to study clustering of absorbers is to search a small field for a large number of quasars and to identify the absorption systems in the quasar spectra. This will allow 3D mapping of the baryonic content of this part of the Universe. The major limitation is that the number of quasars per square degree is large enough to yield interesting conclusions at a magnitude prohibitively large to achieve adequate spectroscopy on 4 m class telescopes.

The possible VLT project would thus include :

– Deep imaging in broad bands with FORS or EMMI at the NTT to select QSO candidates. These images could be used for a parallel programme to detect galaxies at high z.

– Low resolution spectroscopy with FORS to confirm the candidates. This could be part of a programme aimed at determining the luminosity function of AGNs.

– Intermediate resolution ($m_{QSO} < 22.5$) and high resolution ($m_{QSO} < 19$) spectroscopy of the QSOs with FUEGOS and UVES for the brightest to study the absorptions.

– Multi-object spectoscopy in the field with FORS, NIRMOS and ISAAC to identify the associated galaxies.

Table 1 gives the number of QSOs expected in a one-degree field (Hartwick & Shade 1990) and the mean number of absorbers, with $w_{obs,lim} > 0.2$ Å, per line of sight. These numbers are indicative and take into account various observational

Table 1. One degree field, $w_{\text{obs,lim}} > 0.2$ Å

	Nb of QSOs			Nb of absorbers per QSO		
	$m < 21$	22	22.5	C IV	Lyα	Mg II
$0 < z < 2.2$	33	74	129	5		0.7
$0 < z < 2.2$	7	20	32	7	100	0.9

limitations. It is clear that one would like to observe the largest number of QSOs in the field. A compromize between this number and a reasonable amount of observing time should be found. However for a random spatial distribution of the QSOs, a number of hundred QSOs in the field seems adequate leading to primarily targetting $m < 22$ QSOs. The exposure time needed to obtain spectra of S/N ratio of 20 at a resolution of $R \sim 5000$ (thus $w_{\text{obs,lim}} \sim 0.2$ Å) on a $m = 22$ QSO is about 30 hours. This is large but the use of the MOS capabilities reduces the effective observing time requested to achieve the project. In this prospect the instrument providing the largest field, thus FUEGOS, should be used. Only six FUEGOS settings will be needed.

References

Bechtold, J. (1987): *High redshift and Primeval Galaxies*. IAP Colloquium, ed. by Bergeron J. et al., Editions Frontières, Gif sur Yvette, p. 397

Bechtold, J., Crotts, A.P.S., Duncan, C., Fang, Y. (1994): ApJL **437**, L83

Bergeron, J., Boissé, P. (1991): A&A **243**, 344

Cen, R., Miralda-Escudé, J., Ostriker, J.P., Rauch, M. (1994): ApJ **437**, L9

Charlton, J.C., Churchill, C.W., Linder, S.M. (1995): ApJL **452**, L81

Cowie, L.L., Songaila, A., Kim, T.-S., Hu, E.M. (1995): AJ **109**, 1522

Cristiani, S., D'Odorico, S., Fontana, A., Giallongo, E., Savaglio, S. (1995): MNRAS **273**, 1016

Dinshaw, N., Foltz, C.B., Impey, C.D., et al. (1995): Nat **373**, 223

Dinshaw, N., Impey, C.D. (1996): ApJ **458**, 73

Dinshaw, N., Impey, C.D., Foltz, C.B., et al. (1994) ApJ **437**, L87

Fernández-Soto, A., Lanzetta, K.M., Barcons, X. et al. (1996): ApJL **460**, L85

Foltz, C.B., Hewett, P.C., Chaffee, F.H., Hogan C.J. (1993): AJ **105**, 22

Francis, P.J., Hewett, P.C. (1993): AJ **105**, 1633

Haehnelt, M.G., Steinmetz, M., Rauch, M. (1996): ApJ preprint

Hartwick, F.D.A., Shade, D. (1990): Ann. Rev. Astron. Astroph. **28**, 437

Heisler, J., Hogan, C.J., White, S.D.M. (1989): ApJ **347**, 52

Hernquist, L., Katz, N., Weinberg, D.H., Miralda-Escudé, J. (1996): ApJ **457**, L51

Hu, E.M., Tae-Sun Kim, Cowie, L.L., Songaila, A., Rauch, M. (1995): AJ **110**, 1526

Jakobsen, P., Perryman, M.A.C. (1992): ApJ **392**, 432

Jakobsen, P., Perryman, M.A.C., Ulrich, M.H., et al. (1986): ApJ **303**, L27

Lanzetta, K.M., Bowen, D.V., Tytler, D., Webb, J.K. (1994): ApJ **442**, 538

Le Brun, V., Bergeron, J., Boissé, P., 1996, A&A **306**, 691

Lespine, Y., Petitjean, P. (1996): A&A in press

Meiksin, A., Bouchet, F. (1995): ApJL **448**, L88

Miralda-Escudé, J., Cen, R., Ostriker, J.P., Rauch, M. (1996): ApJ in press

Mücket, J., Petitjean, P., Kates, R.E., Riediger, R. (1996): A&A **308**, 17

Petitjean P., Bergeron J. (1990): A&A **231**, 309

Petitjean P., Bergeron J. (1994): A&A **283**, 759

Petitjean, P. (1995): *Science with VLT.* ESO Workshop, ed. by Danziger J., Walsh J., Springer, Heidelberg, p. 339

Petitjean, P., Mücket, J., Kates, R.E. (1995): A&AL **295**, L9

Petitjean, P., Fontana, A., Giallongo, E., Lespine, Y. (1996): in preparation

Rauch, M., Haenelt, M. (1996): MNRAS **275**, L76

Rauch, M., Carswell, R.F., Chaffee F.H. et al. (1992): ApJ **390**, 387

Riediger, R., Petitjean, P., Mücket, J. (1996): in preparation

Robertson, J.G. (1987): MNRAS **227**, 65

Robertson, J.G., Shaver, P.A. (1983): MNRAS **204**, 69P

Sargent, W.L.W., Young, P.J., Boksenberg, A., Tytler, D. (1980): ApJS **42**, 41

Sargent, W.L.W., Steidel, C.C. (1987): ApJ **322**, 142

Sargent, W.L.W., Boksenberg, A., Steidel C.C. (1988): ApJS **68**, 539

Shaver, P.A., Robertson, J.G. (1983): ApJL **268**, L57

Smette, A., Robertson, J.G., Shaver, P.A., Reimers, D., Wisotzki, L., Köhler, T. (1995): A&AS **113**, 199

Smette, A., Surdej, J., Shaver, P.A., et al. (1992): ApJ **389**, 39

Songaila, A., Cowie, L.L. (1996): AJ preprint

Steidel C.C. (1993): *Third Tetons Summer School, The Environment and Evolution of Galaxies*, ed. by J.M. Shull and H.A. Thronson Jr., Kluwer, Dordrecht, p. 263

Tytler, D. et al. (1993): ApJ **405**, 57

Ulrich, M.H., Perryman, M.A.C. (1986): MNRAS **220**, 429

Webb, J.K., Barcons, X. (1991): MNRAS **250**, 270

Wolfe, A.M. (1993): ApJ **402**, 411

York et al. (1991): MNRAS **250**, 24

Searches for High Redshift Clusters

Mark Dickinson

STScI, 3700 San Martin Dr., Baltimore, MD 21218, USA

Abstract. High redshift galaxy clusters have traditionally been a fruitful place to study galaxy evolution. I review various search strategies for finding clusters at $z > 1$. Most efforts to date have concentrated on the environments of distant AGN. I illustrating these with data on the cluster around 3C 324 ($z = 1.2$) and other, more distant systems, and discuss possibilities for future surveys with large telescopes.

1 Finding Distant Clusters

At one time, galaxy clusters served as the most observationally straightforward means of studying galaxy evolution at high redshift. The reason was primarily one of *contrast*: even without spectroscopy, very rich clusters are recognizable as enhancements in the galaxy surface density out to $z \sim 1$, and the properties of the cluster galaxy population can be studied statistically with imaging data alone if a proper "control sample" of field galaxies can be observed in the same manner. In this way, Butcher and Oemler (1978, 1984) provided the first convincing evidence for galaxy evolution, identifying an apparently systematic bluing trend for galaxies in the cores of rich clusters beyond $z \approx 0.3$.

At very large redshifts ($z > 1$), however, even quite rich clusters are no longer so clearly visible against the tremendously numerous population of faint field galaxies, and thus become correspondingly more difficult to discover and study. This is illustrated by figure 1, a cartoon representing the peak surface density contrast in the I–band of identical rich clusters observed at $z = 0.3$ and $z = 1.2$. Because the angular scale changes only slowly with redshift beyond $z = 0.3$, there is very little boost in the galaxy surface density as one moves a cluster further away, while the combination of distance modulus and k–correction dims the cluster galaxies considerably. The result is that a cluster with a contrast $10\times$ over the field galaxy population at $z = 0.3$ barely peaks up over the field at all at $z = 1.2$. The effects of the k–correction are much more severe for early–type galaxies, rendering a rich, elliptical dominated cluster nearly invisible at optical wavelengths for $z > 1$, even assuming reasonable amounts of passive evolution.

On the one hand, thanks to highly efficient multiplexing spectrographs, the tremendous progress which has occurred in the study of high redshift field galaxies has meant that clusters are no longer "needed" to provide large samples of distant galaxies. But at the same time, the higher the redshift, the more interesting a rich cluster becomes from a cosmological viewpoint. Firstly, as the most massive collapsed structures in the universe, their properties and evolution are highly sensitive to the fundamental cosmological parameters, as well as to the power spectrum of mass fluctuations which give rise to large scale structure in

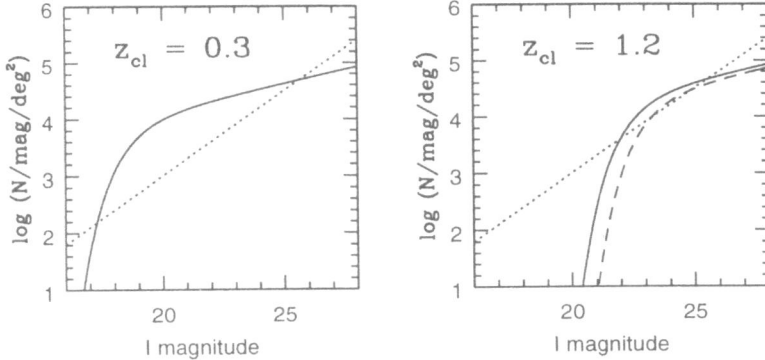

Fig. 1. Schematic plot of surface density contrast for a rich cluster observed at $z = 0.3$ and 1.2. The dotted line sketches I-band field galaxy counts, while the solid curve represents the central galaxy surface density of a rich cluster with a Schecter luminosity function. At $z = 0.3$, the cluster core has a peak contrast of $\sim 10\times$ the background, while at $z = 1.2$ the contrast is $< 2\times$. A k–correction for Sb–type galaxies with mild luminosity evolution has been assumed – the contrast is reduced to virtually nothing if the cluster galaxies are all ellipticals (dashed line) due to the stronger k–correction.

the galaxy distribution. The abundance of massive clusters, and their detailed properties, become increasingly important constraints on cosmological models at larger redshift. Secondly, today's rich clusters are dominated by elliptical galaxies, whose evolutionary history is of particular interest since they may represent the oldest galaxy–sized stellar systems. Tracing their spectrophotometric properties may point back to the earliest epochs of star formation.

The observational challenge in finding and studying distant clusters is one of enhancing their *contrast* against the field. Simply imaging in a single, broad bandpass is not sufficient. In the absence of extensive spectroscopy, multi–color or narrow band techniques must be used to screen out the multitude of mostly foreground galaxies and to isolate candidate cluster members. Here, I consider a variety of means by which this may be achieved. Many of these are costly in terms of observing time: regardless of the method, one must work to very faint apparent magnitudes in order to see distant clusters, requiring large telescope apertures and long exposure times. Thus far, virtually all attempts to find extremely high–z clusters have been "targeted" surveys. Rather than blindly search the sky for contrast enhancements in the galaxy density, targeted searches select likely sites of distant clusters and concentrate their efforts there. Because galaxies cluster, if one already knows where one high–z object is, then that is a good place to start looking for more. High redshift AGN are thus a natural place to begin searching for fainter, more normal companions. In particular, various previous surveys (e.g. Yee & Green 1984, Yates *et al.* 1989, Hill & Lilly 1991) have found that radio–loud AGN at $z \approx 0.5$ are frequently (but not always) situated in rich galaxy clusters. For this reason, my collaborators and I, as well as others, are studying the environments of distant radio galaxies and quasars. Our initial results have been encouraging, and suggest that this may be a fruitful pursuit for future surveys with large telescopes such as the VLT.

2   A Case Study: 3C 324

One means of enhancing the contrast of distant clusters is to search in the near–infrared, where the field galaxy number counts exhibit a shallower slope, and where the k–correction which dims the light of distant early–type galaxies is significantly reduced. Peter Eisenhardt and I have been carrying out a systematic survey of the environments of radio galaxies at $0.8 < z < 1.4$, using deep infrared and optical imaging to search for enhancements in the galaxy surface density around the AGN. Several promising cluster candidates have been found in this manner. In order to confirm these and to study their properties in more detail, Hy Spinrad, Arjun Dey and I have been following these up with Keck multislit spectroscopy. We have also obtained deep *HST* images of two of these clusters so far. Here I will use the environment of 3C 324, a powerful radio galaxy at $z = 1.206$, to illustrate the challenges and promise of these methods.

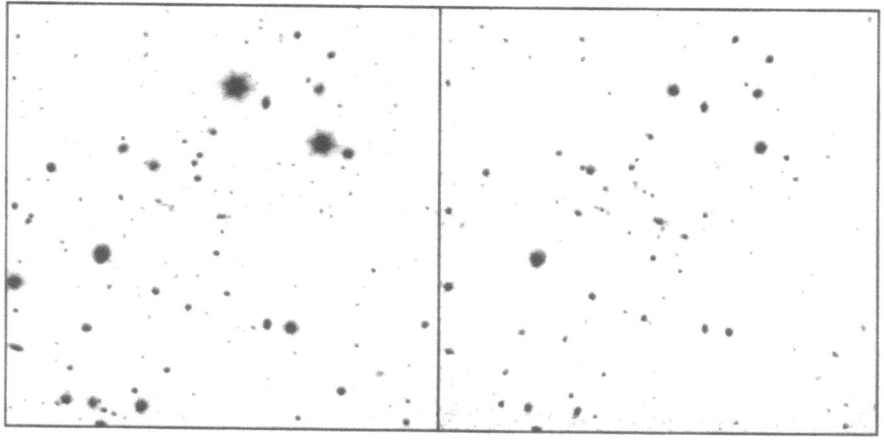

Fig. 2. Deep R (left) and K (right) images of a $2'.5$ field around the $z = 1.206$ radio galaxy 3C 324, which sits at field center. The cluster is evident in the infrared image, but is nearly invisible in the optical data.

Figure 2 shows deep R and K–band images of the field around 3C 324. To the sharp eye, the cluster appears as a density enhancement in the infrared data, but is nearly invisible in the optical image. Figure 3 demonstrates this by showing the radial surface density profile of the cluster in the two images. Optically, the cluster contrast peaks at a factor of only $\sim 2\times$ the background density, whereas in the infrared the contrast reaches a factor of $\sim 6\times$. This is because there are many very red galaxies in this field, which mostly have the expected elliptical morphologies in our deep *HST* images (cf. Dickinson 1995a, b). The k–correction for these ellipticals is killing the cluster contrast optically, but leaving it more prominently visible in the near infrared. Figure 4 demonstrates another means by which distant clusters may be recognized: *color contrast,* wherein a judiciously selected set of broad band filters allows one to distinguish high–z cluster galaxies from the lower redshift field objects by their unusual colors.

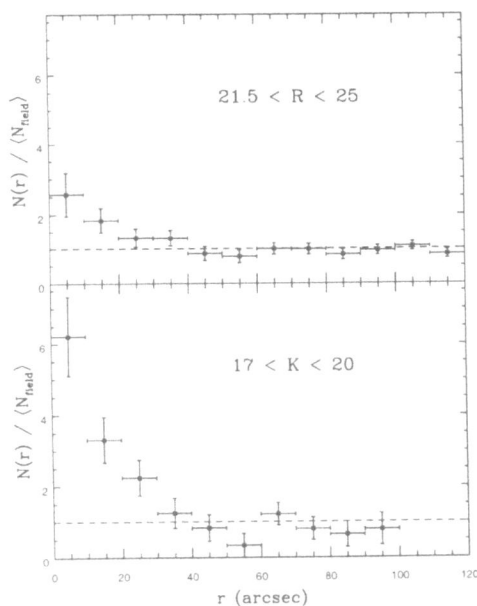

Fig. 3. Radial cluster contrast profile for 3C 324 in the optical and infrared, normalized to the "field" density of foreground/background galaxies. As is evident in figure 2, the cluster contrast is much stronger in the K–band.

Figure 5 shows the results of our spectroscopy around 3C 324, primarily carried out using the Keck LRIS, but with additional redshifts provided by Olivier LeFèvre from the NTT. The vast majority of the galaxies we have observed, particularly at larger separations from the radio galaxy, lie in the foreground or background. However, two sharp spikes in the redshift distribution appear at $z \approx 1.15$ and $z \approx 1.21$. These galaxies are strongly clustered around the radio galaxy: considering only galaxies at radii $< 60''$ from the radio galaxy, these high–z spikes dominate the redshift histogram. Evidently, the "cluster" which appears in the K–band images is comprised of two distinct clumps or sheets of

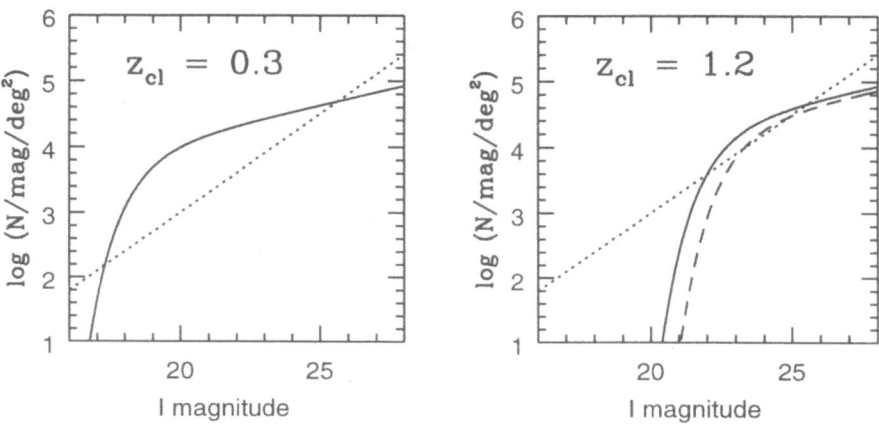

Fig. 4. Histogram of $R - K$ colors for galaxies around 3C 324. The prominent spike at $R - K = 5.9$ largely consists of faint galaxies with elliptical morphologies in the *HST* images.

galaxies separated by ~ 7500 km s^{-1} in their rest frame. Whether this is merely a chance projection, or an indication of supercluster–scale structure at $z \sim 1.2$, is not yet clear.

We believe that one of these two redshift "spikes," probably the one associated with the radio galaxy itself, is genuinely a rich, bound cluster. As part of an x–ray survey of distant radio galaxies, we obtained a Rosat PSPC observation of 3C 324 which detected a faint (6σ) source within $10''$ of the radio galaxy position. At low signal–to–noise, the PSPC resolution was insufficient to show whether the source is resolved, so the question remained whether or not the x–rays arose in the radio galaxy AGN or from the cluster environment. A subsequent 72.1 ksec Rosat HRI exposure clearly shows that the x–ray emission is resolved over a detectable diameter of $\sim 60''$. The redshift distribution shown in figure 4 exhibits spikes at $z < 1$ such as are seen in all faint field galaxy surveys (cf. Cohen *et al.* 1996). However, none of the foreground spikes consists of galaxies particularly concentrated toward the radio galaxy position. Given the close positional agreement between the x–ray source and 3C 324, we regard it as unlikely that the x–rays arise from a foreground group or cluster. If the x–rays indeed originate at $z = 1.2$, the 3C 324 cluster has a bolometric x–ray luminosity $L_X = (8.0 \pm 1.6) \times 10^{44}$ erg s^{-1} (for $H_0 = 50$, $q_0 = 0.5$), comparable to that of the Coma cluster. The presence of a large mass is also supported by the detection of weak shear gravitational lensing centered on the radio galaxy in our WFPC2 images (Smail & Dickinson 1995). This demonstrates the promise of

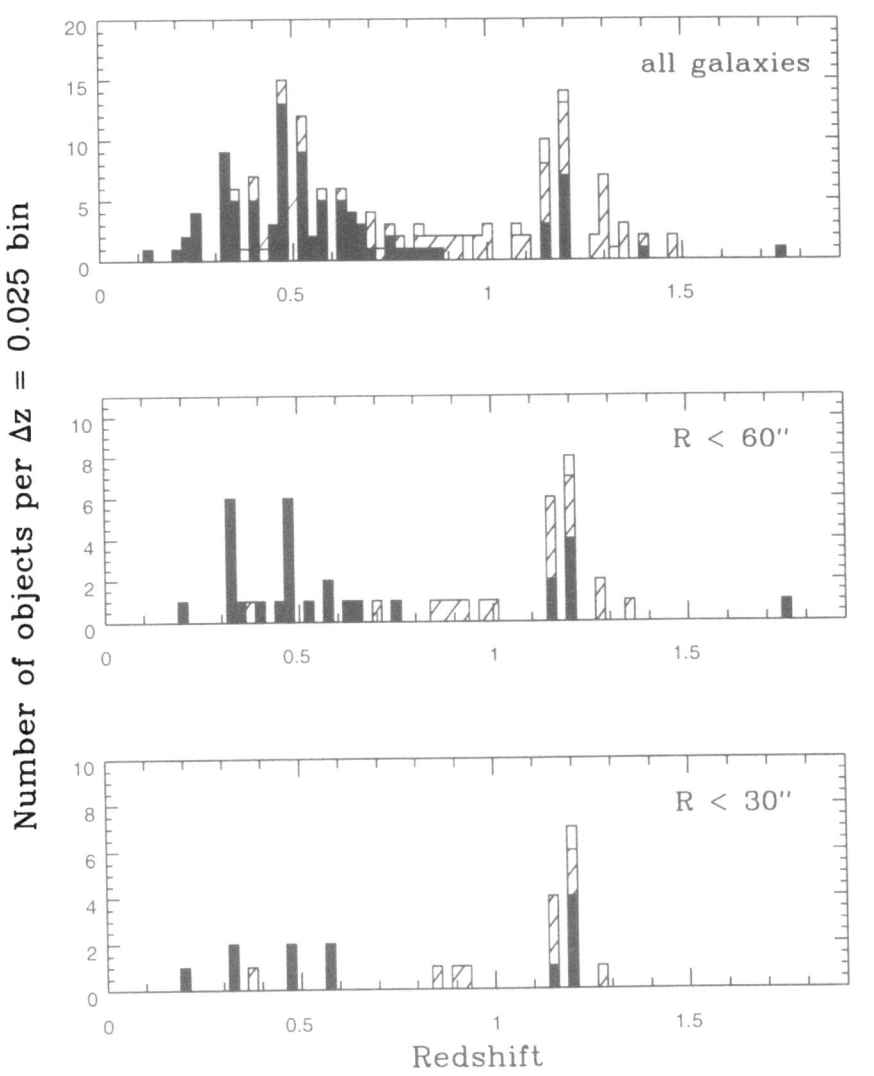

Fig. 5. Redshift histograms of galaxies in the 3C 324 field for samples restricted at several radii from the radio galaxy. The various shadings indicate different redshift "quality classes," with the filled area indicating fully secure redshifts (multiple spectral features) while the hatched area largely represents redshifts based on single emission lines, generally assumed to be [OII]. The "cluster" separates into two structures at $z = 1.15$ and $z = 1.21$. Within $30''$, where extended x–ray emission is found in our Rosat data, these high–z spikes dominate the redshift distribution, suggesting strongly that the x–rays originate at $z = 1.2$.

cluster–hunting at high redshift: the properties of such massive, collapsed structures at $z > 1$ may provide useful constraints on theories of structure formation and evolution.

3   Search Techniques at Higher Redshift

The preceding section demonstrates two methods of contrast enhancement: searches in the infrared, and searches by color contrast. At higher redshifts, it is unlikely that contrast in a single broad bandpass, even in the infrared, will be sufficient to allow recognition of even rich clusters except perhaps in a statistical fashion. More carefully "tuned" methods must be adopted.

Just as optical–infrared color contrast helped us to discover the 3C 324 cluster, various *infrared–infrared* color combinations may be effective for isolating galaxies at $z > 2$. In particular, combined JHK imaging may be effective for detecting galaxies with evolved (ages > 1 Gyr) stellar populations at z 2.5, where the J band lies shortward of 4000Å in the rest frame and the H and K bands roughly measure the rest–frame $B - R$ color. Modelling suggests that older galaxies at $z \sim 2.5$ should have identifiably unique colors in this bandpass combination, separating out from the locus of lower redshift objects.

At $z > 3$, the 912Å Lyman limit passes through the observed U–band. This fact has been exploited to great effect by Steidel and collaborators (cf. Steidel *et al.* 1995 and 1996, and the contributions of Giavalisco and Macchetto to this conference), who have identified and spectroscopically confirmed large numbers of $3 < z < 3.5$ field galaxies by selecting them according to their colors in a specially tuned UGR filter system. This is another means of multicolor selection, using a spectral break in the rest–frame UV rather than the rest–frame optical. Giavalisco (priv. comm.) has used this filter system to identify a density enhancement around the $z = 3.6$ radio galaxy 1243+036, and Lacy & Rawlings (1996) have reported similar results for 4C 41.17 at $z = 3.8$.

The Lyman–break technique has been especially effective for studying field galaxies because it probes a large redshift interval, and hence a large volume of space. For this reason, however, it may actually be less optimal for cluster surveys, where one would ideally like to *restrict* as much as possible the redshift range over which one isolates candidate cluster galaxies. Narrow band techniques searching for line emission may therefore also prove useful and effective, although they are limited to the detection of star–forming galaxies and AGN. But this may not be a strong drawback at high redshift, where many or most galaxies may be actively undergoing star formation. Lyman α, while optically convenient for $z > 2$, is a fragile line to work with and is easily extinguished by dust. Nevertheless, it has been successfully used by Francis *et al.* 1996 (and this volume), Pascarelle *et al.* 1996, and Møller & Warren (1993) to identify galaxies in candidate groups or clusters at $2 < z < 3$. LeFèvre *et al.* (see also contribution by J.–M. Deltorn at this meeting) have also detected and confirmed two Lyα companions to the $z = 3.14$ radio galaxy MGO 0316-257.

In the future, narrow band infrared searches may be particularly effective, probing Balmer and forbidden line emission which is less affected by dust than is Lyα (cf. contributions by Mannucci to this meeting). A particularly "magic" redshift is $z \sim 2.3$, where [OII], [OIII]/Hβ, and Hα are shifted into the J, H and K windows, respectively, and where Lyα is shifted to $\lambda_{\rm obs} > 4000$Å, facilitating narrow band searches and spectroscopic confirmation. At $z \sim 2.3$, the multicolor IR broad band techniques suggested above may also be used to look for older, non–star–forming galaxies. Indeed, Paul Francis has taken advantage of all of these methods in studying the $z = 2.38$ system he described in his contribution to this conference.

4   Future Prospects and the VLT

At present, there is very little known about any of these very distant cluster candidates except that a few galaxies are present at similar redshifts. But these are early days yet, and preliminary detections can, with intensive follow–up studies, lead to more far–reaching results. For 3C 324, the detection of x–ray emission and gravitational lensing gives a first hint at cluster *masses* beyond $z = 1$. Our redshift survey of the 3C 324 field, particularly the unexpected discovery that the "cluster" divides into two distinct redshift–space structures seen in projection, serves as a reminder that extensive spectroscopy is needed to confirm and interpret any individual cluster candidate. This is where telescopes like the VLT will excell, providing the firepower needed to do this efficiently.

For future cluster surveys, the techniques described above will require *wide field* imaging to very faint flux levels in bandpasses ranging from U through K. The wide field imaging aspect must be stressed, especially in this era of increased angular resolution and adaptive optics, which often drives instrument design toward smaller pixels and smaller fields of view. Systems capable of multiplexed imaging (simultaneously observing through several bandpasses, split by dichroics) would be highly desirable. Narrow band capabilities, ideally tunable to any wavelength, are also likely to be particularly useful for cluster work where restricted redshift coverage is desirable, much more so than for field galaxy surveys. Narrow band work is often difficult on very large telescopes because of the large sizes of the optical beams, but should be considered carefully in future instrument designs for the VLT.

For detailed follow–up studies of high–z clusters, spectroscopy is the key. Proposed future VLT instruments such as VIRMOS will be ideal, permitting simultaneous spectroscopy of hundreds of faint galaxies, and extending into the near–IR where [OII], [OIII] and Balmer emission will be redshifted. And while emission lines are useful and important, the success of Steidel *et al.* in measuring redshifts from *UV absorption line* in young, star forming galaxies at $z > 3$ should be kept in mind. In the future, infrared continuum and absorption line spectroscopy may be essential for studying the properties of *older* stellar populations of $z > 1$ galaxies in clusters and in the field.

5 Acknowledgements

I would like to thank my collaborators, particularly Hy Spinrad, Arjun Dey, Peter Eisenhardt and Richard Mushotzky, for permitting me to show data in advance of publication. I also thank the conference organizers for an invigorating meeting in (admittedly frigid) Garching, and for their financial support.

References

Butcher, H.R., and Oemler, A. 1978, ApJ, 219, 18.

Butcher, H.R., and Oemler, A. 1984, ApJ, 285, 426.

Cohen, J.G., Hogg, D.W., Pahre, M.A., and Blandford, R. 1996, ApJ, 462, L9.

Dickinson, M. 1995a, in *Fresh Views on Elliptical Galaxies,* eds. A. Buzzoni, A. Renzini, & A. Serrano, (ASP, San Francisco), p. 283.

Dickinson, M. 1995b, in *Galaxies in the Young Universe,* eds. H. Hippelein, K. Meisenheimer, & H.-J. Röser, (Springer–Verlag), p.144.

Francis, P.J., Woodgate, B.J., Warren, S.J., Møller, P., Mazzolini, M., Bunker, A.J., Lowenthal, J.D., Williams, T.B., Minezaki, T., Kobayashi, Y., and Yoshii, Y., 1996, ApJ, 457, 490.

Hill, G.J., and Lilly, S.J. 1991, ApJ, 367, 1.

Lacy, M., and Rawlings, S., 1996, MNRAS, 280, 888.

LeFèvre, O., Deltorn, J.-M., Crampton, D., and Dickinson, M., 1996, ApJ Letters (in press).

Møller, P., and Warren, S.J., 1993, A&A, 270,43.

Pascarelle, S.M., Windhorst, R.A., Driver, S.P., Ostrander, E.J., and Keel, W.C., 1996, ApJ, 456, L21.

Smail, I., and Dickinson, M., 1995, ApJ, 455, L99

Steidel, C.C., Pettini, M., and Hamilton, D. 1995, AJ, 110, 2519.

Steidel, C.C., Giavalisco, M., Pettini, M., Dickinson, M., and Adelberger, K. 1996, ApJ, 462, L17.

Yates, M., Miller, L., and Peacock, J. 1989, MNRAS, 240, 129.

Yee, H.K.C., and Green, R.F. 1984, ApJ, 280, 79.

Part 6

First Active Objects

High Redshift Radio Galaxies with the VLT

Huub Röttgering & George Miley

Leiden Observatory,
P.O. Box 9513,
2300 RA Leiden,
The Netherlands

1 Introduction

The study of high-z radio galaxies (HZRGs) has matured considerably during the last decade. Large numbers of $z > 2$ radio galaxies have been found and have been studied in considerable detail.

During the mid-eighties there were indications from statistical analyses of radio source counts that there might be a cutoff in the space density of powerful radio sources above a redshift of $z \sim 2$ (e.g. Windhorst 1984). However, it has become clear that a substantial population of radio galaxies with $z > 2$ does exist. The first radio galaxy discovered with a measured redshift beyond 2 was 4C40.36 at $z = 2.269$ (Chambers et al. 1988). This was soon followed by two radio galaxies at even more extreme distances (0902+34 at $z = 3.4$, Lilly 1989 and 4C41.17 at $z = 3.8$, Chambers et al. 1990). At the moment the record holder is 6C0140+326 at $z = 4.41$ (Lacy et al. 1996).

Not only has the maximum distance out to which radio galaxies are observed been extended dramatically, but also the number of known galaxies with $z > 2$ has grown substantially to well above 100 (e.g. Spinrad 1995, Eales and Rawlings 1996, McCarthy et al. 1996, Röttgering et al. 1996a). In Fig. 1 the redshift distribution of 106 $z > 2$ galaxies that are known to the authors (mid-1996) is shown. This histogram is of that given by Spinrad (1995) showing the 79 $z > 2$ radio galaxies that were known to him in mid-1994. The number of known distant radio galaxies has increased by a third over this two year period.

In this review we will address the importance of the VLT for studies of samples of distant radio galaxies. We shall first briefly discuss the importance of radio galaxies for cosmological studies. Then we shall summarize the various emitting components associated with these objects and comment on the possibilities offered by the VLT for detailed observations of these components and the radio/optical alignment effect. Next, the prospects for detecting and studying galaxy clusters around high redshift radio galaxies are briefly discussed. The case of 1138−262 ($z = 2.15$) is discussed as an example of a possible proto-galaxy that might be situated in a cluster-type of environment. Finally, we outline some of the steps that should be taken to optimize use of the VLT in this area.

For the most complete review of distant radio galaxies, as per 1993, we refer the reader to McCarthy (1993).

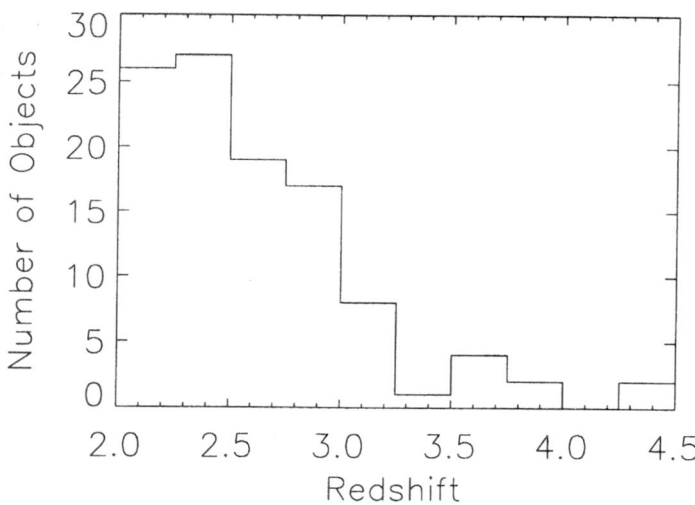

Fig. 1. Redshift distribution of the $z > 2$ radio galaxies known to the authors in mid-1996.

2   Importance

As a probe of the distant universe, distant radio galaxies offer a wide range of distinct components that can be studied. These include stars, AGN, hot (X-ray emitting) gas, ionized gas, neutral (HI) gas, molecular gas, dust and relativistic plasma. Some components have merely been detected (dust), while others have been extensively studied (ionized gas, neutral (HI) gas, relativistic plasma). It is commonly accepted that these objects contain stars and AGN, but their existence have been inferred by indirect methods. Other components have not yet been detected in the highest redshift objects (molecular gas and hot (X-ray emitting) gas), but it is clear that only modest advances in the performance of instrumentation should allow detection of these components.

Radio galaxies have advantages over quasars as cosmological probes in that at least four of these emitting components can be spatially resolved by ground based telescopes. Furthermore, due to the absence of a bright nucleus, the study of components other than the nucleus itself is less challenging.

One of the aims of studying these distant sources is to better understand processes related to galaxy formation. Possibly the best hint that radio source activity in these distant objects is related to galaxy formation is the strong evolution of the radio luminosity function. The number density of the most powerful steep spectrum radio sources (e.g. $P_{2.7} = 10^{27}$ W Hz^{-1} sr^{-1}) increases by a factor 1000 from $z \sim 0$ up to $z \sim 2 - 3$ (Dunlop and Peacock 1993). Such strong evolution is a natural consequence in hierarchical models for structure formation (Haehnelt and Rees 1993).

Finally, radio galaxies are powerful tracers of galaxy clustering at high redshift. There are a number of observational hints that the highest-z radio galaxies are located in relatively dense environments. We shall discuss this in some detail in section 5.

3   Components

In Table 1 we provide an overview of the mass associated with some fo the components of HZRGs. Most of these estimates are highly model-depended and should therefore treated with caution.

Components	Total Mass $M_\odot$	Reference
Stars	10^{12}	4C41.17, $z = 3.8$, Chambers et al. (1990)
Ionized Gas	10^9	1243+036, $z = 3.6$, van Ojik et al. (1996a)
Neutral (HI) gas	2×10^7	0943−242, $z = 2.9$, Röttgering et al. (1995)
Molecular gas	$< 10^{11}$	Evans et al. (1996); van Ojik et al. (1996b)
Dust	3×10^8	4C41.17, $z = 3.8$, Dunlop et al. (1994)
Hot (X-ray emitting) gas	2×10^{12}	3C356, $= 1.079$, Crawford and Fabian (1993)

Table 1. Estimated masses for some of the components associated with distant radio galaxies. A Hubble constant of $H_0 = 50$ km s^{-1} Mpc^{-1} and a density parameter of $\Omega = 1$ is used.

Below we will discuss characteristics of the components that can be studied with the VLT and give some recent results indicating what progress can be made using the VLT.

3.1   Stars

It is clear that characterization of the stellar population in distant galaxies is a prime task for the next generation of optical telescopes. However, the first issue to be addressed is whether there is observational evidence that these systems do indeed contain stars. At low-z, powerful radio sources are located in massive ellipticals. By analogy, one would expect this also to be the case at high redshifts. However, optical imaging studies of the morphology of $z > 2$ galaxies showed that a large fraction of these objects are clumpy and have its main optical axis aligned with the axis of the radio emission (e.g. McCarthy 1993). This is unlike nearby ellipticals and therefore, before the stellar content of these systems can be properly modelled, it has to be established whether these systems do contain stars and, if so, what fraction of the UV/optical/infrared continuum emission is due to stars. We will return to this issue in section 4. Here we will concentrate on observational evidence that these systems do contain stars.

Until recently the best evidence that these systems do contain a massive population of stars came from infrared imaging. The alignment effect and clumpiness is much less prominent at these relatively long wavelengths and the Hubble K-diagram shows a continuity from low to high-z and a relatively low dispersion (e.g. McCarthy 1993). All this can be readily explained as related to the existence of a massive population of relatively old stars (for alternative views see Eales and Rawlings 1996).

The best direct evidence for stars comes from recent KECK spectroscopy of a radio galaxy at $z = 3.54$ (Dey et al. in prep.). In addition to the usual emission lines (Lyα, C IV and He II) it showed very strong UV absorption lines. This absorption spectrum very much resembles that of the two nearby Wolf-Rayet galaxies NGC 1741 (Conti et al. 1996) and NGC 4214 (Leitherer et al. 1996) as well as those of the starforming galaxies at $3 < z < 3.5$ that Steidel et al. (1996) have recently discovered. This similarity clearly indicates that at least one distant radio galaxy contains a massive population of young stars that have been produced during a recent burst of star formation.

The VLT with its proposed optical and infrared instrumentation should allow detailed studies of spectral energy distribution of these systems, including stellar absorption lines. With this information it should be possible to separate the contributions of stellar and non-stellar light to the morphology of the galaxy. This will allow important parameters of these systems to be determined, including stellar content, metallicity, SFR and SF history. The high resolution capabilities of the VLT should allow a number of these parameters to be determined as a function of location within the galaxy. With this information it should be possible to study the distribution of SF over the galaxy and observationally establish the epoch of formation of these galaxies.

3.2   AGN

Since distant radio galaxies all emit powerful radio emission, it is likely that these objects contain, either now or in the past, active nuclei. During this meeting, Cimatti discussed how observations of scattered light will not only contribute to our knowledge of the scattering medium, but also to that of the nature of the AGN.

It is likely that dust absorbs broad UV emission lines associated with the supposed nucleus in the very centre of the HZRGs. Since dust obscuration is less severe in the infrared, searches for broad emission lines and compact central objects should be carried out in this waveband. In addition, the high resolution imaging of radio loud quasars should give detailed information on the galaxies hosting the quasars. A detailed comparison between these host and those of radio galaxies should provide important further constrains on models that attempt to unify quasars and radio galaxies.

3.3   Ionized Gas

The Lyα emission of distant radio galaxies is often spectacular (e.g. McCarthy 1993). It can be as luminous as 10^{44} erg s^{-1} and extend up to 100 kpc ($\sim 10''$ at $z = 2.5$). These halos could well trace the reservoir of gas from which the galaxies are forming.

Studies of the velocity structure of the gas show that there are two distinct components. The inner region has a relatively high velocity dispersion (~ 1000 km s^{-1}) and there is clear evidence that this is induced by hydrodynamical interactions of the radio jet with the gas. In some sources such interaction is directly observed; in 1243+036 ($z = 3.6$) the Lyα gas is displaced both spatially and in velocity at the location of a strong bend in the radio jet (van Ojik et al. 1996a). Further evidence for such strong interaction comes from the correlated distortions of the radio and emission line morphologies seen in samples of HZRGs (van Ojik et al. 1996c). The second gas component is the emission line gas outside the main radio source structure, which has a much lower velocity dispersion (~ 250 km s^{-1}). In the case of 1243+036 we have direct evidence that the gas shows ordered motion, possible due to rotation of a protogalactic gas disk at $z = 3.6$ out of which the galaxy associated with 1243+036 is forming. A gravitational origin of the rotation of such a large disk implies a mass of $\sim 10^{12} \sin^{-2}(i)$ M$_{\odot}$, where i is the inclination angle of the disk with respect to the plane of the sky.

The distinction between the inner and outer halos has been observed using 4-m telescopes with integration times of order 1 night, about the maximum realistically feasible. An interesting challenge for an 8-m telescope would be to determine the maximum extent of the Lyα halos. The largest sizes measured so far are up to 150 kpc. Is this a true cutoff in the distribution or is it possible to trace the Lyα halo out to distances of half a megaparsec?

A very important question concerns the ionisation of the emission line gas. For a number of reasons it is likely that ionisation by a beam from a hidden quasar is playing a role: (i) some HZRGs have a Lyα emitting region that has a cone shaped morphology reminiscent of such a scenario, (ii) the integrated line ratios are well reproduced by nuclear photoionisation models and (iii) there is a tendency for the emission line region to be aligned with the main axis of the radio source (e.g. the optical beam). However, there are a number of problems with this simple picture, the most important being that the outer contours of the Lyα halos are elliptical in shape, and that therefore a significant fraction of halo emission comes from regions perpendicular to the radio axis, ie. not illuminated by the supposed beam. An important constraint on the ionisation mechanisms would be to measure line ratios in these outer regions.

With detailed studies of the morphology, dynamics and ionisation mechanisms of the halos gas, it should be possible to test scenarios for the origin of the gas. It has been suggested that the gas was expelled during a major starburst that the galaxy underwent during its formation. Alternatively, the gas might be indicative of a massive cooling flow that provides a significant fraction of the material from which the galaxy is forming.

3.4   Neutral Gas

There are a number of ways to search for and subsequently study neutral gas associated with distant galaxies. One method is to measure the redshifted 21 cm absorption line against the radio continuum. The only distant radio galaxy for which this has been done is 0902+34 ($z = 3.4$). Uson et al. (1991) found an absorber in this system with a column density of 4.4×10^{22} atoms cm^{-2}, assuming a spin temperature of 10^4 K. The existence of this absorber was confirmed by Briggs et al. (1993) and de Bruyn et al. (1995). It is likely that more cases will be discovered by the new tunable radio receivers of the Westerbork Radio Telescope (see http://www.nfra.nl/nfra/wsrt_info.html).

A second method is studying the deep narrow troughs that often "disfigure" the Lyα profiles. High–resolution spectra show that, in some cases, these features are too sharp to be explained as separate kinematic components of the emission, but that they are definitely due to absorption by neutral hydrogen along the line of sight. We have analysed deep high resolution spectra for a sample of 18 distant radio galaxies (van Ojik 1995; van Ojik et al. 1996c) and H I absorption features appear widespread in the Lyα profiles. 11 radio galaxies out of the sample of 18 have strong ($> 10^{18}$ cm^{-2}) H I absorption. Since, in most cases, the Lyα emission is absorbed over the entire spatial extent (up to 50 kpc), the absorbers must have a covering fraction close to unity. Given the column densities and spatial scales of the absorbing clouds, the typical H I mass of these clouds is $\sim 10^8$ M$_\odot$.

On the source with one of the deepest and best defined HI absorption systems (0943$-$242, $z = 2.9$, see also Fig. 4), we have carried out deep high resolution (1.5 Å) spectroscopy on the C IV and He II line using the AAT telescope. In Fig. 2 we show the resulting spectra. The He II line does not show absorption. This is expected since it is a non-resonant line. The C IV line shows absorption due to the CIV 1548/1551 doublet. We have fitted the line profile with a combination of a Gaussian (for the emission) and two (coupled) Voigt functions (for the absorption). The column density for the absorber is $10^{14.4}$ cm^{-2}. Combined with the measured column density for the HI absorber (10^{19} cm^2), this indicates that the spatially extended absorber is metal enriched. It further shows that such absorption measurement is a good tool to study extended slabs of neutral gas at high redshift. Unfortunately, 0943$-$242 is one of the few objects for which this kind of work is possible with 4-m class telescopes, since it is among the few that have both strong HI absorption and strong CIV emission.

With the availability of the VLT there is the prospect of conducting detailed studies of such systems. Questions that should be addressed concern the dynamics and morphology of the neutral gas. For example, could the neutral gas be in a rotating system or is it the product of colliding proto-galaxies? Such observations should also address the origin of the gas and its ultimate fate as possible food for forming the stellar populations of the galaxy.

3.5   Dust

The existence of dust associated with distant radio galaxies is well established. The most direct method is measuring the dust emission at submillimetre wave-

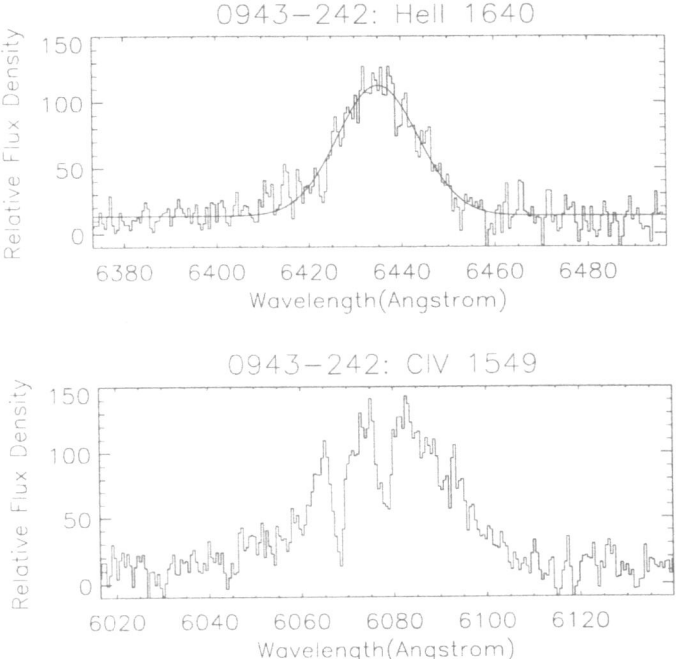

Fig. 2. Parts of the high resolution AAT spectrum (1.5 Å) of the He II 1640 region (upper) and the CIV 1549 region (lower) of the distant radio galaxy 0943−242 at $z = 2.9$). The C IV line shows absorption due to the CIV 1548/1551 doublet.

lengths. In Fig. 3 we show the spectral energy distribution of the radio galaxy 4C41.17 ($z = 3.8$) (from a compilation of Hughes 1996) indicating that this object contains 10^8 M$_\odot$ of dust. Other indirect measurements confirm that HZRGs can indeed contain massive amounts of dust, including (i) the optical/UV polarisation measurements, (ii) the clumpy optical continuum morphologies as compared to those in the infrared and (iii) the Lyα/Hα emission line ratios.

The amount of dust probably greatly varies from object to object. This is exemplified by two objects (TX0211−122, van Ojik et al. 1994 and MG 1019+0535, Dey et al. 1995) out of an estimated 60 HZRGs that have Lyα very much fainter with respect to the high ionisation lines than in typical high redshift radio galaxies. This suggests that these galaxies are undergoing a vigorous starburst producing a copious amount of dust that attenuates the Lyα emission.

An interesting topic for the VLT will be to compare the spatial distribution of Hα to Lyα, thereby estimating the distribution of dust through the whole galaxy. Detailed spectral polarisation measurements with the VLT is another powerful tool for studying the dust distribution. This is discussed by Cimatti in this proceedings.

Finally, it it is interesting to consider whether the thermal infrared instrument VISIR (VLT Imager and Spectrometer for mid InfraRed) that will be mounted on UT2 is sensitive enough to detect dust at these high redshifts. The 6 σ RMS in an 8 hour observation is 0.2 mJy for N-band ($8 - 13$ μm) and 2 mJy for Q-band ($16 - 24$ μm). We have plotted these two limits in Fig. 3. From this it is clear that dust is only detectable if there is a warm component (T $>$ 100 K) comparable in mass to the colder dust. IRAS has established that such warm dust components are a common feature of nearby radio loud AGN (e.g. 3C390.3 Miley et al. 1984). ISO should provide more information. VISIR has the great advantage over ISO in studying such warm dust because of its superb resolution.

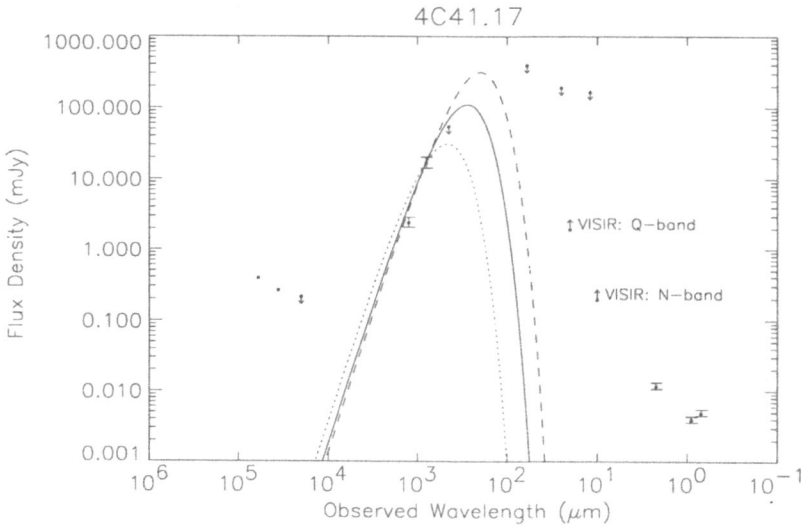

Fig. 3. Spectral energy distributions of the radio galaxy 4C41.17 ($z = 3.8$); the radio emission is for the core only. The dotted, solid, and dashed lines represent isothermal grey-body emission with an emissivity index $\beta = 2$, for dust at temperatures of 30 K, 50 K and 70 K respectively. Data are taken from the compilation by Hughes (1996). Also indicated is the 6 σ RMS in an 8 hour observation for N-band ($8 - 13$ μ m) and Q-band ($16 - 24$ μm) using VISIR, the mid infrared instrument on the VLT.

4   The Alignment Effect

Several years ago it came as a big shock when it was discovered that, unlike the case for nearby radio galaxies, the optical/IR continuum emission radio emission of $z > 0.6$ radio galaxies is roughly aligned with the radio emission (Chambers, Miley and van Breugel 1987; McCarthy et al. 1987). During the last 10 years a large number of explanations of this aligned emission have been proposed, the three most promising being scattering of light from a hidden quasar by electrons or dust (Tadhunter et al. 1989; Fabian 1989), star formation stimulated by the radio jet as it propagates outward from the nucleus (Chambers, Miley and van Breugel 1987; McCarthy et al. 1987; De Young 1989; Rees 1989; Begelman and Cioffi 1989) and nebular continuum emission from the emission line gas (Dickson et al. 1995).

The high resolution imaging capabilities of the HST provide excellent opportunities for studying the nature of the interaction of the jet with the host galaxies. At present we are carrying out an imaging survey on a selected sample of $z > 2$ radio galaxies. Although the alignment effect is clearly present at the kpc scale, the diversity of structures is enormous. Some of the galaxies have a simple cigar shaped morphologies aligned between the radio lobes (e.g. 0943−242, $z = 2.9$, Fig. 4), while others are very complex, showing a number of knots of which some are connected with the radio jet (e.g. 1138−262, $z = 2.15$, Fig. 5).

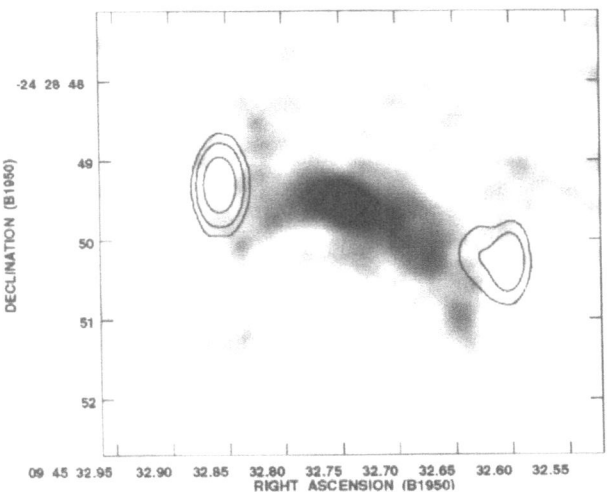

Fig. 4. The grey-scale is the HST image of 0943−242 ($z = 2.9$) through the F702W filter with a total integration time of 5300 sec. The contours show the VLA A-array total intensity radio map at 8.2 GHz with a resolution of 0.25″. The contours are at (0.2,0.8,4) mJy.

At intermediate redshifts we have studied the optical morphologies as observed by HST of a complete sample of 3CR radio galaxies ($1 \lesssim z \lesssim 1.3$) and found that they are highly dependent upon their radio properties (Best et al. 1996). There is a clear evolution of the optical structures as the size of the radio source increases: small radio sources consist of many bright knots, tightly aligned along the radio axis, whilst more extended sources contain fewer (generally no

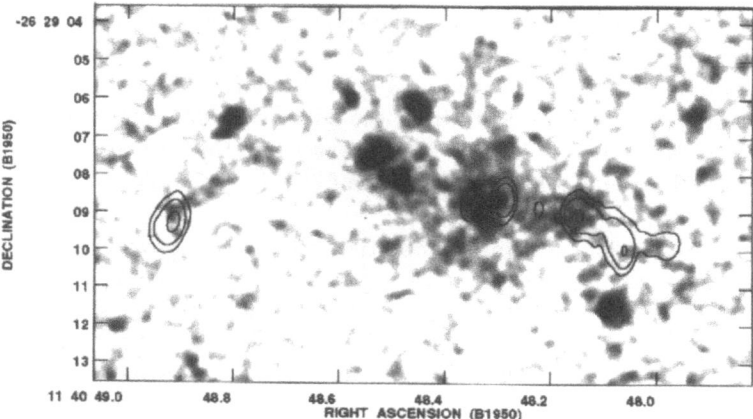

Fig. 5. The grey-scale is the HST image of 1138−262 through the F702W filter with a total integration time of 5300 sec. The contours show the VLA A-array total intensity radio map at 8.2 GHz with a resolution of 0.25″. The contours are at (0.25,1,5) mJy.

more than two) bright components and display more diffuse emission.

The morphologies of the intermediate and high-z sources can be explained as a combination of the three explanations as mentioned above. On the basis of the morphologies at the kpc scales most of the alternative models explaining the alignment effect seem no longer tenable including (i) inverse Compton scattering of CMB photons (Daly 1992), (ii) enhancement of radio luminosity by interaction of the jet with an anisotropic parent galaxy (Eales 1992), (iii) alignment of the angular momentum of the nuclear black hole with an anisotropic protogalactic distribution (West 1994) and (iv) gravitational lensing (Le Fèvre et al. 1987).

If it is indeed correct that the UV/optical continuum can be explained as due to a combination of scattered, stellar and nebular continuum light, then the next step is to properly disentangle these three mechanisms. Techniques for this include polarisation, colour information over a large range of wavelengths and narrow band filtering to separate gas from continuum emission. The superior sensitivity of the VLT will be particularly important for such photon limited studies.

5   High Redshift Clusters

Detection of clusters and groups of galaxies at high-redshift is important both for constraining cosmological models and for providing unique laboratories for studying a diverse range young and forming galaxies.

High-redshift radio sources are important targets for establishing to what extent clustering exists in the early Universe. At low redshifts ($z \sim 0.1$ to 0.5), luminous steep–spectrum radio sources have long been known to be excellent indicators of galaxy clustering (e.g. Miley 1980). It has also been shown that high luminosity radio sources associated with quasars and radio galaxies in the range $0.5 < z < 1$ are located in rich clusters (e.g. Hill and Lilly 1991).

More recent work clearly indicates that at least some of the radio galaxies at $z \sim 1$ are contained in clusters. The most likely explanation of the X-ray emission detected from some of the $z \sim 1$ 3C radio galaxies (e.g. 3C356, $z = 1.079$) Crawford and Fabian 1993) is that it originates in a hot halo of X-ray gas associated with a cluster. At this conference Dickinson presented further evidence from Keck spectroscopy and ROSAT X-ray imaging that 3C324 ($z = 1.206$) is located in a cluster (see also Dickinson et al. 1995).

At $z > 2$ the existence of clusters around HZRGs has not been established. There are important observational indications, however, that they might be in clusters, including (i) strong Faraday polarisation and rotation of the radio emission of some of the HZRGs indicating dense halos of hot electrons (e.g. Carilli et al. 1996), (ii) an excess of companion galaxies detected along the axes of the radio sources (Röttgering et al. 1996b), (iii) deep K-band imaging showing several red companion galaxies possibly at the same redshift as 4C41.17 ($z = 3.8$) (Graham et al. 1994) and (iv) potential companion galaxies around 4C41.17 discovered through imaging below the Lyman limit (Lacy and Rawlings 1996).

The deep imaging work that can be carried out (and is being carried out) with existing 4-m telescopes provides good candidates for cluster galaxies around HZRGs. To establish the existence of a cluster, the redshifts of these candidate cluster galaxies have to be measured. This will be an important task for the VLT.

6   1138−262 at $z = 2.15$: a Young and Forming Galaxy at the Center of Cluster?

Perhaps the ultimate aim of studying distant radio galaxies is to investigate the process of galaxy formation. A number of competing scenarios of galaxy formation have been proposed. Here we mention three classes of models which are currently in vogue and which may be relevant for such studies. First there are cooling flow models in which the galaxy forms during a massive cooling flow. Secondly there are hierarchical models in which dwarf galaxies merge to form the large radio galaxy. Thirdly there are models in which a gas reservoir builds up and undergoes a massive star burst.

This paper is clearly too short to discuss in detail these three different scenarios. Instead what we would like to do here is briefly discuss the remarkable galaxy 1138−262 ($z = 2.15$) in the light of these three scenarios.

A VLA imaging survey of 34 radio galaxies known to be at $z > 2$ has recently been carried out (Carilli et al. 1996) The radio galaxy PKS 1138−262 ($z = 2.15$, see Figs. 5 and 6) was one of the most extreme objects in the sample having (i) the highest rotation measure (RM) and the largest gradient of RM, (ii) the most distorted radio morphology, and (iii) a spectral index that steepens towards the outer knots. Also the optical properties are very peculiar with (i) a very clumpy

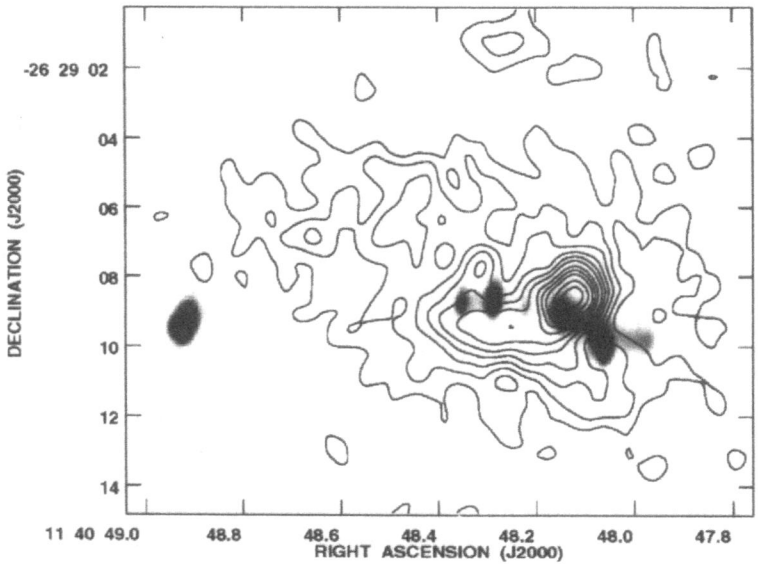

Fig. 6. The contours show an NTT narrow band image of the Lyα emission from 1138−262. The grey-scale represents the VLA A-array total intensity radio map at 8.2 GHz with a resolution of 0.25″.

continuum emission, as seen by HST, and (ii) a distribution of Lyα emission gas that does not follow the optical continuum (Pentericci et al. 1996).

Drawing the analogy between low redshift radio galaxies and 1138−262 we suggest that this source is at the centre of an extreme cooling-flow, as high as 1000 M$_\odot$ yr^{-1}. If this cooling flow could be maintained long enough then it would indeed provide sufficient material from which a large galaxy could be assembled. The extreme clumpiness of this galaxy suggest that that we are witnessing the assembling of this galaxy. Once the subunits have merged, 1138−262 will be a cD type galaxy at the centre of the cluster.

The suggestion that 1138−262 is a forming galaxy is possibly premature. It is clear, however, that detailed studies with the VLT of objects like this will allow the hypothesis that such objects are protogalaxies to be tested. Of particular interest will be a detailed determination of the SED of all the individual clumps of these galaxies. What is their stellar content like? Are they indeed SF regions that mainly contain young stars, or are they fairly old dwarf galaxies that happen to be in a cluster around 1138−262? Such measurements are impossible with current 4-m class telescopes and should be carried out with the next generation of optical telescopes.

7   Strategy

The VLT will be equipped with a broad range of instrumentation, the most important of which are mentioned in Table 2. This table also indicates the major emission components of HZRGs that can studied with each instrument.

	ISAAC	FORS	FUEGOS	UVES	VISIR	CONICA
Stars	*	*		*		*
AGN	*	*				*
Emission line gas	*	*	*	*		*
Absorbing gas	*	*	*	*		
Dust		*			*	
Environment	*	*	*			

Table 2. An overview of the most important instruments that the VLT will be equipped with, together with the components that can be studied with these instruments.

How can we make optimum use of the VLT for these studies? A first concern is the size and quality of samples of distant radio galaxies that are currently available. Preparatory work should go into a number of projects to obtain: (i) significant numbers of $z > 4$ radio galaxies, (ii) complete samples of radio sources that are fully identified and have redshifts, (iii) samples of milli/μ Jansky sources and (iv) samples of radio galaxies near bright stars (for adaptive optics/VLTI).

A final word about instrumentation: we believe that with its first set of instrumentation, the VLT is well equipped to carry out studies of the sort that we have mention here. We have a slight concern about the number of narrow and intermediate band filters that will be available. ESO has always had a good filter set, and we hope that it finds ways to provide an adequate set of filters for the VLT instrumentation. Finally, we believe that for studying the environment of distant radio galaxies a wide field infrared imaging capability is essential. We therefore hope that the present plans to build such an instrument will be pursued further.

Acknowledgements. We would like to thank our collaborators, Malcolm Bremer, Wil van Breugel, Chris Carilli, Arjun Dey, Dick Hunstead, Jaron Kurk, Laura Pentericci, Pat McCarthy, Rob van Ojik, Hy Spinrad and Paul van der Werf.

References

Begelman, M. C. and Cioffi, D. F.: 1989, *ApJ* **345**, L21

Best, P., Longair, M. S., and Röttgering, H. J. A.: 1996, *MNRAS* **280**, L9

Briggs, F. H., Sorar, E., and Taramopoulos, A.: 1993, *ApJ* **415**, L99

Carilli, C. L., Röttgering, H., van Ojik, R., Miley, G. K., and van Breugel, W.: 1996, *Radio Continuum Imaging of High Redshift Radio Galaxies*, APJS: submitted

Chambers, K. C., Miley, G., and van Breugel, W.: 1988, *ApJ* **327**, L47

Chambers, K. C., Miley, G. K., and van Breugel, W.: 1987, *Nat* **329**, 604

Chambers, K. C., Miley, G. K., and van Breugel, W. J. M.: 1990, *ApJ* **363**, 21

Conti, P. S., Leitherer, C., and Vacca, W. D.: 1996, *ApJ* **461**, L87

Crawford, C. and Fabian, A.: 1993, *MNRAS* **260**, L15

Daly, R.: 1992, *ApJ* **386**, L9

de Bruyn, G.: 1995, in M. Bremer, P. van der Werf, H. Röttgering, and C. Carilli (eds.), *Cold Gas at High Redshifts*, Kluwer, in press

de Young, D. S.: 1989, *ApJ* **342**, L59

Dey, A., Spinrad, H., and Dickinson, M.: 1995, *ApJ* **440**, 515

Dickinson, M., Dey, A., and Spinrad, H.: 1995, in H. Hippelein, K. Meisenheimer, and H. Röser (eds.), *Galaxies in the Young Universe. Springer-Verlag*, p. 164

Dickson, R., Tadhunter, C., Shaw, M., Clarck, N., and Morganti, R.: 1995, *MNRAS* **273**, L29

Dunlop, J. S., Hughes, D. H., Rawlings, S., Eales, S. A., and Ward, M. J.: 1994, *Nat* **370**, 347

Dunlop, J. S. and Peacock, J.: 1993, *MNRAS* **263**, 936

Eales, S. A. and Rawlings, S.: 1992, in *Infrared Spectroscopy: future observational directions*, preprint

Eales, S. A. and Rawlings, S.: 1996, *ApJ* **460**, 68

Evans, A. S., Sanders, D. B., Mazzarella, J. M., Solomon, P. M., Downes, D., Kramer, C., and Radford, S.: 1996, *ApJ* **457**, 658

Fabian, A. C.: 1989, *MNRAS* **238**, 41P

Graham, J. R., Matthews, K., Soifer, B. T., Nelson, J. E., Harrison, W., Jernigan, J. G., Lin, S., Neugebauer, G., Smith, G., and Ziomkowski, C.: 1994, *ApJ* **420**, L5

Haehnelt, M. G. and Rees, M. J.: 1993, *MNRAS* **263**, 168

Hill, G. and Lilly, S.: 1991, *ApJ* **367**, 1

Hughes, D.: 1996, in M. Bremer, P. van der Werf, H. Röttgering, and C. Carilli (eds.), *Cold Gas at High Redshifts*, Kluwer, in press

Lacy, M. and Rawlings, S.: 1996, *MNRAS* **280**, 888

Lacy, M., Rawlings, S., Wold, M., Bunker, A., Blundell, K. M., Eales, S. A., and Lilje, P. B.: 1996, in C. Parma (ed.), *IAU Symposium No. 175: Extragalactic Radio Sources*, Kluwer, in press

Le Fèvre, O., Hammer, F., Nottale, L., and Mathez, G.: 1987, *Nat* **326**, 268

Leitherer, C., Vacca, W. D., Conti, P. S., Filippenko, A. V., Robert, C., and Sargent, W. L. W.: 1996, *ApJ* **465**, 717

Lilly, S. J.: 1989, *ApJ* **340**, 77

McCarthy, P., van Breugel, W., Spinrad, H., and Djorgovski, S.: 1987, *ApJ* **321**, L29

McCarthy, P. J.: 1993, *ARA&A* **31**, 639

McCarthy, P. J., Kapahi, V. K., van Breugel, W., Persson, S. E., Athrea, R., and Subramhanya, C. R.: 1996, APJS: submitted

Miley, G. K.: 1980, *ARA&A* **18**, 165

Miller, L.: 1984, in R. Fanti, K. Kellerman, and G. Setti (eds.), *VLBI and compact radio sources*, p. 189, Dordrecht:Reidel

Pentericci, L., Röttgering, H., and Miley, G.: 1996, *1138−262: a young and forming galaxy at the center of a cluster*, AA, in prep.

Rees, M. J.: 1989, *MNRAS* **239**, 1P

Röttgering, H., Hunstead, R., Miley, G. K., van Ojik, R., and Wieringa, M. H.: 1995, *MNRAS* **277**, 389

Röttgering, H., van Ojik, R., Miley, G., Chambers, K., van Breugel, W., and de Koff, S.: 1996a, *Spectroscopy of Ultra-Steep Spectrum Radio Sources: A sample of z > 2 Radio Galaxies*, A & A: in press

Röttgering, H. J. A., West, M., Miley, G., and Chambers, K.: 1996b, *A&A* **307**, 376

Spinrad, H.: 1995, in H. Hippelein, K. Meisenheimer, and H. Röser (eds.), *Galaxies in the Young Universe. Springer-Verlag*, p. 95

Steidel, C. C., Giavalisco, M., Pettini, M., Dickinson, M., and Adelberger, K. L.: 1996, *ApJ* **462**, L17

Tadhunter, C. N., Fosbury, R. A. E., Binette, L., Danziger, I. J., and Robinson, A.: 1987, *Nature* **325**, 504

Uson, J., Bagri, D. S., and Cornwell, D. S.: 1991, *Phys. Rev. Letter* **67**, 3328

van Ojik, R.: 1995, *Ph.D. thesis*, University of Leiden

van Ojik, R., Röttgering, H., Carilli, C., Miley, G., and Bremer, M.: 1996a, *A radio galaxy at z = 3.6 in a giant rotating Lyman α halo*, A&A: in press

van Ojik, R., Röttgering, H., Miley, G., Bremer, M., Macchetto, F., and Chambers, K.: 1994, *A&A* **289**, 54

van Ojik, R., Röttgering, H., van der Werf, P., Miley, G., Carilli, C., Isaac, K., Lacy, M., Jenness, T., Sleath, J., Visser, A., and Wink, J.: 1996b, *A search for molecular gas in high redshift radio galaxies*, A&A: in press

van Ojik, R., Röttgering, H. J. A., Miley, G. K., and Hunstead, R.: 1996c, *The Gaseous Environment of Radio Galaxies in the Early Universe: Kinematics of the Lyman α Emission and Spatially Resolved HI Absorption*, A&A: in press

West, M. J.: 1994, *MNRAS* **268**, 79

Windhorst, R.: 1984, *Ph.D. thesis*, University of Leiden

Keck Spectropolarimetry of High z Radio Galaxies: Discerning the Components of the Alignment Effect

Andrea Cimatti[1,2], Arjun Dey[1,3], Wil van Breugel[1], Robert Antonucci[4], Todd Hurt[4], Hyron Spinrad[5]

[1] IGPP/LLNL, 7000 East Ave, P.O. Box 808, Livermore, CA 94550, USA
[2] Osservatorio Astrofisico di Arcetri, Largo E. Fermi 5, I-50125, Firenze, Italy
[3] NOAO, Tucson, AZ, USA
[4] Physics Department, University of California, Santa Barbara, USA
[5] Astronomy Department, University of California, Berkeley, USA

Abstract. We present the first results of optical spectropolarimetry of high z radio galaxies performed at the W.M. Keck 10m telescope, and discuss the main implications about the components of the UV continuum and the relation to the Unified Model of radio–loud AGN.

1 Introduction

High redshift radio galaxies (HzRGs) are observable up to cosmological distances competitive with the most distant quasars. However, before using them as probes of galaxy evolution, it is crucial to separate the stellar and non-stellar components (see Cimatti 1996 for a recent review). One of the most striking properties of HzRGs is the alignment of the UV continuum with the axis of the radio source (*alignment effect*; see McCarthy 1993). However, the relative importance of the stellar and non-stellar radiation to the *alignment effect* is still unknown, although a significant fraction is recognized to come from scattering of anisotropic radiation emitted by the obscured nucleus, as expected in the unified model of powerful radio sources (di Serego Alighieri, Cimatti & Fosbury 1994). Spectropolarimetry is the most powerful technique to observe at the same time different radiation components, but the 4m class telescopes can reach a sufficient S/N ratio only on the few brightest objects. Therefore, in order to investigate the origin of the *alignment effect* and to test the validity of the unified model of powerful radio-loud AGN, we have started a program of optical spectropolarimetry of HzRGs with the W.M. Keck 10m telescope equipped with the Low Resolution Imaging Spectrometer (LRIS) in polarimetric mode.

2 Observations and Data Analysis

The observations were made at the Cassegrain focus of the 10-m W.M. Keck Telescope using the Low Resolution Imaging Spectrometer (LRIS) in March and July 1995. The spectropolarimeter is a dual beam instrument which uses a calcite analyser and a rotatable waveplate. We used a 300 line/mm grating ($\lambda_{blaze} = 5000$ Å) and a 1 or 1.5 arcsec wide slit. The LRIS detector is a Tek 2048^2 CCD with 24μm pixels which correspond to a scale of 0.214 arcsec pix^{-1}. The dispersion is $\approx$2.5 Å pix $^{-1}$ and the spectral region covered is λ_{obs} ~3970–9010 Å. The observations were made under good seeing conditions ($\approx$0.5-0.7 arcsec). A complete observation includes 4 integrations made at four half-wave plate positions (0°, 45°, 22.°5, 67.°5). Unpolarized and polarized standard stars were observed in order to check and calibrate the instrumental polarization and the polarizance, and to calibrate the position angle zero point offset. The observations and data analysis are described in details in our first papers (Cimatti et al. 1996a, 1996b; Dey et al. 1996).

Table 1. Summary of the results

Galaxy	z	$\Delta\lambda_{rest}$	P	E	$P(\lambda)$	$\theta(\lambda)$	$P \times F_{tot}$
		(Å)	(%)				
3C 441	0.707	2350-5300	2-10	$\perp$	blue	const	const
3C 356a	1.079	1900-4300	7-15	$\perp$	blue	const	const
3C 356b	1.088	1900-4300	2-4	$\perp$	const	const	red
3C 324	1.206	1800-4000	10-13	$\perp$	const	const	red
3C 13	1.351	1700-3800	5-9	$\perp$	const	const	red
3C 256	1.819	1400-3200	10-14	$\perp$	const	const	red

3   General Results and Implications

We observed 6 HzRGs with 0.7< z <1.8. Table 1 and Figure 1 show respectively the main results and one example of our spectropolarimetry. 3C 368 (z=1.132) is not included in Table 1 because the data have been not fully reduced yet. *(1)* High linear polarization of the UV continuum is detected in all the observed galaxies (with the exception of 3C 356b). The perpendicularity of **E** to the UV continuum axis and the constancy of $\theta(\lambda)$ suggest that scattering is the dominant polarization mechanism. *(2)* The detection of the MgIIλ2800 emission line in polarized flux suggests that the incident radiation comes from an obscured quasar nucleus and is emitted anisotropically along the radio axis. In particular, the broad and polarized MgIIλ2800 in 3C 324 and 3C 356a has velocity and equivalent widths consistent with those observed in radio-loud quasars (Cimatti et al. 1996a, 1996b). On the other hand, the always lower or null polarization of the forbidden narrow lines implies that they are emitted isotropically outside the obscuring region. *(3)* In the two galaxies analysed in detail so far (3C 256, Dey et al. 1996; 3C 324, Cimatti et al. 1996a), we observe *spatially extended* polarization along the UV continuum axis, implying that the scattered flux is spatially extended. This result was possible thanks to the good seeing during the observations and the high S/N ratio of our spectropolarimetry. *(4)* These results are in agreement with the requirement of the unified model of powerful radio-loud AGN, where the differences between Type 1 and Type 2 AGN are mainly due to orientation effects and not to intrinsic diversities (Antonucci 1993). Our results also imply that scattered light is a necessary ingredient to explain the *alignment effect*. In particular, we find that scattered light contributes to 50–60% of the total UV continuum in 3C 324 and 3C 356a (Cimatti et al. 1996a, 1996b). *(5)* We detect basically two kinds of $P(\lambda)$: flat and blue. Since the total flux spectra are generally red ($F_\nu \propto \nu^{-1 \div 2}$), the polarized flux spectra $P \times F_\nu$ are either red or flat. These informations will be used to investigate in detail the nature of the scattering particles, the incident spectrum and the importance of unpolarized UV radiation.

4   The Case of 3C 356 ($z = 1.079$)

3C 356 is of particular interest since it exhibits the alignment effect at both optical and near-infrared wavelengths (Lacy & Rawlings 1994). This has led to the suggestion that the alignment effect in 3C 356 is a result of star formation triggered by a jet rather than scattered light from a quasar. However, the location of the nucleus of the radio source is not clear, although Lacy & Rawlings (1994) suggested that b is the core of the double–lobed radio source. Our observations show that both components a and b are polarized with the electric vectors in both cases oriented approximately orthogonal to the optical $a - b$ axis. Component a also shows evidence for broad MgIIλ2800 emission both in polarized and total light, while the narrow forbidden lines are unpolarized. Our observations allow us for the first time to quantify that the non–stellar radiation (scattered + nebular continua) contributes to more than half of the total UV continuum observed flux. We also find that a stellar population $\sim$1–2 Gyr old can account for the rest of the light at 2800 Å, in agreement with the detection in both galaxies (a&b) of the stellar CaII K absorption lines. Although the present data do not clarify unambiguously whether a or b is the nucleus of 3C 356, they suggest that the hidden quasar nucleus is located in a. In fact, if the nucleus were located in b, this would require an excessive and unrealistic luminosity of the hidden quasar in order to explain the luminosity of the scattered radiation coming from a.

References

Antonucci, R. 1993, ARAA, 31, 473.

Cimatti, A. 1996, in "New Extragalactic Perspectives in the New South Africa – Changing Perceptions of the Morphology, Dust Content and Dust-Gas Ratios in Galaxies", ed. D. Block et al., Kluwer Academic Publishers, in press

3C 356a

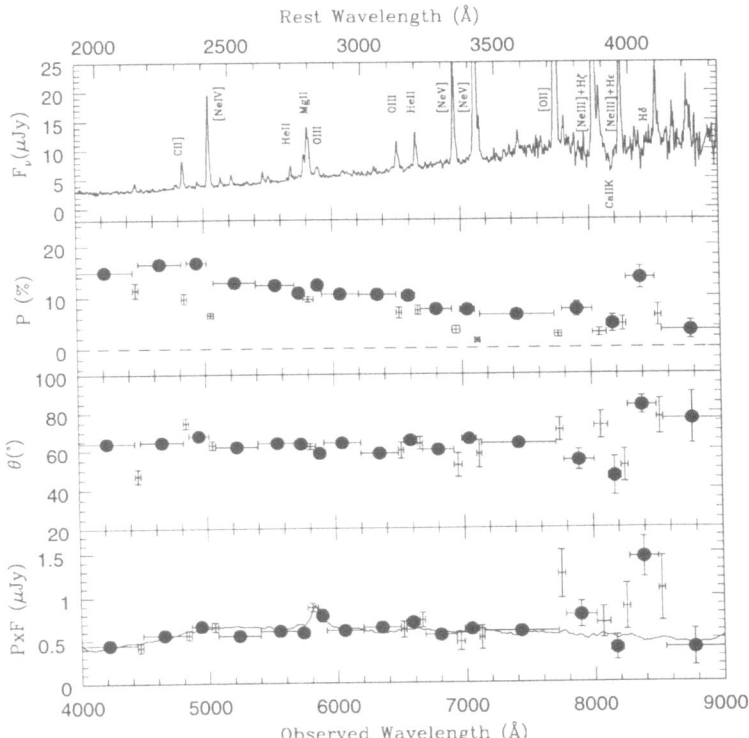

Fig. 1. The spectral and polarization properties of 3C 356a. From top to bottom: the total flux spectrum, the percentage polarization, the position angle of the electric vector and the polarized flux spectrum. Filled circles and crosses indicate respectively continuum, and emission lines with their underlying continuum. The continuum line in the $P \times F$ plot is a radio–loud quasar average spectrum (Cristiani & Vio 1990) scattered by Galactic type dust ($E_{B-V}=0.05$) (see Manzini & di Serego Alighieri 1996 for the dust scattering model).

Cimatti, A., Dey, A., van Breugel, W., Antonucci, R. & Spinrad, H. 1996a, ApJ, in press

Cimatti, A., Dey, A., van Breugel, W., Hurt T., Antonucci, R. 1996b, ApJ, submitted

Cristiani S., Vio R. 1990, A&A, 227, 385

Dey A., Cimatti A., van Breugel W., Antonucci R., Spinrad H. 1996a, ApJ, in press

di Serego Alighieri S., Cimatti A., Fosbury R.A.E. 1994, ApJ, 431, 123

Lacy M., Rawlings S. 1994, MNRAS, 270, 431 (LR94)

Manzini A., di Serego Alighieri S. 1996, A&A, in press

McCarthy P.J. 1993, ARAA, 31, 693

The Importance of IR Polarimetry for the Study of High Redshift Galaxies

Sperello di Serego Alighieri

Osservatorio Astrofisico di Arcetri, Largo E. Fermi 5, I-50125 Firenze, Italy

Abstract. Polarimetry in the IR range, in imaging and in spectroscopy, is very important in the study of distant galaxies to improve our understanding of the early Universe, both of the active galaxies, by disentangling the stellar and the active components, and by clarifying the unification issue for the most luminous AGN, as well as of the normal galaxies, by checking for the presence of a non-stellar component, by providing information on the dust content and properties, and by clarifying the links between galaxy formation and activity. Better coverage of the IR polarimetry is necessary to be able to attack these problems with the VLT instruments.

1 Introduction

The advantage of the VLT over 4m class telescopes is greatest when the data are photon noise limited rather than background limited (e.g. di Serego Alighieri 1995), like in polarimetry. The importance of polarimetry in the optical range for studying the AGN environment has already been discussed (di Serego Alighieri 1995). With FORS the VLT is indeed well equipped for optical polarimetry. Here I would like to concentrate on the IR polarimetry, stressing its usefulness for the study of distant galaxies, both active and normal, and calling for a better coverage of this technique by the VLT instrumentation than presently foreseen.

2 Distant Active Galaxies

Most of the distant galaxies ($z > 2$) known today are radio galaxies (RG). Therefore most of the information that we have on stellar systems in the early Universe comes from RG. This situation is not completely satisfactory, since we would like to have information on the formation and early evolution also of the normal non-active objects, which form the vast majority of the galaxy population today. Indeed the HST (see the Hubble Deep Field) and the Keck 10m telescope (e.g. Steidel et al. 1996) are opening up new important possibilities for studying distant normal galaxies, and the VLT will undoubtably follow this line.

Nevertheless distant active galaxies are likely to continue to play an important role for the study of the early Universe also in the VLT era for a number of reasons. First, it is possible that most galaxies have had an active phase in the past, possibly nearly coinciding with their main star forming phase: only by observing distant active galaxies we can learn about these important links between galaxy formation and activity. Second, active galaxies act as markers and illuminators of their environment, likely to contain normal galaxies and primordial gas clouds, which can only be found and understood by studying the distant active galaxies (e.g. Smail & Dickinson 1995). Finally it is clearly important to study RG to understand activity *per se* and its evolution.

In the recent past polarimetry in the optical range has provided important new information on distant RG, by disentangling the active component and the stellar one, thereby permitting a detailed and safe analysis of both of them. In fact the strong contamination of the light of distant RG by scattered anisotropic radiation coming from the active nucleus has been discovered with imaging polarimetry and specropolarimetry gives the spectral shape of the scattered and stellar components (di Serego Alighieri, Cimatti & Fosbury 1994). By scaling from the results obtained with 4m class telescopes with polarimetry on RG at redshift 1–2 (see fig. 1), we estimate that the VLT should be able to obtain the same results on RG at redshift 3–4. However the strongly polarized rest–frame range between 2000 and 4000Å, which we observe in the optical for objects at $z \sim 1$, would then be pushed in the near IR for the objects observable with the VLT. Therefore we need IR polarimetry with the VLT to extend to the highest redshifts the studies of the active galaxies which we do at redshift 1–2 with the 4m class telescopes. In addition, the Hα line, which is often the strongest permitted line in active galaxies, is pushed in the IR for z>0.5. Therefore IR polarimetry is again needed to observe the broad polarized Hα, foreseen by the scattered quasar model (see fig. 1) and it would give the ratio of narrow to broad Hα, and the amount of (stellar) dilution at Hα, from the ratio of the equivalent width in the total and in the polarized spectra (di Serego Alighieri 1996).

3   Distant Normal Galaxies

Infrared polarimetry is useful also to study distant normal galaxies, a field in which it is easy to foresee a large use of the VLT, as noted before. The main reason is that in these objects it is most important to check that they do not have an active component, otherwise the whole point of studying them breaks down. This component cannot be excluded *a priori* and may very well be hidden from direct view. In the context of the Unified Model for AGN it is not yet clear where the parent population of the radio quiet quasars is, and this population could be very numerous, since the space density of radio quiet quasars is much larger than that of radio loud ones, and the solid angle in which the strong nuclear radiation escapes could be rather small for radio quiet objects. Indeed many of the faint galaxies (the so called *stringy* galaxies) in the Hubble Deep Field bear a striking resemblance with the structure of distant RG, where we know that a large fraction of the extended structure is due to scattering.

The best way to check about the 'normality' of distant galaxies and understand the links between galaxy activity and formation is by investigating the presence or absence of scattered radiation with polarimetry. Again, since the redshift at which this check is important is around 3 and 4, the range of expected high polarization is shifted to the IR. Furthermore we now know that

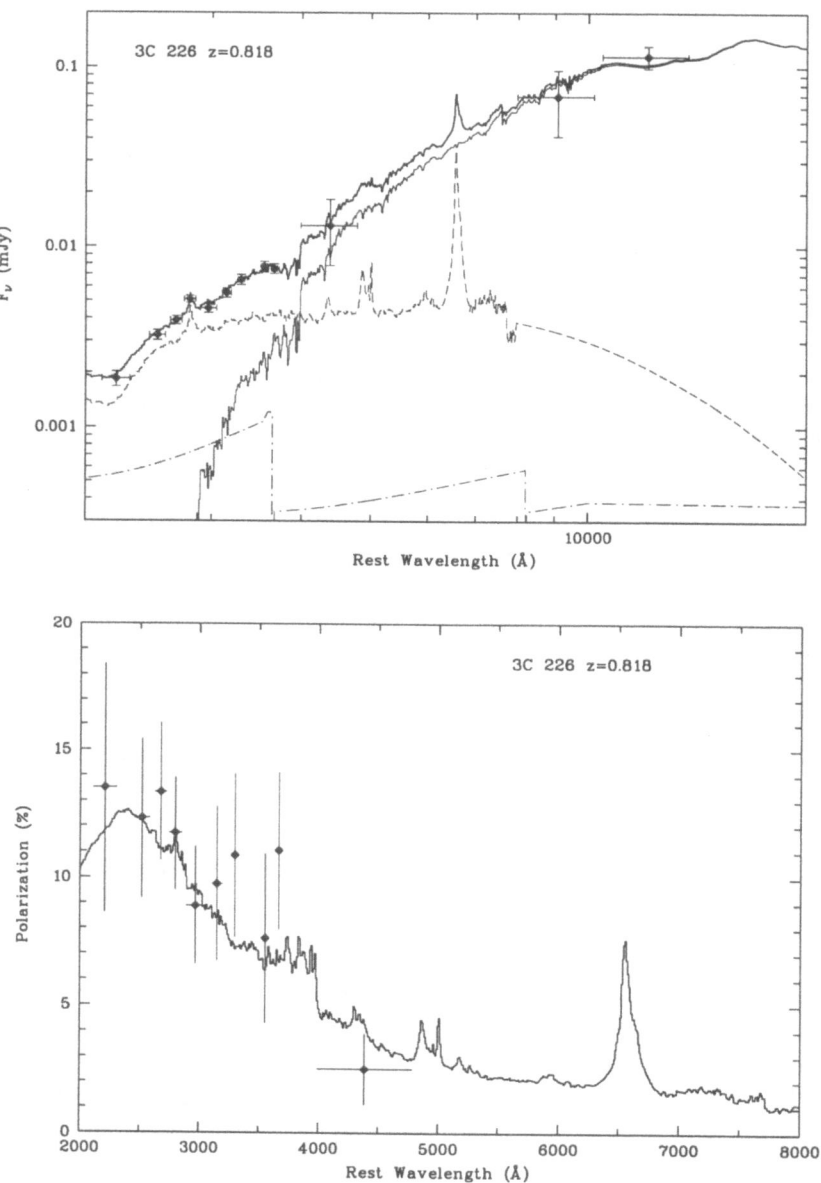

Fig. 1. Model of the SED and polarization of the distant radio galaxy 3C 226. The top panel shows the model of the SED made of a dust scattered quasar (dashed line), a 4 Gyr old stellar population (thin line), and a nebular continuum (dot–dashed line), which fits the points obtained from the spectropolarimetry of di Serego Alighieri et al. (1994) and IR imaging. The bottom panel shows the polarization foreseen by the model together with the data from di Serego Alighieri et al. (1994).

the presence of dust cannot be excluded even in the most distant objects (Ivison 1995) and that its unequivocal detection is a clear sign that the primordial material has been reprocessed in stars and can be used to push back in time the epoch of galaxy formation. Again polarimetry, in the IR because of the redshift, is of great help, since polarization is produced by transmission of radiation through aligned dust grains even in the most normal galaxy.

4   Technical Considerations

In the IR it is even more important than in the optical to use polarization analyzers of the beam splitting type, like Wollaston prisms, since the faster variations in the sky transparency and emissivity make it very difficult to rely on a succession of exposures with a rotating single beam polarizer in the IR (di Serego Alighieri 1995). Suitable birifringent materials for Wollaston prisms exist also for the IR. There are however some practical difficulties that have prevented so far to duplicate the excellent polarimetric capabilities of FORS in the present set of IR instruments for the VLT. These difficulties are mainly linked with the necessity of cooling the optics that work in the IR above about 1.8μm. This is not so much of a problem for imaging polarimetry, since the beam splitting analyzer (e.g. a Wollaston prism) can be placed in the grism wheel where the beam is smallest and where there is enough room, and the slotted mask can go in the aperture wheel which is normally inside the dewar and therefore cooled. The rotation necessary for the polarization measurement can be implemented rotating the whole instrument using the rotator/adapter or having a set of Wollaston prisms and masks at different angles. The situation is more difficult for spectropolarimetry, since a rotation of the whole instrument or a set of slits at different angles would change the part of the object that goes through the slit. This problem is normally solved by using a rotating half–wave plate, but its implementation is difficult in an IR instrument, because its rotating mechanism has to be cold. There may be however other solutions to this problem, which are worth investigating in order to be able to do IR spectropolarimetry with the VLT. In fact instrument builders should realize that with the VLT it will be a waste not to measure all Stokes parameters, particularly in the IR, where the high background forces split exposures anyway.

References

di Serego Alighieri, S., Cimatti, A. & Fosbury, R.A.E., 1994, ApJ **431**, 123

di Serego Alighieri, S., 1995, in *Science with the VLT*, J.R. Walsh & I.J. Danziger eds., (Springer, Berlin, Heidelberg), p. 317

di Serego Alighieri, S., 1996, in *Instrumentation for Large Telescopes*, VII Canary Isl. Winter School of Astrophys., J.M. Rodriguez ed., (Cambridge Univ. Press), in press.

Ivison, R.J., 1995, MNRAS **275**, L33

Smail, I. & Dickinson, M., 1995, ApJ **455**, L99

Steidel, C.C., Giavalisco, M., Pettini, M., Dickinson, M. & Adelberger, K.L., 1996, ApJ **462**, L17

Quasar Surveys

S. Cristiani[1], F. La Franca[2]

[1] Dipartimento di Astronomia, Università di Padova, Vicolo dell'Osservatorio 5,
I–35122 Padova, Italy
cristiani@astrpd.pd.astro.it
[2] Dipartimento di Fisica, Università di "Roma TRE", Via della Vasca Navale 84,
I–00146 Roma, Italy

Abstract. The identification of complete samples of quasars and of absorption lines in QSO spectra provides basic information about the formation and evolution of large and low-mass halos in the Universe. The QSO luminosity function and the clustering properties of quasars and absorbers are instrumental to study the physics driving the energy generation in AGNs, the origin and evolution of perturbations, the geometry of the Universe. QSO surveys to be carried out with the VLT have different requirements according to the immediate goal they are focussing on, but, to exploit the full potential of the various techniques, they have to be integrated in a coherent strategy.

1   The Luminosity Function

The general behaviour of the quasar luminosity function (LF) is fairly well established at redshifts lower than 2.2, for which color techniques provide reliable selection methods. La Franca and Cristiani (1996) have summarized the most recent measurements in this domain, discussing the most significant deviations from the "classical" Pure Luminosity Evolution (PLE) of the QSO population. The number density of quasars rapidly increases from the local neighborhood up to redshifts $\gtrsim 2$. Beyond this, the picture becomes less and less clear: there are evidences that the LF evolution slows down and indications that the number density of moderately bright quasars drops significantly at $z > 4$ (Warren, Hewett and Osmer 1994). The latter result is controversial with some authors suggesting that the luminosity function remains constant at z> 3 for the most luminous quasars, while others suggest that the luminosity function remains unchanged for all quasars up to z~4 (Hawkins and Veron 1995, a conclusion partially revised in Hawkins and Veron 1996). The origin, as well as the existence, of the suggested decline in quasar numbers is also controversial, since it is unclear whether the decline would be intrinsic to the QSO population, or is due to absorption by intervening dust (Warren, Hewett and Osmer 1994, Fall and Pei 1993). The shape of the LF, conventionally assumed to be a double power-law with an evolving break, has also been questioned (Hawkins and Veron 1995).

Most of these results are based on photographic material, typically obtained with Schmidt telescopes. The need for a breakthrough in the selection depth becomes apparent when considering that conventional surveys, limited to $R \lesssim 20 - 21$ are fully sampling the QSO population (i.e. down to $M_B = -23$ QSOs) only up to $z \sim 1.5$. A complete picture up to $z \sim 4 - 5$ will become available only

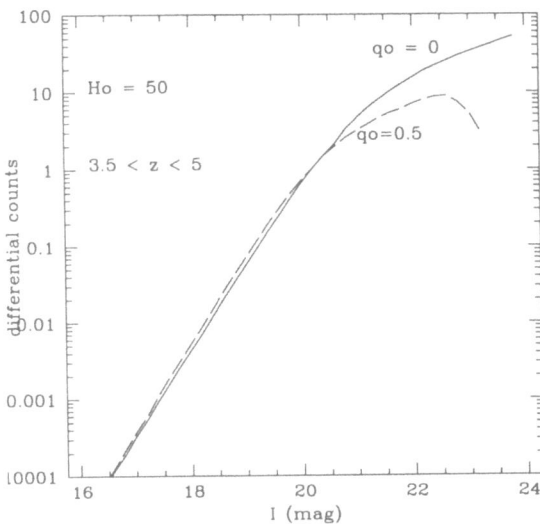

Fig. 1. Expected counts of quasars with $3.5 < z < 5$ as a function of the I magnitude. Quasars with $4 < z < 5$ account for roughly half of the counts

when deep surveys down to $I \sim 23$ will be carried out. The I band is needed for high-z QSO counts, because the R band drifts in the Lyman forest at $z \gtrsim 4$ (for the I band this happens only at $z \sim 5.5$). In this way the K-correction based on the R magnitude for $z > 4$ quasars becomes not only unfavorable but also aleatory, preventing from a reliable reconstruction of the LF.

Fig. 1 shows the counts of high-z quasars as a function of the I magnitude, predicted on the basis of a PLE with a "pessimistic" decline ($\propto (1 + z)^{-2}$) of the LF beyond $z = 2.2$, in agreement with the results of Warren et al. (1995) at $R \lesssim 20$. Depending on the model universe, we expect at $I \leq 22.5$ a surface density of $z > 3.5$ QSOs between 20 and 45 per sq. deg., which would become a factor $3 - 4$ larger in absence of a LF decline. This relatively large uncertainty can be removed quickly with low-resolution spectroscopy at the VLT on a few sq.deg. FUEGOS is the instrument of election, if the improved efficiencies of its new optical design are confirmed. Assuming a 10% success rate, there will be $30 - 50$ candidates per FUEGOS field ($28'$ diameter), i.e. 5 sq.deg. can be covered with 30 FUEGOS pointings of about 3 hours each, which makes less than 2 VLT weeks. This would provide a sample of $100 - 200$ high-z QSOs.

The selection of the candidates has to be carried out with CCD photometry in $BVRIZ$, using similar criteria to those described for galaxies (E. Giallongo, this conference). But, there's the rub for the ESO community, because, if we had to carry out the photometric survey with EMMI, we would need to observe 230 fields, making up about 150 nights ! As already mentioned by D. Tytler, the present ESO capabilities are able to "feed" the VLT at a rate that is one order of magnitude less than it should, showing the dramatic need for a dedicated wide field facility. It is interesting to note that the only QSO survey that approaches the proposed apparent magnitude limits and photometric quality, the KPNO multicolor survey (Hall et al 1996), found an excess of z>3 quasars with respect to previous works.

2   The Large-Scale Structure. Quasar Clustering

The situation of QSO clustering is similar to that of the LF: the picture is much better at $z \lesssim 2.2$, for which a 5σ detection exists (Andreani and Cristiani 1992, Shanks and Boyle 1994), it becomes much less clear at larger redshifts. As a consequence, the issue of the clustering evolution cannot be properly addressed, due to the lack of statistical significance.

An area of 5 sq. deg. would provide a fair sample: in terms of transversal size it corresponds to $300 - 500$ comoving Mpc ($q_o = 0.5 - 0.1$ and $H_o = 50$), so that clustering can be sampled in all directions and not only in the redshift space.

The most straightforward measure of sensitivity to clustering is given by the *numbers of pairs* of quasars within a given separation. Simulations with mock catalogues show that the prospects are not exciting: if the LF decays after $z = 2$ with a $(1+z)^{-2}$ law, even in the case of stable clustering in comoving coordinates, there would be just a 3σ detection of clustering at redshift $3.5 - 5$. The reason is easily understood if we consider that the S/N ratio for the clustering can be expressed (in the case of weak signal) as:

$$S/N = \sqrt{4\pi n_v N} \; \frac{3}{3 - \gamma} \; r_o^\gamma \, l^{1.5-\gamma} \tag{1}$$

where n_v is the density of QSOs per comoving volume density, N is the total number of QSOs in the sample, γ the index of the two-point correlation function modeled as $\xi(r) = (r/r_o)^{-\gamma}$, r_o is the correlation length and l is the radius within which the pairs are counted. l cannot be made too small, otherwise the number of expected pairs becomes also very small. The lower density of QSOs at high z reduces the sensitivity to clustering, even for a relatively deep survey as the one described in the previous section. To improve the sensitivity, the z-threshold has to be lowered. A survey covering the redshift range $2.2 - 5$ (or $2.2 - 4$) would not only detect clustering at a $5 - 6\,\sigma$ level but also be able to discriminate between the standard models customarily used to classify its evolution: $\xi(r, z) = \xi(r)(1 + z)^{-(3-\gamma+\epsilon)}$, with $\epsilon = 0, -1.2, -3$.

Selecting intermediate-z QSOs ($z = 2.2-3.5$) implies either adding a U band to the photometric survey or resorting to a slitless search for Ly$_\alpha$ emission in quasars down to $B \sim 23.5$. The former option is of course not available in the Red Arm of EMMI (once more the need for wide field imagers !), the latter would increase of about 20% the time required by the photometric survey. The number of quasars (and of candidates) would be about doubled, not affecting the total number of FUEGOS pointings, given the number of available fibers.

3   Quasar Absorption Lines

The wealth of lines revealed in intermediate-high resolution spectra of $z \gtrsim 1.5$ quasars (about 10 CIV per unit redshift with $EW > 0.1$, and more than one order of magnitude larger density of Ly_α) can provide unique information about the high-z universe. The limiting magnitude for a survey of quasars suitable for absorption-line work would be determined by the ability of the intermediate-high resolution spectrographs of acquiring spectra with the necessary S/N ($\gtrsim 10$). For UVES it can be placed at $V \sim 21$. There will be $125 - 150$ such quasars with $z > 1.6$ and $50 - 75$ with $z > 2.2$. The sum over all the quasars of the redshift intervals sampled would be $\Delta z(\text{CIV}) \sim 75 - 100$ (for absorbers up to $z = 3$, i.e. $\lambda \leq 620$ nm) and $\Delta z(\text{Ly}_\alpha) \sim 30 - 40$ (for absorbers up to $z = 4.1$). This means $750 - 1000$ CIV absorbers distributed between $z = 1.3$ and 3 and $3500 - 5000$ Ly_α's between $z = 2$ and 4.1, in a 5 sq.deg. field.

To estimate the sensitivity for a clustering signal one cannot apply eq. 1, since the absorptions are detected in different lines of sight separated typically by 20 Mpc in the case of CIV and 30 Mpc for Ly_α. The effect would be to increase the sensitivity on small scales, within the individual line of sight, still allowing to probe correlations on the scales of superclusters, among different lines of sight. In this way it will be possible to study the clustering of CIV lines at a $> 10\sigma$ level and also its evolution (assuming that CIV lines cluster roughly as galaxies, as already known).

If on the one hand for the Ly_α observations it would be possible to use the quasars detected in the surveys described in the previous sections, for the CIV observations one would need U photometry to select the UVx ($z < 2.2$) quasars that would make a considerable part of the CIV sample. Given the relative brightness of these QSOs, requiring a limiting $U \sim 21$ this could be easily accomplished, even with a Schmidt telescope, by stacking a few deep exposures.

Observing with UVES would require about 1/2 night per $V \sim 21$ object, corresponding to about 50 VLT nights. The possibility of a fiber-link between FORS and UVES, allowing to carry out at the same time a high-resolution survey of the absorption lines and a low-resolution survey of galaxies, could make even more interesting the investment of observing time.

References

Andreani P., Cristiani S., 1992, ApJ 398, L13
Fall S. M., Pei Y. C., 1993, ApJ 402, 479
Hall P.B., Osmer P.S., Green R.F., Porter A.C., Warren S. J., 1996, ApJ 462, 614
Hawkins M. R. S., Veron P., 1995, MNRAS 275, 1102
Hawkins M. R. S., Veron P., 1996, MNRAS, in press
La Franca F., Cristiani S., 1996, AJ submitted
Shanks T., Boyle B., 1994, MNRAS, 271, 753
Warren S. J., Hewett P. C., Osmer P. S., 1994, ApJ 421, 412
Warren S. J., Hewett P. C., Osmer P. S., 1995, ApJ 438, 506

Starburst Galaxies at High Redshift

Michael Rowan-Robinson

Astrophysics Group, Blackett Laboratory, Prince Consort Rd, London SW7 2BZ

Abstract. The properties of ultraluminous and hyperluminous galaxies are reviewed and it is argued that the majority of the far infrared radiation from these galaxies is powered by starbursts. The evidence that the starburst population is also seen in the sub-mJy radio source-counts is reviewed and a model for the source-counts from radio to X-rays is presented. Planned surveys at far infrared and submillimetre wavelengths are discussed. The nature of the galaxies in the Hubble Deep Field and their relationship with infrared starburst galaxies is briefly discussed.

1 What Do We Know About Starburst Galaxies from IRAS?

1.1 Ultraluminous Infrared Galaxies

One of the major discoveries of the IRAS mission was the existence of ultraluminous infrared galaxies, galaxies with L_{ir} $10^{12}h_{50}^{-2}L_o$($h_{50} = H_o50$). The peculiar Seyfert 1 galaxy Arp 220 was recognised as having an exceptional far infrared luminosity early in the mission (Soifer et al 1984). Sanders et al (1988) discussed the properties of 10 IRAS ultraluminous galaxies with 60 μm fluxes 5 Jy and concluded that (a) all were unteracting, merging or had peculiar morphologies (b) all had AGN line spectra. On the other hand Leech et al (1989) found that only 2 of their sample of 6 ultraluminous IRAS galaxies had an AGN line spectrum. Leech et al (1994) found that 67% of a much larger sample of ultraluminous galaxies were interacting, merging or peculiar. Lawrence et al 1989 had found a much lower fraction amongst galaxies of high but less extreme infrared luminosity. The incidence of interacting, merging or peculirr galaxies by ir luminosity is summarised in Fig 1 of Rowan-Robinson (1991). The situation on point (b) remains controversial, though, since Lawrence et al (1995) find only a fraction 21% of 81 ultraluminous galaxies in the QDOT sample to have AGN spectra. 21

Rowan-Robinson and Crawford (1989) found that their standard starburst galaxy model gave an excellent fit to the far infrared spectrum of Mk 231, an archetypal ultraluminous ir galaxy. However their models for Arp 220 appeared to require a much higher optical depth in dust than the typical starburst galaxy. Condon et al (1991) showed that the radio properties of most ultraluminous ir galaxies was consistent with a starburst model and argued that these galaxies required an exceptionally high optical depth. This suggestion was confirmed by the detailed models of Rowan-Robinson and Efstathiou (1993) for the far infrared spectra of the Condon et al sample.

Quasars and Seyfert galaxies, on the other hand, tend to show a characteristic mid infrared continuum, broadly flat in νS_ν from 3-30 μm. This component was modelled by Rowan-Robinson and Crawford (1989) as dust in the narrow-line region of the AGN with a density distribution n(r) αr^{-1} . More realistic models of this component based on a toroidal geometry are given by Piers and Krolik (1992), Granato and Danese (1994), Rowan-Robinson (1995), Efstathiou and Rowan-Robinson (1995) . Rowan-Robinson (1995) suggests that most quasars contain both (far ir) starbursts and (mid ir) components due to (toroidal) dust in the narrow line region.

1.2   Hyperluminous Infrared Galaxies

In 1988 Kleinmann et al identified P09104+4109 with a z = 0.44 galaxy, implying a total far infrared luminosity of 1.5×10^{13}, a factor 3 higher than any other ultraluminous galaxy seen to that date. In 1991, as part of a program of systematic identification and spectroscopy of a sample of 3400 IRAS FSS sources, Rowan-Robinson et al discovered IRAS F10214+4724, an IRAS galaxy with z = 2.286 and a far infrared luminosity of $5 \times 1014 \, h_{50}^{-2} L_o$. This object appeared to presage an entirely new class of infrared galaxies. The detection of a huge mass of CO by Brown and vandenBout (1991), $10^{11} h_{50}^{-2} M_o$ confirmed by the detection of a wealth of molecular lines (Solomon et al 1992), and of submillimetre emission at wavelengths 450-1250 m (Rowan-Robinson et al 93, Downes et al 1992), implying a huge mass of dust, $10^9 h_{50}^{-2} M_o$ confirmed that this was an exceptional object. Early models suggested this might be a giant elliptical galaxy in the process of formation (Elbaz et al 92). Simultaneously with the growing evidence for an exceptional starburst in F10214, the Seyfert 2 nature of the emission line spectrum (Rowan-Robinson et al 1991, Elston et al 1994a) was supported by the evidence for very strong optical polarisation (Lawrence et al 93, Elston et al 94b). Subsequently it has become clear that F10214 is a gravitational lens system (Graham et al 1995, Broadhurst and Lehar 1995, Serjeant et al 1995) with a magnification of about 10 at far infrared wavelengths, but not much greater than that (Green and Rowan-Robinson 1995). Even when the magnification of 10 is allowed for, F10214 is still an exceptionally luminous far ir source.

In 1992 Barvainis et al successfully detected submillimetre emission from the z=2.546 'clover-leaf' gravitationally lensed QSO, H1413+117, which suggested that H1143 is of similar luminosity to F10214.

The program of follow-up of IRAS FSS sources which led to the discovery of F10214 has also resulted in the discovery of a further 8 galaxies or quasars with far ir luminosities $10^{13} h_{50}^{-2} L_o$ (McMahon et al 95a). Cutri et al (1994) report a search for IRAS FSS galaxies with 'warm' 2560μm colours, which yielded the z = 0.93 Seyfert 2 galaxy, F15307+3252. Dey and van Breugel (1995) report a comparison of the Texas radio survey with the IRAS FSS catalogue, which yielded 5 galaxies with fir ir luminosities $10^{13} h_{50}^{-2} L_o$.

Finally, inspired by the success in finding highly redshifted submillimetre continuum and molecular line emission in F10214, several groups have studied an ad hoc selection of very high redshift quasars and radio-galaxies, with several notable successes (Andreani et al 1993, Dunlop et al 1995, Isaak et al 1995, McMahon et al 1995b, Ojik et al 1995, Iveson 1995). Most of these detections imply far ir luminosities $10^{13}h_{50}^{-2}L_o$, assuming that the far infrared spectra are typical starbursts. In addition there are 2 PG quasars from the sample detected by IRAS and studied by Rowan-Robinson (1995), which also satisfy this condition.

1.3 Models for Hyperluminous Infrared Galaxies

For a small number of these galaxies we have reasonably detailed continuum spectra from radio to uv wavelengths. The continuum emission from F10214 was the subject of a detailed discussion by Rowan-Robinson et al (1993). Green and Rowan-Robinson (1995) have discussed starburst and AGN dust tori models for F10214 and for F15307. Fig 1 of Rowan-Robinson (1996) shows the continua of these and several other hyperluminous galaxies, with fits using radiative transfer models (generally the standard starburst model of Rowan-Robinson and Efstathiou (1993) or the standard QSO dust model of Rowan-Robinson (1995).

For the remaining objects we have only 60 m or single submillimetre detections and for these we estimate their far infrared luminosity, and other properties, using the standard starburst model of Rowan-Robinson and Efstathiou (1993). In Fig 1 we show the far infrared luminosity against redshift for hyperluminous galaxies, with lines indicating observational constraints at 60, 800 and 1250 μm. Of the sources with luminosities above $5\times10^{14}h_{50}^{-2}L_o$, two are gravitationally lensed (F10214, M = 10; H1413, M = 7.6). The other 3 might be strong candidates for also being lensed.

On the other hand there is overwhelming evidence for a population of galaxies with far ir luminosities in the range $1-50\times10^{13}h_{50}^{-2}L_o$. By analogy with F10214 , we believe that the rest-frame radiation longward of 50 μ comes from a starburst component. The luminosities are such as to require star formation rates in the range 1-50 x $10^3h_{50}^{-2}M_o$, which would in turn generate most of the heavy elements in a $10^{11}M_o$ galaxy in $10^7 - 10^8$ yrs. Most of these galaxies can therefore be considered to be undergoing their most significant episode of star formation, ie to be in the process of 'formation'.

It appears to be significant that a large fraction of these objects are Seyferts, radio-galaxies or QSOs. For the very high redshift ($z3$) objects, this is a selection effect in that quasars and radio-galaxies are the only objects known at such redshifts. However even for the population of objects found from direct optical follow-up of IRAS samples (and omitting objects found in searches biassed to 'warm 6025μm colours, which are biassed towards galaxies with a 3-30 μm dust tori component), out of 10 objects, 5 are QSOs, one is Seyfert 1, 2 are Seyfert 2, and only 2 are narrow-line objects. Thus in a high proportion of cases, this phase of exceptionally high far ir luminosity is accompanied by AGN activity at optical and uv wavelengths.

In the Sanders et al (1989) picture, the far infrared and submillimetre emission would simply come from the outer regions of a warped disk surrounding the AGN. However the weaknesses of this picture as an explanation of the far infrared emission from PG quasars have been highlighted by Rowan-Robinson (1995). A picture in which both a strong starburst and the AGN activity are triggered by the same interaction or merger event is far more likely to be capable of understanding all phenomena (cf Yamada 1994).

2   Sub-mJy Radio-sources as Starburst Galaxies

The bright radio source-counts ($S(1.4GHz) >> 1mJy$) are due to radio-galaxies and quasars undergoing strong evolution, which is approximately of the form of luminosity evolution (eg Condon 1984, Dunlop and Peacock 1990), though some models involve a small amount of density evolution also. Below 1 mJy at 1.4 GHz the slope of the counts steepens again and there is evidence of a new population of blue radio-emitting galaxies (Mitchell and Condon 1985, Windhorst et al 1987, Thuan and Condon 1987, Franceschini et al 1988, Condon 1989, Lonsdale and Harmon 1991).

Benn et al (1993) have carried out spectroscopy of a sample of 112 idensitifi-cations of sources with $S(1.4GHz) > 0.1mJy$ and shown that below 1 mJy most of the sources are starburst galaxies very similar to those seen by IRAS. The luminosity function for these galaxies agrees well with that at 60 μm, shifted by the radio-fir relation

$$S(60\mu m) = 90S(1.4GHz).$$

Fits to the sub-mJy source-counts (Fig 4) then show that strong evolution is required in this population, with luminosity evolution strongly favoured over density evolution (Rowan-Robinson et al 1993a).

3   A New Model for Source Counts and Background Radiation from Radio to X-ray Wavelengths

A new model for extragalactic source counts proposed by Pearson and Rowan-Robinson (1996) defined in the infra-red consisting of a normal spiral + cir-rus component defined by cool 100/60μm colours, a starburst/IR ultraluminous galaxy component defined by warm 100/60μm colours, hyperluminous IR galax-ies ($L_{60} > 10^{12.5}L_\odot$), a 3-30 μm Seyfert/QSO + dust torii component and an elliptical component (negligible in far IR) has been used to predict source counts from radio to X-ray wavelengths ($H_o = 50$, $\Omega = 1$).

K-corrections are found using model source spectra. For the normal galaxies the cirrus model of Rowan-Robinson (1992) is used in the IR, while the model Sbc data of Coleman et al. (1980) and Yoshii and Takahara (1988) is used for the near IR/optical spectrum assuming that 30% of the optical light is re-radiated in the IR. For the starburst/ultra/hyperluminous components, the model spectra of Rowan-Robinson and Efstathiou (1993) are used in the far-IR while at near-IR/optical wavelengths, for $\nu < \nu_B$, the Sab spectrum of Coleman et al. (1980) is used and for $\nu > \nu_B$ a HII galaxy spectrum (Mk36, Neugebauer et al. (1976)) is used. In modelling the optical spectra of starburst galaxies it is assumed that 95% of the optical light is re-radiated in the far-IR. For the Seyfert galaxies the torus model of Rowan-Robinson (1995) is used and the elliptical galaxies are modelled using the spectral energy distributions of Yoshii and Takahara (1988) and Bertola et al. (1982)

The normal spirals and starburst/ultra/hyperluminous populations are respectively modelled using the cool and warm 60 μm galaxy luminosity functions of Saunders et al. (1990). For optical counts an assumption of $L_{IR}/L_B \approx 30\%$ is assumed for the normal spiral galaxies to shift the luminosity function to the B-band. The Seyfert/QSOs use the 12μm parameters of Rush et al. (1993) who represented their sample with the 2 power law luminosity function of Lawrence et al. (1986). the elliptical galaxies use a B-band Schecter function to model their contribution to the B and K band counts.

Pure luminosity evolution is assumed for the starburst, ultra/hyperluminous and AGN components and is of the form used by Boyle et al. (1988) found from fits to optical QSO data. An evolution rate of $(1+z)^Q$ $\{z < 2\}$, $(1+z_*)^Q$ $\{z > 2\}$ is assumed where $Q = 3.1$, $z_* = 2$ and the formation redshift, $z_f = 5$. This form of evolution also fits the data for radio counts of radio galaxies and QSOs, sub-mJy counts of starburst galaxies at 1.4GHz and far-IR counts of IRAS galaxies (Fig 3). The physical motivation behind such evolution would be galaxy-galaxy interaction driven star formation and feeding of black holes in galactic nuclei.

4   Future Surveys at Infrared and Submillimetre Wavelengths

Several surveys are in progress with ISO, at wavelengths from 6.7 - 90 μm in Open and Guaranteed Time. The ELAIS survey (European Large Area ISO Survey) will survey about 15 sq deg at 15 and 90 μm (Oliver et al 1996). It is anticipated that these will find a number of infrared galaxies at z > 1, though not many at very high redshift (Table 1).

Submillimetre surveys have been proposed for the new SCUBA bolometer array on the JCMT and for ESA's Fourth Cornerstone Mission FIRST. It is clearly of interest to compare the performance of these with what could be achieved with large ground-based millimetre arrays like LSA and MMA.

Rowan-Robinson, Pearson, Mobasher, Griffin, Gear, Dunlop, Longair and Blain are collaborating to propose a large survey at 450 and 850 μm with SCUBA on the JCMT. We estimate that with 120 hours of observing time we could cover 1 sq deg to a sensitivity of 11 mJy or 10 sq deg to a sensitivity of 36 mJy at 850 μm, in either case detecting 30-100 sources.

For FIRST it has been proposed that 10% of the observing time over 2 years could be allocated to a survey at 200, 400 and 800 μm. A sensitivity of about 10 mJy could be achieved at 200 and 400 μm for a 10 sq deg survey, and some 6000 galaxies should be detected, a high proportion at z > 1.

Finally, for LSA/MMA, if 10% of the observing time were allocated to a survey over the first 3 years of operation, 7 sq deg could be surveyed to a sensitivity of 0.9 mJy at 1 mm (5-σ), and some 18000 galaxies would be expected to be detected, again most at high redshift.

The performance of the different proposed far infrared and submillimetre surveys is compared in Table 2.

In conclusion, the proposed large millimetre arrays could have great interest for cosmology, because of their capacity to detect large numbers of high redshift galaxies. However such surveys would be complementary to those of ISO and FIRST, because without multi-wavelength information, nothing would be known of the bolometric power of the galaxies detected, their dust mass, star formation rate etc.

5   Starburst Galaxies in the Hubble Deep Field

In December 1995 an area of five square arcminutes was surveyed by the Hubble Space Telescope to an unprecedented depth (Williams et al 1996). This survey, known as the Hubble Deep Field, reaches at least one magnitude deeper than previous optical galaxy surveys.

The morphology of the brighter galaxies in the HDF has been discussed by Abraham et al (1996). Mobasher et al (1996) have analyzed the spectral energy distributions of 1611 galaxies detected in the V and I bands. They find that a high therefore appear to be optical counterparts of the starburst galaxies studied at far infrared wavelengths. The link with the infrared will be tested directly in a survey at 6.7 and 15 μm being undertaken by the author and collaborators with ISO.

References

Andreani P., La Franca F., Cristiani S., 1993, MN 261, L35

Abraham R.G., et al, 1996, MNRAS, in press

Barvainis R., Antonucci R., Coleman P., 1992, ApJ 399, L19

Benn C.R., Rowan-Robinson M., McMahon R.G., Broadhurst T.J., Lawrence A., 1993, MNRAS, **263**, 98

Bertola F., Capaccioli M. & Oke J.B., 1982, ApJ, **254**, 494-499

Boyle B.J., Shanks T. & Peterson B.A., 1988, MNRAS, **235**, 935-948

Broadhurst T. and Lehar L., 1995, ApJ

Brown R.L. and vanden Bout P.A., 1991, AJ 102, 1956

Coleman G.D., Wu C. & Weedman D.W., 1980, ApJS, **43**, 393-416

Condon J.J., 1989, ApJ, **228**, 13

Condon J.J., Huang Z.-P., Yin Q.F., Thuan T.X., 1991, ApJ 378, 65

Cutri R.M., Huchra J.P., Low F.J., Brown R.L., Vanden Bout P.A., 1995, ApJ

Danese L., de Zotti G., Franceshini A., Toffolatti L., 1987, ApJ, **318**, L15

Dey A. and van Breugel W., 1995, in Mass Transfer Induced Activity in Galaxies

Downes D., Radford S.J.E., Greve A., Thum C., Solomon P.M., Wink J.E., 1992, ApJ 398, L25

Dunlop J.S., and Peacock J.A., 1990, MN 247, 19

Dunlop J.S., Hughes D.H., Rawlings S., Eales S.A., Ward M.J., 1995, Nat 1995

Elbaz D., Arnaud M., Casse M., Mirabel I.F., Prantzos N., Vangioni-Flam E., 1992, AA 265, L29

Elston R., McCarthy P.J., Eisenhardt P., Dickinson M., Spinrad H., Januzzi B.T., Maloney P., 1994, AJ 107, 910

Fisher K.B., Strauss M.A., Davis M., Yahil A., Huchra JP., 1992, ApJ, **389**, 188

Franceshini A., Cesarsky C., Rowan-Robinson M., 1996, Memoria della Societa Astronomica Italiana (in press)

Franceshini A., Danese L., de Zotti G., Xu C., 1988, MNRAS, **233**, 175

Graham J.R. and Liu M.C.,1995, ApJ

Granato G.L. and Danese L., 1994, MN 268, 235

Green S. and Rowan-Robinson M., 1996, MN in press

Hacking P.B., Houck J.R., 1987, ApJS, **63**, 311

Hacking P.B., Soifer B.T., 1991, ApJ, **367**, L49

Hacking P.B., Condon J.J., Houck J.R., 1987, ApJ, **316**, L15

Isaak K.G., McMahon R.G., Hills R.E., Withington S., 1995, MN (in press)

Iveson R.J., 1995, MN (in press)

Kleinmann S.G., Hamilton D., Keel W.C., Wynn-Williams C.G., Eales S.A., Becklin E.E., Kuntz K.D., 1988, ApJ 328, 161

Lawrence A., Walker D., Rowan-Robinson M., Leech K.J. & Penston M.V., 1986, MNRAS, **219**, 687-701

Lawrence A., Rowan-Robinson M., Leech K.J., Jones D.H.P., Wall J.V., 1989, MN 240, 329

Lawrence A. et al, 1993, MN 260, 28

Leech K.J., Rowan-Robinson M., Lawrence A., Hughes J.D., 1994, MN 267, 253

Lonsdale C.J., Hacking P.B., 1989, ApJ, **338**, 712

Lonsdale C.J., Hacking P.B., Conrow T.P., Rowan-Robinson M., 1990, ApJ, **358**, 60

Lonsdale C.J., Harmon R.T., 1991, Adv.Space.Res., **11**, 255

McMahon R.G., Omont A., Bergeron J., Kreysa E., Haslam C.G.T., 1995, MN

Mitchell K.J., Condon J.J., 1985, AJ, **90**, 1957

Mobasher B., Rowan-Robinson M., Georgakakis A., & Eaton N., 1996, MNRAS, in press

Neugebauer G., Becklin E.E., Oke J.B., & Searle L., 1976, ApJ, **205**, 29-43

van Ojik R. , Rottgering H.J.A., Miley G.K., Bremer M.N., Macchetto F., Chambers K.C., 1995, AA

Oliver S.J., Rowan-Robinson M., Saunders W., 1992, MNRAS, **256**, 15p

Oliver S. et al, 1995, in Wide-Field Spectroscopy and the Distant Universe, eds S.J.Maddox, A.Aragon-Salamanca, World Scientific(Singapore) p.274

Pearson C.P. & Rowan-Robinson M., 1996, MNRAS, *submitted*

Pier E., and Krolik J., 1992, ApJ 401, 99

Rowan-Robinson M. et al, 1991, Nat, **351**, 719

Rowan-Robinson M., 1992, MNRAS, **258**, 787-798

Rowan-Robinson M., Efstathiou A., Lawrence A., Oliver S. & Taylor A., 1993, MNRAS, **261**, 513-521

Rowan-Robinson M., Benn C.R., Lawrence A., McMahon R.G., Broadhurst T.J., 1993a, MNRAS, **263**, 123

Rowan-Robinson M., 1995, MNRAS, **272**, 737-748

Rowan-Robinson M. et al 1993b, MN 261, 513

Rowan-Robinson M. and Efstathiou A., 1993, MN 263, 675

Rowan-Robinson M., McMahon R.G., 1996, in preparation

Rowan-Robinson M., 1996, in Cold Gas at High Redshift, ed. P.van der Werf

Rowan-Robinson M., 1996, in 'Dynamics of Molecular Cloud Distributions' eds F.Combes and F.Fasoli (Kluwer) p.211

Rush B., Malkan M., A. & Spinoglio L., 1993, ApJS, **89**, 1-33

Sanders D.B., Soifer B.T., Elias J.H., Madore B.F., Matthews K., Neugebauer G., Scoville N.Z., 1988, ApJ 325, 74

Sanders D.B., Phinney E.S., Neugebauer G., Soifer B.T., Matthews K., 1989, ApJ 347, 29

Saunders W., Rowan-Robinson M., Lawrence A., Efstathiou G., Kaiser N., Ellis R.S. & Frenk C.S., 1990, MNRAS, **242**, 318

Serjeant S., Lacy M., Rawlings S., King L.J., Clements D.L., 1995, MN 276, L31

Soifer B.T., Helou G., Lonsdale C.J., Neugebauer G., Hacking G., Houk J.R., Low F.J., Rice W., Rowan-Robinson M., 1984, ApJ 283, L1

Solomon P.M., Downes D. and Radford S.J.E., 1992, ApJ 398, L29

Thuan T.X., Condon J.J., 1987, ApJ, **322**, L9

Windhorst R.A., Dressler A., Koo D.C., 1987, in Observational Cosmology, eds A.Hewit, G.Burbidge, L.Z.Fang, Reidel (Dordrecht), p.573

Yamada T., 1994, ApJ 423, L27

Yoshii Y. & Takahara F., 1988, ApJ, **326**, 1-18

Table 1 : Number of sources expected for ISO survey at 90μm and 12μm covering 25 sq. deg.

λ (μm)	lgν	S (mJy)	lgS	ALL COMPONENTS N	N(z>1)	%(z>1)	NORMAL (CIRRUS) N	N(z>1)	%(z>1)	STARBURST N	N(z>1)	%(z>1)
90	12.5214	15	-1.8239	7362	1548	21	2323	0	0	5003	1513	30
15	13.2996	1.7	-2.7696	3298	592	18	1526	0	0	1120	438	39

λ (μm)	lgν	S (mJy)	lgS	HYPERLUMINOUS N	N(z>1)	%(z>1)	SEYFERT 1 N	N(z>1)	%(z>1)	SEYFERT 2 N	N(z>1)	%(z>1)
90	12.5214	15	-1.8239	3.75	3.41	91	15.6	15.0	96	16.8	16.2	96
15	13.2996	0.9	-3.05	2.77	2.44	88	294	65	22	355	87	25

Table 2 : Comparison of FIR and sub-mm Surveys

Survey	Date	λ	S_{lim} (mJy)	Area	No. sources	Main type	Positional accuracy
IRAS-FSS	1987	60μm	200	whole sky	60000	spirals/starburst	30″
VLA	1995	20cm	2	80% sky	10^6	AGN	1″
ISO-ELAIS	1997	90μm	15	$20^\square$ (215hrs)	7000	starburst	10″
ISO-slew	1997	200μm	1000	$8000^\square$	5000	spirals	30″
SCUBA	1993	850μm	13	$10^\square$ (900hrs)	700	starburst/protogalaxies	3″
FIRST	2007	400μm	10	$10^\square$ (1800hrs)	6000	starburst/protogalaxies	7″
LSA	2010	1mm	0.9	$7^\square$ (2700hrs)	18000	starburst/protogalaxies	0.1″

Figure 1 : Starburst luminosity distribution with redshift

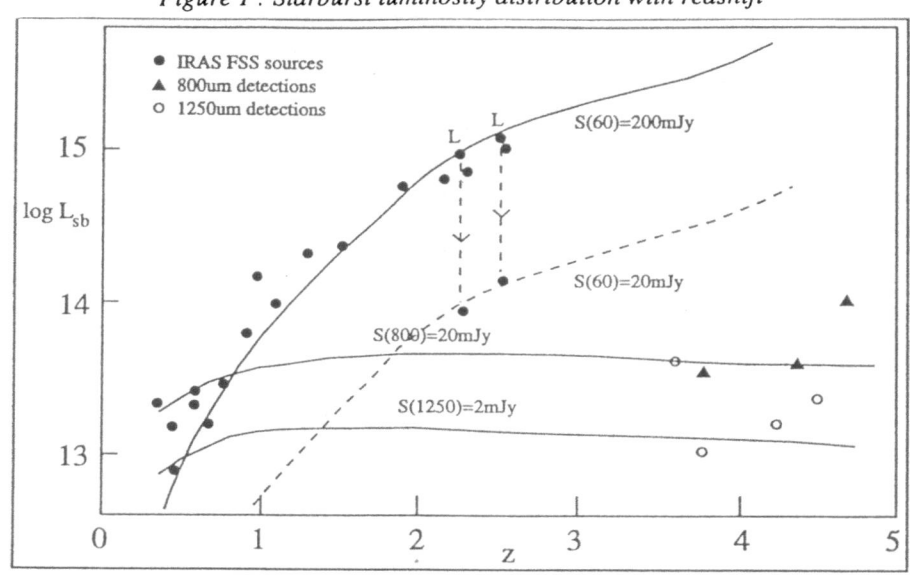

Figure 2 : N-z distribution at 90μm over 25 sq. deg. to a flux limit of S=15mJy

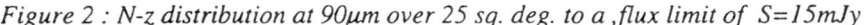

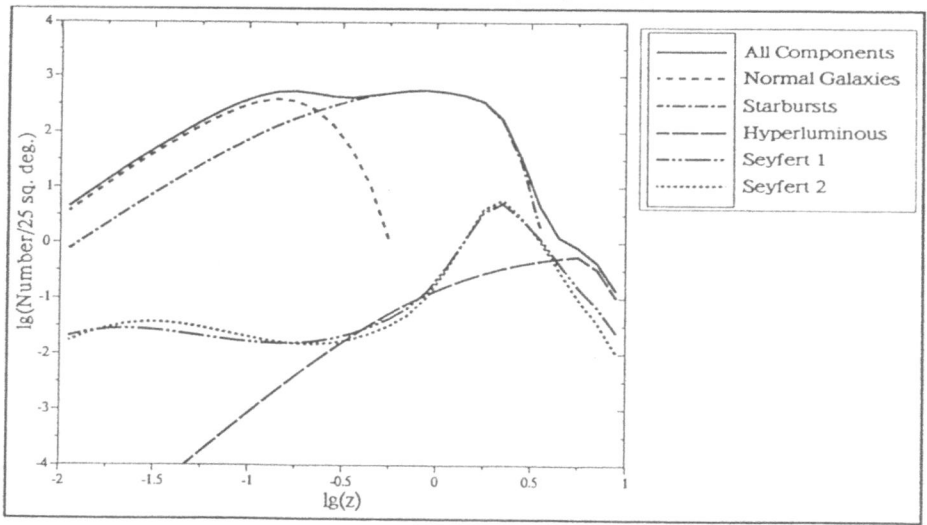

Figure 3 : Normalized differential counts at 60μm

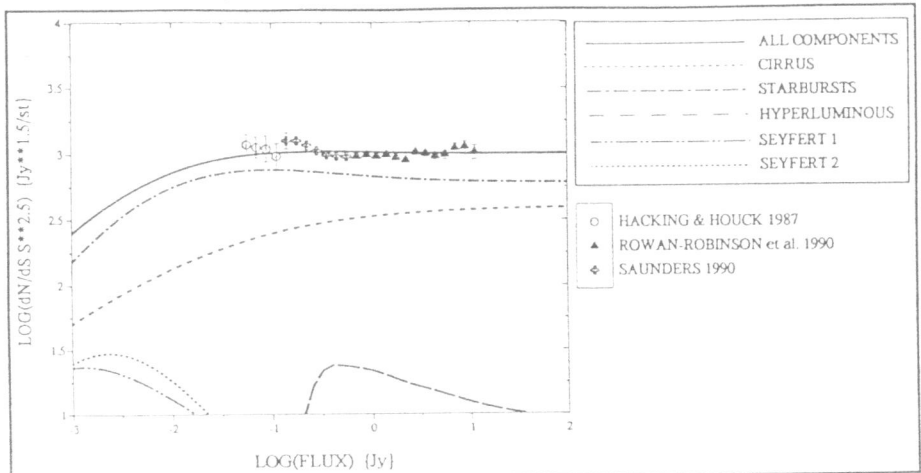

Figure 4 : Differential counts of sub-millijansky sources at 1.4GHz

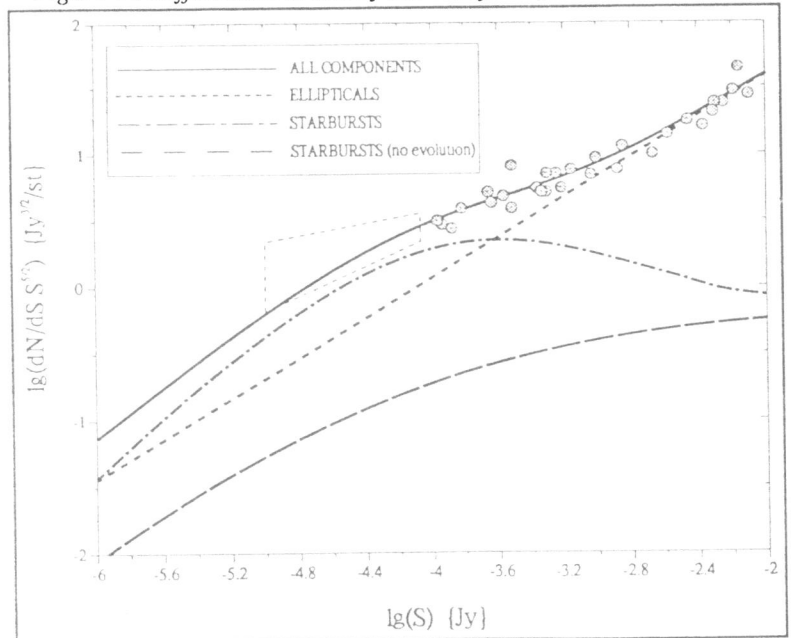

Prospects for ISO Deep Surveys

Alberto Franceschini[1] and Luigi Danese[2]

[1] Universita' di Padova, Dipartimento di Astronomia,
Vicolo Osservatorio 5, I-35122 Padova, Italy
[2] Scuola Internazionale Superiore di Studi Avanzati, Trieste, Italy

Abstract. We review various survey programs currently under execution by the Infrared Space Observatory, and to which an important fraction of the total mission time is devoted. In spite of the small primary collector, ISO is expected to provide results of relevant cosmological impact, especially to test rival scenarios of galaxy and quasar formation.

1 Introduction

Unprecedented collecting power, superb sensitivity, spatial and spectral resolution are the characteristics of the ESO Very Large Telescope. There is clearly no competition with such a monster telescope within one wavelength decade centered in the optical.

One of the lessons from the IRAS mission, however, was that some qualitatively new and otherwise unrecoverable astrophysical information can be gained by a substantial extension of the dynamic range in observing frequency, even though performed with a rather unsensitive instrument, as IRAS was. It is not just a question of complementing the limited spectral coverage of optical observations with data at longer wavelength, but rather to overcome the problem of the selection-wavelength bias. In the presence of source populations with widely scattered spectral properties, observations at one wavelength may totally miss entire classes of objects and astrophysical phenomena. An illustration of this fact is the well-known discovery by IRAS of a new class of galaxies with enormously high far-IR luminosities (comparable to those of quasars), but weak emitters in the optical because of dust extinction. Such relatively bright IR objects may have, according to the current interpretation, a deep relationship with unsolved problems as the formation of quasars and of early type galaxies, and are still the subject of a warm debate.

Thirteen years after IRAS, the Infrared Space Observatory (ISO) is now the major IR space mission of the decade. Of the three planned successors, the japanese IRIS is sheduled for the 2001, the ESA's FIRST will not be issued untill 2007, while the US SIRTF has'nt yet a launch date. Such a unique opportunity is currently used to study all known classes of astrophysical objects in the 3 to 240 μm wavelength domain, where there is still so much to be learned after IRAS.

We have soon realized that ISO has, in addition, attractive features as a surveyor of relatively large sky areas, to faint enough flux limits, at wavelengths mostly unaccessible from ground. This was suggested by its limited angular resolution (from several to several tens arcsec in the mid- to far-IR, due to diffraction), compared with the capability to achieve, with few minutes of integration, orders-of-magnitude improvements in sensitivity with respect to IRAS. In spite of the small primary mirror, it has been estimated that ISO could have provided results of cosmological impact when used in survey mode, particularly in connection with complementary observations at higher spatial resolution from ground in the radio, sub-mm, optical and near-IR. These will be needed to probe into the nature of the selected sources, and to unveil new classes of objects and unsuspected phenomena. After a brief presentation in Sect. 2 of the ISO mission and its present status, we summarize in Sect. 3 a few expected outcomes of these observations and possible tests of cosmological interest. The main planned surveys are mentioned in Sect. 4, and the relationship with the VLT project is discussed in Sect. 5.

2   The Infrared Space Observatory: Current Status

The decision to fly a second generation space IR observatory was taken by ESA on 1983, shortly after the first results by IRAS were reported to the european community. After the IRAS all-sky exploration at four IR wavelengths (12, 25, 60, 100 μm), the natural follow-up mission should have been one dedicated to deepen our knowledge of the IR sky through dedicated studies of selected regions with much greater spatial and spectral resolution, over a wider frequency interval and with much improved sensitivities (Cesarsky and Elbaz, 1996).

As known to everybody, such a space observatory has been studied, developed and tested by ESA during more than a decade, and eventually successfully flown from the Kourou base on November 17, 1995. Crucial steps of the launch campaign have been the release of the payload into orbit 20 min after launch, and the first activation of the valves controlling the flow of the 2100 l of helium cryo-cooler. On Nov 23 there has been the first switch-on of the instruments and on Nov 27 the first light in the form of a nice image of M51. Then the PV phase lasted a total of 65 days, after which on Feb 4, 1996, the routine operation phase has started.

2.1   Overview of the ISO Mission and Instrumentation

Organized as an IR observatory open to the astronomical community, 40% of its total observing Time is Guaranteed (GT) to scientists involved in the development and management of the facility, the remaining 60% is Time Open (OT) to the community.

ISO consists of a large dewar (5.3x3.5 m overall payload size) containing a superfluid helium tank cooling the entire telescope and instrumentation down to 3 K. The cooling system constrains the size of the primary mirror to 60 cm, which dictates its angular resolution (diffraction limited at $\lambda > 5\ \mu m$) and sensitivity limits.

The focal plane is shared by four instruments: a mid-IR camera (CAM), a far-IR photo-polarimeter (PHT), and two spectrographs (SWS and LWS) covering the entire 2 to 240 μm range. The four instruments, fed through a pyramidal mirror in a Ritchey-Chretien configuration, look simultaneously at 4 closeby regions of the sky, whose unvignetted size is 3 arcmin. Whenever an instrument other than CAM is observing, CAM is used in a *Parallel Mode* to image a sky region few arcmin apart from the target with its most sensitive filter (5-8.5 μm). Data are then telemetred to ground with a reduced rate. These data are currently analysed by the CAM Consortium and the ISO Operation Center, and will be published after the end of the mission.

During the very numerous telescope slews of more than 2.3 degrees in the sky, required to point at the several thousands planned target positions (for a total of 32000 observations), ISO is collecting photons with PHT through a broad-band filter centered at 180 μm, with a sensitivity comparable to that of IRAS at shorter wavelengths. A sizeable fraction of the sky ($\sim 10\%$) is expected to be surveyed in this *Serendipity Mode*. These data are analysed by the PHT Consortium.

The two instruments of cosmological interest are CAM and PHT, the spectrographs being limited to bright galactic and closeby extragalactic objects. CAM is a two channel camera feeding two 32x32 array detectors, one for short (SW, 2.5-5.5 μm) and the other for long wavelengths (LW, 5-18 μm). Two filter wheels per channel carry a set of lenses to change the pfov among 1.5, 3, 6 and 12 arcsec, and a set of fixed and CVF filters (up to a resolution of 45). Short exposures of a few minutes allow to easily detect sources at the sub-mJy level, longer ones reaching the few tens μJy limit with the most sensitive filters. At such levels the high zodiacal background requires flat-fielding accuracies attainable only with suitable procedures of micro-scanning and beam-switching, as used for ground observations (Cesarsky and Elbaz, 1996).

PHT includes three subsystems: a far-IR camera, a multi-aperture photometer and a spectrograph. The PHT camera, which is used for deep surveys, features a 3x3 and a 2x2 array working at the diffraction limit at 100 μm (C100) and 200 μm (C200), respectively (Lemke et al., 1994).

PHT and CAM are quite sensitive instruments, thanks to the cryogenic reduction of noise and clean environment. Because of this, both are fundamentally limited by source confusion even at high galactic latitudes. PHT C100, in particular, reaches the expected (extragalactic) confusion limit ($\sim 10\ mJy$ at 90 μm) in 1 minute of integration only. In practice, all deep PHT surveys are limited by confusion. Also CAM is expected to reach confusion at a level of 10-20 μJy in 1-2 hours of integration with sensitive filters (Franceschini et al., 1991).

2.2 Current Status of the Mission

At the time of writing, the ISO satellite and all scientific instruments are in very good shape, all main systems performing close to, or better than, expected. The cryo-cooler consumption rate is lower than expected, which means that the ISO lifetime is expanded from the originally scheduled 18 months to 24 ± 2 months.

This implies not only a good 30% more data than predicted, but also access to the entire sky (the visibility hole centered in Orion, and due to the various pointing constraints, is removed thanks to such an extended lifetime). Also, the pointing performances are better than qualification, with a rms 5" absolute and 1.5" relative pointing accuracy, which ease a lot the CAM data reduction at faint fluxes.

The planned deep surveys will exploit the ISO imagers (CAM and PHT) at their sensitivity limits. Then it has been crucial to establish with great care, during and soon after the PV phase, the performances of these instruments. Good news for CAM were that the high zodiacal background between 5 and 100 μm and the galactic *cirrus* do not appear structured on scales of interest ($< 3'$), and that the sensitivity performances are close to expectation. Some difficulties in the data reduction process have arisen from a transient behaviour of the detectors once illuminated, from memory effects, and from a rate of glitches due to cosmic rays (~ 1/sec on the whole array) larger than expected. Care in the analysis of the temporal behaviour of the signal in the pixels is required to efficiently remove the corresponding spurious signals, which have not to be confused with faint sources.

Also the PHT far-IR camera suffers from a general increase of the noise due to the effect of ionizing cosmic radiation, not present during laboratory tests, with a corresponding reduction of S/N by typically a factor 2 for a given integration time. However, thanks to the high sensitivity of this instrument and to some redundancies in the planned survey rasters, there appears to be no net loss of sensitivity for the deep PHT surveys above the long-wavelength source confusion limits.

Altogether, as emphasized during the recent First ISO Science Meeting in ES-TEC, the mission is already a bright success of the European Space Agency and of european astronomy in general. It is, in particular, up to the task of exploring the distant universe orders-of-magnitude deeper than the IRAS precursor.

3   Science Goals for the ISO Deep Surveys

In addition to the obvious prime motivation of exploring a largely unknown universe at wavelengths of 3 to 200 μm (because of the limited sensitivity, IRAS has detected just a handful of quasars at $z > 0.5$, and discovered just two, see Sect. 3.2 below), the major goals of the ISO survey projects are to provide qualitatively new tests of the enigmatic phases of formation and early evolution of galaxies and active galactic nuclei.

3.1   The Epoch of Galaxy Formation

Studies of the formation and early evolution of cosmic structures on the galaxy to galaxy cluster scale have significantly benefited from the advent of powerful 10m-class telescopes and from the HST refurbishment. In particular, the use of suitable colour selection techniques (Steidel et al., 1995) and exploitation of high-z bright quasars as source of background photons to look for spectral absorption features by intervening clouds (e.g. Wolfe, 1995) have recently unveiled a rich population of moderately luminous galaxies over a wide redshift interval ($0 \lesssim z \lesssim 4$). Their properties are reminiscent of those expected for forming spirals, with rates of star-formation (SF) (as inferred from Lyα or Hα line intensities and UV continuum luminosities) of typically several $M_\odot/yr$, i.e. only within factors of a few of the current SF rate in the Milky Way. Note that building up an $M^\star$ galaxy with such a low SF rate would require an entire Hubble time. The ISM metallicities of the QSO absorbers, estimated from line fitting, are correspondingly a small fraction of solar at high redshifts and slowly grow to half solar by $z \simeq 0.5$ (Petitjean et al., 1996). Altogether, these results explain why it has been so difficult to identify such high-redshift galaxies: the inferred luminosities, colours, and line intensities are not appreciably different from those of local galaxies.

On the other hand, fundamental insights into the past history are gained from studies of local fossil remnants, such as the stellar populations in nearby galaxies, the intra-cluster medium (ICM) and the hot plasma halo component around early-type galaxies. The latter two are observed to include substantial amounts of heavy elements, which have to be processed by stars and released in some sorts of galactic winds or outflows. Analyses of metallicity and abundance ratios of these plasmas imply an enhanced activity of massive stars during time periods ($t \sim 10^8$ yr) short enough to avoid exceeding contributions of Fe-enriched material by type-I SN. Many observed photometric, dynamical and chemical properties of early-type galaxies (e.g. the observed tightness of the fundamental plane, the red colours of high-z cluster ellipticals, and the tight colour-velocity dispertion relationships; see Renzini, 1995) also indicate an early and brief formation event for the bulk of stars.

Though still controversial, this argument suggests that an important phase of galaxy formation, and one of quick and efficient metal production in the universe, could have yet been missed by current observations in the optical, even at very low limiting fluxes. This is particularly surprising since such phase is expected to be very bright, as much as it is short (Zepf and Silk, 1996). Franceschini et al. (1994) have interpreted this problem of *visibility* as an effect of dust in the galaxy's ISM, produced by the earliest stellar generations, enshrouding the main SF phase and degrading the optical emission to the mid- and far-IR. In this picture (which has received some support by the IR background detection of Puget et al., 1996) the SF event was assumed to happen at moderate redshifts ($2 < z < 5$), and, as such, it may be tested by long-λ observations with ISO (see for more details Franceschini et al. 1994).

Another possible solution for the *visibility* problem of the E/S0's formation phase is to consider it confined to very high redshifts (e.g. Tyson, 1988), so that the redshifted Lyman continuum erases the whole optical emission. Franceschini and Gratton (1996) have analyzed element abundances in absorption-line systems associated to the quasar – and interpreted as originating from stellar activity in the host galaxy – and compared them with the observed abundances in the quasar nucleus. This analysis suggested that the putative host galaxy may be older than the quasar by typically 1 Gyr. Since quasars are observed allover the $2 < z < 4$ redshift interval, this could imply that the main formation phase of massive early-type galaxies is confined to very high redshifts, $z > 4 - 5$. Figure 1 illustrates the capabilities of mid-IR CAM observations to test such high-z formation phases.

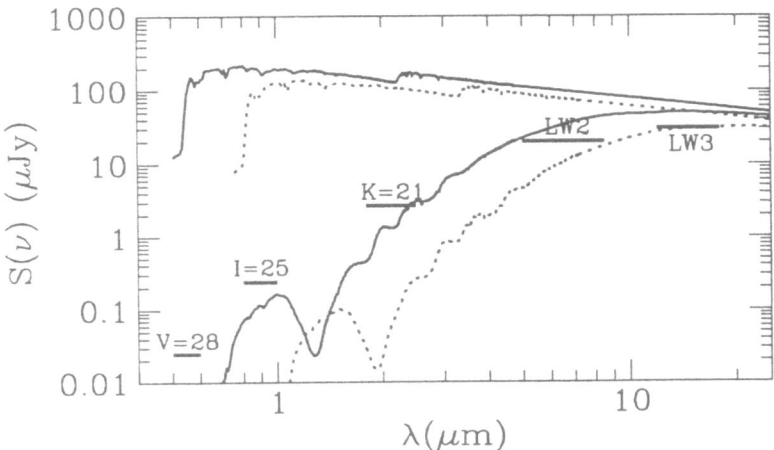

Fig. 1. Comparison of the predicted spectral energy distribution (SED) of a forming elliptical of $t = 0.1$ Gyr by Bressan et al. (1995) with limiting fluxes of deep CAM and optical surveys. The two upper lines are the unreddened SEDs for masses of $3\ 10^{11} M_\odot$ at z=5 (continuous heavy line) and z=8 (dotted line), observed during a strong burst with peak luminosity of $\sim 4\ 10^{13} L_\odot$ (H_0=60, q_0=0.25). The two lower lines include the effect of a E(B-V)=1 extinction by a dusty ISM in the object.

A drastic alternative is one often discussed in the framework of hierarchical structure formation, in which the smaller scales are the first to collapse after decoupling, due to propagation and dumping of seed fluctuations through the recombination. The more massive galaxies are then formed at relatively low redshifts ($z < 1-2$) through coalescence of sub-galactic units and barion collapse in massive dark matter halos forming at late cosmic epochs (Kaufmann et al., 1993). Under this scheme, head-on collision and merging of gas rich systems could produce, via shock formation, the centrally peaked $r^{-1/4}$ profile of ellipticals (Van der Werf, 1996). Luminous IR galaxies found by IRAS at low redshifts could represent such forming ellipticals (Kormendy and Sanders, 1992). If this latter scheme applies, no qualitatively new outcomes are expected from ISO surveys, all main SF events either happening at low-z or being quite faint at all wavelengths.

ISO provides ways to significantly constrain such diverse scenarios of galaxy formation.

3.2   Hyper-luminous IR Galaxies and Quasar Formation

Three of the highest redshift objects discovered by IRAS (IRAS F10214 at z=2.29, IRAS F15307 at z=0.93, and IRAS 09104 at z=0.44) have been recently interpreted as bright examples of a new class of peculiar hyperluminous IR galaxies, characterized by similarly high bolometric emissions ($L \sim$ several $10^{13} L_\odot$; note that the F10214's flux was shown to be amplified by a gravitational lens) and very similar IR SEDs. For all objects a dominant fraction of the emission falls in the far-IR, with the optical flux being very faint: clearly dust emission and absorption shapes the broad-band SEDs.

Proposed alternative interpretations for such objects were either a huge starburst during the formation of a massive spheroidal galaxy (Mazzei and De Zotti, 1994), or a dust enshrouded quasar (Rowan-Robinson et al., 1993). The presence of an active nucleus in all sources is clearly supported by the strong emission lines typical of local type-2 AGNs, the optical-UV polarization, the small size, and the broad emission lines observed in polarized light. Granato et al. (1996) have successfully reproduced the whole broad-band SEDs in terms of the emission from a quasar embedded in a massive torus-like dusty structure observed close to the equator. The same structure, if observed pole-on, also fits the spectrum of the BAL Cloverleaf quasar. The latter exhibits Broad Absorption Lines (BAL) in the spectrum, a situation which may arise from radiation-pressure acceleration of material in the torus polar regions for very high, possibly super-Eddington, accretion rates. All this is suggestive of a possible initial transient phase of formation of a super-massive black-hole, when the gas is still very abundant around the nucleus.

How this phase relates with that of formation of the host galaxy? The detection of a large amount of CO in IRAS F10214 may indicate that SF is ongoing around the nucleus. Also the detection of dust in the sub-millimeter for several high-z quasars may be in keeping with the idea of an initial phase in which a 'young' nucleus is immersed in a 'young' galaxy with large amounts of (already processed) gas not yet locked into stars.

A better statistics is clearly needed for any conclusions to be drawn about this enigmatic class of objects. The ISO surveys are expected to improve it substantially, if we consider that all of them have been found at the IRAS sensitivity limit. We estimate that of order of one thousand UVX AGNs will be discovered in the foreseen 25 sq. deg. sky area covered by various ISO survey activities at $\lambda \simeq 5$ to $100\ \mu m$. The dusty QSOs that will be found, whose number is unpredictable now since we do not know the duration of the dust enshrouded phase, will cast light on the quasar formation process and its relationship with that of the host galaxy.

3.3 Hidden Active Galactic Nuclei, the Unified Model, and Background Radiations in the IR and X-rays

Circum-nuclear dust plays an important role in re-processing the optical-UV emission of local AGNs too. The nuclear and the host galaxy's emissions are roughly comparable for Seyfert 1 galaxies at 2.2 μm, while dust in the host galaxy is responsible for most of the far-IR emission. Growing evidence indicates that observations at mid-IR wavelengths are best suited to evidentiate the emission directly or indirectly powered by the active nucleus.

The phenomenology of various Seyfert types is interpreted as due to the filtering action of a dusty torus present in the nuclear regions, such that objects observed for line-of-sights close to the torus axis or intersecting it appear as type-1 or type-2, respectively. IR bands are then suited, better than the optical ones, to test this unified picture of AGN activity and find the optically unobservables nuclei, which may outnumber the canonical optical objects by large factors ($\sim 3-5$). These obscured AGNs, with a suitable distribution of torus column densities and strong cosmological evolution, may produce the cosmic X-ray background above 3 KeV (HXRB) (Comastri et al. 1995). Such highly obscured sources should be powerful far-IR emitters, and their contribution to the IR background may be substantial. There is an interesting relationship between the HXRB and the IR background to be explored after ISO.

4 ISO Deep Survey Projects

Programs for both Guaranteed Time and Open Time observations have been scheduled to perform extensive surveys. An important fraction ($10-15\%$) of the whole ISO time will be devoted to such observations.

4.1 Observations in the ISO Guaranteed Time

Table I summarizes some information about planned surveys with ISO. The most important programs (Cesarsky et al., 1994, and Taniguchi et al., 1994) will carry out both very deep integrations over restricted areas (of typically some 0.1 sq.deg.) and shallower maps of larger fields (typically some sq.deg.). This will allow coverage of large dynamic ranges in flux, distance and luminosity for the selected sources.

The surveys are performed through long and repeated micro-scans in which any sky pixels are observed by different parts of the camera at different times, to minimize systematic effects (Cesarsky and Elbaz, 1996). The filters used are the most sensitive of CAM (LW2 at $5-8.5\mu m$ and LW3 at $12-18\mu m$) and of PHT (C100 at 90 μm). A 2.4 sq.deg. area with a minimum *cirrus* emission will be surveyed at 160 μm by the japanese-Hawaii team. Limiting sensitivities are estimated to be of the order of few tens of μJy in the mid-IR (set by the flat-fielding noise) and ~ 10 mJy at 90 μm (the confusion limit).

The choice of the areas has favoured a high galactic latitude, the lowest cirrus emission and galactic absorption (as inferred from IRAS and COBE data) and

Table 1. ISO Deep Survey Programs

Title / P.I.	Sky Area	Filters (μm)	Observ. Time (hours)
Guaranteed Time Observations			
ISOCAM Deep Survey Program	1 sq.deg. North	5 − 8.5 (CAM)	20
C. Cesarsky et al.	"	12 − 17 (CAM)	27
	1 sq.deg. South	5 − 8.5 (CAM)	30
	"	12 − 17 (CAM)	40
	"	90 (PHT)	29
Search for PGs and QSOs	2.4 sq.deg. North	5 − 8.5 (CAM)	30
Y. Taniguchi et al.	"	90 (PHT)	25
(Japanese + Univ.Hawaii Team)	"	160 (PHT)	25
Deep Observation of SA57	0.4 sq.deg. North	60 (PHT)	13
H. Norgaard-Nielsen et al.	"	90 (PHT)	13
Measurement of the EBL	0.5 sq.deg. North	3 to 200	6
K. Mattila et al.	0.5 sq.deg. South	3 to 200	6
Open Time Observations			
Large Area Survey, ELAIS	13 sq.deg. North	8 − 15 (CAM)	64.5
M. Rowan-Robinson et al.(1995)	"	90 (PHT)	64.3
	6 sq.deg. South	8 − 15 (CAM)	43.5
	"	90 (PHT)	43.0
Survey in the cosmological window, Puget et al.	1 sq.deg. South	4 − 5 (CAM)	22.5
	"		
Discretionary Time Observations			
Hubble Deep Field	12 sq.arcmin	5 − 8.5 (CAM)	6.4
M. Rowan-Robinson et al.(1996)	"	8 − 15 (CAM)	6.4

the lack of bright sources. Further criteria have been the existence of high-quality data at optical, radio and X-ray wavelengths and a good visibility of the area during the ISO mission.

Two sky fields will receive particular attention: the Lockman Hole ($\alpha = 10\ 49$, $\delta = 57\ 37$) and a southern field (the Marano field, $\alpha = 03\ 13$, $\delta = -55\ 24$). Given the already existing data and planned ISO follow-up activities, the two fields will be among the most deeply and extensively investigated of the sky.

4.2   A Wide Area Survey in the Open Time: ELAIS

It has been agreed to dedicate a significant fraction of the ISO Open Time to make a deep survey over 20 sq. degree, 70% of which in the northern hemisphere, with two wide-bands filters centered at 90 and 15 μm (Rowan-Robinson et al. 1995). The total survey area will be split into 4 independent high-galactic latitude plus 6 other small fields of specific interest for already existing data. At least thousands of sources are expected to be found, from local stars to high-z hyperluminous galaxies.

This survey will provide a major legacy from ISO, which will stimulate follow-up studies in all e.m. bands with ground and space observatories. To have the greatest impact on the community, the intention is to complete the analysis of the survey data within one year after the end of the mission. More details on this project may be found in Oliver (1996) and the WWW page http://icstar5.ph.ic.ac.uk.

4.3   Very Deep Integration in the Hubble Deep Field from the ISO Director's Discretionary Time

ISO will survey the area during July 1996 following a call for the ISO Discretionary Time (Rowan-Robinson et al. 1996). Both CAM LW2 and LW3 filters will be used down to the confusion limit. The aim of these observations, though far from competing with HST as for angular resolution, is to detect highly obscured objects according to the rationale of Sect. 3.

5   Relationship with VLT and Other Ground-Based Observatories

ISO surveys are expected to discover some ten thousands extragalactic sources, including normal and starbursting galaxies, AGNs, primeval objects and forming galaxies, evolved stars, and perhaps some brown-dwarfs. There will obviously be a large demand of photon collecting power to spectroscopically follow them up. However, because of the long-wavelength selection, detection of very extinguished objects is favoured, which will be very difficult to survey spectroscopically. To quantify this point, let us adopt IRAS F10214 as a template. Sources selected by ISO at 15 and 90 μm will be typically 3-4 magnitudes dimmer. Assuming that the same 4 mag scaling applies to the optical flux too, the corresponding optical magnitudes of these ISO selected sources would be close to V=26 and K=20. This is much too faint than allowed by multi-object spectroscopy from ground in the foreseeable future.

A kind of alternative to optical spectra will be to obtain multi-wavelength data over a wide frequency interval – including radio, sub-mm, near-IR and optical-UV data, in addition to the ISO data. The large information content of such spectra will allow to use broad-band features of the sources (in particular, the far-IR dust emission feature) in a similar way as the optical lines are observed by low-resolution spectroscopy. This is most relevant for the identification of high-redshift objects. Note also that various survey programs have deliberately omitted the use of ISO's short-wavelength filters, to optimize the scientific return of space observations by complementing them with appropriate data in the near-IR to be obtained from ground.

A specific mention is finally in order for the planned VLT mid-IR camera and spectrograph, VISIR (Lagage et al. 1994), which is ideally complementary to ISOCAM in terms of spatial resolution (the former improving on the latter by a factor 20). In particular, it is expected that VISIR will be able to resolve a sizeable fraction of all ISO detected sources with long enough integration.

References

Bressan, A., Chiosi, C., Fagotto, F. (1995): *Astrophys. J. Suppl.* **94**, 63

Cesarsky, C.J. et al. (1994): ISO Central Progamme, CCESARSK.IDSPCO

Cesarsky, C.J., Elbaz, D. (1996): *Examining the Big Bang and Background Radiations*, Kafatos and Y. Kondo eds., 109-116

Comastri, A., Setti, G., Zamorani, G., Hasinger, G. (1995): *A&A* **296**, 1

Franceschini, A., Toffolatti, L., Mazzei, P., Danese, L., De Zotti, G. (1991): *Astron. Astrophys. Suppl. Ser.* **89**, 285

Franceschini, A., Mazzei, P., Danese, L., De Zotti, G. (1994): *ApJ* **427**, 140-154

Franceschini, A., Gratton, R. (1996): *MNRAS*, submitted

Granato, G., Danese, L., Franceschini, A (1996): *ApJ Lett.* **460**, L11-14

Kaufmann, G., White, S., Guiderdoni, B. (1993): *MNRAS* **264**, 201

Kormendy, J., Sanders, D. (1992): *ApJ Lett.* **390**, L53

Lagage, P.O., et al. (1994): VLT Instrumentation Plan, Phase A Report

Lemke, D., et al. (1994): Opt. Eng. 33 20

Mazzei, P., De Zotti, G. (1994): *MNRAS* **266**, L5

Oliver, S. (1996): in *Cold Gas at High Redshift*, M. Bremer et al. Eds., Kluwer

Petitjean, P., Riediger, R., Rauch, M., 1996, *A&A* , in press

Puget, J.L., et al. (1996): *A&A Lett.* **308**, L5-9

Renzini, A. (1995): in *Stellar Populations*, IAU Symposium 169, B. Barbuy and A. Renzini Eds., (Kluwer, Dordrecht), in press

Rowan-Robinson, M., et al. (1993): *MNRAS* **263**, 123-130

Rowan-Robinson, M., et al. (1995): ISO OT Proposal

Rowan-Robinson, M., et al. (1996): ISO Discretionary Time Proposal

Steidel, C., Pettini, M., Hamilton, D., 1995, *Astron. J.* **110**, 2519

Taniguchi, et al. (1994): ISO Central Progamme, YTANIGUC.DEEPPGPQ

Tyson, J. (1988): *Astron. J.* **96**, 1

Van der Werf, P. (1996): in *Cold gas at high redshift*, M.N. Bremer et al. Eds., Kluwer, in press

Wolfe, A. (1995): in *QSO Absorption Lines*, G. Meylan Ed., (Springer, Berlin), 13

Zepf, S., Silk, (1996): *ApJ*, in press

Optical Searches for Quasars with $z > 5$

Richard G. McMahon

Institute of Astronomy, Madingley Road, Cambridge CB3 0HA, UK

Abstract. There are now ~65 quasars known with z>4 of which ~75% have been discovered using computer based survey techniques and photographic plate material. In the 8m era, it is expected that techniques based on electronic detectors ie CCDs and IR arrays will become much more commonplace, especially in the regime of surveys for z>5 quasars. Now, that it is practical to fill the unvignetted focal plane of 2–4m class telescopes with optical or near IR array detectors, it is an ideal time to carry out systematic CCD based surveys for optically selected quasars with z>5. In addition surveys for z>5 quasars will generate important bye-products such as low luminosity M-star or brown dwarf candidates.

1   Introduction

The study of quasars at the highest observable redshifts has wide ranging relevance to AGN astrophysics, galaxy formation, and the growth of large scale structure. The astrophysical information extracted from the study of quasars at high redshift encompasses studies of both the quasars themselves and the cosmologically distributed intervening absorption line systems.

The mere existence of quasars at high redshift and the properties of their host galaxies and local environment is a constraint on theories for the formation of galactic scale objects (eg Efstathiou & Rees 1988, Haehnelt & Rees, 1993) The wealth of astrophysical information that can be obtained from quasar absorption lines is covered in the recent ESO meeting on Quasar Absorption Lines (Meylan, 1995). The power of this type of observations is demonstrated by the fact that we can observe the properties of intervening HI clouds detectable over a staggering ~10 decades of column density ($\sim 10^{12} - 10^{21}$ atoms cm^{-2}).

Before I summarise the current status of searches for z>4 quasars it is worthwhile to remind you how high redshift quasar surveys have progressed over the last 30 years. I use the record redshift as an indicator of progress purely out of convenience. Following each record breaker there usually followed a phase when the new survey technique was exploited and thereafter a new survey approach was developed.

Within two years of the discovery of 3C273 redshifts in excess of 2 were reached(Schmidt 1965). However, it was the early 70's before redshifts in excess of 3 were reached. After the discovery of the z=3.52 quasar OQ172 (Wampler et al. 1973) the record remained unbroken until the discovery of the z=3.78 quasar PKS2000−330 (Peterson et al. 1983). In 1986, Hazard, McMahon and Sargent(1986) discovered, the z=3.80 optically selected quasar Q1208+1011 using a UK Schmidt Telescope objective prism plate. Prior to this the highest

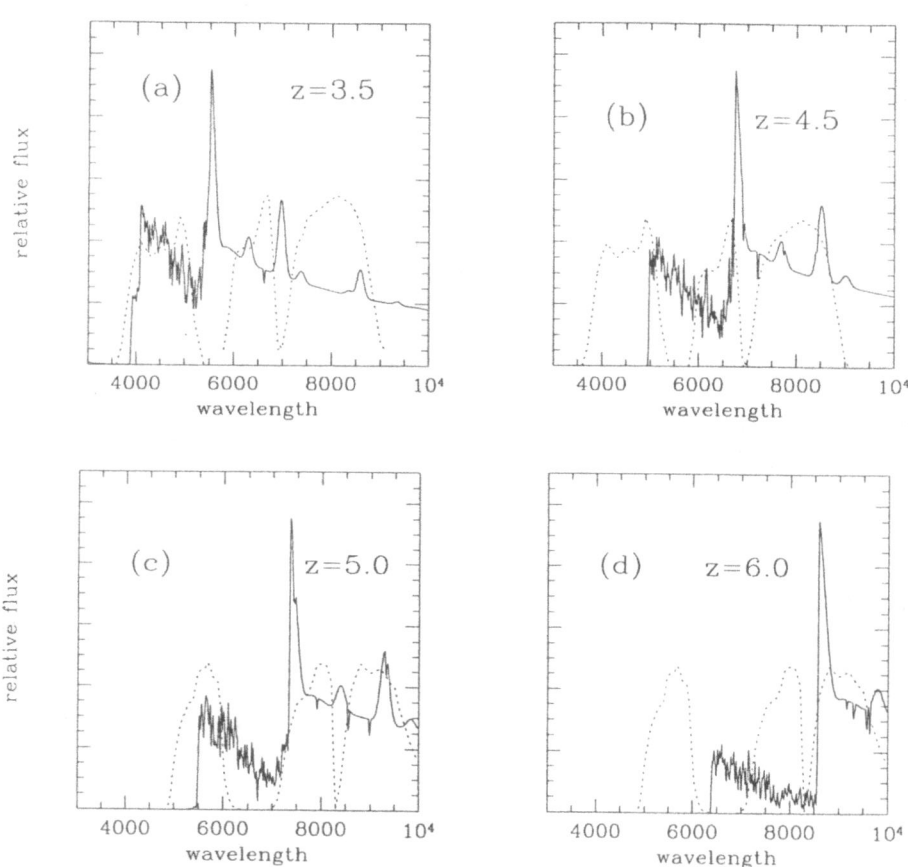

Fig. 1. Model quasar spectra computed as described in section 2.1. The dotted curves on (a), (b) show the UKST B$_J$, R and I emulsion/filters responses. The dotted curves in (c), (d) show the response functions for our CCD based V–I, I–Z survey.

redshift quasars had all been radio selected. Since then the majority of high redshift quasars have been optically selected. As shown in table 1, only 2 of the ~65 quasars that are now known with z>4 are radio selected.

2   The Status of z > 4 Quasar Surveys

This first quasar to be discovered with a redshift over 4 (Warren et al. 1987) was discovered using quantitative computer based analysis techniques to analyze digital photometric data based on APM measurements of UK Schmidt Telescope

Table 1. THE HIGHEST REDSHIFT QUASARS April 1996

By Technique

Survey Technique	Number of quasars	reference
Photographic Surveys		
APM + UKST + B+R	16	1
APM + UKST + B+R+I	13	1
APM + POSS-2 + B+R	1	1
APM + UKST + B+R+I	3	2
APM + Radio + POSS1	1	3
COSMOS + UKST + Variability	4	4
STSCI + POSS2 B+R+I	9	5
UKST + Prism + Eyeball	1	6
Sub Total	48	
CCD surveys		
CCD + Grism + Palomar 200inch	9	7
CCD + colours + Palomar 200inch	1	8
CCD + colours + KPNO 4m	1	9
Sub-Total	11	
Non-optical surveys		
Radio	2†	3,10
X-rays	1	11
Serendipity	2	12,13
Grand Total	**63**	

Notes: † includes one quasar from Hook et al(1991) who also used photographic plate material.
(1) my own work with Irwin and Hazard (eg Irwin, McMahon & Hazard(1991)) (2) Warren, Hewett & Osmer (1994) (3) Hook et al (1995) (4) Hawkins & Veron (preprint) (5) Kennefick, Djorgovski & Carvalho (1995) (6) Hazard & McMahon (unpublished) (7) Schneider, Schmidt & Gunn(1994a) (8) Schneider, Schmidt & Gunn(1991) (9) Hall et al. (1996) (10) Shaver, Wall & Kellerman(1996) (11) Henry et al (1991) (12) Schneider, Schmidt & Gunn(1994b) (13) McCarthy et al(1988)

plates. Prior to this work most optically selected quasars had been selected by visual inspection of multicolour or objective prism photographic plates. The majority (>90%) of the known z>4 quasars have been discovered using well defined quantitative digital techniques. This move towards computer based techniques in the discovery of quasars has been fundamental to the success of optical surveys for high redshift quasars.

Table 1 summarises the various survey techniques that have been used to discovery quasars with z>4. Whilst the highest redshift(z=4.9) quasar PC1247+3406 (Schneider, Schmidt &Gunn 1991) was discovered the majority of the other quasars have been discovered using photographic plates taken with the 1.2m UK or Palomar Schmidt telescopes (see Table 1 for references) In the 5-10 year timescale the Sloan Digital Sky Survey will transform the area of large area surveys for bright quasars in the Northern Hemisphere. There is clearly a need for a similar initiative in the Southern Hemisphere.

In the 8m era, it is expected that techniques based on electronic detectors ie CCDs and IR arrays will become much more commonplace, especially in the regime of surveys for z>5 quasars. Now, that it is practical to fill the unvignetted focal plane of 2–4m class telescopes with optical or near IR array detectors, it is an ideal time to carry out systematic CCD based surveys for optically selected quasars with z>5.

2.1   Principal Survey Techniques

Optical surveys for z>4 quasars are primarily based on the selection of quasar candidates on the basis of either emission lines or red B−R colours. The principle behind both techniques is shown by Figure 1 which shows a model quasar spectrum at a series of redshifts. The quasar spectrum is based on a power law with spectral index $(S_\nu \propto \nu^{-0.7})$ with an emission line spectrum based on the Francis et al.(1991) composite spectrum. Absorption due to the intervening Lyman-α forest and metal line systems has been monte-carlo simulated on the basis of an absorption line population described by the parameters derived by Williger et al.(1994) for the forest and by Storrie-Lombardi et al.(1994) for the optically thick lyman-limit absorption systems

The strongest emission line the observed spectra is red-shifted Lyman-α $(\lambda_{rest}=1216\text{Å})$. The detection of this line is the basis of the CCD surveys carried out by Schmidt, Schneider & Gunn with the Palomar 200inch telescope.

One of the most conspicuous features of the spectra of high redshift quasars compared with lower redshift quasar is the absorption shortward of Lyman-α due to the Lyman-α forest and higher column density optically thick lyman-limit systems. Whilst the lyman-limit systems are quite rare compared with the Lyman-α forest lines they effect the observed colours of quasars very strongly. It is this absorption which is the physical basis of multicolour surveys for z>4 quasars. By selecting stellar objects on the basis of red B−R colours alone we have discovered 17 quasars with z>4.

Optical colours can also be used to supplement non-optical surveys for high redshifts quasars. To complement survey work on optically selected quasars, Isobel Hook and I, are carrying out a survey for high redshift radio selected quasars based on the 6cm radio surveys carried out the the Greenbank and Parkes radio telescopes. Rather than carry out an indiscriminate redshift campaign on the stellar optical counterparts we have used the colours derived from APM scans of POSS1 O/E and UKST B/R plates to select red star-like optical identifications. This program has so far resulted in one radio radio selected quasar with z>4 (Hook et al, 1995). This program is also using CCD based colours for the 'blank fields'. Shaver, Wall & Kellerman(1996) are carrying out a similar program using the Parkes 2700Mhz survey. Radio identification are important because whilst radio loud quasars have lower observed space densities than optically selected quasars, radio samples can be identified independent of optical selection effects.

2.2   Quasar Candidate Rates and Science from the Chaff

It is important to note that whilst surveys for quasars with z>4 have been highly successful, the efficiency measured in terms of the ratio of quasar candidates to confirmed high redshift quasars is quite low. For instance the highly successful Palomar Transit CCD transit grism survey produced 1655 emission line candidates (Schneider, Schmidt & Gunn 1994). 1052 of these were confirmed with followup spectroscopy as emission lines, primarily due from low redshift emission line galaxies galaxies(z<0.45). The survey produced 90 quasars with z>2.7 ie ~5% efficient. This emission line galaxy sample is a very valuable bye-product.

The operational efficiency of our (Irwin, McMahon & Hazard 1991) survey work based on multicolour selected quasar candidates is similar ie ~5%. Our candidates are dominated by compact low redshift (z<0.5) red galaxies and late type M-stars. In fact it was one of these late type M-stars BRI0021−0214 (Irwin. McMahon & Reid 1991) which was one of the first objects to be studied with the HIRES echelle spectrograph on the Keck 10m Basri & Marcy 1995). Similarly, Warren et al.(1991) have identified distant halo Carbon stars from their quasar survey. Therefore, it is worth emphasizing the non-quasar content of quasar surveys to time allocation committees.

3   Searches for z > 5 Quasars

The multi-colour B−R, R−I approach can be extended to higher redshifts using CCDs by shifting the survey bands to redder wavelengths eg V−I, I−Z. My colleagues and I are carrying out such a survey using the NAOJ Mosaic CCD (Sekiguchi et al. 1992) on the 4.2m William Hershel Telescope. Our first survey observations shall took place in April 1996, the week after this meeting.

The basic principles behind the technique are shown in Figures 1 and 2. In the case of our z>4 quasar survey we initially carried out a B−R, R−I survey but subsequently we were able to simplify this to a B−R survey. This was because the B−R colours of quasars with z>4.25 alone were sufficient to efficiently isolate them from the more common M-stars. similar observed colours. It is possible

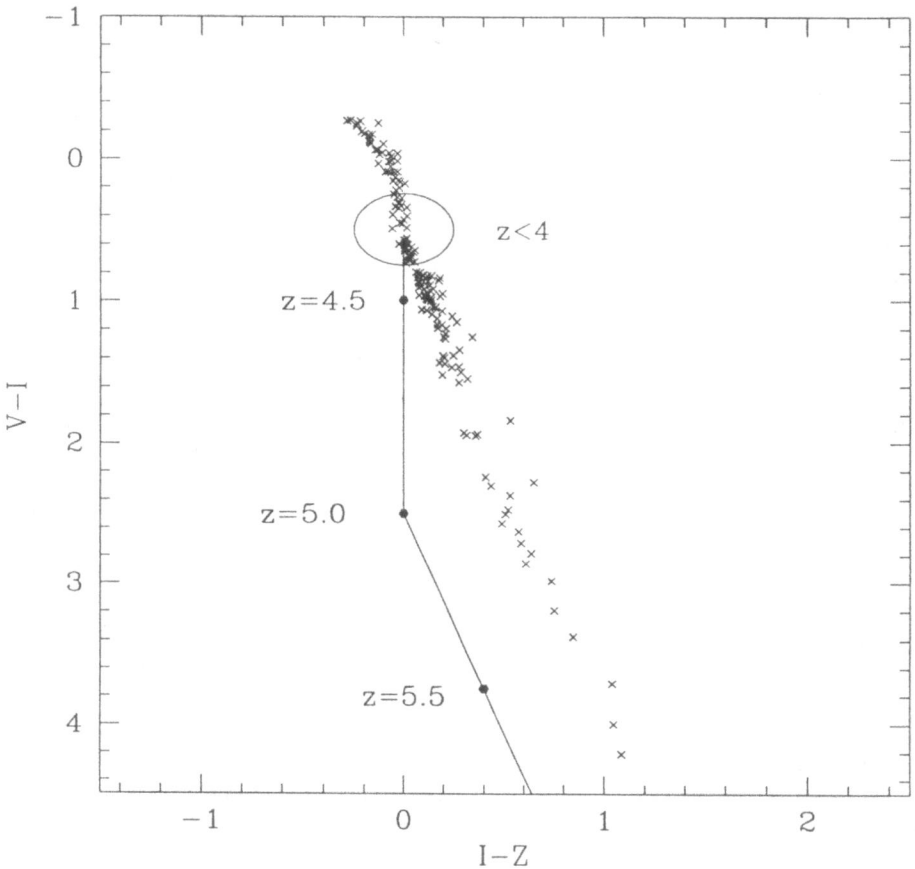

Fig. 2. Expected colours for high redshift quasars based on the simulations similar to the quasar spectra shown in Fig 1. The crosses are colours for stars calculated using the Gunn & Stryker(1991) stellar atlas. The large ellipse shows the region populated by z<4 quasars and the solid line is a schematic representation of how the observed colours higher redshift quasars change with redshift

that similarily, z>5 quasars may be discover using a single colour such as R–Z. If a single colour is sufficient this would be exciting result.

The optimum survey strategy depends crucially on three factors; the relative surface densities of faint low mass M-stars, the space density of z>5 quasars and the cosmological evolution of the intervening absorption line population. All three are highly uncertain and the answers to all three questions are important astrophysically.

4   Searches for z > 7 Quasars

It seems to reasonable to assume that due to absorption by intervening HI that quasars with z>7 will have very little flux at rest frame wavelengths shortward of 912Å; λ_{obs}=7300Å and more than 2 magnitudes of absorption across the Lyman-α emission line (λ_{obs}=9730Å) so that we must move to the NIR wavebands (see Figure 1d). In the last few years there have been substantial improvements in the manufacture of arrays that work at wavelengths longward of $1\mu m$. It is now feasible to consider building wide field cameras that fill the unvignetted focal planes of 2-4m class telescopes.

The Institute of Astronomy is currently building a NIR wide field imaging camera system based on a mosaic of 4 Rockwell 1024^2 HgCdTe arrays. Compared with previous generation of NIR cameras based on single NICMOS 256^2 arrays this camera represents a leap by a factor of 64. A similar leap in field of view will not occur in the near future and hence it is an ideal time to initiate a search for z>7 quasars.

It is intended that the camera will be ready for astronomical observations towards the end of 1997. Initially we propose to use the camera on the 2.5m Isaac Newton Telescope on LaPalma. At the prime focus of the INT the pixel size of $18.5\mu m$ produces an image scale of 0.45"/pixel and a field of view of 15.4' x 15.4'. A number of survey programs will be carried out with the system and whilst it may not be the primary program due to the speculative nature the J and H survey material in conjunction with optical CCD data will be used to search for quasars with z>7. Even a firm upper limit would be of great cosmological interest.

Qualitatively, a search for z>7 quasars in feasible. The highest redshift quasar known PC1247+3406 (z=4.9) (Schneider, Schmidt &Gunn 1991) has JHK magnitudes of 18.0, 17.6 and 17.0 (McMahon et al, in prep). The quasar PC1247+3406 was discovered in a survey covering $3deg^2$. At these redshift objects quasars dim as $\sim(1+z)^3$ so going from z=5 to z=10 produces 2 magnitudes of dimming in an unabsorbed continuum band. Therefore a survey over an area of $\sim20deg^2$ to H$\sim$20 would be in the right ball park ie 5-10minute exposures on a 2.5m class telescope. Nevertheless with the camera system descibed above it would take $\sim$10 nights per waveband. This is quite reasonable when you factor in the 8m science programs that would the quasars would generate.

The uncertainties in the expected numbers of quasars is large. The aim of any survey program would to determine the space density or set the best limits possible.

5   Final Remarks: VLT Specific

Since this meeting is intended to focus on VLT and 8m science issues I will finish with a few specific remarks. Most the quasar work on Keck has been based on small number of targets. One needs new targets particularly in the South. Our own APM UKST surveys could be extended South of -20 declination and $\sim$30 more z>4 quasars would result. Efficient CCD based surveys will need wide field CCD mosaic cameras on a 2m class telescope. Since high redshift quasar survey produce many candidates the multi-object capabilities of FUEGOS could be used in the filtering process. Finally the optical follow-up of sources detected at non-optical wavelengths should be considered ie radio and X-ray wavelengths.

Acknowledgements I thank the organizers for a thoroughly enjoyable meeting and my colleagues Cyril Hazard, Isobel Hook, Mike Irwin, Isobel Hook and Lisa Storrie-Lombardi for their contributions to the APM quasar surveys and follow-up programs.

References

Basri, G., Marcy, G.W. (1995) AJ, **109**, 762-773.

Efstathiou, G., Rees, M.J. (1988), MNRAS, **230**, 5p–11p.

Francis, P.J., et al. (1991), ApJ, **373**, 465–470.

Gunn, J.E. & Stryker, L.L. (1983), ApJS, **52**, 121–153.

Haehnelt, M.G., Rees, M.J. (1993), MNRAS, **263**, 168–178.

Hall, P.B. et al. (1996), ApJ, **462**, 614.

Hazard, C., McMahon, R. G. & Sargent, W.L.W., (1986) Nature, **322**, 38-40.

Henry, J.P. et al. (1994), AJ, **107**, 1270-1273.

Hook, I.M., McMahon, R.G., Patnaik, A., Browne, I.W., Irwin, M.J. & Hazard, C. (1995), MNRAS, **273**. 63p–67p.

Irwin, M.J., McMahon, R.G. & Hazard, C., (1991), *The Space Distribution of Quasars*, ASP Conference Series, Vol. 21, p117 Ed. D. Crampton.

Irwin, M.J., McMahon, R.G. & Reid, N, (1991), MNRAS, **252**, 61p.

Kennefick, J.D., Djorgovski S.G., & Carvalho, R.R. (1995) AJ, **110**, 2553.

McCarthy, P.J., et al. (1988), ApJL, **328**, L29-L32.

Meylan, G. (editor) (1995) *QSO absorption lines* (Springer, Berlin, Heidelberg)

Peterson, B.A., Savage, A., Jauncey, D.L., Wright, A.E. (1982), ApJ, **260**, L27–L29.

Schmidt, M. (1965), ApJ, **141**, 1295.

Schneider, D.P., Schmidt, M., & Gunn, J.E., (1991), AJ, **102**, 837

Schneider, D.P., Schmidt, M., & Gunn, J.E., (1994a), AJ, **107**, 880.

Schneider, D.P., Schmidt, M., & Gunn, J.E., (1994b), AJ, **107**, 1245,

Sekiguchi, M, et al. (1992), PASP, **104**, 744–751.

Shaver, P.A., Wall, J.V. & Kellerman, K.I., (1996), MNRAS, **278**, L11–L15.

Storrie-Lombardi, L.J., McMahon, R.G., Irwin, M. & Hazard, C. (1994), ApJL, **427**, L13–L16.

Wampler, E.J., Robinson, I.B., Baldwin, J. & Burbidge, E.M. (1973), Nature, **243**, 336–337.

Warren, S.J., ,Hewett, P.C., Osmer (1994) ApJ, **421**, 412–433.

Warren, S.J., Irwin, M.J., Evans, D.W., Liebert, J., Osmer, P.S., Hewett, P.C. (1993), MNRAS, **261**, 185–189.

Warren, S.J., Hewett, P.C., Irwin, M.J., McMahon, R.G., Bridgeland M.T., Bunclark, P.S. & Kibblewhite, E.J., (1987), Nature, **325**, 131-133.

Williger, G.M., Baldwin, J.A., Carswell, R.F., Cooke, A.J., Hazard, C., Irwin, M.J., McMahon, R.G. & Storrie-Lombardi, L.J. (1994), ApJ, **428**, 574–590.

Dust and the Search for High Redshift Quasars

Rachel L. Webster[1], Michael J. Drinkwater[2], Paul J. Francis[1], Frank J. Masci[1], and Bruce A. Peterson[3]

[1] School of Physics, University of Melbourne, Parkville, Victoria 3052, Australia
[2] Anglo-Australian Observatory, Coonabarabran, NSW, 2357, Australia,
[3] Mount Stromlo and Siding Springs Observatory, Private Bag, Weston Creek PO, ACT, 2611 Australia
E-mail: rwebster@physics.unimelb.edu.au

Abstract. Quasars provide a unique probe of the matter distribution along high redshift lines-of-sight, both through absorption line and gravitational lens studies. However if the selection of quasars is biased, then the related statistical descriptions of the matter distribution may also be biased. Recent work on a completely identified sample of radio sources shows that many quasars are significantly reddened. Arguments are presented which favour dust as the explanation of the reddening. Possible locations for the dust are explored, and the consequences of biased quasar selection briefly discussed.

1 Background

Two major areas of astronomical research are facilitated by the study of samples of high redshift quasars: the distribution and nature of matter at high redshift, and the physics of the central engine of active galactic nuclei. In both cases, useful phenomenological guidance for physical models relies on an unbiased selection of background sources.

Absorption line studies, which can be used to determine the evolution and clustering of matter and its metallicity might be affected if dusty absorption systems redden a background quasar, reducing its probability of selection using standard optical techniques. Statistical studies of gravitational lenses provide our strongest direct probes of clumped matter distributions. However the lines-of-sight for strong lensing generally pass through galaxies which may also be dusty. In such cases, lensed quasars may be preferentially magnified into a magnitude-limited sample, but have a low probability of selection using standard optical techniques. Finally, active galactic nuclei (AGN) are almost certainly asymmetric, and angle of orientation will directly effect the relative components of emission observed. Recent HST images verify that at least some AGN have dusty tori surrounding their nuclei. Thus the effects of dust local to the active galactic nucleus must be seriously considered.

Our discussion of quasars, and in particular radio quasars is in the context of the unified models of AGN (eg. Antonucci (1993), Urry & Padovani (1995)). Although some of the details of the unified models are controversial, there is general agreement that the orientation of the AGN to the line-of-sight will strongly affect the type of emission observed. For radio-loud objects, if the line-of-sight is directly down the radio jet, then the object will be classified as a BL Lac; somewhat off-axis orientation will result in a radio quasar, and a line-of-sight perpendicular to the radio axis would result in classification as a radio galaxy.

There are a range of components known to exist in radio AGN: jets, broad-line regions (BLR), narrow-line regions (NLR), tori, host galaxies, and also reflection screens, presumably of either dust or electrons, which are responsible for the reflected and polarised broad-line regions seen in nearby radio galaxies or Seyfert II galaxies (eg. Cimatti et al. (1993)). We would like to emphasize the importance of the last-mentioned component, as this may also affect observations of sources viewed at angles near the line-of-sight to the radio jet.

In order to design a search technique for AGN at any wavelength, a clear understanding the the expected emission at that wavelength is required. Many quasar selection programs have used criteria based on either a blue colour excess or images taken in relatively blue passbands. However it is possible that either the region local to the nucleus or the host galaxy might be dusty. Dust and its associated gas would attentuate optical and UV emission, as well as soft X-ray emission from the AGN. Hard X-ray, radio and IR emission should be relatively unaffected by dust. Thus in addition to variations with orientation angle, and in some cases, magnification by gravitational lensing, dust might also attenuate emission at several wavelengths.

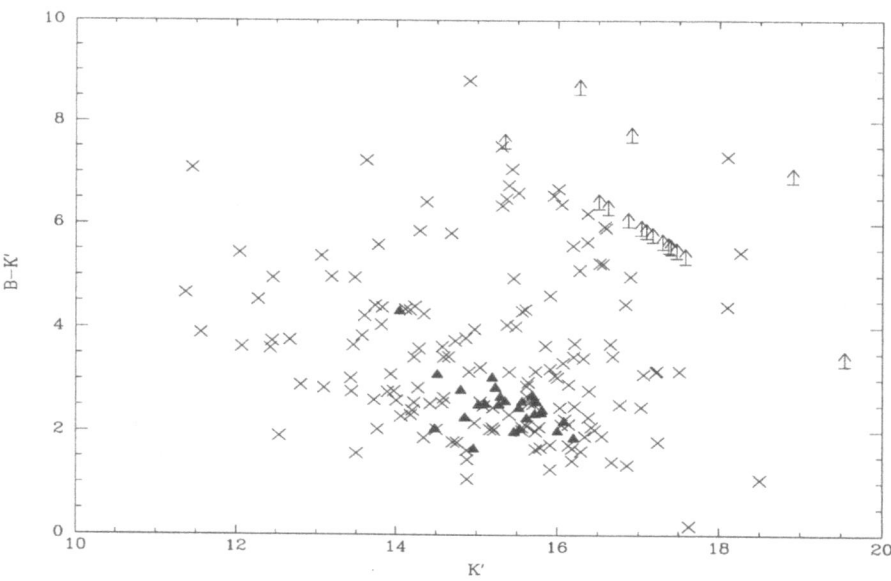

Fig. 1. Plot of K magnitude against colour, $B_J - K$ for PHFS quasars which are marked with crosses. Sources with only a lower limit on their B_J magnitude are marked by arrows; triangles are similar data for a sample of LBQS quasars.

2   Evidence for Red Quasars

We have undertaken a survey of a complete sample of 323 flat-spectrum ($S_\nu \propto \nu^\alpha$ where $\alpha > -0.5$) radio sources with fluxes $> 0.5\,Jy$ at $2.7\,Ghz$. The data paper on the Parkes Half-jansky Flat-spectrum Sample (PHFS) will be published shortly (Drinkwater et al. (1996)). Accurate radio positions have been used to identify 312 sources with optical or near-IR counterparts. Of these sources, 260 have redshifts and a further 20 have smooth spectra showing no identifiable lines. About ~ 35 of the sources have extended images in the optical or near-IR and are therefore classified as radio galaxies. The selection criteria mean that the sample is strongly biased toward the selection of quasars. K-band magnitudes have been obtained for ~ 180 of these sources. A rough estimate of the colour of the sources is given by the K-band magnitudes and B_J magnitudes from COSMOS. The range of 8 magnitudes in colour for PHFS quasars can be compared to a range of less than 3 magnitudes for a sample of optically-selected quasars from the Large Bright Quasar Sample (LBQS: Foltz et al. (1993)).

Figure 1 shows that many of the radio-selected quasars have red optical colours; indeed the LBQS colour distribution can be characterised by a peak at $B_J - K \sim 2.5$, while the PHFS quasars have a peak at $B_J - K \sim 2.5$ with a substantial tail to $B_J - K \sim 8$. Figure 2 shows the distribution of colour as a function of redshift. This distribution is likely to be biased at high redshift as many faint red sources do not yet have redshifts. However the available data does not show a trend of increasing redness with redshift.

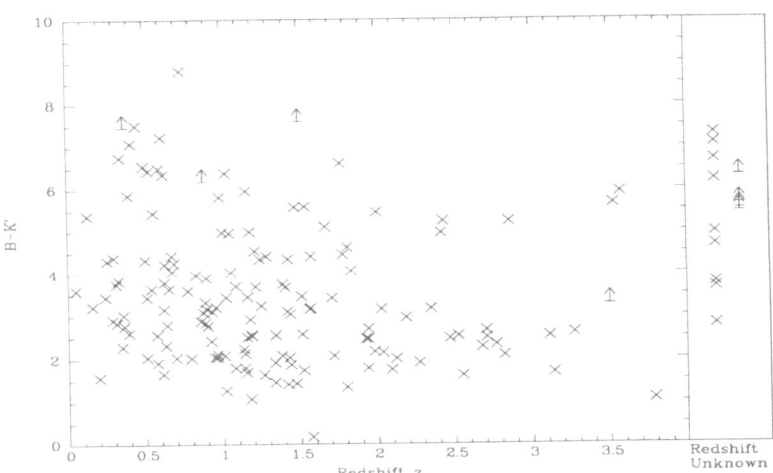

Fig. 2. Plot of redshift against colour, $B_J - K$. Those sources without redshifts are plotted on the right of the diagram, showing that we have incomplete redshift information for some of the reddest sources in the sample.

3   What Causes Redness?

Three different physical processes have been suggested for the observed reddening of the sources. Since early near-IR observations of optically faint radio sources
(Rieke, Lebofsky & Kinman (1979)), it has been known that some BL Lacs are
quite red. In the unified models, BL Lacs are AGN viewed directly down the
the radio jet, so that beamed emission is doppler-boosted. In one version of the
unified models, flat-spectrum radio quasars are FR II galaxies viewed directly
down the jet, while BL Lacs are their their counterparts for FR I galaxies (Urry
& Padovani (1995)). Thus it has been suggested that the reddening of radio
quasars may be due to a doppler-boosted optical component of the jet (Sergeant
& Rawlings (1996)). Since the models of jet emission have many free parameters,
it is difficult to test this model against the data. We believe that some of our
sources are too red to be fitted by synchrotron models for the beamed emission
(Masci (1996)), however this possibility will not be discussed further here.

Bright radio sources are often found in large elliptical galaxies. A strong
correlation has been observed between the K magnitude and the redshift of these
galaxies (the K-z relation), which has been interpreted as a population of low
mass stars in the galaxy acting as a standard candle (Lilly & Longair (1984)). It
has been suggested that the reddening of the PHFS sources is due to emission
from the underlying galaxy dominating the spectrum at near-IR wavelengths.

Finally, the reddening might be due to dust located between the observer
and the BLR of the quasar. In at least a few cases, the AGN must be observed
through a dusty galaxy in the line-of-sight; however an alternative possibility is
that the dust is located in the AGN host galaxy, or very near to the AGN central
region, such as in the NLR clouds (Netzer & Laor (1993)).

3.1   The Underlying Galaxy?

The discovery of a fairly good correlation between K and z for high redshfit
radio galaxies prompted suggestions that the K magnitudes could be used as
a robust standard candle for cosmological studies. However spatially resolved
spectroscopy of these galaxies has shown that in some cases, the K-band emission is due to a non-thermal stellar component (Lacy et al. 1995), considerably
weakening the argument that the underlying galaxies of these radio sources are
a uniform population of giant elliptical galaxies. Nonetheless it has been suggested that in the case of the redder PHFS quasars, we are simply seeing the
host galaxy of the quasar in the near-IR.

A simple way to consider this proposition is to plot the PHFS sources on the
K-z diagram, and consider the location of the reddest sources with respect to
the K-z relationship derived for radio galaxies. If the hypothesis is correct, then
the reddest sources should be dominated by the underlying galaxy and lie on
the K-z relation. If the quasar central emission is dominating the observed flux,
then the red sources will be brighter than the K-z relation.

The upper panel of Figure 3 shows the K-z plot for the PHFS quasars, with
the relationship for radio quasars shown by the solid line (Eales & Rawlings

(1993)). The lower panel shows the offset of each quasar from the K-z relation as a function of colour. If the underlying galaxy hypothesis is correct, then we expect all the sources to lie in the lower lefthand corner of this diagram. At least some of the sources do not lie in this region. We therefore conclude that not all the reddening can be explained by light from the host galaxy.

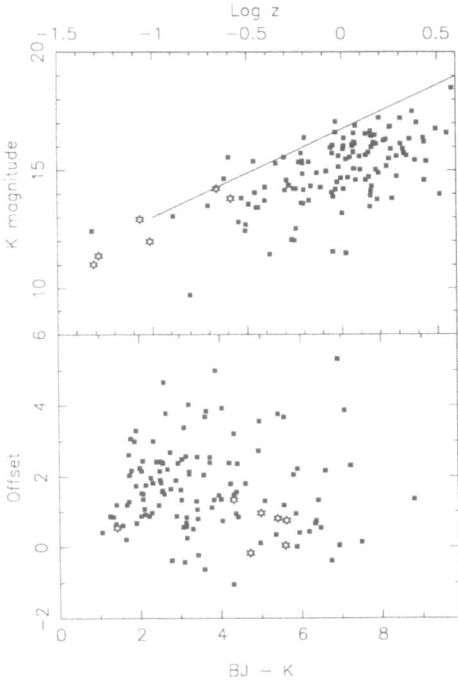

Fig. 3. The upper panel shows the K-z data for PHFS quasars; the solid line represents the relationship for radio galaxies. The lower panel shows the offset from the solid line in K magnitudes, for each of the PHFS quasars, as a function of colour.

3.2   The Case for Dust

Spectral identification of the faint PHFS sources is timeconsuming; however low resolution spectroscopy at the Anglo-Australian Telescope covering the wavelength range $\sim 4000 - 10,000$ Å has been fairly successful. Particular care has been taken to ensure that the observations are not biased by atmospheric dispersion, though it is unclear whether older observations taken from the literature are spectrophotometric. We make the simplest assumptions: intrinsically each quasar has spectrum similar to the blue envelope of the spectral distribution, and that deviations from that spectrum are due to reddening by dust. Using models of SMC and Milky Way dust, we are able to reproduce our observed spectra with dust optical depths of a few for the reddest quasars. While this modelling provides a consistency check, it does not prove that the reddening is due to dust.

A much stronger argument can be made by considering the distribution of the equivalent widths of the broad emission lines with colour. If the emission lines are due to gas photoionised by the UV photons from the continuum source of the quasar, then the equivalent width provides a direct measure of the relative number of UV photons. Thus an anticorrelation between equivalent width of the emission lines and $B_J - K$ should be observed if the continuum source is deficient in UV photons. Figure 4 shows that the equivalent widths of the MgII lines are not correlated with colour. This is strong evidence that the source of reddening is located beyond the broad line region of the AGN. If dust is the source of reddening, this conclusion is not surprising, as dust within the BLR would be vapourised (Netzer & Laor (1993)).

The suggestion that the NLR region of quasars may contain dust is not a new one (eg. Netzer & Laor (1993), Netzer et al. (1995)). Netzer & Laor (1993) calculate that the covering factor in dusty NLR clouds may be $\sim 30\%$, which is a little lower than, but not inconsistent with our estimates of numbers of reddened quasars in our sample (Webster et al. (1995)). A simple observation could be made which would provide the another test for the synchrotron beaming model: if radio quasars are oriented close to the line-of-sight, and a significant fraction of the near-IR emission is beamed at relativistic velocities, then as the quasar increases in brightness, the redness of the spectrum should also increase. This prediction would only be valid if the beamed emission varied more strongly than the continuum emission from the central engine. Published data on quasar variability suggests the opposite: as a radio quasar brightens, its colour becomes bluer. This data includes a few quasars from PHFS (Netzer et al. (1996)).

3.3   Dust in the Line-of-Sight

Calculations which assume that the dust content of galaxies along the line-of-sight to a background quasar remains constant with redshift, predict a strong statistical increase in observed reddening with redshift (eg. Fall & Pei (1993)). This trend is not observed in our dataset. However, if the amount of dust in galaxies is allowed to evolve as a function of redshift, then it is possible to match the lack of evolution in the observed reddening with a modest evolution in the length scale of the dusty component of the galaxies. If the scale length is $\propto (1+z)^\alpha$ then $\alpha = -2 \pm 1$ (Masci & Webster (1996)). However in order to scale the number of observed red quasars to the model predictions, the scale length of the dust component of the galaxies needs to be $\sim 170 \pm 40 h^{-1}\,kpc$, much greater than could be reasonably assumed from observations of reddening around nearby galaxies. Recent work by Pei & Fall (1995) in which calculates a selfconsistent model of reddening by dusty $L\alpha$ systems, predicts that by $z \sim 3$, $27 - 44\%$ of quasars will be too red to be selected by standard techniques in the B-passband.

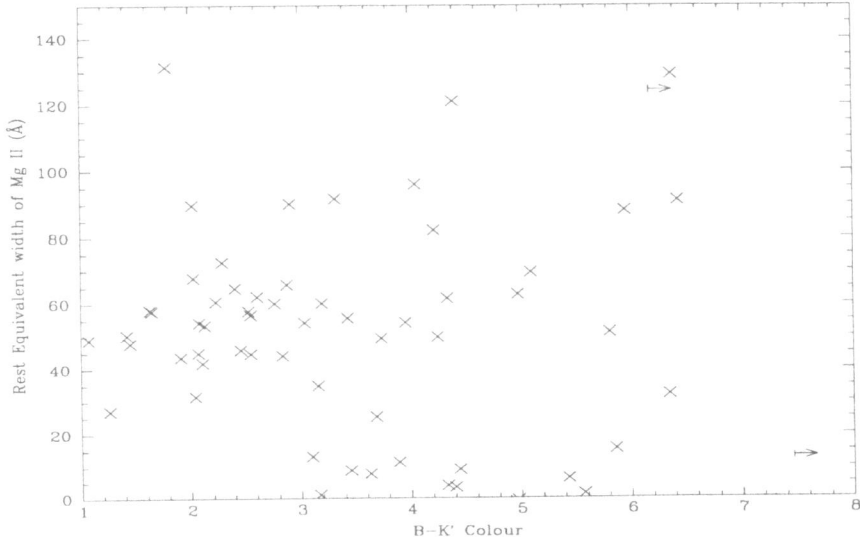

Fig. 4. Plot of equivalent width of MgII against colour, $B_J - K$ for PHFS quasars.

In summary, it is hard to rule out a doppler-boosted synchrotron component, using direct physical arguments. There may be cases where the underlying galaxy is important, but a red galaxy cannot account for the observed reddening in all cases. Similarly, there are definitely dusty intervening objects in the line-of-sight; however if all the observed reddening were due to line-of-sight dust, galaxies would require a very large dust envelopes. Thus for most red quasars, we favour a model where dust is located in the environment of the quasar, either near the AGN, or in the inner regions of the host galaxy.

4   Additional Results

We are able to make extensive comparisions between our sample and the LBQS and have found the following:

1. The mean PHFS spectrum is identical to the mean LBQS spectrum in the optical; this is a strong argument against bias due to an enhanced continuum due to synchrotron emission. We note that our spectra are low resolution, and thus we are unable to comment on the actual shapes of individual lines, though the distribution of equivalent widths is similar.
2. The redshift distributions of the two samples are essentially the same, if allowance is made for the high redshift cutoff in the LBQS which occurs as $Ly\alpha$ moves through the B_J passband in which the LBQS is selected.

3. Since quasars in the PHFS are viewed face-on (most are compact flat-spectrum sources), this may mean that quasars selected by the LBQS are viewed face-on as well.

4. If the optical and near-IR part of the spectrum is identical for these two samples of quasars, then the physics underlying the emission at these wavelengths is also the same.

5. The last points suggest that if dust in the NLR clouds is responsible for the observed reddening, then radio-quiet quasars are likely to be effected in the same way as radio-loud quasars. These reddened quasars would be missing from optically-selected samples.

5 Conclusions

– Reddened quasars are not selected by standard optical techniques; hence incidences of strong gravitational lensing and absorption by intervening systems will be underestimated. Estimates of the effects of dust on $Ly\alpha$ absorption systems are discussed by Fall in these proceedings.

– The observed reddening in radio quasars cannot all be due to dust in the line-of-sight to the quasar. Most of the dust must be located either in the host galaxy or in the immediate environs of the AGN.

– Radio-quiet and radio-loud quasars may be identical apart from the radio jet emission, including some form of aspect dependence.

References

Antonucci R. (1993), ARAA, **31**, 473

Drinkwater M. J., et al. (1996): MNRAS, in press

Cimatti, A., di Serego Alighieri, S., Fosbury, R. A. E., Salvati, M. & Taylor, D. (1993): MNRAS, **264**, 421

Eales, S. A. & Rawlings, S. (1993): ApJ, **411**, 67

Fall, S. M. & Pei, Y. C. (1993): ApJ, **402**, 479

Foltz, C. B., Hewett, P. C., Chaffee, F. H., & Hogan, C. J. (1993): AJ, **105**, 22

Lacy M., Rawlings S., Eales, S., & Dunlop J. S. (1995): MNRAS, **273**, 821

Lilly, S. J. & Longair, M. S. (1984): MNRAS, **211**, 833

Masci F. J. (1996): in preparation

Masci F. J., & Webster R. L. (1996): in preparation

Netzer H., & Laor A. (1993): ApJ, **404**, L51

Netzer H., et al. (1995): ApJ, **448**, 27

Netzer H., et al. (1996): MNRAS, **279**, 429

Pei, Y. C. & Fall, S. M. (1995): ApJ, **454**, 69

Rieke, G. H., Lebofsky, M. J. & Kinman, T. D. (1979): ApJ, **232**, L15

Sergeant, S. & Rawlings, S. (1996): Nature, **379**, 304

Urry C. M. & Padovani P. (1995): PASP, **107**, 803

Webster, R. L., Francis, P. J., Peterson, B. A., Drinkwater, M. J. & Masci, F. J. (1995): Nature, **375**, 469

Radio-Selected High Redshift Quasars

Peter Shaver[1], Jasper Wall[2], Ken Kellermann[3], Carole Jackson[4],
and Mike Hawkins[5]

[1] ESO, Karl-Schwarzschild-Str. 2, 85748 Garching bei München, Germany
[2] Royal Greenwich Observatory, Madingley Road, Cambridge CB3 0EZ, England
[3] National Radio Astronomy Observatory, Edgemont Road, Charlottesville,
   VA 22903-2475, U.S.A.
[4] Institute of Astronomy, Madingley Road, Cambridge CB3 0HA, England
[5] Royal Observatory, Blackford Hill, Edinburgh EH9 3HJ, Scotland

Abstract. Radio-based searches for high-redshift quasars offer significant advantages: large surveys with accurate positions, simple selection effects, relatively unrestricted redshift range - and radio emission is unaffected by dust obscuration. We have recently completed such a search, finding the highest-redshift radio quasar known and showing that the redshift cutoff is real, and not due merely to dust obscuration. Future searches will use one of the most prominent features in quasar spectra as a signature: the far-infrared bump which is redshifted into the millimeter waveband. The VLT will play an important role in this work, not only in identifying the quasars themselves, but also in studying the faint galaxies and absorption systems along the line of sight.

1 Introduction

Radio-loud quasars are a small fraction of the total, but radio-based searches have several advantages. There are now large radio surveys with high positional accuracy covering most of the sky. Combining these with digital optical sky surveys provides an extremely efficient approach to the search for quasars. A large fraction of flat-spectrum radio sources are quasars; they have relatively simple selection effects, and are compact, hence easy to identify. Compared to other search techniques, radio-based searches using flat-spectrum sources are relatively unrestricted in redshift range, and they are unaffected by dust obscuration. This is important with regard to the redshift cutoff, and obtaining unbiased samples for studying the evolution of damped Lyα absorbers.

High redshift quasars are important for several reasons. Their very existence sets constraints on the epoch of formation of the first galaxies, and the properties of these rare objects provide information on the physical and chemical conditions at those early times. They also provide the maximum path length for studying the intervening medium using absorption lines and gravitational lensing.

Radio-loud quasars held the record redshift until the mid-1980s when new optical selection techniques dedicated to high-redshift quasars were introduced. These searches have been successful in pushing the known redshifts to almost 5, but it becomes more difficult to find quasars at $z > 5$ because searches must be made in the red or infrared where the sky background is high and sensitive arrays small. A promising approach is to *combine* the best of radio and optical selection techniques, as described below.

2   A Search for $z > 5$ Quasars

The search was based on a sample of 878 flat-spectrum Parkes radio sources from a survey region covering 40% of the sky (flat spectrum here means spectral index between 11 and 6 cm $\alpha > -0.4$, where $S_\nu \propto \nu^\alpha$), coupled with an optical criterion (absence from the B-band) to select $z > 5$ quasars. Hence, there were two critical assumptions, that the high-redshift quasars would still have flat radio spectra, and that they would be obscured in the optical B-band.

Fig. 1. (*below, left*) Faction of flat-spectrum radio-loud quasars (solid histogram) and BL Lac objects (dashed histogram) for which the *observed* spectral index at 6 cm would be flatter than -0.4, as a function of redshift, based on the quasi-simultaneous measurements of Gear *et al.* (1994).

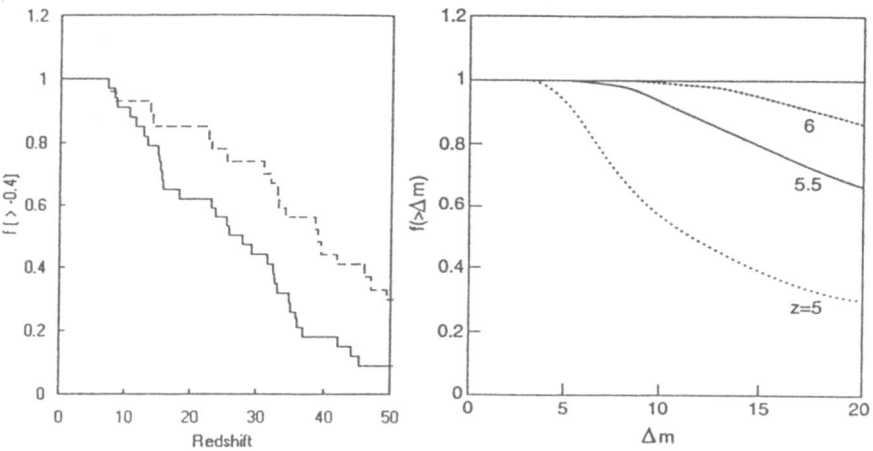

Fig. 2. (above, right) Expected fraction of quasars at redshifts 5, 5.5, and 6 for which the absorption in the B band due to the Lyα forest and Lyman limit is greater than Δm, as a function of Δm (based on calculations by Møller (private communication; *cf.* Møller & Warren (1991)). These extrapolations assume $dn/dz \propto (1 + z)^{2.75}$ and $f(N)dN \propto N^{-1.60}$.

The flat spectrum is an efficient criterion: of the flat-spectrum sources in the Parkes catalogue which were already optically identified, 83% are quasars. But quasar radio spectra which are flat at the usual centimeter wavelengths often become steeper at higher frequencies; at sufficiently high redshifts this steep portion of the spectrum will be redshifted into the window between 11 and 6 cm where the sample is selected, so very high-redshift quasars could be excluded in the selection process. However, analysis of the quasi-simultaneous radio spectra of a sample of quasars and BL Lac objects measured by Gear *et al.* (1994) indicates that this should not become important until $z \sim 10$ (fig. 1). The second assumption, that $z > 5$ quasars will be obscured in the optical B band by Lyα forest and Lyman limit absorption, is supported both by observed quasar spectra at $4 < z < 5$ and by simulations based on extrapolations of the known statistics of intervening hydrogen absorbers to higher redshifts. Fig. 2 shows the absorption in the B band expected at $z \sim$ 5-6; 10 magnitudes is typical even at $z = 5$.

Of the 878 flat-spectrum sources, only 229 were listed in the catalogue as unidentified, and these were our potential candidates for $z > 5$ quasars. We obtained radio positions accurate to better than 1 arcsec for these unidentified sources; about half were from new VLA or Australia Telescope measurements, and the others from previously published and unpublished measurements. These accurate positions were cross-correlated against the COSMOS catalogue of the southern sky, which is based on the UK Schmidt B_J survey, with a limiting magnitude of $\sim$ 22.0-22.5. The scatter in the radio-optical positional differences was 0.9 arcsec rms for both quasars and galaxies, and 3 arcsec was taken as the cutoff for identifications. This cross-correlation produced another 128 identifications. To identify the remaining sources we made CCD observations using EFOSC on the ESO 3.6m telescope in the B, Gunn-i and Gunn-z filters. Any $z > 5$ quasar would be expected to show up as a very red stellar object coincident with the radio position, which is present in Gunn-z but not in B.

All of the unidentified sources have now been identified, either as galaxies, or as stellar objects which are present in the B band (and are therefore not at $z > 5$). For the 24 stellar objects below the plate limit the B magnitude criterion was supplemented by spectroscopy and/or colour measurements. The identification content of the complete sample is shown on a plot of optical magnitude *vs.* radio flux density in Fig. 3. It is noticeable that the quasars tend to lie well above the detection limit, and there is a large population of very faint flat-spectrum radio galaxies (quasars account for 84% of the objects identified on Schmidt plates, but only 25% of those identified from the CCD observations). Only one source, PKS 1251-407, was identified with a very red stellar object, although because it was still faintly visible in B it was not considered a $z > 5$ candidate. Indeed, it was found to have a redshift of 4.46, making it the highest-redshift radio-loud quasar known (Shaver, Wall, & Kellermann, 1996). Objects at higher redshifts could have been found with similar ease. Thus, as all sources are identified, there are none left which could be quasars at $z > 5$.

It is most unlikely that there are any $z > 5$ quasars amongst the sources previously identified on the Schmidt plates. Over half of the catalogued identifications have been checked using new radio positions, and they are found to be highly reliable. Their very presence on B_J Schmidt plates makes them too bright to be at $z > 5$: even the most luminous quasars known would be fainter than the UKST plate limit if they were located at $z > 5$ behind just 5-6 magnitudes of absorption (which is at the low end of the distribution in fig. 2). About half of the identified quasars in the Parkes catalogue already have measured redshifts, and all are below 4. And finally, none of the $B - R$ colours obtained for a random sub-sample of almost 200 of these objects is as red as expected for a $z > 5$ quasar. We conclude that none of the 878 sources in our original sample is a quasar at $z > 5$.

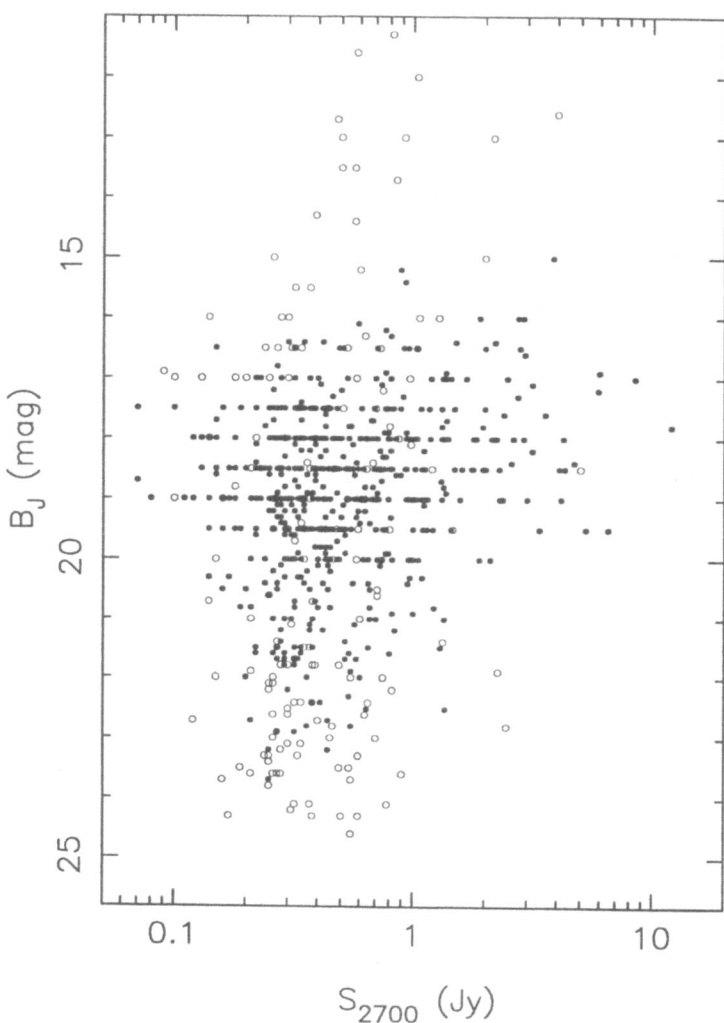

Fig. 3. Optical B_J-magnitude plotted against 11 cm flux density for the quasars ($\bullet$) and galaxies (o) in the sample

3   The Redshift Cutoff

We can therefore place an upper limit on the space density of flat-spectrum radio-loud quasars at $z > 5$ *which is independent of optical magnitude*. The limiting flux density of the Parkes catalogue varies over the sky, so in computing the space density we consider only those areas for which the Parkes catalogue is complete to 0.25 Jy at 11 cm. In the comoving volume covered by this 3.8 steradian area and the redshift range $5 < z < 7$, the space density corresponding to one object is 0.007 Gpc^{-3}, and this can be directly compared with measured space densities in lower redshift bins of quasars with radio luminosities corresponding to flux densities ≥ 0.25 Jy at $z = 7$ ($H_0 = 50$ km s^{-1} Mpc^{-1} and $q_0 = 0.5$). In determining these lower redshift space densities we used a restricted declination range of the Parkes catalogue ($+2.5° > \delta > -45°$), in which the fraction of quasars with measured redshifts is high (87% for $S_{11} > 0.5$ Jy). As we can only use sources with known redshifts, the space densities we compute at $z < 5$ are *lower* limits, to be compared with the *upper* limit at $z > 5$.

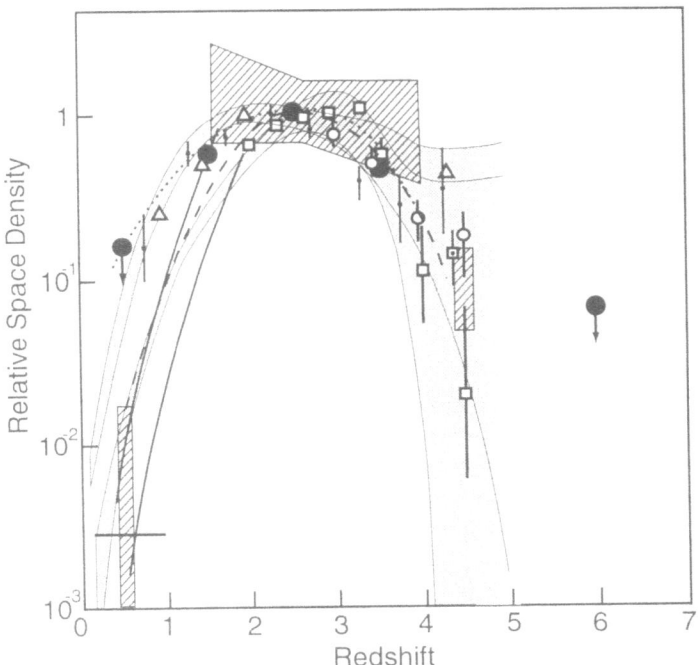

Fig. 4. Space densities, normalized to $z \sim 2-3$ and plotted as a function of redshift, for the Parkes flat-spectrum radio-loud quasars (⊙), the optically-selected quasar samples of Warren *et al.* (1994) (⊔), Schmidt *et al.* (1995) (⊙), Kennefick *et al.* (1995) (⊔), Hawkins & Véron (1996) (●), and Irwin *et al.* (1991) (combined and normalized with Miller *et al.* (1993) (△)), the flat and steep spectrum sources studied by Dunlop & Peacock (1990) (shaded areas), and the flat-spectrum radio-loud quasar sample of Hook (1994) (dotted line). The solid lines represent lower-redshift luminosity functions (Boyle, 1991; Hewett *et al.*, 1993) as used by some of the above authors. Also shown are normalized ultraviolet background radiation intensities (hatched areas) at $z \sim 0.5$ (Kulkarni & Fall, 1993), $z \sim 2-4$ (Bechtold, 1995), and $z \sim 4.5$ (Williger *et al.*, 1994; Storrie-Lombardi, 1994).

These space densities, normalized to $z \sim 3$, are plotted against redshift in fig. 4, The decrease at high redshift is obvious. Considering that 24 quasars above a given radio power are found at $z < 4$ and none at $5 < z < 7$, the turnover is significant at the 99.95% level. If there were no fall-off in the space density of quasars at high redshift, we would have expected to find at least 15 quasars at $5 < z < 7$ in our sample.

This absence of high-redshift quasars in our sample is very unlikely to be due to obscuration, as Ostriker and Heisler (1984) had proposed for optically-selected samples: (1) Radio emission is unaffected by dust, and all sources are already accounted for - there are simply none left to be $z > 5$ quasars. This excludes the possibility of obscuration by intervening galaxies or dusty Lyman limit systems too faint to be seen directly. (2) The possibility that some of the sources are in fact at $z > 5$ but have been *misidentified* with visible foreground galaxies that obscure them cannot explain the turnover, because it would require accurate positional coincidences of *all* of the $z > 5$ quasars with observed foreground galaxies. (3) Another possible source of obscuration at high redshifts is Thomson scattering by an ionized intergalactic medium, which affects all frequencies $\ll mc^2/h$ equally. However, even if all the baryons in the Universe were distributed in an ionized intergalactic plasma with Ω_B at the improbably high value of 0.5, the effect on the results shown in fig. 4 would still be small. (4) The possibility that some sources have been *completely* missed (also in the radio) due to both optical obscuration and radio free-free absorption in intervening objects is highly implausible, because a large free-free optical depth would imply a small size for any reasonable excitation parameter, hence a very small probability of a line-of-sight coincidence. (5) Obscuration by large amounts of matter *intrinsic* to the quasar (both dust, as proposed by Webster *et al.*, 1995 (but note Boyle & di Matteo (1995) and Serjeant & Rawlings (1996)), and ionized gas) is conceivable, but this does not affect the conclusions drawn here. We are concerned with a particular observed phenomenon, *i.e.* quasars similar to those seen at $z \sim 1 - 2$, and the space density of such objects decreases at high redshifts (there must obviously be precursors at still higher redshifts, at least in the form of mass concentrations, but that is a different issue). Furthermore, even if there are "hidden" quasars enshrouded in dust and ionized gas, should quasars be short-lived objects we would expect such hidden quasars to exist at *all* redshifts, so the redshift turnover would be unaffected. For all of these reasons, it is concluded that the observed turnover in the space density of flat-spectrum radio-loud quasars is real.

Can it now be argued that this result applies to *all* quasars, and not just to the flat-spectrum radio-loud quasars? (1) It is unlikely to be just *coincidence* that the radio and optical quasar populations turn over together, as is clear from fig. 4. (2) The radio-loud fraction of quasars may decrease with redshift up to $z \sim 1$, largely due to the steep-spectrum quasars (La Franca *et al.*, 1994; Schmidt *et al.*, 1995), but above that redshift it appears to be roughly constant (Hooper *et al.*, 1995, 1996). There is no evidence that the flat-spectrum radio-loud quasar population turns over before the radio-quiet population, and no theoretical reason to expect it. (3) If the turnover in optically-selected samples were due to obscuration, at least some radio quasars should also be affected. So, if the apparent space density of optically-selected quasars were down by a factor of, say, ten at $z \sim 4.5$ due to obscuration, there would be only a 10% chance of our having found PKS 1251-407 rather than a blank field. (4) Measurements of the ultraviolet background using the "proximity effect" (*e.g.* Bechtold, 1995) in the three redshift ranges as shown in fig. 4 reveals a redshift dependence similar to that of the quasar space densities, suggesting that there is a causal relationship and supporting the existence of a "quasar epoch".

We conclude that the turnover probably *does* apply to all quasars, and is not strongly affected by dust obscuration. In that case, the "delay" from the Big Bang to the "quasar epoch" is then an upper limit on the timescale for galaxy formation. Finally, we note that, just as radio-selected quasars first established the strong *increase* in space density with increasing redshift, they are also needed to confirm the *decrease* beyond the peak.

4   Future Prospects and the VLT

Having complete identifications, we can now measure *all* redshifts for this large sample of radio-loud quasars. For the first time it will thus be possible to trace the evolution of the space density over the entire "quasar epoch", with a single well-defined sample. This sample will also provide the basis for an unbiased study of the cosmic evolution of damped Lyα systems. Until now, such studies have been based on optically-selected samples, resulting in large uncertainty due to a possible selection effect: if the damped systems, especially those of highest column density, contain significant dust, they could obscure the background quasar and so not appear in the observed sample. The use of radio-based samples avoids this problem.

The approach described above can be expanded simply by searching to lower radio flux densities, steeper spectral indicies, and still larger samples provided by the new radio surveys. Large millimeter telescopes will provide a new approach. One of the dominant features in quasar spectra is the far-infrared bump due to dust emission. Redshifting this peak into the millimeter waveband provides a strong signature of high-redshift objects, which could be found easily by cross-correlating with surveys at other wavelengths. Such searches would therefore probe the "dark ages" beyond redshift 5, and may also find any "dust-enshrouded" quasars.

The VLT will play an important role in many aspects of this work, which has stretched the 3.6m telescope to its limits. With its high sensitivity and good image quality the VLT will be essential for the identification and classification of very faint objects - the faint galaxies and the low end of the quasar luminosity function. Infrared imaging and spectroscopy with the VLT will play a large role in this work. The VLT will be also be essential in studying the faint galaxies located both along the line of sight and physically associated with the quasars, and the absorption spectra of these radio quasars, particularly the damped Lyα systems. And finally, there will be a strong synergy between the VLT and a large millimetre array in studying the objects at the highest redshifts, both individually and statistically. They will have similar sensitivity and resolution, and as the VLT will favour dust-free objects while a large millimetre array will favour dust-rich objects, they will be highly complementary.

References

Bechtold. J. (1995): in *QSO Absorption Lines* (ESO Astrophysics Symposia; ed. G. Meylan; Springer-Verlag, Berlin), p. 299

Boyle, B.J. (1991): in *Relativistic Astrophysics, Cosmology, and Fundamental Physics* (ed. J.D. Barrow *et al.*; Ann. N.Y. Acad. Sci.), 14

Boyle, B.J., di Matteo, T. (1995): *Mon. Not. R. astr. Soc.* **277**, L63

Dunlop, J.S., Peacock, J.A. (1990): *Mon. Not. R. astr. Soc.* **247**, 19

Gear, W.K., Stevens, A., Hughes, D.H., Litchfield, S.J., Robson, E.I., Teräsranta, H., Valtaoja, E., Steppe, H., Aller, M.F., Aller, H.D. (1994): *Mon. Not. R. astr. Soc.* **267**, 167

Hawkins, M.R.S., Véron, P. (1996): *Mon. Not. R. astr. Soc* (in press)

Hewett, P.C., Foltz, C.B., Chaffee, F.H. (1993): *Astrophys. J.* **406**, L43

Hook, I.M. 1994, Ph.D. thesis, University of Cambridge; Hook, I.M., McMahon, R.G., Patnaik, A.R., Browne, I.W.A., Wilkinson , P.N., Irwin, M.J., Hazard, C. (1995): *Mon. Not. R. astr. Soc* **273**, L63

Hooper, E.J., Impey, C.D., Foltz, C.B., Hewett, P.C. (1995): *Astrophys. J.* **445**, 62

Hooper, E.J., Impey, C.D., Foltz, C.B., Hewett, P.C. (1996): *Astrophys. J.* (in press)

Irwin, M., McMahon, R.G., Hazard, C. (1991): in *The Space Distribution of Quasars* (ASP Conf. Ser. 21; ed. D. Crampton), p. 117

Kennefick, J.D., Djorgovski, S.G., de Carvalho, R.R. (1995): *Astron. J.* **110**, 2553

Kulkarni, V.P., Fall, S.M. (1993): *Astrophys. J.* **413**, L63

La Franca, F., Gregorini, L., Cristiani, S., de Ruiter, H., Owen, F. (1994): *Astron. J.* **108**, 1548

Miller, L., Goldschmidt, P., La Franca, F., Cristiani, S. (1993): in *Observational Cosmology* (ASP Conf. Ser. **21**, eds. G. Chincarini, A. Iovino, T. Maccacaro, D. Maccagni), p. 614

Møller, P., Warren, S. (1991): in *The Space Distribution of Quasars* (ASP Conf. Ser. 21; ed. D. Crampton), p. 96

Ostriker, J.P., Heisler, J. (1984): *Astrophys. J.* **278**, 1

Schmidt, M., Schneider, D, Gunn, J. (1995): *Astron. J.* **110**, 68

Schmidt, M., van Gorkom, J.H., Schneider, D.P., Gunn, J.E. (1995): *Astron. J.* **109**, 473

Serjeant, S., Rawlings, S. (1996): *Nature* **379**, 304

Shaver, P.A., Wall, J.V., Kellermann, K.I. (1996): *Mon. Not. R. astr. Soc* **278**, L11

Storrie-Lombardi, L.J. (1994): Ph.D. thesis, University of Cambridge; Storrie-Lombardi, L.J., McMahon, R.G., Irwin, M.J., Hazard, C. 1996, *Astrophys. J. Suppl.* (in press)

Warren, S., Hewett, P., Osmer, P. (1994): *Astrophys. J.* **421**, 412

Webster, R.L., Francis, P.J., Peterson, B.A., Drinkwater, M.J., Masci, F.J. (1995): *Nature* **375**, 469

Williger, G.M., Baldwin, J.A., Carswell, R.F., Cooke , A.J., Hazard, C., Irwin, M.J., McMahon, R.G., Storrie-Lombardi, L.J. (1994): *Astrophys. J.* **428**, 574

This work is based on data collected at the European Southern Observatory, the Australia Telescope National Facility, and the National Radio Astronomy Observatory (a facility of the National Science Foundation operated under cooperative agreement by Associated Universities, Inc. It was supported in part by Nato Collaborative Research Grant CRG 931543.

Millimetre Emission from High Redshift Radioquiet Quasars

A. Omont[1], R.G. McMahon, J. Bergeron, P. Cox, S. Guilloteau, E. Kreysa, F. Pajot, E. Pécontal, P. Petitjean, P.M. Solomon, L.J. Storrie–Lombardi

[1]Institut d'Astrophysique de Paris, C.N.R.S. UPR 341, 98 bis Bd Arago, F–75014 Paris

Abstract. The detection at millimetre wavelengths of continuum radiation emitted by dust in high z quasars is favored by the steepness of the far infrared-submm spectrum in the rest frame. About 10 radioquiet quasars at z > 2, the majority of them at z > 4, have now been detected at 1.25mm with the IRAM 30m telescope, with fluxes in the range 3–15 mJy. Millimeter lines of CO are slightly more difficult to detect, but already within the reach of sensitive interferometers for several of these sources. The J = 5–4 line has recently been detected in BR1202–0725 at redshift 4.69.

Since the discovery of FIRAS10214+4724 by Rowan–Robinson et al. (1991), the number of identified high-redshift far-infrared hyperluminous sources ($L_{FIR} \sim 10^{13} L_{\odot}$) has increased considerably reaching ~20 by now. About half of them have been identified from the IRAS Faint Source Catalogue (Rowan–Robinson 1996). A comparable number is associated with bright radio-quiet QSOs, mostly at z≥4 (Omont et al. 1996b). A few are bright radiogalaxies (Dunlop et al. 1994, Chini and Krügel 1994, Ivison 1995). Most of the objects that are not IRAS sources have been detected in the millimetre or sub-millimetre ranges where the rest-frame far-infrared emission is redshifted. The very large millimetre-submillimetre spectral index, ~3–4, measured for a few of them, shows that the emission most certainly originates in dust. This has been convincingly demonstrated in FIRAS10214+4724 and H1413+117 ("The Cloverleaf") where strong gravitational amplification allows detailed studies of the energy distribution.

The far-infrared emission of nearby ultraluminous IRAS galaxies ($\sim 10^{12} L_{\odot}$) is believed to be related at least partly to intensive star formation activity. The presence of a strong starburst activity is well proved in FIRAS10214+4724 and H1413+117 and is likely to be a common characteristic of all far-infrared hyperluminous sources. In addition however, strong nucleus activity is detected in several far-infrared hyperluminous sources, including the two latter (see e.g. Rowan–Robinson 1996, Goodrich et al. 1996, Sanders and Mirabel 1996 and references herein). It is probably related to the large central accumulation of interstellar gas generally observed. The relative importance of the starburst activity and of the AGN in the ultimate origin of the far-infrared luminosity is still mostly unknown, although the latter could dominate in many sources.

The case of the increasing number of known luminous radioquiet QSOs at very high redshift appears particularly interesting since: i) there is a good chance that there is a high star formation rate ; ii) the detectability of mm emission from a thermal dust-like spectrum increases for z>1 because of the steep slope of the submm/far-infrared continuum ($S_\nu \propto \nu^\alpha, \alpha \sim 4$, as observed in FI-RAS10214+4724), due to the ν^2 Rayleigh–Jeans law combined with a ν^2 dust emissivity. As a result the observed flux increases almost linearly with redshift beyond z=1, as shown in Figure 1 of McMahon et al. (1994) ; iii) the detectability of a 3mm CO line remains approximately constant beyond z=1, because the redshift increase brings in the 3mm range CO lines with higher J and hence larger luminosities. Since several years we have thus carried out a systematic programme with the IRAM radiotelescopes to detect millimetre emission of dust and CO from high redshift QSOs. The target QSOs are primarily drawn from the APM optical survey for QSOs with z>4 (Irwin et al. 1991). The present paper is aimed at summarizing the results which are published in detail in McMahon et al. (1994) and Omont et al. (1996a, 1996b).

As concerns the search of dust emission, we have performed a sensitive ($\sigma \sim 1.5$ mJy) systematic study of the 1.25mm continuum emission of ~ 22 radio-quiet QSOs at z$\gtrsim$4, with the IRAM 30m telescope equipped with bolometer arrays. Five radio-quiet QSOs at z>4 have been detected at a 5-σ level in addition to the initial detection of the z=4.7 QSO BR1202−0725 reported in McMahon et al. (1994). The detected fluxes range from 2.5 to 10 mJy. All the reported detections were independently confirmed at the 3-σ level on at least three different nights. In addition 10 other QSOs from the Cambridge APM survey sample and 6 others QSOs from the literature were searched for millimetre emission but not detected with 2-σ upper limits of 3–4 mJy.

From this systematic study of about half of the known optically selected z>4 QSOs, some general trends of their millimetre emission can be inferred. All the QSOs we have detected pertain to the APM sample and are among those which have the largest UV rest-frame luminosities. The detection rate within the APM sample is 6 out of 16 observed, compared with zero in the remaining 6. Two of the four APM broad absorption line QSOs observed were detected and four of the seven weak lined APM QSOs were detected, whereas none of the five strong lined APM QSOs were detected. Thus there is evidence for enhanced millimetre emission from luminous QSOs with weak broad emission lines or broad absorption lines.

There is one clear case known of strong lensing amongst the six millimetre detected objects with z>4. In light of the fact that both previously known objects with confirmed strong millimetre emission at z>2 are gravitationally lensed, i.e. H1413+117 and IRAS F10214+4724, sensitive high resolution observations of these z>4 QSOs are required to determine whether gravitational lensing effects need to be taken into account.

Assuming that the millimetre wave continuum emission is due to dust emission (McMahon et al. 1994, Isaak et al. 1994), the very large amount of dust implied, $\sim 10^8\ h^{-2}\ M_\odot$, means that the host galaxies of these QSOs have undergone a substantial phase of star formation. If the gas-to-dust ratio in these galaxies is similar to that in lower redshift objects, the total gas mass would be $\sim 10^{11}\ M_\odot$. We have begun to explore the 1.25 mm emission of bright radio-quiet QSOs in the redshift range 1.5 to 3.5, using criteria which seem to favor millimetre detections, established from our z>4 detections. One source was detected at z = 2.70. We have observed three QSOs with z>3 that were previously studied at 1.25mm by Andreani et al. (1993) who reported detections at a level higher than 3σ. We have been unable to confirm any of these reported detections. In particular we have a 3σ upper limit of 3.2 mJy for the z=3.19 QSO PC2132+0126 for which Andreani et al. reported a flux of 11.5±1.7mJy. Either this source has substantially varied during the period between the two sets of observations or the single channel bolometer observations were affected by systematic errors.

Sensitive observations with the IRAM interferometer of the bright radio-quiet QSO BR1202-0725 at z=4.69 (Omont et al. 1996a) reveal that its continuum millimetre emission is extended on scales of several arcseconds. The map of the emission at 1.3mm presents two prominent peaks ; one at the QSO position, the other 4" North–West.

The 3mm CO(J=5–4) line is detected at both positions, with similar redshifts and widths. Each of these lines is detected at $\sim(5$–$6)\sigma$ level. As both detections are independent because the two positions are spatially resolved, CO(5–4) emission is definitely detected at about the 8σ level. The coincidences with the two 1.35mm peak positions and the absence of signal elsewhere further strengthen this conclusion. From the total integrated flux of the line (3.1±0.9 Jy km s^{-1}), one deduces a mass of molecular gas $M(H_2) \sim 6 \ 10^{10} \ M_\odot$, assuming no gravitational amplification and a conversion factor between $M(H_2)$ and the CO luminosity comparable to that at smaller redshifts (Solomon et al. 1992). The J=7–6 and 4–3 lines have also been tentatively detected with fluxes consistent with the 5–4 line. The J=5–4 line has also been detected by Ohta et al. (1996) with characteristics in agreement with our results.

Two interpretations are possible for the second millimetre source: a massive hyperluminous infrared companion which could be related with Lyman α emission and an absorption system close to the QSO ; or a gravitational lens with rather constraining conditions to explain the absence of a second optical image.

Regardless of the precise interpretation of the maps, the detection of carbon monoxide and dust shows the presence of large masses of molecular gas and dust in one of the most distant galaxies known, and shows that conditions conducive to huge bursts of star formation existed in the very early Universe.

References

Andreani, P., La Franca, F., Christiani, S. (1993), MNRAS **261**, L35

Chini, R., Krügel, E. (1994), A&A **288**, L33

Dunlop, J.S., Hughes, D.H., Rawlings, S., Eagles, S., Ward, M. (1994), Nature **370**, 347

Goodrich, R.W. et al. (1996), ApJL **456**, L9.

Irwin, M.J., McMahon, R.G., Hazard, C. (1991): *The Space Distribution of Quasars*, ASP Conference Series **21**, (ed.) D. Crampton, 117

Isaak, K.G., McMahon, R.G., Hills, R.E., Withington, S. (1994), MNRAS **269**, L28

Ivison, R.J. (1995), MNRAS **275**, L33

McMahon, R.G., Omont, A., Bergeron, J., Kreysa, E., Haslam, C.G.T. (1994), MNRAS **267**, L9

Ohta, K., Yamada, T., Nakanishi, K., Kohno, K., Akiyama, M., Kawabe, R. (1996a), Nature **382**, 428

Omont, A., Petitjean, P., Guilloteau, S., McMahon, R.G., Solomon, P.M., Pécontal, E. (1996a), Nature **382**, 428

Omont, A., McMahon, R.G., Cox, P., Kreysa, E., Bergeron, J., Pajot, F., Storrie–Lombardi, L.J. (1996b), A&A, in press

Petitjean, P., Pécontal, E., Valls–Gabaud, D., Charlot, S. (1996), Nature **380**, 411

Rowan–Robinson, M. et al. (1991), Nature **351**, 719

Rowan–Robinson, M. (1996): *Cold Gas at High Redshift*, (eds) M. Bremer, H. Rottgering, P. van der Werf & C. Carilli (Kluwer, Dordrecht, in the press)

Sanders, D.B., Mirabel, I.F. (1996), Ann. Rev. Astron. Astrophys. **34**, in press

Solomon, P.M., Downes, D., Radford, S.J.E. (1992), ApJ **398**, L29

Can Photoionization Solve the Overcooling Problem?
A Quick Look at the Cosmological History of Baryons

Alain Blanchard[1], Domingos Barbosa[1], and Simon Prunet[2]

[1] Observatoire astronomique, 11, rue de l'université, F-67 000, Strasbourg, France.
[2] Université de Paris-Sud, Institut d'Astrophysique Spatiale, Bâtiment 121, F-91405 Orsay cedex, France

Abstract. In a simple, strictly homogenous cosmological model, the IGM would consist of neutral primeval gas. However, the Gunn-Peterson test illustrates that the IGM, if it does exist, should be highly ionized. The existence of such an IGM, containing most of the baryons, is very likely, although it remains undetected. Present day constraints are at the edge of eliminating purely photoionized medium, leading to consider other heating mecanisms. In the context of structure formation, the possible existence of a hot IGM is of considerable importance in order to understand the evolution the thermal history of baryons, and may be the solution to the "overcooling problem". Better constraints on the physical IGM, achievable with the VLT, will greatly help us to understand the mecanism of galaxy formation.

1 Introduction

Because astronomical information arises mainly from light, coming from baryons, our knowledge of the baryonic content of the universe is quite good. Most of the visible baryons are stars. The knowledge of the galaxy luminosity function and of the M/L ratio for the various stellar population of galaxies does allow to estimate the total amount of visible baryons present in the universe (including gas in late galaxies) in term of the critical density:

$$\Omega_{*+gas} \approx 0.005$$

This number is substantially smaller than what the standard nucleosynthesis predicted:

$$\Omega_{bbn} \approx 0.04 h_{50}^{-2}$$

indicating that there is a lot of unseen baryons. The presence of X-ray gas in clusters of galaxies may tell us what actually are these remaining baryons: the gaseous component represents the major baryonic component. In the Coma cluster, considered here as the best studied case, the ratio between stars and gas is:

$$\frac{\Omega_{gas}}{\Omega_*} \approx 10 - 15 h_{50}^{-3/2}$$

If we assumed this ratio to be universal, this leads to an estimate of the total baryonic content of the universe in a gaseous form:

$$\Omega_{gas} \approx 0.04 - 0.06 h_{50}^{-2}$$

this number is in quite good agreement with nucleosynthesis, but does conflict with $\Omega_0 = 1$ universe (White et al, 1993).

The above argument may have to be revised anyway because of some recent results: primordial nucleosynthesis may predict a smaller or larger Ω_{bbn} than we have quoted, depending on the deuterium issue. An other caviat in the argument is coming from the MACHO and EROS experiments: althought the halo seems unlikely to be fully baryonic, it may content two to five times more baryons than the disk itself (Chabrier et al, 1996). If we assume Ω_* to be two or three times higher, than Ω_{gas} has to be increased accordingly. In the following we discussed the baryon issue without envisaging such a possibility (in most cases, this does not change the general picture).

2   The Intergalactic Medium at High Redshift

In the absence of structure formation and of any specific reheating mecanism, the baryonic component of the universe should cool down adiabatically at redshift smaller than ~ 200, reaching a very low temperature today. The Gunn-Peterson test however reveals the high degree of ionization of the intergalactic medium. The absence of any positive detection of absorption by neutral hydrogen up to redshift ~ 5, allows one to set a severe constraint on the physical state of the intergalactic medium. The present situation is summarized on figure 1: we have used the best observed limit on the Gunn-Peterson optical thickness: $\tau \leq 0.05$ found recently (Giallongo et al, 1994). In a temperature-density diagram this limit translates in a curve, provides the UV background is specified. We have plotted the constraints corresponding to two different values of the intensity of the UV background: $J_{21} = 1$ and $J_{21} = 0.3$ the latter case being favored by detailled analysis of the proximity effect (Williger et al., 1994).

The ionization state of the IGM unambigously probes that it has to be heated.

The UV flux coming from QSO's is one component of this heating. It is not clear whether QSO's does provide the bulk of the UV flux of another source (like early galaxies) is required to explain the flux deduced form the proximity effect. One important question is whether the IGM is purely photoionized or whether an additional source of heating is required. This question can not be answered from one single constraint. In the temperature-density diagram a photoionized medium appears as a line. This is not trivial: for a given density of the IGM, as the UV flux is increased the ionization fraction grows but the temperature remains the same (actually there is a weak sensitivity to the spectral index). It follows from this curve that photoionization cannot heat the IGM at temperature higher than a few 10^4K. The measured $\tau_{GP} \leq 0.05$ does allow to set an other constraints. If the IGM was purely photoionized with $J_{21} = 0.3$, one should

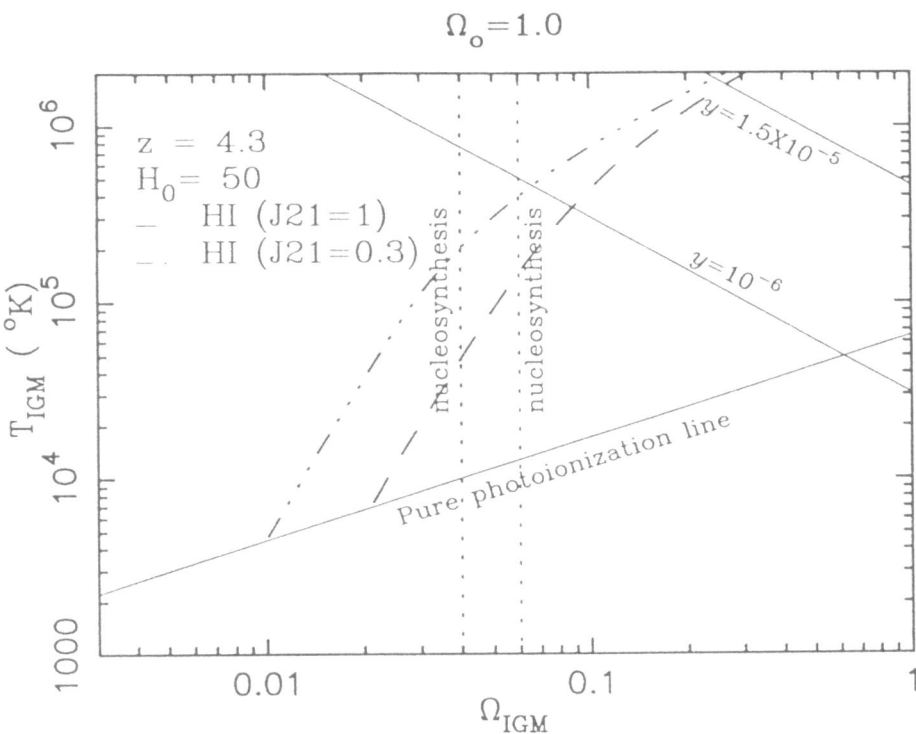

Fig. 1. Temperature-Density diagram where various constraints have been represented. The doted line corresponds to ithe limit implied by the optical Gunn-Peterson limit of $\tau_{GP} \leq 0.05$ and to a flux $J = 10^{-21} \mathrm{ergs^{-1} cm^{-2}}$, while the doted-dashed line correspond to $J = 10^{-21} \mathrm{ergs^{-1} cm^{-2}}$ the IGM should lie above this line. The full line corresponds to a purely photoionized medium.

conclude that the density parameter of the IGM is less than 1%. Photoionization cannot therefore heat the IGM at a temperature higher than a few 10^4K. If we keep the conclusion that Ω_{IGM} is substantially higher, as indicated by the amount of gas in clusters, we see from the figure 1, that the IGM cannot be purely photoinized in order to satisfy the GP limit and should have a temperature of the order of 10^5K.

3   The Overcooling Problem

Structure formation theoreticians have recently focussed on the role and history of baryons. A important criteria proposed in order to differentiate galaxies from clusters is the cooling criteria: when gas falls in a potential it is shocked heated (or adiabatically compressed) up to a temperature which does allow the gas to be in hydrostatic equilibrium. This temperature is currently called the virial temperature of the gas. When the gas reaches this temperature it has a caracteristic cooling time (its constrast density within a structure being at least 200 times the mean density). Clusters typically represent structure for which the cooling time exceeds the age of the universe, while for galaxies this typical time scale is much shorter. It is tempting therefore to think that the cooling criteria can be used as a criteria for star formation: if the gas is able to cool, it will contract in a runaway fashion, which is can end up only by star formation (as there is not so much cooled gas in the universe).

However this simple scheme meets a fundamental problem: at high redshift the bulk of the baryons lies is small potentials with temperature in the range $10^4 - 10^6$K in which cooling is extremely efficient. As a result most of the baryons should have been cooled by now. This is in clear contradiction with two basic facts that we have met: known stars represent only a small fraction of baryons predicted by nucleosynthesis, and most of the baryonic content of clusters is still in the gas phase. Both facts suggest that only 10% of the baryons were actually turn into stars during the cosmic history. This problem was first pointed out by Blanchard et al. (1990) and Cole (1991). It is clear that the solution implies that the gas have undergone some reheating.

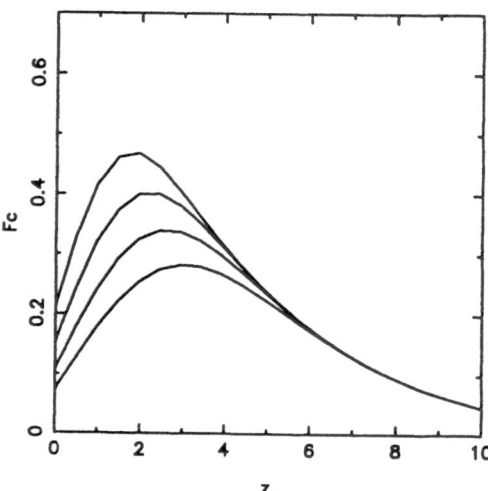

Fig. 2. Instantaneous fraction of gas able to cool at various redshift. The different curves correspond to different values of Ω_b, from 0.08 (top curve) to 0.01 (bottom curve).

Within a simple spherical model it is possible to evaluate the virial temperature. The knowledge of the mass function allows to estimate the fraction of gas able to cool at any redshift by considering the mass range in which cooling is likely to occur. More details on this type of calculations can be found in Blanchard et al (1992). The non-photoionized case is presented in figure 2. This figure represent the instantaneous fraction of gas able to cool within structure at arbitrary redshift. The dependance on the total baryonic content of the universe is rather moderate (the instantaneous maximum fraction is changed by less than a factor of 2 when Ω_b is changed by a factor of 8). This fraction has been evaluated as in Blanchard et al. (1992) by assuming a cooling function for an IGM with no metals).

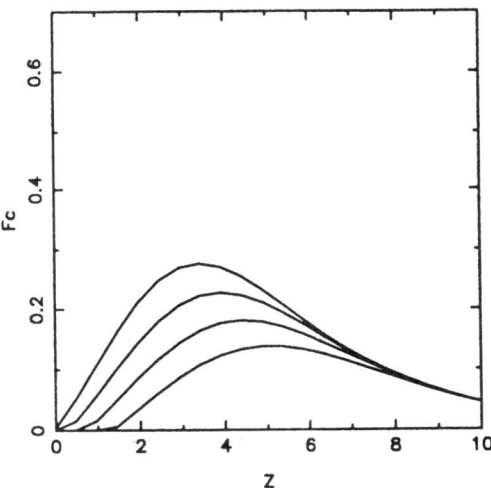

Fig. 3. Same quantity than in figure 2, but here the cooling curve is taken in the photoionized case.

It has been suggested that photoionization can strongly suppress cooling in small potentials and therefore small galaxy formation (Efstathiou, 1992). Let us examine whether photoionization can act efficiently to suppres the overcooling problem. The first effect of photoionization is to heat the IGM to a temperature of the order of 10^4k and therefore this will prevent the formation of small potentials with virial temperature smaller than this. This is not relevant to the overcooling problem. An other important point is that the photoionization substantially alter the cooling rate of the gas (which is normally dominated by colisional exitation cooling). It is therefore interesting to examine whether this significantly reduces the fraction of gas able to cool.

In figure 3 the same quantity is evaluated in the case of a fully ionized. It is clear that although the photoionization reduces the instantaneous fraction, the change is at most a factor of two. This is unlikely to be a solution to the overcooling problem.

4   Conclusion

What we know from baryons clearly indicate that most of the bulk of the baryons lies in an IGM with a temperature greater than 10^4 K may be in the range of $10^5 - 10^6$K. Photoionization hardly explain such a high temperature. The overcooling problem cannot be solved easily by the suppression of the colisional exitation cooling. Althought photoionization reduces the amplitude of the cooled fraction, the effect seems too small to offer a solution to the overcooling problem. Detailed numerical simulations lead to similar conclusions (Navarro, Steinmetz, 1996). The simplest solution which agrees quite well with present day information is the existence of a hot IGM heated by some other mecanism as suggested by Blanchard et al (1992): in such a case the high temperature of the gas prevents it to fall in most of the potential at high redshift. If the temperature of the IGM is gretaer than a few 10^5K the overcooling problem is solved.

Aknowledgments: D.B. is supported by the Praxis XXI CIENCIA-BD/2790/93 grant attributed by JNICT, Portugal.

References

Blanchard, A., Valls-Gabaud, D., Mamon, G. (1990): in *"Particle Astrophysics, the Early Universe and Cosmic Structures"*, Proceedings of the XXth Rencontres de Moriond in Astrophysics", ed. J.-M. Alimi, A. Blanchard, A. Bouquet, F. Martin de Volnay and J. Trân Thanh Vân, Editions Frontières, Gif-sur-Yvette, p. 403.

Blanchard, A., Valls-Gabaud, D., Mamon, G. (1992): A&A **264**, 365–378

Chabrier, G, Secretain, L., Méra D. (1996): astro-ph/9606083

Cole, S, (1991): ApJ **367**, 45

Efstathiou, G., (1992): MNRAS **456**, 43p

Giallongo, E., D'Odorico, S., Fontana, A., McMahon, R. G., Savaglio, S., Cristiani, S., Molaro, P., Trevese, D. (1994): ApJ **425**, L1-L4

Navarro, J.F., Steinmetz, M. (1996): astro-ph/9605043

White, S., Navarro, J..F., Evrard, A.E., Frenk, C..S. (1993): Nat **366**, 429-433

Williger, G. M., Baldino, J. A., Carswell, R. F., Cooke, A. J., Hazard, C., Irwin, M. J., McMahon, R. G., Storrie-Lombardi, L. J. (1994): ApJ **428**, 574-590

Part 7

Concluding Remarks

Which Second Generation VLT Instrumentation?

Guy Monnet and Roberto Gilmozzi

European Southern Observatory,
D-85748 Garching bei München, Germany

Abstract. Getting a competitive "cosmological" instrumentation package at the VLT Observatory (with eventually contribution from smaller telescopes than the 8m's) is one of our most important goals. Both present and future instrumental capabilities are analyzed in that context, with special emphasis on providing a full complement of observational modes to probe deeply and efficiently the far Universe.

1 Present VLT Instrumentation

Presently planned VLT Instruments have been thoroughly presented during the Workshop. Figure 1 shows their location according to the current VLT planning. Their ability to tackle cosmological studies can be summarized as follows.

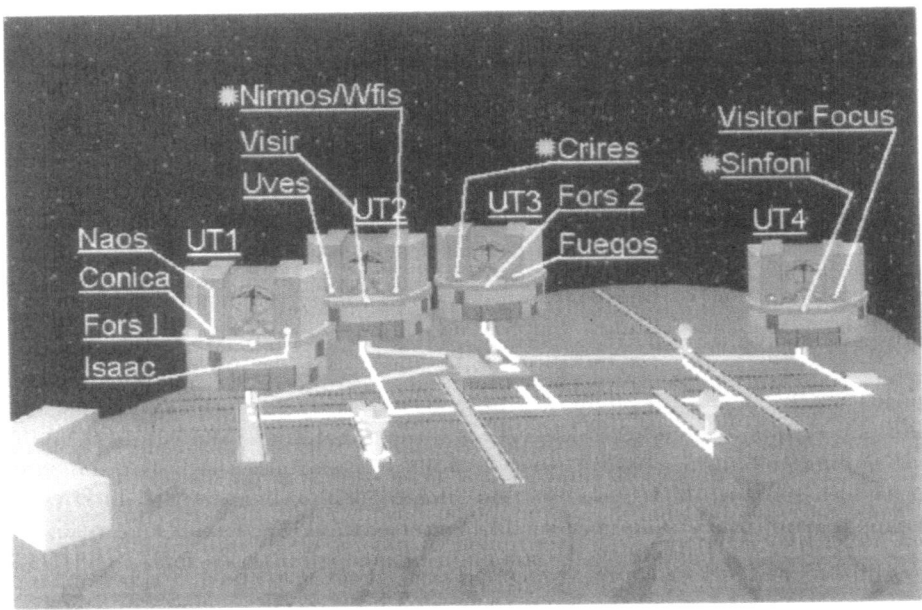

Fig. 1. Schematic view of the VLT Observatory at Paranal. The location of the foreseen instruments is indicated. An asterisk marks instruments which are under study but have not been approved yet.

1.1   Local Cosmological Tests

Our early capability should be basically adequate for such local tests (to use Tytler's terminology, this Workshop) and especially abundance determinations with UVES, ISAAC and FORS.

FUEGOS, with the relatively low multiplex advantage of 80 offered by its multi-fiber mode, but with a very substantial maximum spectral resolution of 45,000 will be highly efficient for such precise measurements in the resolved stellar population of nearby dwarf galaxies.

1.2   Early Galaxy Evolution

ISAAC, especially with its deep 1-2.4 μm imaging and long slit capability, will access far galaxies in a very large distance range, potentially from $z = 0$ to 20. Note that, with spectral resolution around 3,000, it provides "software" suppression of the strong OH airglow lines. The CONICA 1-5 μm imager, coupled to an adaptive optics "bonnette", will give near-diffraction limited images, quite comparable in spatial resolution with present visible images from HST, but will be limited by the availability of natural guide stars. FORS, while presently limited to 19 slitlets, could efficiently screen out potential $z > 3$ (proto)galaxies, especially if a wide-field imager is available for prior deep surveys of potential candidates. UVES, as a 100,000 spectral resolution echelle spectrograph with performances similar to the Keck HIRES instrument, will play a significant role in the probe of intergalactic gas clouds from quasar absorption lines. We are thus anticipating a good ability for significant contributions in this field.

1.3   Single Object Spectrographic Follow-Up

Single object follow-up can be carried out, on a variety of spectrographic instruments, in their long-slit mode, from 0.35 to 2.4 μm, and even up to 24 μm (VISIR), if the object is bright enough. Additionally, FUEGOS in its "ARGUS" mode provides in the visible band simultaneous spectroscopic coverage of about 660 adjacent pixels on a small field (between 18 and 5 arcsec square), with a variety of spectral resolutions between 1,500 and 45,000. CONICA adds a sub-0.1 arcsec spectroscopic capability, either in a slit mode in the $J\,H\,K\,L\,M$ bands with spectral resolutions of 200 to 1,500, or as scanning Fabry-Perot Interferometry in the K band at a resolution of 1,800 in roughly half an arcmin fields.

2   Future VLT Instrumentation

Vigorous efforts are nevertheless needed to become competitive in three key areas of observational cosmology, namely: (i) the evolution of "normal" galaxies back to at least $z = 2$; (ii) the evolution of active/forming objects to the largest possible redshifts and (iii) individual objects follow-up in the near-infrared.

2.1   Galaxy Evolution at $1 < z < 2.4$

In this interval all emission and absorption lines from the objects are shifted in the near-infrared. Two studies presently being made in competition between the AUSTRALIS and the VIRMOS Consortia address this issue with huge multi-object spectrographs, multiplex capabilities in the hundred(s) of objects, and fields of acquisition of the order of 15-25 arcmin. The goal is to cover the J-band, and as much as the H-band as feasible, within the limits set by thermal background emission in the relatively warm environment of Paranal. Both on-going studies have been presented at the Workshop; interestingly they both feature a near-infrared "ARGUS" mode, with up to 6,000 spatial pixels on their half arcmin square field. Note that this whole domain relies heavily on the availability and performances of the present 1K x 1K Rockwell arrays, eventually to be enlarged to 2K x 2K in the near-future.

2.2   Distant Objects Surveys

The same studies cover the need for even larger multi-objects deep pencil beam surveys for very distant objects ($z \gg 2$), where UV rest frame lines are now shifted in the visible-near UV transmission window of the atmosphere. Spatial multiplex capabilities near to one thousand appear feasible in that wavelength range, especially with the new generation of low noise 2K x 4K CCDs. In a more modest range, the multi-slit capability of one of the FORS first generation spectrographs could conceivably be upgraded to close to one hundred; this potential modification will be explored with the VIC Consortium in the near future.

2.3   Distant Objects Follow-Up

As the HST Deep Field amply shows, considerable sub-structuring is present in myriads of (presumably) far objects. Deep spectrography of individual objects, producing a full data cube and with spatial resolutions in the 0.1-0.2 arcsec range is required to glimpse at their physical nature. We are presently looking at a near-Infrared (1 to 2.4 μm) cryogenically cooled three-dimensional spectrograph, coupled to a moderate order ($\sim$36) Adaptive Optics system to provide that facility. Dubbed SINFONI (for SINgle Faint Object Near-ir Investigation), this project in collaboration with the MPI-E Garching is presently at the feasibility phase level. Figure 2 shows one example of the simulation work currently under way.

2.4   Wide-Field Capability

A major weakness of the ESO (Paranal + La Silla) Observatory, and especially in this field of observational cosmology, is the lack of a decent wide-field capability there. Such a facility is crucial for at least three purposes:

   (i) building statistically valid infrared (resp. visible) samples for multi-object near IR (resp. visible) instruments. If the two multi-slit VIRMOS proposals are

SINFONI simulation

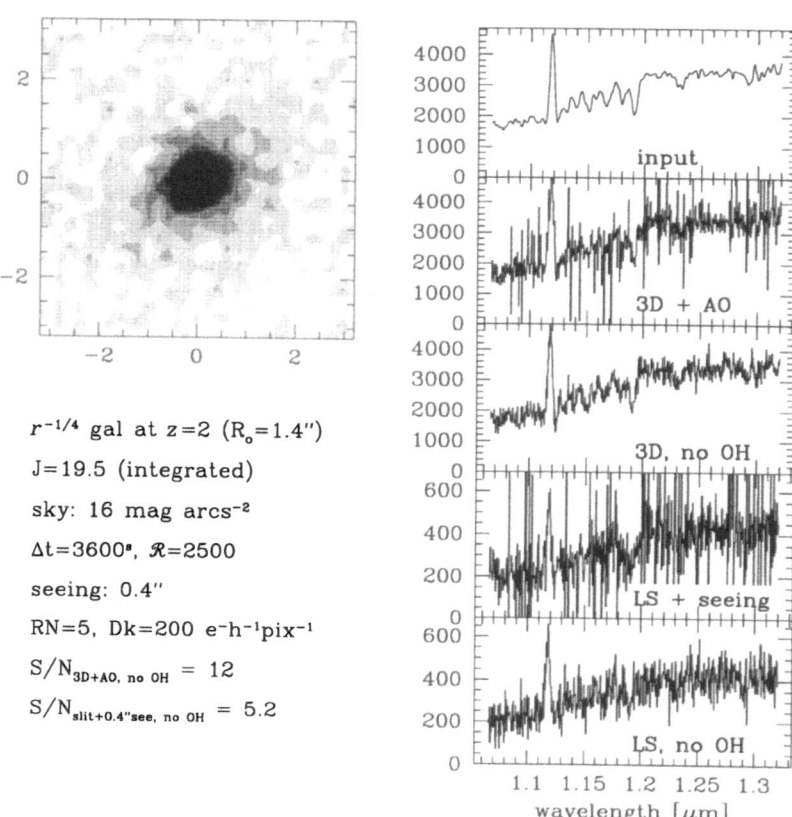

$r^{-1/4}$ gal at $z=2$ ($R_o=1.4''$)

J=19.5 (integrated)

sky: 16 mag arcs^{-2}

$\Delta t=3600^s$, $\mathcal{R}=2500$

seeing: 0.4''

RN=5, Dk=200 e$^-$h^{-1}pix^{-1}

$S/N_{3D+AO,\ no\ OH} = 12$

$S/N_{slit+0.4''see,\ no\ OH} = 5.2$

Fig. 2. Simulation of an observation of an elliptical galaxy at $z = 2$ with SINFONI ("3D") and with conventional long slit spectroscopy ("LS"). The top left panel shows the 32×32 pixel field of view (axes in arcsec). The right hand panels show the simulated spectra. Panels with "no OH" in their label indicate that software OH suppression has been applied. The relevant parameters used are reported in the lower left. A 2.4× gain in S/N is achieved in this particular simulation.

selected, it is foreseen that the instruments themselves, in their imaging mode, will provide this deep imagery over their 14 x 14 arcmin field. In the same vein, the single multi-fiber AUSTRALIS Consortium proposal comes with a suggested separate and even larger IR and Visible wide field facility, again using the VLT itself.

(*ii*) selecting relatively rare targets in the visible, e.g. large z candidates, for single or multi-object follow-ups in the visible or in the infrared. This calls for a "true" wide-field instrument, with at the very least a half a degree field, but it can be envisioned at intermediate size telescopes in the 2.2 to 3.6 m range. We are presently looking at a number of possibilities at various La Silla telescopes, as well as at a possible Paranal option: in any case, an 8K x 8K camera project is starting. Initially, at least, it will be installed at the MPI-ESO 2.2 m, where it will provide a 30 arcmin square field.

(*iii*) in its own sake, especially for wide-field gravitational shearing measurements, in large structures as well as around nearby galaxies, to weigh dark matter distribution at all scales.

3   Conclusion

The high quality and breathtaking variety of the contributions presented at this Workshop amply confirms that the observational cosmology field is indeed an exciting one, where the ESO VLT Observatory must absolutely play a major role in the nearest possible future. It is already a tribute to the importance of the field that the vast majority of the present (and, even more, of the future) VLT instruments address its needs, incorporated as major science drivers.

Developing the initial instrumentation package at 11 of the 12 VLT foci (one being reserved for visitor instruments) is a long term venture, however, that will carry us easily up to the year 2005. This means that priorities in time must be setup in the "wish" list of cosmological instrumentation briefly presented above. One of the very helpful result of this Workshop will be using your input to come to enlightened choices on this issue, e.g.:

• What is the proper balance between getting large samples of $z = 1$ to 2.5 galaxies and the study of individual objects (up to $z = 5$)?

• What is the most essential wide-field visible imaging facility to select $z > 2.3$ galaxies: a 40 arcmin field at a 3m class telescope or a 15 arcmin field at an 8m telescope?

• Would the choice be the same if the goal was to produce gravitational shear maps?

• And finally: is there any other instrumental approach that we have not yet considered which would further improve our capabilities in this field?

Part 8

Poster Papers

Galaxy Evolution at Low Redshift?
Inferences from Optical and IR Counts

E. Bertin[1,2] and M. Dennefeld[1]

[1] IAP, Paris
[2] Sterrewacht Leiden

A homogeneous catalog of galaxies covering about 400 square degrees has been selected from reprocessed IRAS data with a 4σ detection limit of 120 mJy at 60 μm, thanks to low cirrus contamination and to a larger than average satellite coverage. This highly complete sample of galaxies is therefore intermediate between the Faint Source Catalogue (FSC, large area, but limiting flux > 220 mJy) and the small, deep survey (6 square degrees, 50 mJy) by Hacking & Houck (1987). Schmidt plates in a 120 square degrees subarea have been digitized, yielding optical counterparts (as well as normal field galaxies) down to $B_j = 21$ and R = 19.5. We use both IR and optical counts to put constraints on galaxy evolution at the low redshifts corresponding to this survey.

1 IR Galaxies Counts and Optical Counterparts

Number counts at 60 μm down to the completeness limit (around 110 mJy) are compared to various models of luminosity evolution in Fig. 1. The raw counts have been corrected for the various biases (including Eddington bias and confusion) appearing in the original catalogue thanks to Monte-Carlo simulations. Taking into account these corrections provides a smaller excess of faint sources over no-evolution models than previously claimed (Lonsdale *et al.* 1990, Gregorich *et al.* 1995). Adopting a pure luminosity evolution scenario $L \propto (1+z)^Q$, a best fit is obtained for $Q = 3.2 \pm 0.2 \pm 0.3$, where the second uncertainty comes from normalisation of the model at high fluxes. A somewhat faster evolution might occur below 150 mJy, although low S/N makes the estimation uncertain. Background fluctuations measurements are consistent with this result and do a priori exclude luminosity evolution models with $Q > 4$. The identification rate of optical counterparts, corrected for "chance superpositions", is still high at these low IR fluxes and confirms the reality of the excess. The colours of these candidates are much redder than optically selected field galaxies at faint magnitudes, and follow the average trend expected for bright, non-evolving Sbc galaxies (Bertin, Dennefeld and Moshir, 1996).

2 Optical Counts

From the digitized Schmidt plates, carefully calibrated with CCD frames over the whole magnitude range, number counts have also been produced for galaxies,

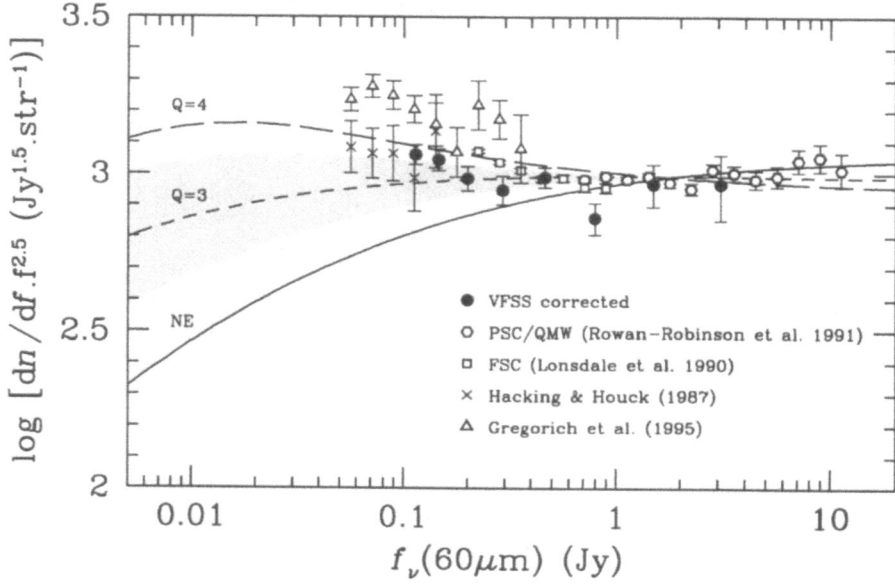

Fig. 1. Corrected number counts at 60 μm in the Very Faint Source Survey (this work), compared to previous studies, a no-evolution model (NE), and pure luminosity evolution models (see text). The grey zone indicates the model-dependant domain allowed by the level of IRAS background fluctuations.

irrespective of their association or not with an IRAS source. Optical counts show no excess over no-evolution models up to B_J around 20-21, implying a "high" normalisation of the luminosity function (Bertin & Dennefeld 1996).

3   Consequences

We therefore observe an excess of faint IR galaxies, while no excess is seen in the optical in the magnitude range corresponding to their optical counterparts. Can these facts be reconciled?

The colours and magnitudes of the IR galaxies in excess suggest they are typical M^* galaxies (not dwarfs), an analogy supported by spectroscopic identifications of counterparts of mJy radio-sources (Benn et al.1993) linked to the IR thanks to the "universal" FIR/radio correlation (e.g. Helou *et al.* 1985). They are probably undergoing a burst of star formation as usual in IRAS galaxies, and further illustrated for the brighter FSC counterparts by Clements et al. (1996). On the one hand, starburst galaxies represent only a small fraction of the optically luminous galaxies in the corresponding luminosity range in the local universe. On the other hand, starbursts in IR galaxies are most of the time burried in the dusty central regions so that no large effect on the optical luminosity is expected (nor observed). An excess in the IR number counts is therefore compatible with the "no excess" seen in the optical.

The VLT provides a unique opportunity to study in details the spectroscopic properties of this complete sample of distant starburst galaxies. New counts at fainter IR levels with ISO should also bring new light in the near future.

References

Benn C.R., Rowan-Robinson M., McMahon R.G., Broadhurst T.J., Lawrence A., 1993, MNRAS 263, 98

Bertin E., Dennefeld M., 1996, A&A, in press

Bertin E., Dennefeld M., Moshir M., 1996, submitted to A&A

Clements D.L., Sutherland W.J., Saunders W., Efstathiou G.P., McMahon R.G., Maddox S., Lawrence A., Rowan-Robinson M., 1996, MNRAS, in press

Gregorich D.T., Neugebauer G., Soifer B.T., Gunn J.E., Herter T.L., 1995, AJ 110, 259

Hacking P.B., Houck J.R., 1987, ApJS 63, 311

Helou G., Soifer B.T., Rowan-Robinson M., 1985, ApJ 298, L7

Lonsdale C.J., Hacking P.B., Conrow T.P., 1990, ApJ 358, 60

Rowan-Robinson M., Saunders W., Lawrence A., Leech K., 1991, MNRAS 253, 485

Novae as Distance Indicators: Simulation of VLT Observations of Fornax

Massimo Della Valle[1] and Roberto Gilmozzi[2]

[1] Astronomy Dept., Universitá di Padova, Italy
[2] European Southern Observatory, Garching bei München, Germany

Abstract. In this contribution we study the possibility to calibrate the absolute magnitude at maximum of a sample of type Ia SNe occurred in the Fornax Cluster using the Maximum Magnitude *vs.* Rate of Decline Relationship of Classical Novae.

1 Introduction

Type Ia SNe are commonly regarded as reliable standard candles (e.g. Branch and Tammann 1992). They occur both in spirals and ellipticals but only in the case of spirals it has been possible to provide a direct calibration of the SN absolute magnitude at maximum via Cepheids (e.g. Saha et al. 1996). On the other hand, recent results may suggest the existence of systematic differences among SNeI – both in their spectral evolution and in their absolute magnitude at maximum. It is therefore of paramount importance to provide **direct** calibrations of the absolute magnitude at maximum of type Ia SNe which occurred in ellipticals.

In this contribution we study the possibility to carry out this task by using Novae as calibrators. Novae are, potentially, very useful distance indicators. At maximum light they are more luminous than the longest period Cepheids; they are observable in both spiral and elliptical galaxies; and they are mainly located in the bulges of spirals whereas Cepheids are concentrated in spiral arms where crowding and absorption problems are more important. We have simulated a VLT observation of 25 novae, distributed along the magnitude at maximum vs. rate of decline relationship recently studied by Della Valle and Livio (1995), in a giant elliptical of the Fornax cluster like NGC 1380. In addition to the possibility of determining the distance to the Fornax Cluster itself (Sandage and Tammann 1990), the giant ellipticals of this cluster have produced in the past some spectacular SN events like the type Ia SN 1992A. This object has been particularly well investigated (Kirshner et al. 1993) both photometrically and spectroscopically and the calibration of its absolute magnitude at maximum would be of great astrophysical impact.

2 The Data Set

We have distributed 25 novae arranged in different rate of decline and magnitudes at maximum according to the Della Valle and Livio (1995) recipe. The simulated

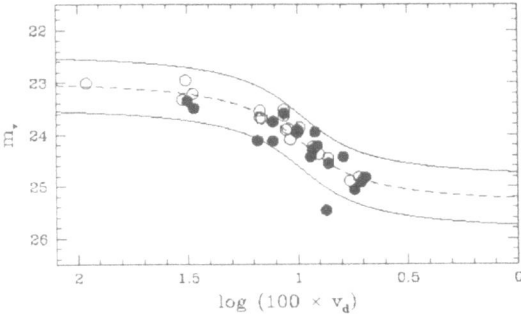

Fig. 1. Original (open circles) and recovered (filled circles) $m_V - v_d$ relation.

image contains a $r^{-1/4}$ galaxy whose magnitude and dimensions closely match the typical giant elliptical in Fornax plus 500 stars distributed with a standard luminosity function. The simulation corresponds to a FORS V band image of 30m duration in 0.8″ seeing. We have assumed a distance to Fornax of 25 Mpc corresponding to a distance modulus of (m–M)=32. Frames were simulated at maximum light and at 2, 4, 8 and 16 days past maximum. The observing strategy will call for observations (of several galaxies) at similar intervals to insure that at least 25 novae will be accessible during the observing "window". One interesting aspect of this simulation is that there was no "cross-talk" between the simulation process and the recovery of the novae: one of us created the simulated images, the other performed the photometry independently.

3   Conclusions

We have analyzed the frame with daophot, recovering 18 (filled circles in the figure) out of the 25 original novae (open circles). We note that by adopting the observational strategy described above we have lost one half of the fast objects ($\sim 10\%$ of the total). A second, more severe source of incompleteness is due to the objects occurring in the inner part of the galaxy ($\sim 20\%$). However the best fit to the black circles gives a distance to the Fornax of 27 Mpc, which corresponds to a distance modulus of 32.15 ± 0.10 (1 σ). For a photometric accuracy on the order of ±0.1 mag, which is quite normal for CCD photometry of type Ia SNe at maximum, we should be able to constrain the accuracy of the calibration of the absolute magnitude at maximum of such objects to below 0.2 mag.

References

Branch, D., Tammann, G. 1992, *AAR*, **30**
Della Valle, M., Livio, M. 1995, *ApJ*, **452**, 704
Kirshner et al. 1993, *ApJ*, **415**, 589
Saha et al. 1996, *ApJ*, in press
Sandage, A., Tammann, G. 1990, *ApJ*, **365**, 1

The Search for Clustering at $z > 0.8$

Jean-Marc Deltorn[1], Olivier Le Fèvre[1], David Crampton[2], Mark Dickinson[3]

[1] DAEC, Observatoire de Paris, Section de Meudon, 92195 Meudon, France.
[2] Dominion Astrophysical Observatory, Victoria, Canada.
[3] Space Telescope Science Institute, Baltimore, USA.

Abstract. We present here the succesful identification of high redshift galaxy clusters around two powerful radio-galaxies, 3C265 (z≃0.81) and 3C184 (z≃1), as well as the observations of a cluster candidate associated with the radio-galaxy 0316-257 at z≃3.14.

In the first two cases, the identification is based on (1) a significant excess of galaxies observed in projection from I band photometry, with a peak density 12σ and 8σ above the mean galaxy background of 3C265 and 3C184 respectively; and (2) 11 and 12 galaxies with redshifts within $1000 km s^{-1}$ from 3C265 and 3C184. These are among the first examples of clusters of galaxies spectroscopically confirmed at $z \simeq 1$ around bright radio-galaxies, and increase the small list of large scale structures securely identified at $z > 0.8$.

At higher redshift Lyman α imaging may provide candidates at the same redshift of a nearby radio-galaxy: we report here the identification of 3 galaxies with confirmed redshifts at z≃3.14 around 0316-257.

1 Introduction

Despite their important implications in such different fields as the study of large scale structures formation or the birth and evolution of galaxies in dense environments, only a handful of clusters have been securely identified at z>0.8. Knowing that at lower redshifts (z<0.5) approximately one third of the bright radio-galaxies and QSOs lie within rich clusters (Hill & Lilly, 1991; Yee & Ellinson, 1993), we use z>0.8 radio-galaxies as sign-posts to identify very high redshift clusters. We present the first results of a survey of radio-galaxies environments, using both near infra-red and optical imagery, as well as multi-slit spectroscopy, to identify high redshift clusters.

2 Summary

V and I exposures were obtained for a 10′ by 10′ field centered on 3C265 at the CFHT using the Multi-Object Spectrograph imaging mode. Similarly we observed 3C184 in the R and I bands. Additional images were obtained using a narrow band filter centered on the redshifted [OII] line. Galaxies within the completeness limits of the samples, and showing a flux excess in the redshifted [OII] filter were preferentially chosen for spectroscopic analysis. Subsequent multi-slit spectroscopy was performed on MOS and the resulting spectra processed using the MULTIRED package implemented in IRAF (Le Fèvre et al., 1995).

We compute the projected number density excess of galaxies having V-I>1 (for the 3C265 field), and R-I>0.2 (for 3C184), which should predominantly lie at z>0.5 (Bruzual & Charlot, 1993). We find a significant excess of 3.8 (12σ above the background) for 3C265 and of 3.4 (8σ above the bakground) for 3C184. The number of galaxies brighter than $m_3 + 2$ and within $0.5h^{-1}Mpc$ is respectively 20 and 10. The galaxy excess in redshift space appears also to be significant : 12 galaxies were found at the same redshift as 3C265 (out of 58 identified objects) with $< z > = 0.8102$; and 11 at the same redshift as 3C184 (out of 56) with $< z > = 0.9954$ (cf. fig. 1). The expected number of field galaxies collected in a $1000kms^{-1}$ bin (I<22), is expected to be 0.1 and 0.03 for 3C265 and 3C184 resp.

V, I exposures and narrow band imaging at 500.7 nm, with a bandwidth of 96 nm (corresponding to $Ly\alpha$ redshifted to z=3.142) of a 10' by 10' field around the radio-galaxy 0316-257 were obtained. Object with excess flux in the 500.7 filter were preferentially chosen for spectroscopy. 2 objects (a galaxy and an AGN) were found at the same redshift as the radio-galaxy (ie $z = 3.1378$ and $z = 3.1351$), leading to a comoving density of galaxies with $V < 23.8$ and $Ly\alpha$ flux $> 10^{-16}erg/cm^2/sec$ in the vicinity of 0316-257 approximately 25 times higher than the expected background density.

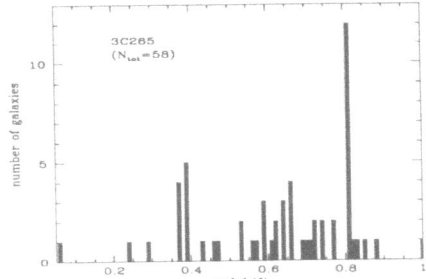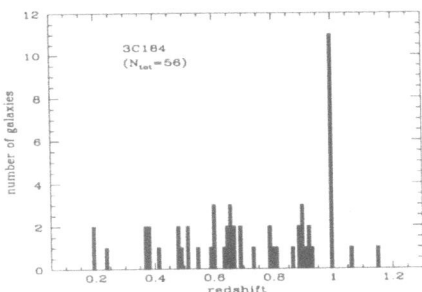

Fig. 1. Redshift distributions in the fields of 3C265 (12 objects within a $1000kms^{-1}$ bin, out of 58) and 3C184 (11 objects in a $1000kms^{-1}$ bin, out of 56).

References

Bruzual, G., Charlot, S., ApJ, **405**, 538
Hill, G. J., Lilly, S. J., 1991, ApJ, **367**, 1
Le Fèvre, O., Crampton, D., Lilly, S. J., Hammer, F., Tresse, L., 1995, ApJ, **455**, 60
Le Fèvre, O., Deltorn, J.M., Crampton, D., Dickinson, M., submitted to ApJL
Yee, H. C. K., Ellingson, E., 1993, ApJ, **411**, 43

Emission-Line Profile Studies of QSOs at $z \simeq 3$

Matthias Dietrich, Martin Kümmel, and Stefan Wagner

Landessternwarte Heidelberg, Königstuhl, D-69117 Heidelberg, Germany

Quasars are the only population of objects which can be observed up to redshifts of z $\simeq$ 5.0. This allows the study of their spatial distribution and physical properties as a function of time. The luminosity function of the quasar population with z < 2.2 became well established through extensive surveys done by a number of groups (e.g. Boyle et al. 1988) while for larger redshifts it is not documented in a comparable way. However, the quasar space density shows only little variations in the redshift range of 2 - 3 with a maximum around z $\simeq$ 2.2 (Schmidt et al. 1991). At this time quasars were nearly 30 times more frequent than at present times, indicating a strong evolution and suggests that the physical properties of quasars change with time.

The study of elemental abundances as a function of redshift is of major interest. C, N, and O abundances can be determined from diagnostical emission lines. The line fluxes can be calculated using photoionization codes.

A comparison of properties of quasars on either side of the maximum of the number density distribution (z $\sim$ 2.2) will give constraints on the physical evolution of AGN. This will eventually lead to the possibility of quasar-dating which will allow us to determine ages of the most highly redshifted quasars.

Before one can start to compare properties of quasars at high redshifts with corresponding counterparts at lower redshifts these objects have to be determined first. The quasar catalogue provided by Hewitt & Burbidge (1993) supplemented with 50 quasars with z > 4 (Shaver 1996) has been used to display the distribution of quasars (Fig. 1).

Pre-VLT observational techniques limit the investigation to the brightest high

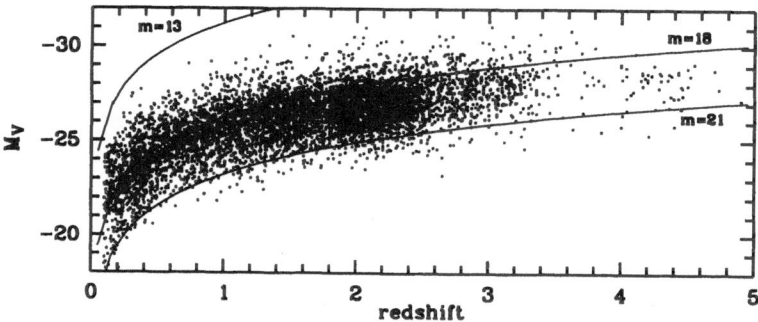

Fig. 1. The quasar distribution in the z-M_V plane. The apparent magnitudes have been transformed to M_V (H_0=50 km s^{-1} Mpc^{-1}, q_0 = 0.5). The solid lines show the location of quasars with an apparent magnitude of 13^m, 18^m, and 21^m.

redshifted quasars. Nevertheless first results have been obtained by various groups studying quasars at epochs prior to the maximum of quasar distribution (e.g. Wills et al. 1995).

We studied a small sample of 9 QSOs with redshifts of $z \simeq 3.0$ which we have observed at Calar Alto/Spain (Dietrich & Wilhelm-Erkens 1996). Within these quasars three objects show unusually narrow line profiles (FWHM$\simeq$2400$\pm$250 km s^{-1}). The profiles of these narrow-lined quasars can be scaled with a single factor to match with the line profile of Lyα (Fig. 2). The relative line ratios as a function of velocity are constant. This can be taken as a strong hint that the conditions of the line emitting gas are similar for different locations in velocity space. A single ionization parameter U might be sufficient for photoionization models for NL QSOs. However, the Lyα profiles of the broad-lined quasars differ in comparison to the other line profiles. While the blue wing of Lyα is comparable to SiIV+OIV]λ1402, CIVλ1548, and CIII]λ1909, the red wing is characterized by strong excess emission (Fig. 2). If the flux is ascribed NVλ1240 line emission the relative flux ratio NVλ1240 / Lyα is 0.07$\pm$0.02 for the NL QSOs and 0.23$\pm$0.06 for the BL QSOs. A similar high line ratio like for the BL QSOs has been found for QSOs at lower redshifts (Laor et al. 1994) and might be due to a higher than solar N abundance and metallicity.

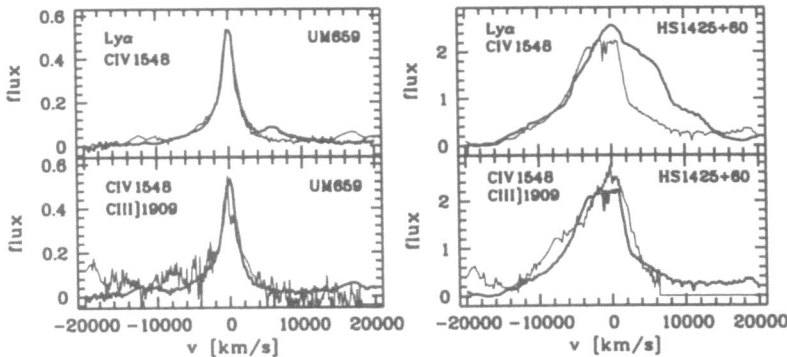

Fig. 2. The scaled line profiles of Lyα (thick), CIVλ1548, and CIVλ1548 (thick), CIII]λ1909 are shown for UM 659, a narrow-lined quasar (left panel) and for HS 1425+60 (right panel), a broad-lined quasar.

References

Boyle B.J., Shanks T., & Peterson B.A., 1988, MNRAS, 235, 935

Dietrich M. & Wilhelm-Erkens U., 1996, A&A, submitted

Hewitt A. & Burbidge G., 1993, ApJS, 87, 451

Laor A., Bahcall J.N., Jannuzi B.T., Schneider D.P., et al., 1994, ApJ, 420, 110

Shaver P.A., 1996, in 17th Texas Symp. on Relativistic Astrophysics and Cosmology, eds. Böhringer, Morfill, Trümper, p.87

Schmidt M., Schneider D.P., & Gunn J.E., 1991, in 'The Space Distribution of Quasars', ASP Conf.Series 21, ed. D. Crampton, p.109

Wills B.J., Thompson K.L., Han M., et al., 1995, ApJ, 447, 139

Redshift and Photometric Survey of the X–ray Cluster of Galaxies Abell 85

Florence Durret[1,2], Paul Felenbok[2], Daniel Gerbal[1,2], Jean Guibert[3],
Gastao Lima-Neto[4], Catarina Lobo[1,5], Vincent Pislar[1] and Eric Slezak[6]

[1] I.A.P., CNRS, 98bis Bd Arago, F-75014 Paris, France
[2] DAEC, Observatoire de Paris, F-92195 Meudon Cedex, France
[3] Observatoire de Paris, 61 Av. de l'Observatoire, F-75014 Paris, France
[4] Universität Potsdam, An der Sternwarte 16, D-14882 Potsdam, Germany
[5] Centro de Astrofísica da Universidade do Porto, P-4100 Porto, Portugal
[6] O.C.A., B.P. 229, F-06304 Nice Cedex 4, France

1 Galaxy Spatial and Velocity Distribution

We obtained a catalogue of **4232 galaxies** in the rich cD cluster of galaxies A 85
(z=0.0556) by scanning a bj Schmidt plate with the MAMA measuring machine
in a region of ±1°; magnitudes were recalibrated with six CCD fields in V and R.
We also obtained 421 redshifts, which added to the 150 in the literature[2,6] gave
a catalogue of **551 galaxies**.
A kernel analysis of the photometric catalogue shows a regular structure with
a small elongation towards the south in the central regions coinciding with the
presence of the South blob observed in X-rays.
Galaxies with velocities in the interval $13350 \leq v \leq 19950$ km/s were taken to be
in A 85, leading to a sample of **308 galaxies in the cluster**. Their bi-weight
mean velocity is 16458 km/s with a dispersion of 1185 km/s.

2 Clustering Patterns

A hierarchical test based on the binding gravitational energy of galaxies members of the cluster was performed to search for possible substructure in A 85[7].
It reveals several interesting features: from left to right on Fig. 1, Groups 1 and 3
are probably just background galaxies, while Group 2 has a velocity close to the
lower limit of that of the cluster and is tight in projection; all the other galaxies
form the actual A 85 cluster, Group 6 being around the cD galaxy.

3 Luminosity Function in the V Band

The luminosity function in the V band was derived directly, for the brightest
galaxies (V<18) with velocities belonging to the cluster, and for fainter galaxies, for all the objects of our CCD photometric catalogue, to which we have

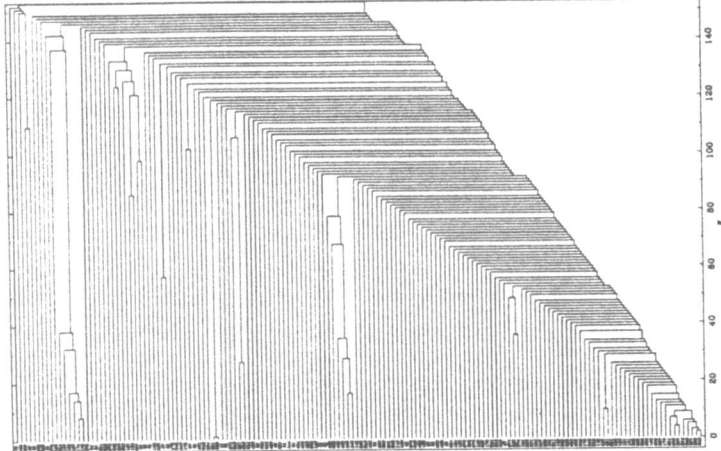

Fig. 1. Hierarchical test performed on the 154 brightest members of A 85.

subtracted the background counts in the V band taken from the ESO-Sculptor survey[1,3]. The best fit is obtained for a power law of index $\alpha = -1.5 \pm 0.1$, comparable to that obtained in central regions of clusters, such as Coma[5].

4   X–ray Analysis

Wavelet analyses[8] of the ROSAT PSPC images of A 85 show a main structure, a south blob already detected with Einstein[4] and a weaker extension south west of the center. We applied a new analysis technique (Pislar et al. in preparation), based on the comparison of two images: 1) Two A 85 fields, where we subtract the background components and correct for exposure, vignetting and quantum efficiency; and 2) a model image including: bremsstrahlung emission, a density distribution (β-model), galactic absorption, convolution with a gaussian (in space and energy).

References

[1] Arnouts, S., de Lapparent V., Mathez G. et al. 1996: A&AS in press
[2] Beers T.C., Forman W., Huchra J.P., Jones C., Gebhardt K. 1991: AJ 102, 1581
[3] Bellanger C., de Lapparent V., Arnouts S., et al. 1995, A&AS 110, 159
[4] Gerbal D., Durret F., Lima-Neto G., Lachièze-Rey M. 1992, A&A 253, 77
[5] Lobo C., Biviano A., Durret F., Gerbal D., Le Fèvre O., Mazure A., Slezak E. 1996, A&A submitted (and poster at this meeting)
[6] Malumuth E.M., Kriss G.A., Van Dyke Dixon W., Ferguson H.C., Ritchie C. 1992, AJ 104, 495
[7] Serna A., Gerbal D. 1996, A&A in press
[8] Slezak E. Durret F. Gerbal D. 1994, AJ 108, 1996

CFRS: The Minimal Ages for Starburst Galaxies

Héctor Flores[1,2] and Francois Hammer[1]

[1] Observatoire de Paris, Section de Meudon, DAEC. , 92195 Meudon, France.
[2] Dpto. Astronomia,Universidad Católica de Chile, Chile.

Abstract. In the frame of the analyses of the Canada France Redshift Survey (CFRS), we have calculated the minimal starburst age for the CFRS galaxies showing $[O_{II}]$ $3727\mathring{A}$ in emission.

We have compared the UV continuum in the region 3050Å - 3950Å, with current model of stellar population synthesis for galaxies with $1.2 \geq z \geq 0.45$. This comparison has been done with two starburst model spectra: a) single burst of finite length, and b) exponentially decreasing star formation rate, produced from the evolutionary synthesis code of Bruzual and Charlot (Bruzual and Charlot (1996)), we have also done including the effect of reddening due to dust for each spectra. We have performed different color-color diagrams with the indices defined in CFRS XIV (Hammer et al. (1996)). From the analysis of the diagram $D(4000)$ v/s $D(3538)$, we conclude that the best model is with a burst of finite length, with solar and sub-solar-metallicities. Preliminary results are : The presence of starts at intermediates ages in most of galaxies, and the decrease of the predicted extinction with redshift and metallicity.

1 Introduction: Overview and Data Basis

The C.F.R.S. has produced a unique sample of 591 field galaxies with $I_{AB} <$ 22.5, from z=0.1 to z=1 (Lè Fevre et al. (1995)). Data have been taken at C.F.H.T. during a set of several observational runs (Lè Fevre et al. (1995)). Same observational strategy was used during all these runs, in such a way that all CFRS spectra are an homogeneous set of data.

Line measurement: Line measurements have been made using the software MEASURE implemented at the Meudon Observatory by D. Pelat (Rola and Pelat (1994)). Several emission and absorption lines have been measured in the 591 galaxies including H_{α}, H_{β}, H_{δ}, Ca_{II}, Mgb, NaD, $[O_{II}]3727\lambda$, $[O_{III}]5007\lambda$ and$[S_{II}]6725\lambda$ when available in the spectral window.

Continuum indices: Six continuum indices have been calculated which, added to the photometric colors, provide a fair representation of the galaxy restframe continuum properties. Unlike broad-band photometric colors, each of the continuum indices is related to one or several physical properties in the galaxies.

2 Summary

We have developed a method for determining the minimal age for starburst galaxies which is based on the UV region of the spectra, also we have found that the sub-metallic model is highly representative for our sample (specially , for object identified as sub-metallic ones (D(4000) deficient)). From the color-color diagram we found that the burst cte. model is more representative for our sample and that the extinction decreases as a function of z, and this result confirm the similar result found with our modeling. We see that from $z = 0.6$ to $z = 0.9$ A_v vary in $-0.3mag$(Fig 1b). We have also found too, one remarkable trend in the ages : 30% of the sample show a minimal burst age $> 0.2Gyr$. (figure 1b). From the plot A_v v/s z , it's noticely a weak correlation between $< A_v >$ and z. The figure 1b show too, the relation between $L(O_{II})$ and A_v for two M_B domains. It's remarkeable a certain correlation, overall for $M_B < -21$. The Figure 1c shows the comparation between the SFR caculated from O_{II} (Kennicutt (1992)) and the calculated from BC95 burst cte. model (maximal for our method), this one shows that only 17% the objects with $W_o(O_{II}) > 30Å$ have the $SFR_{BC} > SFR_{O_{II}}$.

As an interesting perspective, we will try to see which is the best model for each individual CFRS spectra, and to go deeper on the composite model behavior.

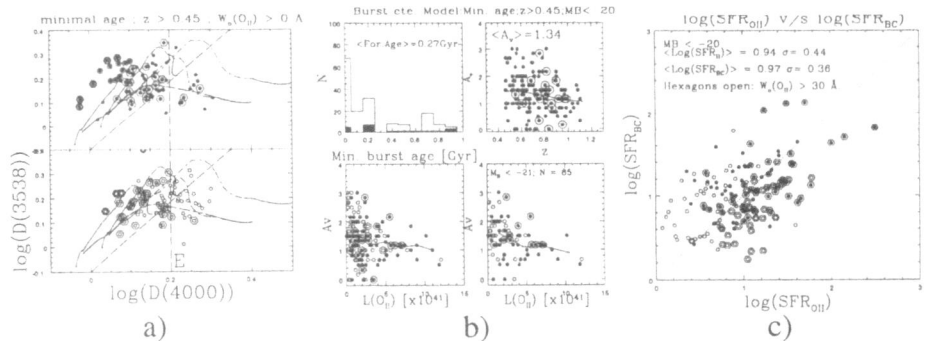

a)   b)   c)

Fig. 1.

References

Bruzual G., ApJ., **273**, 105, (1983).
Bruzual G. and Charlot S., ApJ, **405**, 538 (1993).
Bruzual G. and Charlot S., in Preparation (1996).
Hammer F. et al., in Preparation (1996).
Kennicut R., ApJ. **338**, 310,(1992).
Lè Fevre et al.,ApJ, **455**, 60, (1995) (CFRS II).
Rola C. and Pelat D., A&A , (1994).

Object Detection and Classification in CADIS

R. Fockenbrock, E. Thommes, H. Hippelein, K. Meisenheimer and H.-J. Röser

Max-Planck-Institut für Astronomie, Königstuhl 17, D-69117 Heidelberg, Germany

Abstract. The Calar Alto Deep Imaging Survey (CADIS) employs multi-color photometry in 15 broad and narrow bands between $\lambda = 400\,\text{nm}$ and $2200\,\text{nm}$ to enable an almost complete object classification down to a limiting magnitude of $R = 23.5$ ($B = 24.0$, $K = 20.5$; see Meisenheimer et al. 1996). Additional Fabry-Pérot observations detect emission line galaxies in three wavelength intervals around $\lambda = 700$, 820, and 920 nm. We outline our strategy to obtain a complete object list from object detection on each of the ≥ 24 images obtained through the filters and the Fabry-Pérot. Accurate photometry at the common object position with exactly matching resolution on each of the > 100 individual frames generates a "photometric spectrum" of each object. Automatic classification algorithms are mandatory to deal with the $\geq 50\,000$ objects expected in the entire survey area of 0.3 square degrees.

1   Object Detection

The observations for CADIS are carried out at Calar Alto Observatory in Spain. Up to now we obtained 4 to 18 CCD-frames per filter or Fabry-Pérot wavelength. Each frame has been reduced by correcting electronic *bias* (from overscan), *flatfield, interference-fringes* (if necessary) and eliminating the *cosmics*. After determing the shift between all frames of one filter with respect to a common reference-frame, the sum of all frames is build. Object detection with FOCAS gives us an output with a set of information on each object detected in any filter/wavelength. The complete list of objects is obtained by merging these wavelength-dependent object-lists into one OBJECT-list (where positions within 1" are regarded as one single object).

2   Object Photometry

The OBJECT-list is applied to each individual input CCD-frame by retransforming the coordinates of this big table, taking into account instrumental distortion. From the position information of the OBJECT-list and the photons detected on every input frame we derive the flux of each object. In order to allow for varying seeing PSF_s our photometry uses a Gaussian weighting function (G_w), which is chosen in such a way that the effective PSF ($= PSF_s \otimes G_w$) is the same for all frames (Meisenheimer + Röser 1986). This results in a photometry table for every individual input frame. According to changing weather and seeing conditions, each photometry table is calibrated with respect to standard-stars. After that we combine the complete photometric information for each object contained in our OBJECT-list into one table, providing us with a photometric spectrum for each object. The biggest table we got up to now, has a size of 75 MB. It contained roughly 1/3 of the complete CADIS filter set, and about 1/2 of the number of expected objects (due to a restricted field of view).

3 Object Classification

The procedure of classifying the emission line objects is described in detail by Thommes et al. (1996). For the non-emission line objects we will apply template fitting for reconstructing the most likely spectrum (Belloni 1994). Although, at present, we have to classify each object interactively by assigning the most likely redshift, we envisage that an overwhelming fraction of the objects can be classified automatically by comparing them with an appropriate set of templates.

4 Results

So far the full exposure time could only be reached for the B,R filters and several Fabry-Pérot wavelengths. We find 35.9 objects/$\square'$ in R, 37.1 objects/$\square'$ in B and 9.4 objects/$\square'$ per Fabry-Pérot wavelength. Taking into account the occurence of objects in more than one filter, we get about 50 independent objects/$\square'$. Thus, for one CADIS field ($100\square'$) we expect 5000 objects.

5 Summary and Prospects

Object detection and classification in the multi-wavelength imaging survey CADIS is the central preposition for a scientific evaluation of the recorded data. We established a strategy to obtain a complete OBJECT-list from object detection on each of the more than 24 wavelength-settings, which we will get for every field on sky. Accurate photometry at a common object position with exactly matching resolution on each of the more than 100 individual frames will generate a photometric spectrum of each object. An automatic algorithm for classification is under construction at the moment. The complete strategy gives us a powerful instrument to handle the expected > 50.000 objects in CADIS. So, basic preparations for doing very interesting and cosmological relevant science have already been done. Currently we are looking forward to the next round of observation, which should yield some 250 hours of integration time. Enjoy the future!

References

Meisenheimer, K., Röser, H.-J. (1986): *The Optimization of the Use of CCD Detectors in Astronomy (proceedings)* (ESO-OHP, Garching bei München), p.227

Belloni, P., et al. (1994): A&A, Vol.297, p.61

Meisenheimer, K., et al.(1996):
The early universe with the VLT (proceedings) (Springer, Berlin, Heidelberg), p.??

Thommes, E., et al. (1996): *The early universe with the VLT (proceedings)* (Springer, Berlin, Heidelberg), p.?? , and references therein.

Simulations of High Redshift Galaxies Colors at the NTT-SUSI2 and VLT-FORS

A. Fontana[1], S. Charlot[2], S. Cristiani[3], S. D'Odorico[4], E. Giallongo[1], R. Gilmozzi[4], G. Marconi[1]

[1] Osservatorio Astronomico di Roma, Italy
[2] Institut d'Astrophysique du CNRS, 98 Paris, France
[3] Dipartimento di Astronomia, Università di Padova, Italy
[4] European Southern Observatory, Garching, Germany

Abstract. We show here that a deep multi-band photometry in the visible range may be used to select candidate galaxies in the redshift range $z = 3 - 4.5$, which may be excellent targets for the first VLT spectroscopic observations with FORS. The new NTT imager SUSI–2 will be the instrument of choice for these preparatory surveys.

1 The Colors of High Redshift Galaxies

To derive the expected colors of high z galaxies we have used the population synthesis model of Bruzual & Charlot (1996, in preparation), adding the average IGM absorption Madau (1995, ApJ, 141, 18).

An extended set of simulations has been carried on, including the effects of dust and emission lines. A simple example is shown in fig.1.

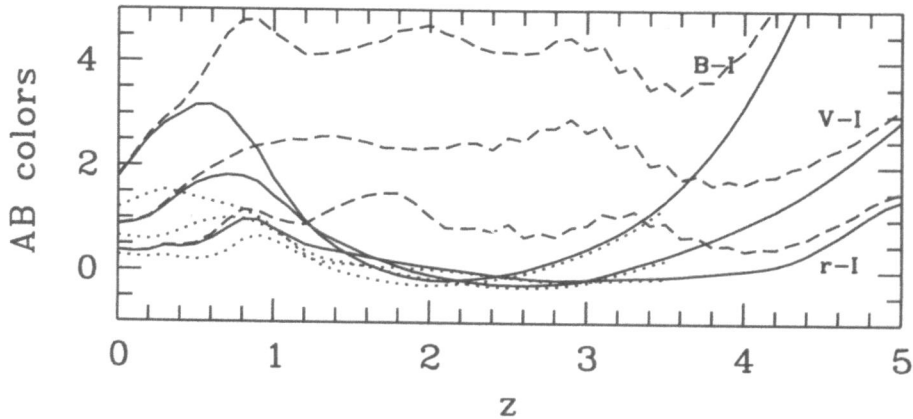

Fig. 1. Color evolution of three different galaxy templates. Solid line: a model with an exponential SFR timescale $\tau = 1$ Gyr formed at $z_f = 7$; dotted: $\tau = 5$ Gyr, $z_f = 4$; dashed: $\tau = 0.1$ Gyr, $z_f = 7$. All models have a Salpeter IMF and no dust.

As it can be seen, galaxies with an ongoing SF are virtually indistinguishable at $z > 2$, if $\tau \geq 1$ (note that a $\tau = 1$ Gyr model yields the colors of a local E0, while a $\tau = 5$ Gyr model those of a local Sb). On the other hand, two models which can be indifferently used to generate the colors of local E0 (i.e. $\tau = 1$ Gyr and $\tau = 0.1$ Gyr) have a completely different evolution at $z > 0.5$.

2   The Identification of Galaxies at $z > 3$

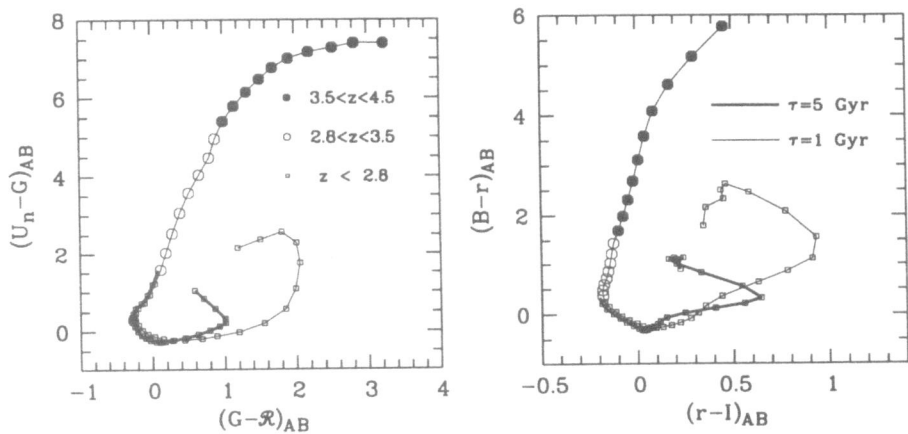

Fig. 2. Color evolution of two templates galaxies in the $U_n GR$ (left panel) and in the BrI system (right panel). Points are separated by $\Delta z = 0.1$

The colors of star–forming galaxies at $z \geq 3$ are independent of the detail of the stellar models used, since are dominated by the effects of the IGM absorption, whose statistical properties are well known. In fig. 2 is shown an example of galaxy color evolution in the two–color $U_n GR$ system used by Steidel et al. (1996 ApJ, 462, L17). Galaxies at $z > 3.5$ become confused with the lower redshift evolved population, at the $U_n - G \leq 3$ limits attainable for fainter objects.

To extend the search at larger redshifts, a four-bands system can be used, with the standard B, V, Gunn r, I filters. In particular, objects at $3.5 < z < 4.3$ are selected through the flat $r - I$ color. As shown in fig. 2, the contamination from evolved population at lower redshift may be minimized with this color selection.

SUSI–2, the new high resolution wide field imager for the NTT, is particularly suited for the preparatory imaging surveys. Its main features are a relatively large field of view ($5.5' \times 5.5'$), an high UV sensitivity and a high spatial sampling to match the excellent image quality attainable at the NTT.

The spectro–imager FORS at the VLT (whose field of view matches the SUSI–2 one) will then be used to confirm the redshift identification with spectroscopic follow–up and to refine the redshift identification of the faintest candidates through intermediate–band photometry.

Spectral Classification of Galaxies with $z \lesssim 0.5$

Gaspar Galaz[1] and Valérie de Lapparent[1]

CNRS, Institut d'Astrophysique de Paris
98bis, Bv. Arago 75014 Paris, France

Abstract. Using the ESO-Sculptor faint galaxy redshift survey data, we have extensively tested the Principal Component Analysis (PCA) method to perform the spectral classification of galaxies with $z \lesssim 0.5$. This method allows us to separate all galaxies in an ordered and *continuous* spectral sequence, which is strongly correlated with the Hubble type. The PCA method works independently of any set of templates, and can also unveil the activity and abnormal spectral features of the sample of galaxy spectra. We study the stability of the approach on the full redshift catalogue in order to search for possible relationships between spectral type, activity and redshift.

We present the spectral classification for some of the spectroscopic data gathered for the ESO-Sculptor redshift survey (Galaz & de Lapparent). The data was obtained with both the 3.6m and the 3.5m NTT ESO telescopes (de Lapparent *et al.*, 1993; Bellanger *et al.*, 1995a; Bellanger *et al.*, 1995b; Arnouts *et al.*, 1996), using multi-object spectroscopy. For $z \approx 0.15$, it is very difficult to determine objectively the morphology of galaxies because their angular diameter is $\leq 7"$. The spectroscopic data provides a wealth of information for classification purposes: it constraints the characteristic properties of the most representative stellar populations, and then provides clues on their possible evolution.

We base our classification mainly on the Principal Components Analysis (Murtagh & Heck, 1987; Connolly *et al.*, 1995) (PCA), which allows to separate all galaxies in an ordered and *continuous* spectral sequence strongly correlated with their morphological type (*i.e.*, the Hubble sequence). We check our results with a simple χ^2 test, using selected Kennicutt templates (Kennicutt, 1992).

Briefly, the PCA technique searches for an orthonormal base in the M-dimensional space, containing a cloud of N spectra, each of them with M wavelength bins. This new base has the following property: The Euclidean distance of each spectrum to all the new axes defining the base is a minimum. Each vector of this new base is called a Principal Component (PC). The PCA technique is interesting when there are some systematics in the input data. In that case, one can *reconstruct all* the data points (*i.e.*, spectra), with only a few PC's. Obviously, in such a case the reconstruction is not perfect, but it may serve to separate the different physical contributions to the spectra and the different PC's tell us about the dominant sources of variation in the data. The first PC is the average spectrum, with an elliptical-like or early shape. The 2nd PC has an spiral-like or late shape and in general the 3rd PC has a very weak continuum and strong emission lines. Therefore, the first 3 PC's contain important information about the systematics of our data. On the other hand, higher PC's are dominated by the noise, and allow us to study its contribution to the data.

We found that we need no more than 3 PC's (resp. 4 PC's) to reconstruct 90% (resp. 97%) of the integrated each flux of each spectrum, in the average. Theses 3 PC's may serve to discriminate the old and the young populations of galaxies, giving also a measure of the stellar activity. The method constitutes a powerful tool as an objective, unconstrained (*i.e., unsupervised*) and reproducible classification system. It is therefore a promising technique, allowing (a) to describe the general spectral features of a database, (b) to detect peculiar objects (for instance the so-called "E+A" galaxies), and (c) to unveil a possible new type of galaxy. Figure 1 shows the projections onto the first two PC of 277 galaxies observed in photometric nights. The analysis was done for $3700 \leq \lambda \leq 5250$ Å, using flux calibrated spectra. Galaxies with small values of α_1 and negative α_2 values are early types and galaxies with small values of α_1 and positive values of α_2, are late types. High values of α_1 and $\alpha_2 \sim 0$ are intermediate spiral types.

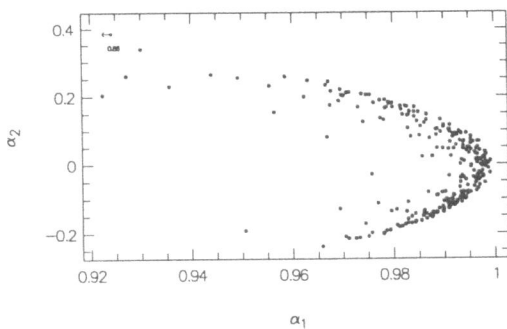

Fig. 1. Projections onto the first two PC of galaxies observed in photometric nights. The analysis was done for $3700 \leq \lambda \leq 5250$ Å.

References

Arnouts, S., de Lapparent, V., Mathez, G., Mazure, A., Mellier, Y., Bertin, E., and Kruszewsky, A. 1996, *A&A Supp.*, in press.

Bellanger, C., de Lapparent, V., Arnouts, S., Mathez, G., Mazure, A., and Mellier, Y. 1995a, *A&A Supp.*, 110, 159.

Bellanger, C., and de Lapparent, V. 1995b, *ApJ* 455, L103.

Connolly, A. J., Szalay, A. S., Bershady, M. A., Kinney, A. L., and Calzetty, D. 1995, *AJ*, 110, 1071.

de Lapparent, V., Bellanger, C., Arnouts, S., Mathez, G., Mellier, Y., and Mazure, A. 1993 *The Messenger*, 72, 34.

Galaz, G. and de Lapparent, V., submitted for publication to *A&A*

Murtagh, F., and Heck, A., "Multivariate data Analysis", 1987, Reidel Publishing Co.

Kennicutt, R. C. 1992, *ApJ* Supp. Series, 79, 255.

The Galaxy Environment of QSOs as a Function of Redshift

Klaus Jäger[1], Klaus J. Fricke[1], and Wolfram Kollatschny[1]

Universitäts-Sternwarte, Geismarlandstraße 11, D-37083 Göttingen, Germany

Abstract. We discuss the importance of galaxy environment studies of QSOs to the investigation of the origin of QSO activity and stress briefly results of present observing campaigns and the benefits of future observations with the VLT/FORS at ESO.

1 QSO Environment Studies

The study of the galaxy environment of QSOs may be fundamental to the investigation of the origin of QSO activity (eg. Osterbrock (1993), Fricke and Kollatschny (1989)). In the activity–interaction picture the fuelling of the central engine with gas and dust is thought to be triggered by gravitational interactions between QSO hosts and nearby galaxies. Such a scenario should reflect itself in a galaxy density enhancement near the QSOs and possibly in an enhanced activity (e.g. starbursts) of this galaxy population. Therefore the scientific goals of our observations are

- the investigation of QSO host galaxies and tidal interaction features,
- the search for companion galaxies very close to the QSO hosts (< 100 kpc),
- the detection of galaxy groups and clusters around QSOs in general,
- the search for differences in the environments of radio–loud and radio–quiet QSOs due to the different host galaxies (elliptical–and spiral galaxies), and
- the investigation of the evolution of the environments of QSOs as a function of redshift.

Until now environmental studies of QSOs exist mostly up to $z = 0.6$ (eg. Ellingson and Yee (1994), Yee and Ellingson (1993)). They show, that at $z < 0.6$ radio–loud QSOs tend to lie in galaxy clusters and radio–quiet QSOs are found in loose groups at best. From a few recent observations at $z > 1$ (Hutchings et al. (1995), Hutchings (1995)) both radio–loud and radio–quiet QSOs are found to reside in compact groups of galaxies, which seems to imply that there may be an evolution of the cluster environment of radio–quiet QSOs at redshifts below $z = 1$. Altogether the nature and evolution of QSO environments is still not clear in detail, because several studies provide contrary results. This might be due eg. to the different statistical methods which have been applied for the detection of galaxy clustering around QSOs.

The search for a number excess in the galaxy density around QSOs is complicated by the discrimination of faint ($m_v > 20$) galaxies against stellar images and

by the dependence of the background galaxy and stellar counts on galactic latitude. We explored such effects within a pilot study based on ESO 3.6m EFOSC
observations (Jäger (1994)) and tested different methods to analyse the frames.
We investigated the environment of $z < 0.5$ – QSOs by direct imaging (B, V, R)
and multiobject spectroscopy. To cope with some difficulties mentioned above
and to determine the galaxy number counts in the fields we specified the Point
Spread Function and furthermore we made a statistical calculation of the number
counts of the star contamination in each field as a function of galactic latitude,
stellar magnitude and color. We detected in 75% of all cases an increase of the
galaxy density towards the central QSOs and/or very close companion galaxies
with blue colors indicating enhanced activity. The results of our statistical and
photometric clustering analysis based on the imaging data were confirmed by
the spectral analysis. 50 % of our randomly investigated galaxy spectra show
the same redshift as the central QSOs, and all of them have emission lines. The
results are in accordance with the general picture that AGN activity is connected
with gravitational interaction.

2   Benefits from ESO/VLT–Observations

Because of the faintness of the galaxies in the QSO fields spectroscopic observations of companion galaxies with 4m–class telescopes are limited to the brightest
field members even at lower redshifts and environment studies of $z > 0.5$–QSOs
are generally restricted to deep imaging observations only. For a reliable as-
sessment of the environment of QSOs up to $z \approx 2$ we will require follow up
spectroscopy with the **VLT**(FORS/MOS Mode). We need spectra for the confirmation of

- the redshift of close companions of the QSOs,
- the state of activity and star formation in the companion galaxies,
- the dynamics within the galaxy clusters,
- the role of gravitational lensing effects, and
- the general evolution of galaxy clustering

Along with spectroscopy, deep imaging observations at limiting magnitudes of
≈ 27 with 1h exposure times will reveal the environment structure of QSOs at
higher redshifts. We expect additional spinoff for a related program concerning
the identification of QSO absorbers (Lindner et al., this workshop).

References

Ellingson, E., Yee, H.K.C. (1993): ApJ Suppl. **92**, 33
Fricke, K., Kollatschny, W. (1989): A&A Suppl. **77**, 75
Hutchings, J.B., Crampton, D., Johnson, A. (1995): AJ **109**, 73
Hutchings, J.B., (1995): AJ **109**, 928
Jäger, K. (1994): Diploma Thesis, Göttingen
Osterbrock, D.E. (1990): ApJ 404, 551
Yee, H.K.C., Ellingson, E. (1993): ApJ **411**, 43

Detailed Kinematics of Extended Lyman-α Regions Around Radio Galaxies at $z > 2$

Anton M. Koekemoer[1], W.J.M. van Breugel[2] and J. Bland-Hawthorn[3]

[1] Institut d'Astrophysique de Paris, 98bis Bd. Arago, Paris F-75014, France
[2] Institute for Geophysics and Planetary Physics, Lawrence Livermore
National Laboratories, P.O. Box 808, Livermore, CA 94550, U.S.A.
[3] Anglo-Australian Observatory, P.O. Box 296, Epping, N.S.W. 2121, Australia

Abstract. Radio galaxies in the early universe are often associated with large Lyα emitting regions extended along the radio axis. Detailed study of the morphology and kinematics of this gas can offer valuable insights into the formation and evolution of radio sources and their impact on the environment. Here we present results from Fabry-Perot kinematic imaging of the extended Lyα region around the radio galaxy 2104−242. The derived physical properties of the gas are discussed in the context of its origin, dynamics and energetics, and its relationship with the radio source.

1 Introduction

High-z radio galaxies often display associated regions of emission-line gas, extended on scales $\gtrsim 100$ kpc and with complex kinematics: e.g., 4C 41.17 has velocity shears ~ 2000 km s^{-1} (Chambers et al. 1990), while 3C 435A shows an arc along the radio lobe (Rocca-Volmerange et al. 1994). We present Fabry-Perot observations of the radio galaxy 2104−242 ($z = 2.491$; McCarthy et al. 1990), which has Lyα extended $\sim 12''$ along the radio axis (~ 135 kpc for $H_0 = 50, q_0 = 0$).

2 Lyα Observations of 2104−242

We observed 2104−242 in Lyα during Oct. 1994 using TAURUS-2 on the Anglo-Australian Telescope, obtaining 33 frames spaced at 0.88 Å. Fig. 1 shows the Lyα and 3 cm radio emission; Fig. 2 shows the kinematics which is complex, with multiple peaks across ~ 1500 km s^{-1}. The total Lyα volume is $\sim 1.6 \times 10^5$ kpc^3 (for reasonable 3-d geometries) and luminosity $L_{\rm Ly\alpha} = 9.8 \times 10^{44}$ erg s^{-1}; this yields the electron density $n_{\rm e}$ and filling factor f in two scenarios:

- Lyα clouds distributed uniformly, e.g. resulting from a violent merger or cooling flow. We find $n_{\rm e}^2 f \sim 0.05$; comparison with van Breugel et al. (1985) suggests $f \sim 10^{-5}$, in which case $n_{\rm e} \sim 70$ cm^{-3} and $M_{\rm HII} \sim 3 \times 10^9$ M$_\odot$.
- Lyα emission from a relatively thin interface (e.g. ~ 1 kpc) between the radio lobe and a confining medium. If $n_{\rm e}$ is to remain $\lesssim 100$ cm^{-3} then $f \gtrsim 10^{-3}$.

The gas velocity structure implies a dynamical timescale $\tau_{\rm dyn} \sim 1 \times 10^9$ yr and dynamical mass $M_{\rm dyn} \sim 2.6 \times 10^{12}$ M$_\odot$ (agreeing with the host galaxy magnitude). However, $\tau_{\rm dyn}$ is $\sim 10^2$ times larger than the radio source age, suggesting that the gas is unsettled, e.g. due to a merger or disruption by the radio plasma.

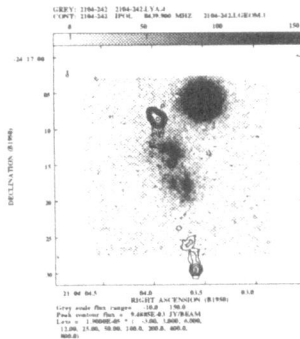

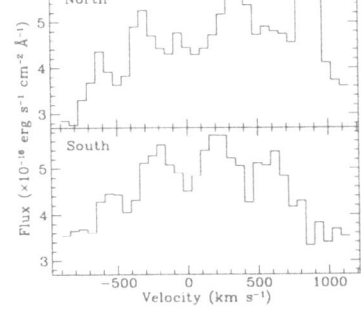

Fig. 1. Lyα (greyscale) and 3 cm radio continuum (contours) of 2104−242. The radio core is identified with the K-band continuum core. The bright object to the N-NW is a foreground star.

Fig. 2. Kinematic structure of the northern and southern Lyα lobes of 2104−242 (integrated over each lobe). Note the multiple line-of-sight velocity components, typically separated by ∼ 500 km s^{-1}.

3 Discussion

The *kinetic energy* of the gas is $E_{\rm K} \sim 1.3 \times 10^{58}$ erg (assuming isotropic velocities and uniform density clouds), requiring $\lesssim 1\%$ of the radio energy if $L_{\rm radio} \sim 1.1 \times 10^{46}$ erg s^{-1} (McCarthy et al. 1990) during $\tau_{\rm radio} \gtrsim 3 \times 10^6$ yr.

The gas *ionization* requires continuous energy input, as the Lyα radiative timescale is $\sim 10^2 \times$ less than the radio source lifetime. Two possibilities are:

- Nuclear photoionization, requiring an isotropic ionizing flux $\sim 1.8 \times 10^{57}$ s^{-1}. For a typical AGN spectrum (e.g. $F_\nu \propto \nu^{-1.5}$, $\log U = -2.5$), the observed nuclear flux at 1200 Å is too low by $\sim 10^2$, consistent with obscuration.
- Energy from the radio plasma through cloud-cloud collision shocks (Sutherland et al. 1993). With shocks ~ 500 km s^{-1} (from line-of-sight velocities), a collective shock area $\sim 2.5 \times 10^2$ kpc^2 (outer extent of the Lyα region), and $n_{\rm e} \sim 70$ cm^{-3}, then $L_{\rm Ly\alpha} \sim 1.5 \times 10^{45}$ erg s^{-1}, agreeing with that observed.

Thus, the Lyα emission is explained by either nuclear photoionization or shocks from radio plasma / gas interactions, while the kinematics strongly suggest disturbance by the radio plasma. However, both scenarios involve uncertainties due to the gas density and filling factor, which will be addressed by observations of other emission lines and detailed hydrodynamical models of jet / gas interactions.

References

Chambers, K., Miley, G., van Breugel, W. (1990): ApJ **363**, 21
McCarthy, P., Kapahi, V., van Breugel, W., Subrahmanya, C. (1990): AJ **100**, 1014
Rocca-Volmerange, B., Gilles, A., Ferruit, P., Bacon, R. (1994): A&A **292**, 20
Sutherland, R., Bicknell, G., Dopita, M. (1993): ApJ **414**, 510
van Breugel, W., Miley, G., Heckman, T., Butcher, H., Bridle, A. (1985): ApJ **290**, 496

AGN Host Galaxies at Intermediate Redshift

Wolfram Kollatschny[1], Alexander Goerdt[1]

Universitäts-Sternwarte, Geismarlandstraße 11, D-37083 Göttingen, Germany

Abstract. We analyse the stellar component of nearby Seyfert host galaxies using population and evolution synthesis methods. We propose to study the host galaxy properties of AGNs as a function of redshift with the ESO VLT.

1   Synthesis Studies of Seyfert Galaxies

We are investigating the relation between nonthermal nuclear activity and host galaxy properties for a sample of nearby Seyfert galaxies i.e. the stellar population in the galaxy, circumnuclear starburst activity, kinematics etc.

So far we have obtained deep two dimensional spectra of a complete sample of face on Seyfert galaxies with redshift of less than 0.01 for having high spatial resolution and low internal obscuration. Furthermore we have taken spectra of a control sample of non active galaxies. For the analysis of the nuclear and extra-nuclear galaxy spectra we have developed a computer code. We synthesize the spectra with stellar spectra of different libraries. Absorption due to dust is included and we consider a central nonthermal component. By using isochrone spectra instead of individual stellar spectra we determine the age distribution of the stellar population in the galaxies.

A first description of our numerical code and of synthesized spectra is given in Goerdt and Kollatschny (1996). We detected an enhancement of a young ($< 10^9$ years) stellar component in the vicinity of the Seyfert nuclei in comparison to non active galaxies of the same Hubble type.

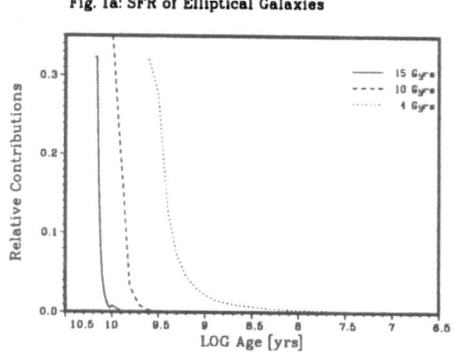

Fig. 1a: SFR of Elliptical Galaxies

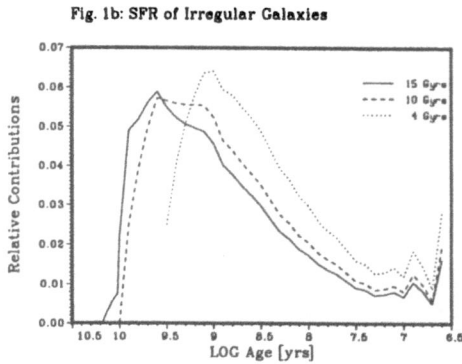

Fig. 1b: SFR of Irregular Galaxies

Fig. 2a: Spectra of Elliptical Galaxies at Different Ages

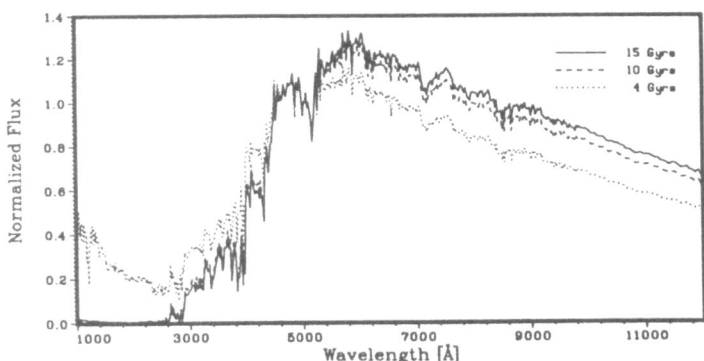

Fig. 2b: Spectra of Irregular Galaxies at Different Ages

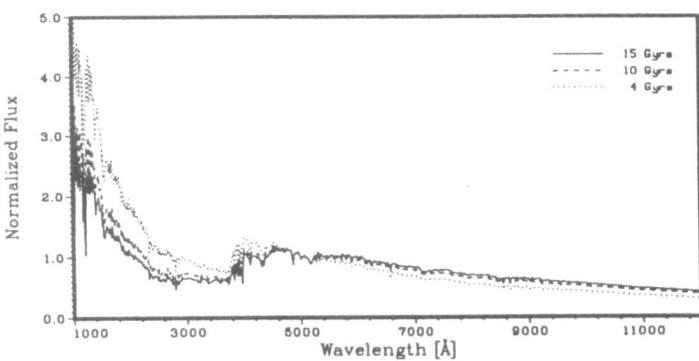

2   ESO VLT Project

It is known that normal galaxies show strong evolutionary effects between z=1 and z=0. In Figs.1a,b we plotted star formation rates as function of time for two different Hubble types (e.g. Sandage (1986)). The constructed theoretical galaxy spectra of the elliptical and irregular galaxies at ages of 4,10,15 Gyrs are shown in Figs.2a,b. There are strong differences in these spectra.
With the ESO VLT we intend to study the relation of nuclear activity with host galaxy properties as a function of redshift until $z = 1$.

Acknowledgement: This work has been supported be the Deutdche Forschungs-gemeinschaft Ko 857/14-1

References

Goerdt, A., Kollatschny, W. (1996): Proceedings of Conference on 'The Impact of Stellar Physics on Galaxie Evolution', Crete, in press
Sandage, A. (1986): A&A **161**, 89

Void Hierarchy – a Guiding Principle to the Study of Faint Structures in Voids

Ulrich Lindner[1], Jaan Einasto[2], Maret Einasto[2], and Klaus J. Fricke[1]

[1] Universitäts-Sternwarte, Geismarlandstraße 11, D-37083 Göttingen, Germany
[2] Tartu Astrophysical Observatory, EE-2444 Tõravere, Estonia

Abstract. We introduce Void Hierarchy as an important property of the Large–Scale Structure in the Universe and demonstrate how it can be used to interpret observations. Moreover the void hierarchy constraints any realistic galaxy and structure formation scenario.

1 Hierarchy of Voids

Voids were defined as low density regions or, alternatively, as regions completely devoid of a certain type of object. Mean void diameters listed in Table 1 demonstrate the dependence of the void size on the type of object used in the (second) void definition. Both definitions imply that voids are not completely empty. Thus, the question is meaningful whether the distribution of galaxies in voids is homogeneous or reveals any structure. For example, it was concluded that Blue Compact Galaxies (BCG) from the Second Byurakan Survey (SBS) or other peculiar galaxies occur isolated within voids (Pustil'nik *et al.* 1995). Such questions are very relevant concerning scenarios of large scale structure and galaxy formation, but they are not conclusively answered up to now.

Table 1 Mean diameters of voids surrounded by different types of object

type of object	mean void diameter
rich clusters (Abell/ACO–Catalogue)	$100\ h^{-1}$ Mpc
poor clusters (Zwicky–Catalogue)	$37\ h^{-1}$ Mpc
bright ($M \leq -20.3$) elliptical galaxies	$30\ h^{-1}$ Mpc
galaxies brighter than $M = -20.3$	$23\ h^{-1}$ Mpc
galaxies brighter than $M = -19.7$	$16\ h^{-1}$ Mpc
galaxies brighter than $M = -18.8$	$13\ h^{-1}$ Mpc

Using the second void definition we have studied the properties of voids surrounded by galaxies from three different luminosity (absolute magnitude M) limited samples. Three void catalogues have been compiled. Comparisons of voids from different catalogues revealed that voids form a hierarchical system (cf. Lindner *et al.* 1995, A&A 301, 329) as it is visualized in Fig. 1a). In this hierarchical concept apparently isolated galaxies in voids may have faint close neighbors which are not detected because of selection effects as it is shown in Fig. 1b).

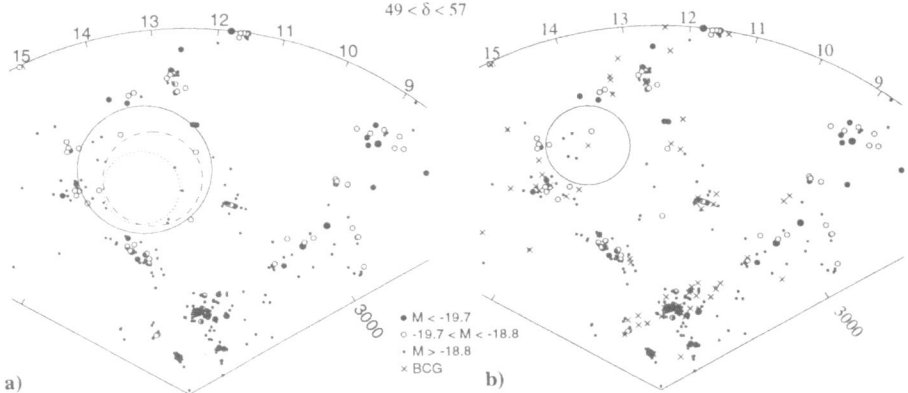

Fig. 1. Wedge diagrams of a slice of the Universe 6000 km s^{-1} deep and bordered by about $9^h < \alpha < 15^h$ and $49° < \delta < 57°$. a) The three circles indicate an example of hierarchically interlaced voids defined by galaxies of different luminosity limit. b) Additionally BCGs from SBS are shown (crosses). The circle indicates the distance to the nearest bright ($M < -19.7$) neigboring galaxy.

2   Conclusions

By now the concept of void hierarchy is established only for galaxies brighter than $M = -18.8$ in the nearby Universe (up to distance $60h^{-1}$Mpc). The study of the radial distribution of fainter galaxies in voids along with nearest neighbor tests (Lindner *et al.* 1996) suggests that this hierarchy continues to fainter magnitudes and therefore contradicts a homogeneous distribution of dwarf galaxies in voids claimed by some theories of galaxy formation (e.g. Dekel & Silk 1986). With second generation instruments attached to the VLT (e.g. VIRMOS) it will be possible to confirm the hierarchy of voids towards fainter luminosity limits and for more distant regions of the Universe. The void hierarchy itself will be helpful to devise new concepts for the study of the large scale structure in the Universe.

References

Dekel, A., Silk, J., 1986, ApJ 303, 39

Lindner, U., Einasto, J., Einasto, M., Freudling, W., Fricke, K.J., Tago, E., 1995, A&A 301, 329

Lindner, U., Einasto, M., Einasto, J., Freudling, W., Fricke, K.J., Lipovetsky, V.A., Pustil'nik, S.A., Izotov, Y., Richter, G.M., 1996, A&A in press

Pustil'nik, S.A., Ugryomov A.V., Lipovetsky, V.A., Thuan, T.X., and Guseva, N.G., 1995, ApJ 443, 499

Implications for Optical Identifications of QSO Absorption Systems from Galaxy Evolution Models

Ulrich Lindner, Uta Fritze – von Alvensleben, and Klaus J. Fricke

Universitäts-Sternwarte, Geismarlandstraße 11, D-37083 Göttingen, Germany

Abstract. We have made an attempt to compile all currently available data on optically identified QSO absorber systems (Lindner *et al.* 1996) to establish the status quo of absorber galaxy data as a basis for the investigation of galaxy evolution. We present a first comparison with results from our galaxy evolutionary synthesis models to demonstrate the potential power of this kind of approach and to guide future observations to identify absorber galaxies.

1 Introduction and Preliminary Results

Our chemical and spectral synthesis model (Fritze – von Alvensleben *et al.* 1994) describes the evolution of various types of galaxies (E to Sd) with appropriate star formation histories and supplies us with time dependent values of luminosities from UV to NIR, colors and metallicities of model galaxies. Adopting any cosmological model characterized by H_0, Ω_0, Λ_0 and the redshift of galaxy formation z_{form}, these results can be transformed into redshift dependent quantities. Apparent R magnitudes as a function of redshift z calculated with this model are plotted in Fig. 1.

Pioneering work in the optical identification of QSO absorption systems was done by Bergeron & Boissé (1991) and up to now, there are more than a dozen of publications reporting on photometric data, equivalent widths and impact parameters for absorbing galaxies. All available data on apparent R magnitudes are plotted in Fig. 1. Different symbols correspond to different authors. Absorbing galaxies with spectroscopically confirmed redshifts (i.e. $z_{gal} = z_{abs}$) are marked by filled symbols, whereas open symbols indicate absorber candidates.

Accounting for the luminosity ranges from the luminosity functions of the various galaxy types (cf. $2\sigma_R$ bars in the lower right corner of Fig. 1) we can state that virtually all observational data points fall between the curves for E– and Sd–models and, accordingly, we can establish global agreement between our galaxy evolution models and observational data up to $z \approx 2$.

Most of the absorber galaxies appear to be early through intermediate type spirals (Sa–Sbc) but many ellipticals and some late type spirals seem to be present, too. The presence of intermediate and late type galaxies among QSO absorbers would imply that a considerable fraction of these galaxies do have extended gaseous halos metal rich enough to cause detectable absorption.

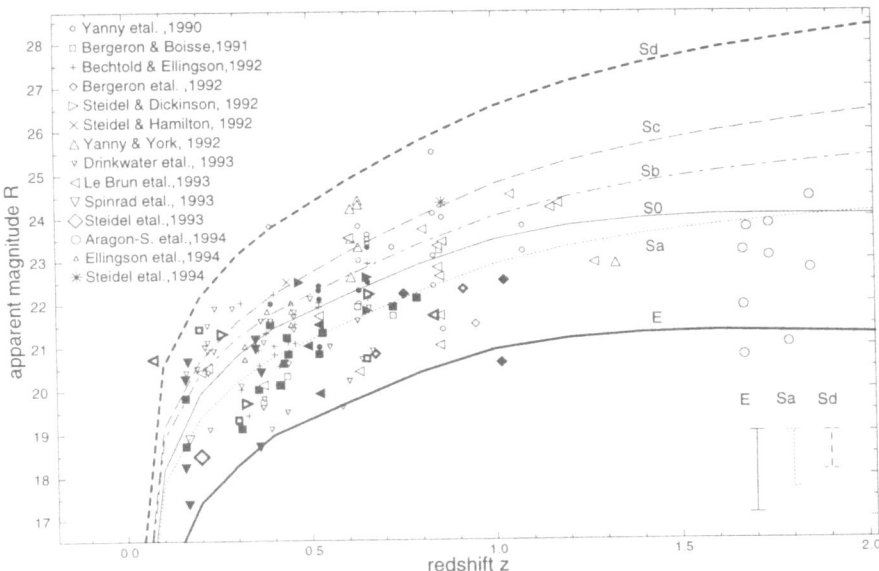

Fig. 1. Apparent R magnitude as a function of redshift for observational data and results from our galaxy evolution model using $z_{form} = 5$ and cosmological parameters $H_0 = 50$ km s^{-1} Mpc^{-1}, $\Omega_0 = 1$, and $\Lambda_0 = 0$.

Varying the cosmological parameters we find that e.g. model galaxies for $H_0 = 50$ km s^{-1} Mpc^{-1} and $\Omega_0 = 0.1$ are much too faint as compared to observations leading us to exclude this compination of cosmological parameters.

2   Outlook and VLT Perspective

Unfortunately, data are very inhomogeneous at present. Measurements in different passbands, i.e. colors, from present day instruments are needed. FORS attached to the VLT will provide deeper imaging ($24 < R < 26$) to detect intermediate and late type spiral galaxies in the redshift range $1 < z < 2$ and the MOS unit of FORS will allow to spectroscopically confirm absorber candidates in the magnitude range $22 < R < 24$ (cf. Fig. 1).

Acknowledgement
This work was supported by Verbundforschung Astronomie/Astrophysik through BMFT grant FKZ 50 0R 90045

References

Bergeron, J., Boissé, P., 1991, A&A 243, 344
Fritze – von Alvensleben, U., Gerhard, O.E., 1994, A&A 285, 751
Lindner, U., Fritze – von A., U., Fricke, K.J., 1996, A&A in press (astro-ph/9604090)

The Extended Luminosity Function in Clusters of Galaxies: a Tool to Search for Environmental Effects

Catarina Lobo[1,2], Andrea Biviano[3,4], Florence Durret[1,5], Daniel Gerbal[1,5], Olivier Le Fèvre[5], Alain Mazure[6], Eric Slezak[7]

[1] Institut d'Astrophysique, CNRS, 98bis Bd Arago, F-75014 Paris, France
[2] CAUP, Rua do Campo Alegre 823, P-4150 Porto, Portugal
[3] Istituto T.E.S.R.E., CNR, via Gobetti 101, I-40129 Bologna, Italy
[4] ESA Villafranca, ISO Science Team, Apto. 50727, CAM IDT, E-28080 Madrid, Spain
[5] DAEC, Obs. Paris, Univ. Paris VII, CNRS (UA 173), F-92195 Meudon Cedex, France
[6] IGRAP, LAS, Tv. du Siphon, Les Trois Lucs, B.P. 8, F-13376 Marseille Cedex, France
[7] Observatoire de la Côte d'Azur, B.P. 229, F-06304 Nice Cedex 4, France

Abstract. We have derived the luminosity function (LF) in the Coma cluster from an imaging catalogue of 7023 galaxies. For $13.5 < V \leq 21.0$ ($-22.2 < M_V \leq -14.7$) the best fit is a gaussian (bright galaxies) plus a steep Schechter function ($\alpha = -1.8$). A power law fit in the range $V > 16.5$ ($M_V > -19.2$) gives a flatter LF in the groups surrounding the giant central galaxies than in the overall cluster. These differences may be produced by dynamical processes related to the environment.

1 The LF in Different Regions of the Cluster

From our catalogue cluster membres are determined by redshift, colour-magnitude relation[6], or by statistical subtraction of field galaxies (CFRS[4] or ESFGRS[1] surveys).

In region O - see Fig. - the best Maximum-Likelihood fit to the LF in the range $13.5 < V \leq 21.0$ ($-22.2 < M_V \leq -14.7$) is given by a Gauss plus a Schechter functions ($\chi_r^2 \sim 0.6$) with parameters $M_V^* = -22.7 \pm 0.4$ ($V^* \simeq 13.0$), $\alpha = -1.80 \pm 0.05$, $\mu = -20.4 \pm 0.2$ ($V_\mu \simeq 15.3$) and $\sigma = 1.1 \pm 0.3$.

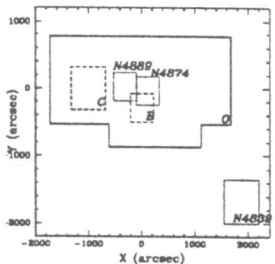

We also fit a power law $\frac{dN}{dL} \propto L^\alpha$ to the LFs obtained in different regions of the cluster for $16.5 < V \leq 21.0$ ($-19.24 < M_V \leq -14.74$) - see Fig. and Table 1.

Table 1. Power law fit results

Region	Power law slope α $\pm 1\sigma$ error	N_{gal}	surface area $arcmin^2$
O	-1.81 ± 0.03	1025	1422
B	-1.71 ± 0.11	60	52
$N4874$	-1.58 ± 0.10	57	51
$N4889$	-1.51 ± 0.13	32	51
$N4839$	-1.74 ± 0.11	54	117
C	-1.88 ± 0.10	77	116

2   Summary of the Results

Here we briefly summarize our results - see details in submitted paper[5].
Our large sample of galaxies allowed us to deduce with great statistical quality
the Schechter index $\alpha = -1.8$ in the Overall region. This steep value indicates
the presence of a huge number of faint galaxies.
The power law slope varies according to the intra–cluster local environment :
around NGC 4874 and NGC 4889 the faint end of the luminosity function is
flatter ($\alpha \sim -1.5$). This may be due to an excess of bright galaxies[3] which can
be explained by a dynamical scenario of formation of giant galaxies by a cannibal-
ism process (see also results for A85 presented by Durret et al., this conference).
Our result for the B region is in contradiction with previous works[2] that produced $\alpha = -1.42 \pm 0.05$ (same power law fit to 15.5<R<23.5). Differences could
be due to the simplistic V-R=1 transformation we used to compare results. The
α index for the group surrounding NGC 4839 is the same as the one determined
for the Overall region. Again from the dynamical point of vue, this apparent
discrepancy between the LFs around groups can be due to different histories[3].

References

[1] Arnouts S., de Lapparent V., Mathez G. et al. (1996) A&AS, in press
[2] Bernstein G.M., Nichol R., Tyson J.A. et al. (1995) AJ **110**, 1507
[3] Biviano A., Durret F., Gerbal D. et al. (1996) A&A in press
[4] Lilly S.J., Le Fèvre O., Crampton D., Hammer F., Tresse L. (1995) ApJ **455**, 50
[5] Lobo C., Biviano A., Durret F. et al. (1996) A&A submitted
[6] Mazure A., Proust D., Mathez G., Mellier Y. (1988) A&AS **76**, 339

A High Redshift Line Emitting Object Discovered in the Near Infrared

Filippo Mannucci[1], Steven V.W. Beckwith[2] and David J. Thompson[2]

[1] CAISMI–CNR, Florence, Italy
[2] Max-Planck-Institut für Astronomie, Heidelberg, Germany

During a near-infrared search for high-redshift forming galaxies (see Thompson et al, 1996, AJ, submitted; or Mannucci et al., this volume) we discovered a complex line emitting object in the field of the B2 0149+33 quasar at z=2.431. This object was selected because of its excess in a narrowband image at the QSO Hα wavelength (2.248 μm) with respect to the K' band image.

The narrow- and broadband images of this field were obtained at the Calar Alto 3.5m telescope, using the MAGIC NICMOS3 cameras. Total exposure times of 91 min for the narrowband image and 26 min for the broadband image were used. The 3σ flux limit for this field, derived from the sky noise and the number of pixels within the photometric aperture, is 1.9×10^{-16}erg cm^{-2} s^{-1}.

The emission line object has an angular separation of 50.1 arcsec from the QSO, corresponding to rougly 400 kpc at the QSO redshift It has a magnitude in the K' filter of 18.52, and is resolved in both images. The excess in the narrowband image indicates a line flux of 6.5×10^{-16}erg cm^{-2} s^{-1}, corresponding to a 10σ detection. If this galaxy is at the same redshift as the quasar, then its Hα flux corresponds to a restframe line luminosity of 2.8×10^{43} erg s^{-1} in a cosmology with $H_0 = 50\,km s^{-1}\,Mpc^{-1}$ and $\Omega_0 = 1$. The rest frame equivalent width is 428 Å.

Near infrared spectra of this object were obtained in the J, H and K bands with the CGS4 spectrograph during two UKIRT runs in November 1995 and January 1996. The slit was oriented through both the quasar and the candidate.

The resulting K spectra of both the QSO and the candidate from are shown in the figure. For both data sets, the bright Hα line from the quasar can be fitted by two gaussians (core + wings), with FWHM of about 2300 and 9300 Km/sec, similar in the two runs. However, the two candidate spectra are very different: in the January 1996 run, a bright line with FWHM=2600 Km/sec was detected, while in the November 1995 the line is much fainter and its FWHM is only 1200 Km/sec.

The width of the line in the January 96 spectrum indicates that an AGN is probably present. The differences between the two spectra has been tentatively attributed to a to a slight offset in the slit alignement between the two observing runs, which resulted in the sampling of different regions of the (spatially resolved) candidate. In January 1996 the AGN was in the slit, while in November 1995 only the outer regions were sampled. These outer region show a narrower line, possibly indicating off-nuclear star formation.

In case this line is the Hα+[NII] complex at the QSO redshift, the Hβ and

[OIII] lines should be observable in the H band, the [OII] line in the J band, and Lyα in the optical. In our J spectrum, the [OII] lies in a region of enhanched noise from the sky lines and was not detected; the Hβ region is not covered by our H spectrum, and no emission is detected at the expected position of the [OIII] lines.

An optical spectrum covering the rest-frame UV region will be crucial to confirm the redshift of this object, either using the Lyα emission or the prominent stellar and interstellar absorption lines.

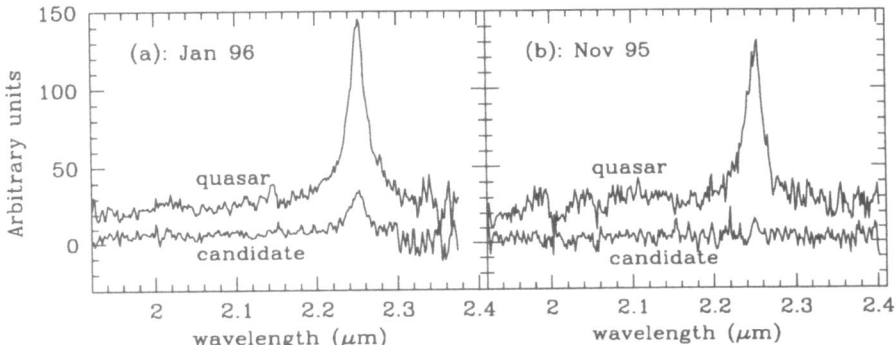

Figure caption: K-band spectrum of quasar B2 0149+33 and of the emission-line object in its field. Panel (a) shows data from the January 1996 UKIRT run, the panel (b) data from the November 1995 run. Coincidence with the quasar's emission line wavelength, as well as faintness of the galaxy and lack of plausible emission lines redward of Hα make it likely that this object lies at $z = 2.43$, at the same redshift as the quasar. In the Jan 96 data the candidate line has a FWHM of 2600 Km/sec, while in the Nov 95 data its FWHM is 1200 Km/sec.

Looking for High Redshift Forming Galaxies in the Near Infrared

Filippo Mannucci[1], David J. Thompson[2] and Steven V.W. Beckwith[2]

[1] CAISMI–CNR, Florence, Italy
[2] Max-Planck-Institut für Astronomie, Heidelberg, Germany

We present the results from an infrared search for high-redshift forming galaxies that makes use of narrowband near-infrared images. A full discussion of the method and the results can be found in Mannucci & Beckwith, 1995, ApJ 442, 569 (MB95) and Thompson, Mannucci & Beckwith, 1996, AJ, submitted (TMB96).

Most of the stars in spheroidal galaxies must have formed at high redshift, ($z > 2$). To discover the population of these high-redshift star-forming galaxies is of considerable interest because it would give information on the process of galaxy formation, on the nature of dark matter, and on cosmology. To date, the bulk of these forming stars have still escaped identification.

We are looking for these galaxies trying to detect their optical emission lines redshifted into the near-infrared bands. Using the Calar Alto 3.5m telescope, we have obtained narrow- and broadband images of fields centered around known QSOs at a redshift placing one of the strong, restframe optical emission lines (Hα, Hβ, [O III], or [O II]) into the narrowband filter. Our approach consists of looking for objects which are relatively brighter in the narrow filter, thus indicating a substantial flux in an emission line. If there is any tendency toward clustering, the QSO will pinpoint regions of overdensity.

A total of 30 fields were imaged, most in the K' band, with a total area coverage of 276 sq.arcmin. Filters with 1% and 3% resolution were used. The most sensitive limits reach $1.4 \times 10^{-16}\,\mathrm{erg\,cm^{-2}\,s^{-1}}$, while the area weighted survey line flux limit was $3.4 \times 10^{-16}\,\mathrm{erg\,cm^{-2}\,s^{-1}}$ (3σ). The total comoving volume sampled by the survey is $1.4 \times 10^5\,\mathrm{Mpc^3}$ (for $H_0 = 50\,\mathrm{km\,s^{-1}\,Mpc^{-1}}$, $\Omega_0 = 1$). Considering only Hα emission in the K' band ($2.05 < z < 2.65$), the survey covers a comoving volume of $3.0 \times 10^4\,\mathrm{Mpc^3}$ to a volume-weighted average star formation rate of $112\,\mathrm{M_\odot\,yr^{-1}}$

One emission-line object was discovered near a QSO at z=2.431, and the presence of an emission line was spectroscopically confirmed (see Mannucci et al., this volume). More observations are needed to study the nature of this object.

Assuming the candidate is not a forming galaxy, we can put upper limits on the comoving density of these objects at various redshift ranges and compare them with model expectations. We conclude that it is unlikely that most of the star formation in elliptical galaxy takes place at redshift lower than 3.5

We present this comparison using two rest-frame quantities directly linked to the models of galaxy formation: the comoving volume density of PG and their

SFR (derived from the line flux).

The expected integral comoving density of PGs are derived by 3 classes of models described in MB95 and TMB96. The comparison must be made in various redshift bins. Two of the resulting plots are presented in the figure.

In the lower redshift bin ($2.0 < z < 3.5$, left panel) our data sample enough volume at a low enough SFR limit to exclude two of the models (labelled *constant* and *bursts* in the figure below) at these redshifts, while the expectations of the third (*hierarchical*) model are only partially sampled. This means that a PG population can be present at low redshift ($2 < z < 3.5$) only if either they are small (low-surface brightness) systems, or for some reason (for example dust) they are not very efficient in emitting lines. At these low redshifts the simplest model, i.e, normal star formation in dust-free galactic size clouds, can be rejected.

As in the case of the Lyα based searches for PGs, adding dust would help to explain the non detection, but in this case a much larger amount of dust would be needed to quench the optical lines instead of the Lyα line.

Dust-free PGs in this redshift range could escape detection only if they are divided into small pieces or in extended low-surface brightness systems (the *hierarchical* models in the figures below).

Our data are also extended enough in the higher redshift bin (right panel) to sample the relevant regions of the parameter space: we expect detections if the *bursts* or (marginally) *constant* models apply. PGs could escape the detection if star formation starts very early ($z > 10$) and goes on until $z \simeq 3.5$, fading out after this epoch. The *hierarchical* model is again out of reach.

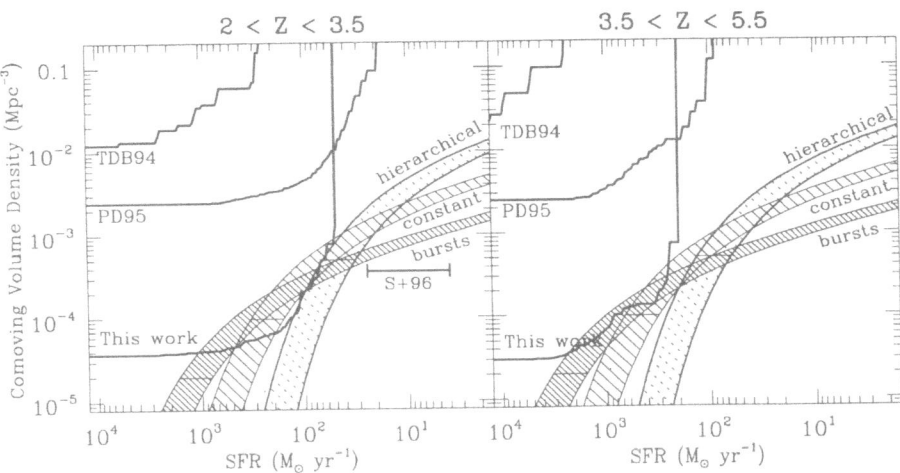

Figure caption: Limits and expectations about forming galaxy comoving density vs. SFR, in two redshift ranges ($2 < z < 3.5$ and $3.5 < z < 5.5$) The thick lines are the upper limits to the PG volume density from three surveys (TDB94: Thompson et al, 1994, AJ 107, 1; PD95: Pahre and Djorgovski, 1995, AJ 449, L1) The short horizontal line in the left panel marks the comoving density and SFR of the population of star-forming galaxies detected by Steidel et al. (1996) at $3.0 < z < 3.5$.

Neoclassic Cosmological Tests with QSOs

G. Mathez[1], Y. Mellier[1,2], J.-P. Picat[1], and L. Van Waerbeke[1]

[1] Laboratoire d'Astrophysique, Observatoire Midi–Pyrénées, 14 Av. E. Belin, F-31400 Toulouse, France
[2] Institut d'Astrophysique, 98bis Bd Arago, F-75014 Paris, France

Abstract. A new cosmological test has been devised, based on QSO samples. Its sensitivity to the parent cosmology (Ω and Λ) and to the various other parameters has been checked with numerical simulations. In particular it is found that increasing the upper redshift limit is largely as efficient as increasing the sample size.

1   A New V/V_{max} Test

This test applies for quasars in a sample complete for magnitudes $m < m_{lim}$. As usual, quasars are assumed to undergo a strong evolution, either a pure density evolution (PDE; Schmidt 1968) or a pure luminosity evolution (PLE; Mathez 1976), according to a law depending on a single parameter k. So, absolute magnitudes M, maximum redshifts z_{max} and the *global* luminosity function $\Phi(M)$ (computed from the entire sample, see Kassiola and Mathez 1991) must all be computed according to the cosmological and evolution hypotheses (e.g. M is brought back to the present cosmological epoch $z = 0$ in case of PLE).

There are 3 free parameters: the density parameter Ω_0, the cosmological constant Λ_0 and one evolution parameter k. It is well known that under uniformity conditions the mean $< V/V_{max} >$ equals $1/2$, independently of the GLF Φ and within any absolute magnitude bin $[M_1, M_2]$. Unfortunately, the V/V_{max} test in this version is *not* a cosmological test because it is always possible to find an evolution parameter k for any given cosmology such that $< V/V_{max} >= 1/2$. Fortunately, one can show that individual V/V_{max} ratios have a uniform distribution, and that the individual V/V_{max} values and the absolute magnitudes M are INDEPENDENT, and we can apply a uniformity test of individual $V/V'_{max}s$ within each absolute magnitude bin and constrain more strongly the three parameters Ω, Λ, k, in a similar way, either for PDE or for LDDE.

2   Maximum Likelihood

The sample is divided in N absolute magnitude bins. Each bin contains n_0 QSOs. In each bin i we compute the Kolmogorov-Smirnov probability $\mathcal{P}_i$ that the V/V_{max} distribution is uniform. The bins are statistically independent then we can compute the likelihood of the whole sample for a given set of parameters (Ω, Λ, k_L) from the product of the $\mathcal{P}_i$'s. Varying k at given (Ω, Λ) gives some maximum likelihood characterizing the cosmological model, so we get an entropy map (Fig. 1) for a grid of cosmological models (Van Waerbeke et al. 1996).

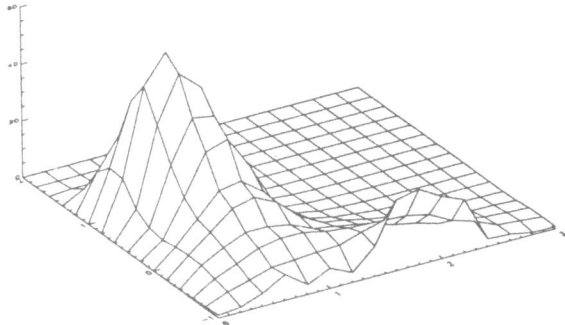

Fig. 1. Likelihood map in the (Ω_0, Λ_0) plane, from Boyle et al. 1990 sample

3   Simulations

A catalogue of redshift and apparent magnitudes is computed from the following method (see Mathez et al. 1996):

- choose a model (cosmology + evolution functional form)
- determine the *global* luminosity function $\Phi(M)$ from the sample
- given limiting magnitude and absolute luminosity, compute the available volume $V_a(M)$
- sort in luminosity according to the observed luminosity pdf in a sample, that is the product $\Phi(M) \times V_a(M)$
- given $V_a(M)$, sort in redshift ensuring homogeneous spatial distribution.

These simulations allow one to test the efficiency of the test according to various parameters or biases, from the point of view of the precision on cosmological parameters determined from our V/V_{max} test. The cutoff towards high redshift seems to be a crucial parameter: increasing it from 2.2 to 3.5 could allow the same gain to be obtained in the precision on Ω as increasing the sample size by a factor around 5. In order to construct quasar samples with redshift cutoffs as high as possible, we proposed (Mathez et al. 1995) to take spectra of all compact and stellar objects brighter than some given limiting (infra)red magnitude.

References

Boyle, B.J., Fong, R., Shanks, T., Peterson, B.A. 1990 MN 243, 1

Kassiola, Mathez, G. 1991, A&A 230, 255

Mathez, G., 1976, A & A 53, 15

Mathez, G., Mellier, Picat, J.P. 1995 *in 'Science with the VLT'*, 1994 ESO symposium, Walsh and Danziger eds., p. 397

Mathez, G., Van Waerbeke, L., Mellier, Y., Bonnet, H., Lachièze–Rey, M., 1996, A&A (in press, SISSA atro-ph 9508029)

Schmidt, M., 1968, ApJ 151, 393

Van Waerbeke, L., et al. 1996, A&A (in press, SISSA atro-ph 9508028)

Chemical Evolution at High Redshift

Francesca Matteucci[1], Paolo Molaro[2], and Giovanni Vladilo[2]

[1] Dipartimento di Astronomia, Universita' di Trieste, Via G.B. Tiepolo 11, I-34131 Trieste, Italy
[2] Osservatorio Astronomico di Trieste, Via G.B. Tiepolo 11, I-34131 Trieste, Italy

Abstract. Observations of high-redshift Damped Ly$-\alpha$ systems (DLA) suggest that the relative abundances of elements might be roughly solar, although with absolute abundances of more than two orders of magnitude below solar. The result comes from observations of the [SII/ZnII] ratio, which is a reliable diagnostic of the true abundances, and from DLA absorbers with a small dust depletion and negligible HII contamination. In particular, in two DLA systems nitrogen is detected at remarkably high levels (Vladilo et al. 1995, Molaro et al. 1996, Green et al. 1995). Here we compare the predictions from chemical evolution models of galaxies of different morphological type with the abundances and abundance ratios derived for such systems. We conclude that solar ratios and relatively high nitrogen abundances can be obtained **only** in the framework of a chemical evolution model relative to a galaxy having one of its first intense bursts of star formation, which, in turn, triggers O-enriched galactic winds, together with a primary origin for nitrogen in massive stars. This model is the most successful in describing the chemical evolution of dwarf irregular galaxies and in particular of the peculiar galaxy IZw18 (Kunth et al. 1995). Thus, solar ratios at very low absolute abundances, if confirmed, seem to favour dwarf galaxies rather than spirals as the progenitors of at least some of the DLA systems.

On the other hand, other three DLA observed by Pettini et al. (1995) and Lipman (1995) are better fitted by models reproducing the solar neighbourhood and therefore are likely to be protospirals.

We conclude by stressing that abundance ratios between elements produced by stars in different mass ranges, such as N/O and α/Fe, represent extremely useful tools to interpret the nature of high-redshift objects.

References

Dufour R.J., Garnett D.R., Shields G.A., 1988, ApJ 332, 752
Green R.F., York D., Huang K., Bechtold J., Welty D., Carlson M., Khare P., Kulkarni V., 1995, Proc. *ESO Workshop on QSO Absorption Lines*, ed. G. Meylan, Springer Verlag, 85
Kunth D., Matteucci F., Marconi G., 1995, A&A 297, 634
Lipman K., 1995, PhD thesis, Cambridge University
Matteucci F., Molaro P., Vladilo G. 1996, A&A, in press
Molaro P., D' Odorico S., Fontana A., Savaglio S., Vladilo G., 1996, A&A 308, 1
Pettini M., Lipman K., Hunstead R.W., 1995, ApJ 451, 100
Renzini A., Voli M., 1981, A&A, 94, 175
Skillman E.D., Kennicutt R.C., 1993, ApJ, 411, 655
Vladilo G., D'Odorico S., Molaro P., Savaglio S., 1995, in *QSO Absorption Lines*, ed. G. Meylan, Springer Verlag p. 103

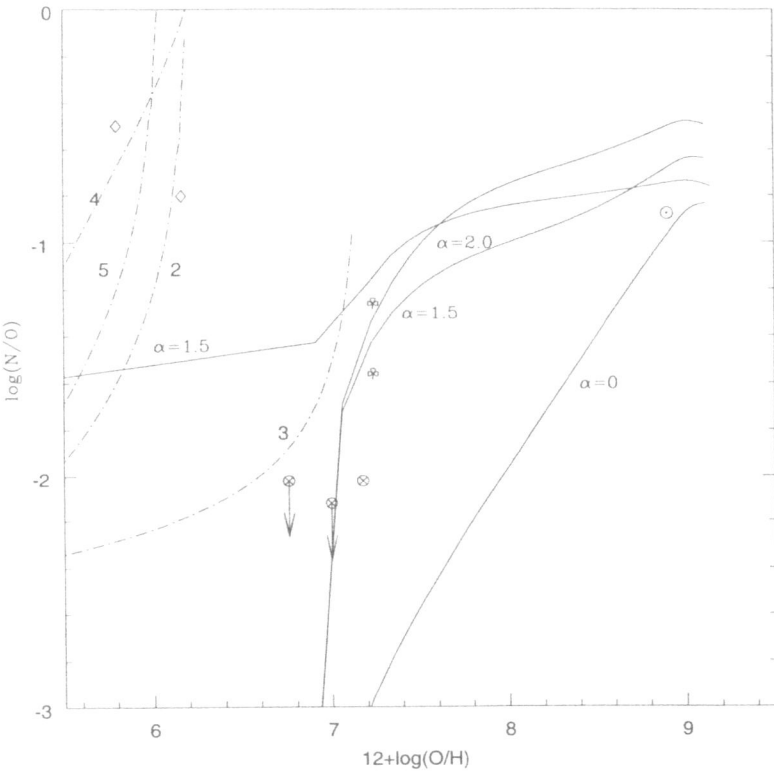

Fig. 1. log(N/O) vs. 12+log(O/H) as predicted by different models of chemical evolution (see Matteucci et al. 1996 for more details). In particular, the continuous lines refer to a model for the solar neighbourhood with different prescriptions concerning the nucleosynthesis of N. The value of the α parameter in Renzini and Voli (1981) is indicated on each curve; $\alpha=0$ means only secondary N in low and intermediate mass stars, $\alpha=1.5$ and 2.0 indicate different amounts of primary N produced during the third dredge-up according to the efficiency of convection. The line with a plateau and labelled $\alpha=1.5$ represents a model where N is assumed to be primary in massive stars (in all the other models is secondary) plus primary N from intermediate mass stars with $\alpha=1.5$. The dash-dotted lines are the predictions of starburst models with only one burst, but with different intensities for the differential galactic wind. In particular, model 3 is the best model for IZw18 (Kunth et al. 1995), whereas models 4, 5 and 2 are similar to model 3 but with higher efficiencies of galactic wind. The symbols refer to measurements of N/O and O/H ratios in the sun and in the DLA. In particular, the rhomb with the highest N/O is the system observed by Molaro et al. (1995) whereas the other rhomb is the DLA system observed by Green et al. (1995). The clovers refer to different measurements for IZw18, the one with the highest N/O is from Skillman and Kennicutt (1993), whereas the other is from Dufour et al. (1988). The crossed circles are the DLA systems discussed in Pettini et al. (1995) and Lipman (1995).

Measurement of q_0 with Type Ia Supernovae

R. Pain

LPNHE, CNRS-IN2P3 and Universités Paris VI & VII, Paris, France

Abstract. Our search for high-redshift Type Ia supernovae discovered, in its first years, a sample of nine supernovae. The spectra and light curves indicate that almost all were Type Ia supernovae at redshifts $z = 0.35$ to 0.5. These high-redshift supernovae could provide a distance indicator to measure the cosmological parameter q_0. In this poster, observation strategies, analysis and preliminary results on q_0, were presented.

1 Introduction

The results presented in this poster paper were obtained by the Supernova Cosmology Project [1]. A more detailled description of this work can be found in our contributions to the NATO Advanced Study Institute on Thermonuclar Supernovae (S. Perlmutter et al., A. Kim et al., G. Goldhaber et al. and R. Pain et al., 1996) held in Aiguablava, Spain on June 20-30, 1995.

2 Data Set and Data Analysis

We present results concerning seven supernovae found during 1994 and followed with photometry and spectroscopy (Perlmutter et al. 1994, 1995a). We observed light curves for all of the supernovae in at least one filter (usually R band), and spectra for all of the host galaxies and three of the supernovae.

The data analysis involves several stages, first to reduce the observed image data to individual photometry points and then to compare these points with nearby SN Ia light curve templates (Leibundgut et al. 1991) to determine the luminosity distance and hence q_0.

For our preliminary estimate of q_0, we have applied a width-brightness correction (Phillips et al. 1993, Hamuy et al. 1995 and Riess, Press, and Kirshner 1995) to all seven of the high-redshift supernovae, on the assumption that they are all SNe Ia with negligible extinction and that the zero point of the width-brightness relation has not evolved in the ~4 billion years back to $z \sim 0.4$.

[1] S. Perlmutter, S. Deustua, S. Gabi, G. Goldhaber, D. Groom, I. Hook, A. Kim, M. Kim, J. Lee, R. Pain, C. Pennypacker and I. Small, Lawrence Berkeley National Laboratory and Center for Particle Astrophysics, University of California, Berkeley; A. Goobar, University of Stockholm; R. Ellis and R. McMahon, Institute of Astronomy, Cambridge University; B. Boyle, P. Bunclark, D. Carter, K. Glazebrook and M. Irwin, Royal Greenwich Observatory; H. Newberg, Fermi National Accelerator Laboratory; A.V. Filippenko and T. Matheson, University of California, Berkeley; M. Dopita and J. Mould, MSSSO, Australian National University; W. Couch, University of the new South Wales

Given the error bars, our current measurements of q_0 do not yet clearly distinguish between an empty $q_0 = 0$ and closed $q_0 > 0.5$ universe. The data do, however, indicate that a decelerating $q_0 \geq 0$ is a better fit than an accelerating $q_0 < 0$ universe. This is an important conclusion since it limits the possibility that $q_0 = \Omega_0/2 - \Omega_\Lambda$ is dominated by the cosmological constant Λ.

3   Conclusion

The data presented in this paper show that scheduled discovery and follow up of batches of pre-maximum high-redshift supernovae can be accomplished. Since then, we discovered 11 additional high-redshift SNe (Perlmutter et al. 1995c). Other groups have started similar searches and the prospects for high-redshifts Type Ia SNe to become a well-studied distance indicator are very good.

The observations described in this paper were primarily obtained as visiting/guest astronomers at the Isaac Newton and William Herschel Telescopes; the Kitt Peak National Observatory 4-meter and 2.1-meter telescopes and Cerro Tololo Interamerican Observatory 4-meter telescope; the Keck Ten-meter Telescope; and the Siding Springs 2.3-meter Telescope. This work was supported in part by the Physics Division, E. O. Lawrence Berkeley National Laboratory of the U. S. Department of Energy under Contract No. DE-AC03-76SF000098, and by the National Science Foundation's Center for Particle Astrophysics, University of California, Berkeley under grant No. ADT-88909616. The author would like to thank Gerard Fontaine of CNRS-IN2P3 and Bernard Sadoulet of CfPA, Berkeley for encouraging his participation to the project.

References

G. Goldhaber et al., 1996, astro-ph 9602124 to appear in Thermonuclear Supernovae, eds R. Canal, P. Ruiz-Lapuente, & J. Isern, (Dordrecht: Kluwer), in press

M. Hamuy, M. M. Phillips, J. Maza, N. B. Suntzeff, R. A. Schommer, and R. Aviles. *Astron. J.*, 109:1, 1995.

A. Kim et al., 1996, astro-ph 960213 to appear in Thermonuclear Supernovae, eds R. Canal, P. Ruiz-Lapuente, & J. Isern, (Dordrecht: Kluwer), in press

B. Leibundgut, G.A. Tammann, R. Cadonau, and D. Cerrito. *Astro. Astrophys. Suppl. Ser.*, 89:537, 1991.

R. Pain et al., 1996, astro-ph 9602125 to appear in Thermonuclear Supernovae, eds R. Canal, P. Ruiz-Lapuente, & J. Isern, (Dordrecht: Kluwer), in press

S. Perlmutter et al., IAUC 5956 and 5958, 1994

S. Perlmutter et al. *Ap.J.*, 440:L41, 1995a.

S. Perlmutter et al., IAUC 6263, 1995b

S. Perlmutter et al., IAUC 6270, 1995c

S. Perlmutter et al., 1996, astro-ph 9602122 to appear in Thermonuclear Supernovae, eds R. Canal, P. Ruiz-Lapuente, & J. Isern, (Dordrecht: Kluwer), in press

M. M. Phillips. *Ap.J.*, 413:L105, 1993.

A. G. Riess, W. H. Press, and R. P. Kirshner. *Ap.J.*, 438:L17, 1995.

Spectrophotometric Investigations of Blue Compact Dwarf Galaxies with the VLT

Polichronis Papaderos, Klaus J. Fricke

Universitäts-Sternwarte, Geismarlandstraße 11, D-37083 Göttingen, Germany

Abstract. A recent analysis of a sample of Blue Compact Dwarf Galaxies by means of surface photometry implies that the structural properties of their LSB stellar component (i) are related to those of their starburst component and (ii) differ from those of other classes of dwarf galaxies. VLT can largely improve our knowledge of the nature of this LSB component through observational projects proposed below.

1 Introduction

Blue Compact Dwarf Galaxies (BCDs) are low–luminosity ($M_B \gtrsim -18$), gas–rich ($M_{gas}/M_{tot} \sim 0.3$) and metal–deficient ($Z_\odot/50 \leq Z \leq Z_\odot/3$) extragalactic systems undergoing brief ($\lesssim 10$ Myr) bursts of star formation being separated by longer (1–3 Gyr) quiescent periods (Thuan 1991). Since conditions, such as high star formation rate and low metallicity, are assumed to have prevailed during the early epoch of the galaxy formation, BCDs are nearby laboratories to study processes which otherwise can be monitored only in high–z environments. Among other issues related to BCDs important questions are:

(i) What is the physical origin of the cyclic starburst phenomenon in BCDs? Since BCDs have been found in their majority to be isolated, or even populate low–density environments (Hopp 1994, Lindner et al. 1996), the concept of interaction-triggered activity has to be dismissed.

(ii) How do BCDs appear during quiescent interburst phases and how are they related to other types of dwarf galaxies? Answering these questions is a requirement for understanding the evolutionary status of dwarf galaxies, in general.

2 Relations Between the Starburst and the LSB Component

An analysis of the structural properties of 12 BCDs as obtained by decomposition of their surface brightness profiles (Papaderos et al. 1996a) reveals that (i) the extent of the starburst component is on average half of the extent of the LSB component and (ii) the fractional area of a BCD wherein star formation occurs tends to increase with decreasing luminosity of the LSB component (Papaderos et al. 1996b). These relations imply that the global star formation in BCDs is partly dependent on the structural properties of the old LSB stellar population onto which it is superimposed.

At equal B luminosity, the LSB component of BCDs is by ~ 1.5 mag brighter in its central surface brightness and by a factor ~ 2 smaller in its exponential scale length than in dwarf irregulars (dIs) and dwarf ellipticals (dEs). Moreover, contrary to dIs/dEs, the central surface brightness of the LSB component in BCDs does not appear luminosity–dependent, being nearly constant around 21.5 mag arcsec^{-2} (Papaderos et al. 1996b). The results of this pilot study require a more detailed analysis of the nature of the LSB component in BCDs and in other dwarf galaxies.

3   Observational Programs for the ESO/VLT

FORS (imaging & MOS mode), FUEGOS and UVES attached to the VLT are the appropriate instruments to carry out observational projects devoted to BCDs and to faint dwarf galaxies in general, such as:

(i) Surface photometry: Structural properties of the LSB component and colour gradients can be obtained with greater accuracy and out to larger scales.

(ii) Multi–object spectroscopy of knots of star formation: High S/N spectra which can be obtained within short exposures will be interpreted by means of spectral synthesis codes. These will in turn give results on the temporal and spatial occurrence of starbursts and on the metallicity distribution within the starburst component. Moreover, spectra of superb quality will enable us to detect and investigate the formation of faint, high–velocity components in emission lines as those detected in few nearby BCDs (cf. Izotov et al. 1996).

(iii) Even more important, within reasonable integration times absorption–line–spectra of sufficient quality can be obtained at intensity levels as faint as ~ 25 mag arcsec^{-2}. They will allow (a) to derive the age and metallicity distribution within this LSB component and (b) to determine the velocity dispersion in order to quantify a possible contribution of dark matter to the mass of BCDs.

(iv) Search for BCDs and inactive dwarf galaxies in low–density regions and comparison of their spectrophotometric properties with objects of equal absolute luminosity in clusters in order to investigate any environmental biases.

References

Hopp, U. 1994, in "Dwarf Galaxies", eds. G. Meylan, P. Prugniel, 37

Izotov, Y.I, Dyak, A.B., Chaffee, F.H., Foltz, C.B., et al. 1996, ApJ 458, 524

Lindner, U., Einasto, M., Einasto, J., Freudling, W., Fricke, K.J. 1996, A&A in press

Papaderos, P., Loose, H.–H., Thuan, T.X., Fricke, K.J. 1996a, A&AS in press

Papaderos, P., Loose, H.–H., Fricke, K.J., Thuan, T.X. 1996b, A&A in press

Thuan, T.X. 1991, in "Massive Stars in Starbursts", eds. C. Leitherer, N.R. Walborn, T.M. Heckman, C.A. Norman, 183

This work has been partly supported by Deutsche Agentur für Raumfahrtangelegenheiten (DARA) GmbH grant 50 OR 9407 6.

A Deep 20 cm Radio Mosaic of the ESP Galaxy Redshift Survey

I. Prandoni[1], L. Gregorini[1], P. Parma[1], R.H. de Ruiter[2], G. Vettolani[1], M.H. Wieringa[3] and R.D. Ekers[3]

[1] Istituto di Radioastronomia del CNR, Via Gobetti 101, 40129 Bologna, Italy
[2] Osservatorio Astronomico di Bologna, Via Zamboni 33, 40126 Bologna, Italy
[3] Australia Telescope National Facility, P.O. Box 76, Epping NSW 2121, Australia

Abstract. In the last two years we used the Australia Telescope Compact Array to make a deep radio mosaic of the Eso Slice Project (ESP) galaxy redshift survey. We surveyed the entire region covered by the ESP galaxies at 20 cm with $\sim 70\,\mu$Jy sensitivity (1σ), getting a radio detection rate of $\sim 15\%$ (3σ threshold).

The main interest in making these radio observations lies in the unique possibility of studying the existing correlations between optical (line activity, colors, morphologies, luminosities, etc.) and radio properties of galaxies. Such an analysis is most reliable when a complete and homogeneous optical sample, like the present one, is available.

As a further result, we are producing a new catalogue consisting of all the radio sources detected above 6σ. This homogeneous and fairly deep sample will allow better studies of the sub–mJy population.

1 The ESP Galaxy Redshift Survey

Vettolani et al. (1993) have made a deep redshift survey as an ESO Key–Project. Redshifts for 3348 galaxies brighter than $b_J \leq 19.4$ have been obtained in two areas of $22° \times 1°$ and $5° \times 1°$ near the South Galactic Pole. The survey has a typical depth of $z = 0.1$ (with 10% of objects at $z > 0.2$) and fully samples the optical luminosity function down to $B = -15$.

Further data available include galaxy b-r colors, galaxy morphologies and K–corrections.

2 ATCA Radio Observations

The Australia Telescope Compact Array (ATCA) was used to image at 20 cm the entire area ($22° \times 1°$ plus $5° \times 1°$) of the optical survey.

The ATCA supports a mosaic observing mode which allows efficient coverage of large areas by interleaving short observations of a grid of pointings.

The observing campaign (34 blocks of 12^h) started in Nov.'94 and has been completed in January '96. Data reduction has been completed too.

We have got 16 big mosaiced radio maps which cover the 27 sq. degr. region surveyed. The spatial resolution is $16'' \times 8''$. The noise level is $\sim 70\,\mu$Jy and is fairly uniform (as needed for statistical studies) over all the maps.

3   First Results of the Radio Survey

About 15% of the ESP galaxies have been detected at 20 cm above a 3σ-threshold of 0.2 mJy. Typically radio detected ESP galaxies are associated with point-like radio sources. Only a few of them show extended radio emission.

The distance distribution of radio detections covers the entire range covered by the total sample of optical galaxies (see Fig. 1) showing that our radio survey is deep enough for a reliable statistical study of the radio–optical properties of the ESP galaxies.

As expected, a large fraction ($\sim 60\%$) of the galaxies detected shows one or more emission lines, confirming the presence of a correlation between radio emission and star formation, traced by the OII line (Kennicut 1983, Kennicut 1992).

We are also searching for all the radio sources present in the region surveyed above a 6σ-threshold. In preliminary analysis on an area of 1.3 sq. degr., we found 120 radio sources above 0.4 mJy. A large fraction of them ($\sim 40\%$) are sub–mJy objects.

This leads us to expect a total number of ~ 2500 radio sources in the entire area surveyed (27 sq. degr.) and ~ 1000 sub–mJy sources. This new catalogue will therefore be especially useful in studying the sub–mJy population which is still poorly understood.

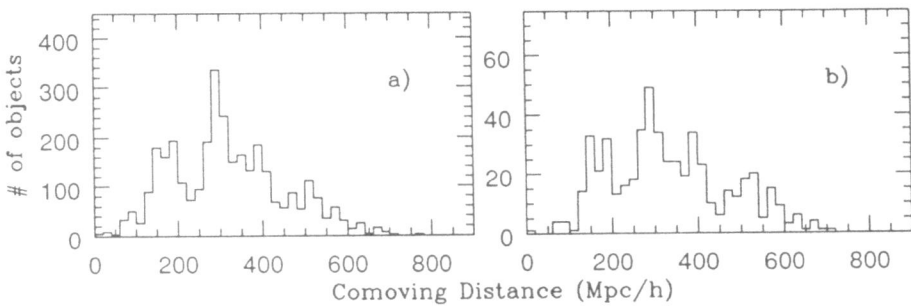

Fig. 1. Distance distribution for a) all ESP galaxies, b) only radio detected ones.

References

Heydon–Dumbleton, N.H., Collins, C.A., MacGillivray, H.T. (1988) in *Large–Scale Structures in the Universe*, eds. W. Seitter, H.W. Duerbeck and M. Tacke (Springer–Verlag), 71

Kennicut, R.C. (1983), A&A, **120**, 219

Kennicut, R.C. (1992), ApJ, **388**, 310

Lund, G. (1986) in *OPTOPUS – ESO Operating Manual No. 6* (ESO, Garching bei Munchen).

Vettolani et al. (1993): IAU Symposium 161, *Astronomy from Wide Field Imaging*, H.T. MacGillivray ed. (Reidel), in press

A New Model of Spectral Evolution: PEGASE Application to Galaxy Counts

Brigitte Rocca-Volmerange and Michel Fioc

Institut d'Astrophysique de Paris, 98 bis Bd. Arago, 75014 Paris, France

1 The Model

A new model of spectral evolution of galaxies (Fioc & Rocca-Volmerange 1996) named **PEGASE** (in french, **P**rojet d'**E**tude des **GA**laxies par **S**ynthèse **E**volutive) has been developped. It uses the principle of isochrone synthesis and allows to compute synthetic spectra of galaxies on timescales as short as 1 Myr for starbursts or as long as 20 Gyrs for Hubble sequence galaxies. Its wavelength range extends from the far UV (200 Å) to the NIR (3 μm). A new library of stellar spectra, mainly observational, has been built. Isochrones are computed for 2 sets of tracks with solar metallicity (Padova and Genova) up to the EAGB completed by the TPAGB phase (Groenewegen et al., 1993) and the PAGB phase from Schönberner (1983) and Blöcker (1995). Nebular emission (continuum and lines), extinction in disks and in spheroids computed from the metallicity of the ISM and ejecta from stars are also considered.

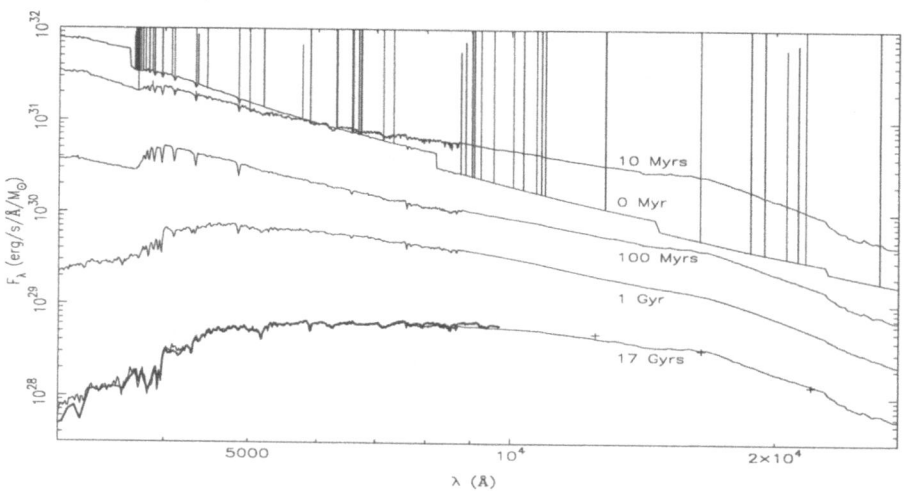

Fig. 1. Evolution of an instantaneous burst. The thick line and the crosses refer to an E1 template (nucleus) provided by Arimoto which is well fitted by a 17 Gyrs old burst.

Star formation scenarios have been fitted on spectra (see Fig. 1), UV, optical and NIR colors of nearby galaxies of the Hubble sequence. Best fits are obtained for 16 Gyrs old E/S0, 13 Gyrs old early-spirals and late-spirals and irregular galaxies 10 Gyrs old or less.

2   Galaxy Counts

K- and e-corrections are derived from these synthetic templates (Rocca-Volmerange & Fioc 1996) and may be used to predict galaxy counts in various bands. Comparison with observed counts (see Fig. 2) in an open cosmology shows that despite the very good agreement of the K-counts and of the blue counts till $b_J = 25$, the excess of blue galaxies still remains at fainter magnitudes. The normalisation of the LF agrees with the blue bright counts of Bertin et al. (1996).

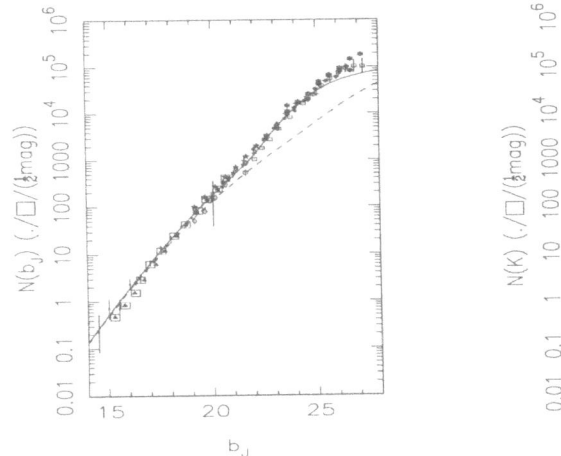
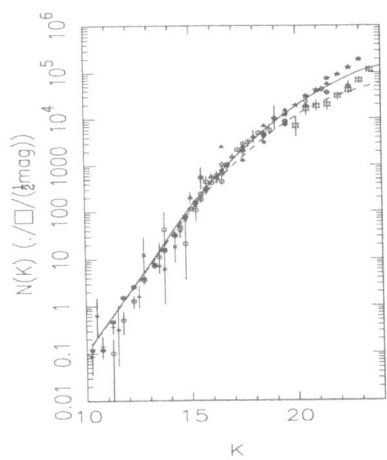

Fig. 2. Galaxy counts for $q_0 = 0.05$, $H_0 = 50$km/s/Mpc in b_J and K computed with Martzke et al. (1994) LF. Solid: with evolution, dashed: no evolution.

3   Access to the Codes and Data

The codes to compute synthetic spectra, the stellar library, the synthetic templates and the k- and e-corrections will be soon avalaible on a AAS CDROM (Leitherer et al. 1996) and by anonymous ftp at *ftp.iap.fr* in the directory *pub/from_users/pegase*.

An Hα Search for Star Forming Galaxies at z=1.5.

Jari Rönnback[1], Nils Bergvall[2] and Göran Östlin[2]

[1] European Southern Observatory, Karl-Schwarzschild-Str. 2,
   85748 Garching bei München, Germany
[2] Astronomiska observatoriet, Box 515, 75120 Uppsala, Sweden

Abstract. Almost exclusively, searches for actively star forming galaxies at high redshifts have been carried out on the redshifted Lyα emission line in the optical window, corresponding to redshifts above z=1.8. The low detection rate argues for searches of star forming galaxies at intermediate redshifts, perhaps representing the major epoch of galaxy mergers. Here we report on the results from an Hα search of starbursting massive galaxies in the vicinity of 3 QSOs at z=1.5.

1 Introduction

Despite the recent detections (Steidel et al., 1996; Fontana et al., 1996) of star forming galaxies at high redshifts ($z \geq 3$), our understanding of galaxy evolution at intermediate redshifts is still very limited. Strong efforts have been made to search for Lyα emission in the field or from QSO Lyα or metal line absorbers at $z \geq 1.8$. Very few detections have been made using this method (Djorgovski et al., 1993; Phare et al., 1995), however. It is normally assumed that dust is important to explain the weak emission but the recent observations by Steidel et al. (1996) do not support this assumption. Searches in the optical for strong emission lines long-ward of Lyα allows a redshift coverage out to z=1.2 (e.g. Le Févre et al., 1995). The intermediate region ($1.2 \leq z \leq 1.8$) has been unaccessible until recently when the sensitivity of near-IR detectors has improved. This redshift epoch is interesting because it could represent the major epoch for disk galaxy buildup through mergers. This is supported by recent observations (Schade et al., 1996), showing that there is a strong luminosity evolution of disk galaxies in the epoch z=0.5-1.2.

2 Observations

Here we present the results from a search for Hα emission from galaxies at z∼1.5 using IRAC2 at the ESO 2.2-m telescope. In order to increase the probability of finding galaxies (see e.g. Bachall and Choski, 1991) we chosed to center on QSO fields. Off-line observations were carried out in the J window and for the on-line observations we used a narrow-band filter with a passband of $\Delta\lambda = 400$Å, corresponding to a velocity window of 7500 km s^{-1}. Three QSO fields were observed.

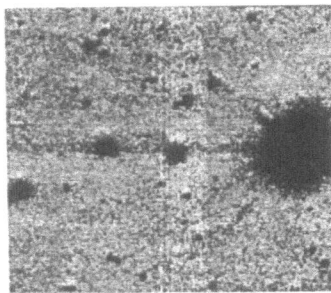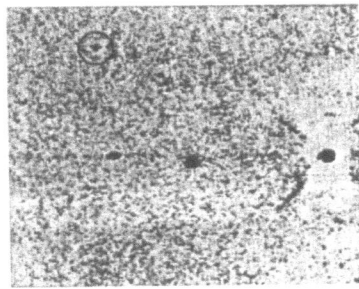

Fig. 1. Left) J image of the field around PKS 0743-67 (centre) and right) continuum subtracted narrow band image of the same field. The ring indicate the position of the Hα source. The source to the left of PKS 0743-67 is a ghost from the bright star.

3   Result

Our observations resulted in at least one candidate detection. The object is situated in the field of QSO PKS 0743-67. Fig. 1 shows the broadband image in J and the on-off line image, where only the Hα emission is seen. The object is also found to be non-circular, indicating that what we see is really a galaxy. The projected size (after seeing corrections) is 1.0 arcsec, corresponding to ∼ 5 kpc (H_0=80, q_0=1/2). The equivalent width of Hα is 80 Å, characteristic of a star forming galaxy. The total Hα luminosity is 3 10^{-19} Wm^{-2}, corresponding to roughly 20 $M_\odot$ year^{-1} (assuming a standard Salpeter IMF). Spectroscopic follow up observations using ISAAC at the VLT would be rewarding. When combined with spectral evolutionary models this would allow us to explore the star formation history of the S+E galaxy population in clusters of galaxies. With ISAAC, the expected Hα flux limit is ∼ 10^{-17} erg cm^{-2} s^{-1}, corresponding to star formation rates ≤ 7% of what we can obtain with IRAC2. We will thus be able to map the distribution of galaxies with normal star formation properties. Images of higher resolution will allow us to study galaxy properties and merging frequencies among the cluster members.

References

Bachall, Choksi: 1991, ApJ 380, L9

Djorgovski, S., Thompson, D., Smith, J.D., 1993, in B. Rocca-Volmerange, M. Dennefeld, B. Guiderdoni and Tran Than Van, eds., First Light in the Universe, Stars or QSOs, Frontiers, 67

Fontana, A., Cristiani, S., D'Odorico, S., Giallongo, E., Savaglio, S.: 1996, MNRAS, in press.

Le Févre, O., Crampton, D., Lilly, S.J., Hammer, F., Tresse, L.: 1995, ApJ 455, 60

Pahre, M.A., Djorgovski, S.G., 1995, ApJL, 449, L1

Schade, D., Lilly, S.J., Le Févre, O., Hammer, F., Crampton, D.: 1996, preprint

Steidel, C., Giavalisco, M., Pettini, M., Dickinson, M., Adelberger, Kurt L., 1996, preprint

RXJ1347-1145: a Galaxy Cluster with Exceptional Properties

Sabine Schindler[1,2]

[1] Max-Planck Institut für extraterrestrische Physik, 85740 Garching, Germany;
[2] Max-Planck Institut für Astrophysik, 85740 Garching, Germany

Abstract. We report on the exceptional properties of the most X-ray luminous and gravitational lensing cluster RXJ1347-1145. Besides its high luminosity it shows a relatively high metallicity, a high temperature and a cooling flow with a huge mass accretion rate.

1 Introduction

The ROSAT source RXJ1347-1145 was identified as the most X-ray luminous cluster and as a gravitational lens (Schindler et al. 1995) in the ESO Key Programme "Redshift Survey of ROSAT Clusters" (Böhringer 1994; Guzzo et al. 1995). Follow-up X-ray observations with the ROSAT/HRI and ASCA reveal a number of interesting properties.

2 Results

We confirm that RXJ1347-1145 (Fig. 1) at $z = 0.451$ is with a luminosity $L_X(bol) = 2 \times 10^{46}$ erg/s the most luminous X-ray cluster discovered to date. The mass of the cluster within 1.7 Mpc is $9.8 \times 10^{14} \mathcal{M}_\odot$. A comparison of the central X-ray mass and the mass determined from a simple gravitational lens model shows a discrepancy of a factor of 2-3 with the X-ray mass being smaller. A spatial analysis of the HRI image yields a small core radius of $r_c = 8.4 \pm 1.8$ arcsec (57 ± 12kpc) and $\beta = 0.56 \pm 0.04$. The temperature of the intra-cluster gas derived by ASCA spectra is relatively high $T = 9.3^{+1.1}_{-1.0}$ keV. We detect a strong FeK line corresponding to a metallicity of 0.33 ± 0.10 in solar units, which is an unexpectedly high value for a distant and hot cluster. There are several hints that the cluster contains an extremely strong cooling flow. With the usual assumptions we find a cooling flow radius of 29 arcseconds (200 kpc) and derive formally a mass accretion rate of more than 3000 $\mathcal{M}_\odot$/yr indicating that this may be the largest cooling flow detected so far.

To find these extreme properties in this distant cluster which can be taken as an indication of a well relaxed and old system is of high importance for the theory of formation and evolution of clusters.

For a detailed analysis of the X-ray data see Schindler et al. (1996).

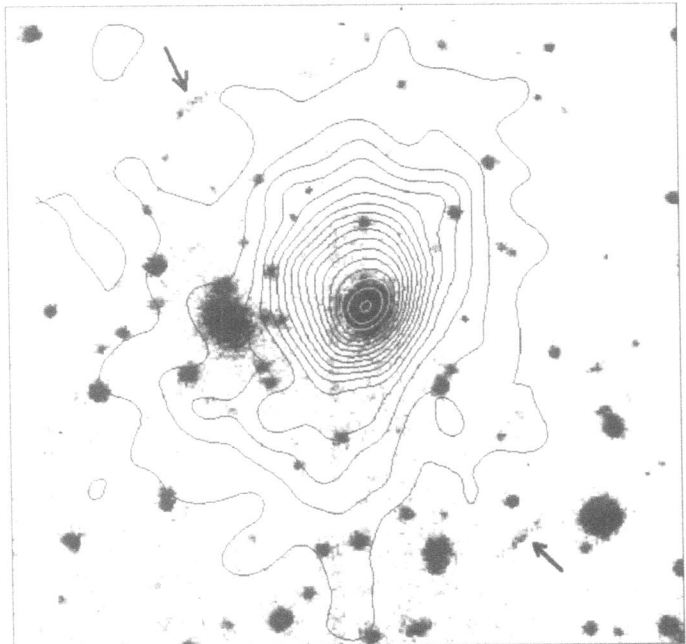

Fig. 1. X-ray contours of the HRI data superposed on an R image. The two images are aligned in such a way that the positions of the X-ray maximum and the central galaxy correspond. The X-ray image is smoothed with a Gaussian filter of $\sigma = 2.5$ arcsec. The contours are linearly spaced with Δcountrate $= 0.032$ counts/s/arcmin^2 the highest contour line corresponding to 0.54 counts/s/arcmin^2. The positions of the gravitational arcs are marked by arrows. The size of the image is 1.4×1.4arcmin^2 (North is up, East is left).

References

Böhringer H., 1994, in: Studying the Universe with Clusters of Galaxies, Böhringer H., Schindler S. (eds.), MPE Report 256, p. 93

Guzzo L., Böhringer H., Briel U., et al., 1995, in: 35th Herstmonceux Conference: Wide–Field Spectroscopy and the Distant Universe, Maddox, S.J., Aragón-Salamanca, A. (eds.). World Scientific, Singapore, p. 205

Schindler S., Guzzo L., Ebeling H., Böhringer H., Chincarini G., Collins C.A., De Grandi S., Neumann D.M., Briel U.G., Shaver P., Vettolani P., 1995, A&A 299, L9

Schindler S., Hattori M., Neumann D.M., Böhringer H., 1996, A&A, submitted

A Survey for Hα Emission at $z = 2.2 - 2.4$

Paul P. van der Werf[1], M.N. Bremer[1], A.F.M. Moorwood[2], H.J.A. Röttgering[1] and G.K. Miley[1]

[1] Leiden Observatory, P.O. Box 9513, NL–2300 RA Leiden, The Netherlands
[2] European Southern Observatory, Karl-Schwarzschild-Str. 2,
     D–85748 Garching bei München, Germany

1 Introduction

Emission-line searches for high-z starburst galaxies carried out to date have mostly targeted the Lyα line. The remarkable lack of success of these searches (Thompson et al. 1995, Thompson & Djorgovski 1995) is most likely a result of absorption by dust, which, even if present in only small amounts, will strongly suppress the resonantly scattered Lyα emission. Indeed, most of the high redshift starburst galaxies recently identified by Steidel et al. (1996) have *no* Lyα emission. Dust absorption is largely avoided by searching for Hα in stead of Lyα. Here we report the first results of a new large-area survey for Hα emission at $z = 2.2 - 2.4$.

2 Observations and Results

We used the near-IR camera IRAC2B at the ESO/MPI 2.2 m telescope at La Silla, with a pixel scale of $0\rlap{.}''72$. The data consist of in total 8 to 13 hour integrations in 2% narrow-band filters and 4 to 6 hours in K′ per field and the total survey covers about 42 □′ to an r.m.s. noise of typically $10^{-16}\,\mathrm{erg\,s^{-1}\,cm^{-2}}$. We selected objects with significant excess narrow-band flux as emission-line candidates.

The limits of our survey are compared to other surveys and to representative luminosity functions in Fig. 1, which shows that this survey for the first time probes all reasonable (evolved) Hα luminosity functions to some extent. Since our survey emphasizes area coverage more than sensitivity, we probe principally the most luminous portion of the Hα luminosity function at $z = 2.2 - 2.4$ and expect a significant detection rate in the case of strong luminosity evolution. However, in the entire survey only three serendipitous emission line objects were found. This result argues strongly against luminosity evolution proportional to $(1 + z)^3$ out to $z_e = 2$, but allows a less steep luminosity evolution, e.g., as $(1 + z)^3$ out to $z_e = 1$ or as $(1 + z)^2$ out to $z_e = 2$. The detection rate significantly exceeds that expected for active galactic nuclei (AGNs) of similar magnitudes, so that the Hα emission is most likely powered by star formation, with implied SFRs of the order of $80\,\mathrm{M_\odot\,yr^{-1}}$. Since only part of the luminosity function is probed, our survey implies a lower limit to the total star formation density

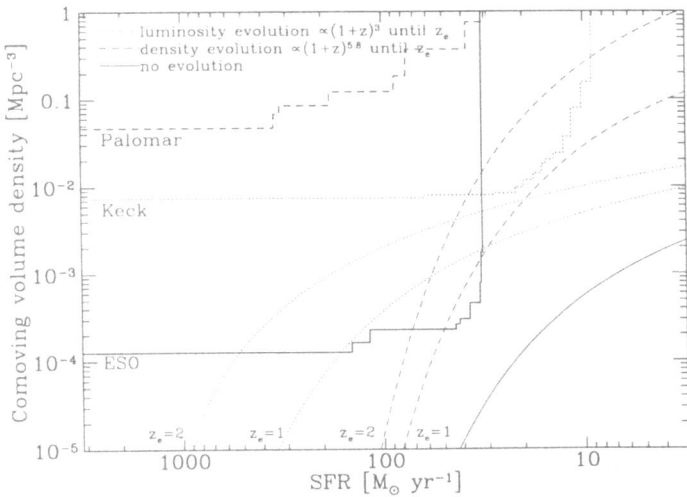

Fig. 1. Limiting comoving volume densities at 90% confidence level implied by our survey (ESO) and the Palomar (Thompson et al. 1994) and Keck (Pahre & Djorgovski 1995) surveys (for $H_0 = 75\,\mathrm{km\,s^{-1}\,Mpc^{-1}}$ and $q_0 = 0.1$, as assumed throughout this paper). The curves indicate integrated Schechter-type luminosity functions based on the local Hα luminosity function (Gallego et al. 1995), labeled "no evolution", and evolved in luminosity or in density using the evolution derived for low to moderate redshift IRAS galaxies (Rowan-Robinson 1996). Star formation rates (SFR) have been calculated from Hα luminosities using the conversion factor given by Kennicutt (1983).

at $z = 2.2 - 2.4$ of $0.05\,\mathrm{M_\odot\,yr^{-1}\,Mpc^{-1}}$, at least a factor 2.5 higher than the local value (Gallego et al. 1995). This survey shows the potential of future work with 8 m class telescopes, which will allow a determination of the Hα luminosity function at the peak of the AGN era, giving a reliable census of the star formation properties of the universe in this important epoch.

Acknowledgements. The research of Van der Werf has been made possible by a fellowship of the Royal Netherlands Academy of Arts and Sciences.

References

Gallego, J., Zamorano, J., Aragón-Salamanca, A., & Rego, M. 1995, ApJ, 455, L1

Kennicutt, R.C. 1983, ApJ, 272, 54

Pahre, M.A., & Djorgovski, S.G. 1995, ApJ, 449, L1

Rowan-Robinson, M., 1996, in Bremer, M.N., Van der Werf, P.P., Röttgering, H.J.A., & Carilli, C.L. (eds.), Cold gas at high redshift, Kluwer, Dordrecht, p. 61

Steidel, C.C., Giavalisco, M., Pettini, M., et al. 1996, ApJ, 462, L17

Thompson, D., & Djorgovski, S.G. 1995, AJ, 110, 982

Thompson, D., Djorgovski, S., & Beckwith, S.V.W. 1994, AJ, 107, 1

Thompson, D., Djorgovski, S., & Trauger, J. 1995, AJ, 110, 963

Neutral Nitrogen in Damped Lyman α Systems

G. Vladilo[1], P. Molaro[1], F. Matteucci[2], and M. Centurión[3]

[1] Osservatorio Astronomico di Trieste, Via Tiepolo 11, I-34131 Trieste, Italy
[2] Dipart. di Astronomia, Univ. di Trieste, Via Tiepolo 11, I-34131 Trieste, Italy
[3] Instituto de Astrofísica de Canarias, E-38200 La Laguna, Tenerife, España

Abstract. Nitrogen abundance determinations in high-redshift galaxies are an important tool for understanding the first stages of galactic chemical evolution. The most intense NI lines, i.e. the triplets at rest wavelengths 1134 Å and 1200 Å, fall in the optical range for $z_{abs} \gtrsim 2$ and can be observed at high spectral resolution in Damped Lyman α (DLA) systems, where nitrogen is expected to be mainly in neutral form. In practice, only a very few detections have been obtained until now due to the difficulty of disentangling these lines from the Lyman alpha forest in the spectrum of the background QSO. Here we present a new NI detection in a DLA, namely the system at z_{abs}=3.025 towards QSO 0347-383. An overview of the available nitrogen measurements in DLA systems shows a dispersion of the N/O ratios which suggests that this class of QSO absorbers include objects with different chemical evolution histories.

1 The DLA System at $z_{abs} = 3.025$ towards QSO 0347-348

As a part of a search for NI absorption in DLA systems we observed QSO 0347-383 at high resolution ($R \simeq 19500$) with the Cassegrain Echelle Spectrograph (CASPEC) at the ESO 3.6m telescope (La Silla, Chile). For the DLA system at $z_{abs} = 3.025$ we identify both NI triplets (Fig. 1) and the transitions OI (1302 Å), SiII (1193 Å, 1304 Å), and FeII (1144 Å). This is the first time that all the 6 transitions of the two NI triplets are detected in a DLA system. Column densities and broadening parameters (b values) were determined by means of a χ^2 minimization of instrumentally broadened Voigt profiles. All the NI lines were fitted simultaneously. The wide range of oscillator strengths of the 6 transitions and the moderate degree of saturation of the NI lines allowed us an accurate determination of the column density, $\log N(\text{NI}) = 14.70 \pm 0.03$ dex, and of the broadening, $b(\text{NI}) = 16.73 \pm 0.90$ km s^{-1}. The OI 1302 Å line is strongly saturated, and this leaves $b(\text{OI})$ poorly constrained with a large error in $N(\text{OI})$. However, given the similar ionization potentials of NI and OI, one expects both atomic species to sample the same material and therefore, thanks to the similar atomic weights, to have $b(\text{OI}) \simeq b(\text{NI})$ even in the extreme case of pure thermal broadening. By adopting the b value of the NI lines and its error for fitting the OI 1302 Å line, we derive $\log N(\text{OI}) = 16.5 \pm 0.3$ dex. Ionization corrections and dust depletion effects are expected to be negligible for these species in DLA systems (see discussion in Molaro et al. 1996). Therefore we derive $\log(\text{N/O}) = -1.8 \pm 0.3$ dex and $\log(\text{O/H}) + 12 = 7.8 \pm 0.3$ dex by taking $N(\text{HI})$ from Pettini et al. (1994).

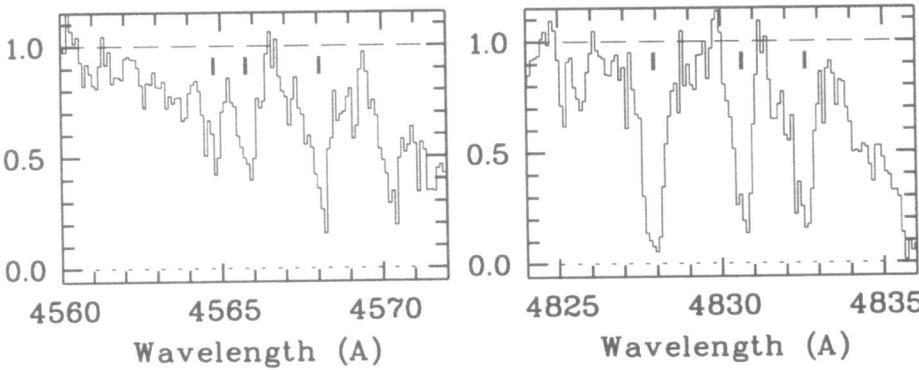

Fig. 1. NI 1134 Å and 1200 Å triplets at z_{abs}=3.025 in the spectrum of QSO 0347-348.

2   Nitrogen Abundances in DLA Systems

Previous measurements of NI in DLA systems include the z_{abs}=3.390 system towards QSO 0000-2619, with log(N/O)=−0.5 and 12+log(O/H)= 5.8 (Molaro et al. 1996) and the z_{abs}=1.78 system towards MC3 1331+170, with log(N/O)= −0.8 and 12+log(O/H)= 6.1 (Green et al. 1995). A marginal detection and two upper limits are also available: log(N/O)= −2.0 and 12+log(O/H) = 6.8 for the DLA at z_{abs}=2.538 towards QSO 2344+124 (Lipman 1995); log(N/O) < −2.2 and 12+log(O/H)= 6.8 for the z_{abs}= 2.28 system towards QSO 2348-147 (Pettini et al. 1995); and log(N/O) < −2 and 12+log(O/H)= 6.3 for the z_{abs}=2.844 system towards Q1946+7658 (Lipman 1995).

By comparing these DLA abundance ratios with the values of log(N/O) and 12+log(O/H) predicted by galactic chemical evolution models (Matteucci et al. 1996) one can see that the N/O ratio measured in the present work and the upper limits reported by Pettini et al. (1995) and by Lipman (1995) are consistent with conventional models of galactic chemical evolution. On the other hand, the N/O ratios measured by Molaro et al. (1996) and by Green et al. (1995) can only be reproduced by models with a primary production of nitrogen, one intense starburst and differential galactic winds. The different chemical evolution models required to explain the dispersion of the observed N/O ratios suggest that DLA systems include objects with different nature.

References

Green R.F., York D.,Huang K., Bechtold J.,Welty D., Carlson M., Khare P.,Kulkarni V. (1995): *ESO Workshop on QSO Absorption Lines*, ed. G. Meylan, Springer Verlag, 85
Lipman K. (1995): PhD thesis, Cambridge University
Matteucci F., Molaro P., Vladilo G. (1996): A&A in press
Molaro P., D' Odorico S., Fontana A., Savaglio S., Vladilo G. (1996): A&A 308,1
Pettini M., Smith L.J., Hunstead R.W., King D.L. (1994): ApJ 426, 79
Pettini M., Lipman K., Hunstead R.W. (1995): ApJ 451, 100

Colours, Luminosity Functions, and Clustering Properties of Faint Starburst Galaxies

Stefan J. Wagner and Martin W. Kümmel

Landessternwarte Heidelberg, Königstuhl 12, 69117 Heidelberg, Germany

1 Source Counts in the Radio and the Optical Regime

Differences between measured source densities and expected values calculated by extrapolating the local luminosity function can be interpreted in two ways. Either they show the evolution of known source classes or they indicate a different, new population of sources at certain flux levels. Log N - log S diagrams at optical and radio wavelengths show excess populations at faint flux levels which are interpreted as new source components. In the radio regime deep source counts at 1.4 GHz (e.g. Oort M.J.A. and Windhorst, R.A. 1985, A&A **145**, 405) revealed a new population at fluxes below a few milli-Jansky which can not be explained by the evolution of radiogalaxies. Source counts in the B-band (Metcalfe N. et al. 1995, MNRAS **273**, 257) exceed theoretical predictions from no evolution models at magnitudes fainter than 21st. Because of their blue colours these objects are called *faint blue galaxies*.

2 Correlating the Different Populations

It has been suspected that both the milli-Jansky population of radio sources and the faint blue galaxies arise from a population of galaxies at moderate redshift undergoing a short starburst phase. In the literature only few optical identifications of the faint radio sources are given. Windhorst and Dressler (1987, in Hewitt A. et al. (eds.), *Observational Cosmology*, IAU Symp. 124) completely identify a sample of mJy-sources in a survey down to $r = 26.5$ mag. The colours of their counterparts are in general "fairly blue". In a later paper Windhorst et al. identified a sample of micro-Jansky sources at 8.4 GHz with the HST (1995, Nature Vol. **375**, 471). The average (V-I) colour of 1.5 ± 0.7 mag is similar to that of a much larger sample of HST field galaxies. Because existing surveys are restricted on small, deep fields the number of objects is too small for detailed statistical examination. The limitation on small samples causes the selection of very special classes of objects. Therefore, it is often impossible to understand all selection effects.

3 Radio Sources at the North Ecliptic Pole

In order to investigate the optical properties of the mJy-population of radio sources we performed a survey in the B and the R band at the North Ecliptic

Pole (NEP) which extends over 1 deg^2. The area around the NEP was chosen because of existing deep surveys in the radio regime and from satellite borne instruments (IRAS, ROSAT). We combined our optical survey with a survey at 1.5 GHz which is sensitive down to flux levels of 1 mJy (Kollgaard R.I. et al. 1994, ApJS **63**, 311). 75 radio sources are within the boundary of the optical survey. With 56 sources below 10 mJy we have a large sample of radio objects with fluxes in the region of the excess population. We investigated our optical survey for the counterparts of the radio sources. After selecting the counterparts we classified them according to their morphologies and derived the luminosity function and their colour distribution.

4   Optical Survey at the NEP

The optical data were obtained with the 3.5 m telescope at Calar Alto, Spain. In both filters we used a TEK CCD in the prime focus of the telescope with a resolution of 0.40 arcsec/pixel. The limiting magnitudes of the B- and the R-survey are 24.5 mag and 23.5 mag, respectively. Down to these levels we detect 80.000 and 100.000 sources in B and R. Given the positional uncertainties of 1" and 2" in the radio and the optical range, respectively, we decided to search for the optical counterparts in a radius of 5" around the radio position. In the case of multiple counterparts we assumed the object with the smallest distance from the radio position to be the right one. Within the search radius we find at least one counterpart for 52 radio sources. 16 radio sources could only be detected in B, 5 are detectable only in R. We expect 12 identifications to be false by accidental coincidence of the radio and the optical position.

5   Optical Luminosity Function and Colour Distribution

The luminosity function in the B-band, where most of the radio sources (46) could be detected, is dominated by objects close to the completeness limit of the survey. 70% of the sources were classified as galaxies or extended objects, the other 30% appear to be pointlike. The colour distribution of the radio objects is comparable to the colour distribution of normal field galaxies as given for example in Metcalfe et al. (1995). There are no particularly blue objects in our sample. The situation could be changed by adding the colours of the 16 objects that could be detected only in the B-survey, but even then an association of mJy-sources with blue galaxies is not possible in our sample. To test the cluster properties of the radio sources we compared the object densities around the radio positions with the object densities around arbitrary positions in our survey. We find weak overdensities around the location of the radio objects on distances smaller than 40 arcseconds. A significant fraction of the mJy-sources is still unidentified at our survey limit. In order to complete and extend such studies of evolutionary properties of faint radio galaxies, deeper multiband and spectroscopic observations with larger telescopes are required.

Large-Scale Structure at $z \sim 3$

G.M. Williger[1], A. Smette[2], C. Hazard[3,4], J.A. Baldwin[4,5] & R.G. McMahon[4]

[1] MPI für Astronomie, Heidelberg, Germany
[2] Kapteyn Institute, Groningen, Netherlands
[3] Univ. Pittsburgh, USA
[4] IOA, Cambridge, UK
[5] CTIO, La Serena, Chile

Abstract. We have made a survey of Lyα absorption systems toward a region of high QSO density near the South Galactic Pole using nine lines of sight spanning Lyα redshifts $2.15 < z < 3.37$. Such a survey makes an unprecedented probe of structure in the Lyα forest along a direction perperdicular to the line-of-sight, at an intermediate scale, and can be used to test predictions of Lyα forest clouds associated with filamentary structures. We find clear evidence for correlations in the Lyα forest on up to 30 arcminute separations throughout the redshift range of the sample. The velocity separation of the correlated systems is $50 < \Delta v < 100$ km s^{-1} at $2.6 \lesssim z < 3.25$, and $100 < \Delta v < 200$ km s^{-1} at $2.3 < z \lesssim 2.6$.

1 Data Analysis

We used 2Å resolution spectra originally obtained from a survey for C IV systems (Williger et al. 1996). We find 377 Lyα forest lines detected to a rest equivalent width of $W_0 = 0.1$Å. We constructed randomised samples of Lyα lines assuming power law redshift distributions and by making random shifts in redshift space to lines binned in $50h^{-1}$ Mpc intervals along each line of sight ($H_0 \equiv 100$ km s^{-1} Mpc^{-1}). The procedure followed is very similar to one described in Williger et al. (1996), which insures that our results are not affected by the incompleteness that affects the lines with equivalent widths below $W_0 = 0.5$Å; this procedure assumes that the incompleteness is slowing varying on these scales. A variety of statistical tests (2 point correlation, nearest neighbour, Kolmogorov-Smirnov test comparing redshift density to a power law in $(1 + z)$) show no anomalies in the Lyα forest in these simulated Lyα lines samples as compared to the literature.

The distribution of equivalent widths as a function of redshift appears to behave as a stochastic variable, as in Carbone & Savaglio (1996) – with redshift intervals containing relatively more lines of high equivalent width being correlated between different lines of sight separated by up to 30 arcmin.

If we ignore spatial separations on the sky and only consider redshift separations, and calculate the distribution of line separations, we find an overabundance of Lyα lines with velocity separations $50 < \Delta v < 100$ km s^{-1} at $2.9 < z < 3.25$ (Fig. 1), significant at the 3σ level. We also find a deficit of separations at $150 < \Delta v < 200$ km s^{-1} with similar significance. At $2.3 < z < 2.6$, a similar overabundance exists at $100 < \Delta v < 200$ km s^{-1}.

We interpret these results as indicating large structures at $z \sim 3$, spanning $\sim 6h^{-1}$ Mpc in the local frame. These may be attributed to filamentary or sheetlike geometries, which will provide useful tests for cosmological structure formation models.

References

Carbone, V. & Savaglio, S. 1996, MNRAS, in press.
Williger, G., Hazard, C., Baldwin, J. & McMahon, R. 1996, ApJS, in press

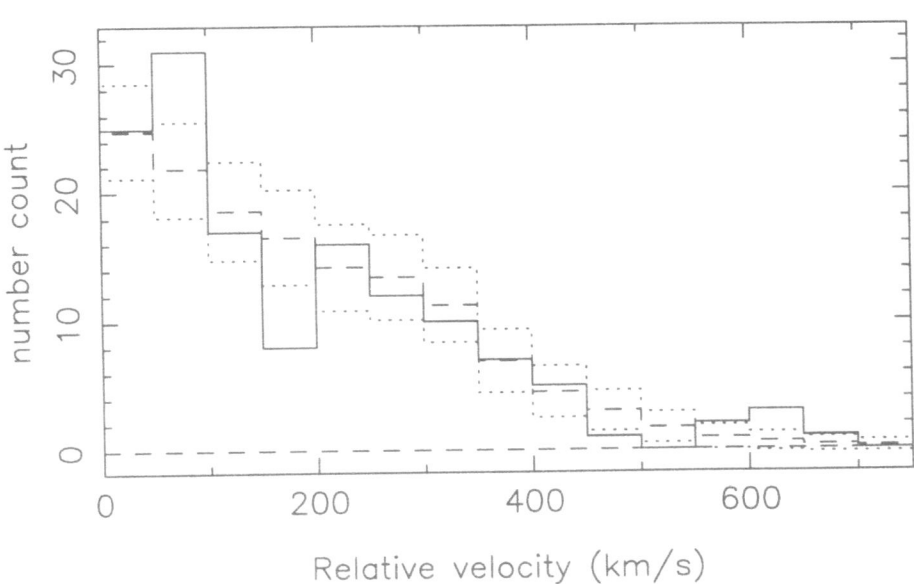

Fig. 1. The distribution of line separations in redshift space for all of the Lyα lines in the sample to rest equivalent width 0.1Å, for $2.9 < z < 3.25$. The solid line is the observations, the dashed line is the mean of 200 Monte Carlo simulations and the dotted line is the 1σ dispersion about the mean. The feature at $50 < \Delta v < 100$ km s^{-1} is significant at the $> 3\sigma$ level.

Author Index

ESO ASTROPHYSICS SYMPOSIA
European Southern Observatory

Series Editor: Philippe Crane

Springer
and the
environment

At Springer we firmly believe that an international science publisher has a special obligation to the environment, and our corporate policies consistently reflect this conviction.

We also expect our business partners – paper mills, printers, packaging manufacturers, etc. – to commit themselves to using materials and production processes that do not harm the environment. The paper in this book is made from low- or no-chlorine pulp and is acid free, in conformance with international standards for paper permanency.

Springer